Advances in Experimental Medicine and Biology

Volume 1348

Advances in Experimental Medicine and Biology provides a platform for scientific contributions in the main disciplines of the biomedicine and the life sciences. This series publishes thematic volumes on contemporary research in the areas of microbiology, immunology, neurosciences, biochemistry, biomedical engineering, genetics, physiology, and cancer research. Covering emerging topics and techniques in basic and clinical science, it brings together clinicians and researchers from various fields.

Advances in Experimental Medicine and Biology has been publishing exceptional works in the field for over 40 years, and is indexed in SCOPUS, Medline (PubMed), Journal Citation Reports/Science Edition, Science Citation Index Expanded (SciSearch, Web of Science), EMBASE, BIOSIS, Reaxys, EMBiology, the Chemical Abstracts Service (CAS), and Pathway Studio.

2020 Impact Factor: 2.622

More information about this series at https://link.springer.com/bookseries/5584

Jaroslava Halper
Editor

Progress in Heritable Soft Connective Tissue Diseases

Second Edition

Editor
Jaroslava Halper
Department of Pathology, College of Veterinary
Medicine, and Department of Basic Sciences,
AU/UGA Medical Partnership
The University of Georgia
Athens, GA, USA

ISSN 0065-2598 ISSN 2214-8019 (electronic)
Advances in Experimental Medicine and Biology
ISBN 978-3-030-80616-3 ISBN 978-3-030-80614-9 (eBook)
https://doi.org/10.1007/978-3-030-80614-9

This Springer imprint is published by the registered company Springer Nature Switzerland AG
The registered company address is: Gewerbestrasse 11, 6330 Cham, Switzerland

Contents

Editor and Contributors

About the Editor

Jaroslava Halper, M.D., Ph.D., is Professor of Pathology at The University of Georgia where she is affiliated with the Department of Pathology, College of Veterinary Medicine, and with the Department of Basic Sciences in the Medical Partnership. She is a board-certified anatomic pathologist who received M.D. degree from the Faculty of Medicine, University of Toronto. She trained in pathology at Albert Einstein College of Medicine (in the Bronx, New York) and at Mayo Clinic in Rochester, Minnesota. She obtained her Ph.D. in experimental pathology from the University of Minnesota at Mayo Clinic where she worked on the identification of transforming growth factor e (known better as granulin) with Harold Moses as her major professor. She continued her work on granulins, growth factors, and tissue repair when she first arrived at The University of Georgia. With time, her research interests have evolved into probing basic problems in tendons. More recently, she has conducted groundbreaking research on equine degenerative suspensory ligament desmitis (DSLD). Her group has established that DSLD, a chronic progressive disorder, is a systemic condition characterized by inappropriate accumulation of proteoglycans in connective tissues, particularly in tendons and ligaments.

Contributors

Jessica M. Beall Department of Animal Science, University of California Davis, Davis, CA, USA

Aude Beyens Center for Medical Genetics Ghent, Department of Dermatology, Department of Biomolecular Medicine, Ghent University Hospital, Ghent University, Ghent, Belgium

Bert Callewaert Center for Medical Genetics Ghent, Department of Biomolecular Medicine, Ghent University Hospital, Ghent University, Ghent, Belgium

Marco Castori Division of Medical Genetics, Fondazione IRCCS-Casa Sollievo della Sofferenza, San Giovanni Rotondo, Italy

Brianne K. Connizzo Department of Biomedical Engineering, Boston University, Boston, MA, USA

Pauline De Kinderen Center for Medical Genetics, Antwerp University Hospital/University of Antwerp, Antwerp, Belgium

Harry Dietz McKusick-Nathans Institute of Genetic Medicine, Johns Hopkins University School of Medicine, Baltimore, MD, USA

Carmela Fusco Division of Medical Genetics, Fondazione IRCCS-Casa Sollievo della Sofferenza, San Giovanni Rotondo, Italy

Nicole L. Gonzales Department of Animal Science, University of California Davis, Davis, CA, USA

Jaroslava Halper Department of Pathology, College of Veterinary Medicine, and Department of Basic Sciences, AU/UGA Medical Partnership, The University of Georgia, Athens, GA, USA

Chloe Taejoo Hwang Department of Pathology, College of Veterinary Medicine, The University of Georgia, Athens, GA, USA

John S. Ikonomidis Division of Cardiothoracic Surgery, University of North Carolina, Chapel Hill, NC, USA

Jeffrey A. Jones Division of Cardiothoracic Surgery, Medical University of South Carolina and Ralph H. Johnson Veterans Affairs Medical Center, Charleston, SC, USA

Tomoki Kosho Department of Medical Genetics, Shinshu University School of Medicine, Matsumoto, Japan

Shireen R. Lamandé Murdoch Children's Research Institute and Department of Paediatrics, University of Melbourne, Royal Children's Hospital, Parkville, VIC, Australia

Bart Loeys Center for Medical Genetics, Antwerp University Hospital/ University of Antwerp, Antwerp, Belgium
Department of Human Genetics, Radboud University Medical Center, Nijmegen, The Netherlands

Ilse Luyckx Center for Medical Genetics, Antwerp University Hospital/ University of Antwerp, Antwerp, Belgium

Naomichi Matsumoto Department of Human Genetics, Yokohama City University Graduate School of Medicine, Yokohama, Japan

Josephina A.N. Meester Center for Medical Genetics, Antwerp University Hospital/University of Antwerp, Antwerp, Belgium

Lucia Micale Division of Medical Genetics, Fondazione IRCCS-Casa Sollievo della Sofferenza, San Giovanni Rotondo, Italy

Michael J. Mienaltowski Department of Animal Science, University of California Davis, Davis, CA, USA

Noriko Miyake Department of Human Genetics, National Center for Global Health and Medicine, Tokyo, Japan

Monica Y. Pechanec Department of Animal Science, University of California Davis, Davis, CA, USA

Lore Pottie Center for Medical Genetics Ghent, Ghent University Hospital, Ghent, Belgium
Department of Biomolecular Medicine, Ghent University, Ghent, Belgium

Jennifer Hope Roberts Department of Pathology, College of Veterinary Medicine, The University of Georgia, Athens, GA, USA

Jeffrey W. Ruberti Department of Bioengineering, Northeastern University, Boston, MA, USA

Seyed Mohammad Siadat Department of Bioengineering, Northeastern University, Boston, MA, USA

Patrick Sips Center for Medical Genetics Ghent, Ghent University Hospital, Ghent, Belgium
Department of Biomolecular Medicine, Ghent University, Ghent, Belgium

Brandon Sloan Division of Cardiothoracic Surgery, Department of Surgery, Medical University of South Carolina, Charleston, SC, USA

Chavaunne T. Thorpe Comparative Biomedical Sciences, The Royal Veterinary College, University of London, London, UK

Lut Van Laer Center for Medical Genetics, Antwerp University Hospital/ University of Antwerp, Antwerp, Belgium

Joe D. Velchev Center for Medical Genetics, Antwerp University Hospital/ University of Antwerp, Antwerp, Belgium

Aline Verstraeten Center for Medical Genetics, Antwerp University Hospital/University of Antwerp, Antwerp, Belgium

Jason B. Wheeler Division of Cardiology, Cleveland Clinic, Cleveland, OH, USA

Danae E. Zamboulis Institute of Life Course and Medical Sciences, Faculty of Health and Life Sciences, University of Liverpool, Liverpool, UK

Sanford M. Zeigler Division of Cardiothoracic Surgery, Department of Surgery, Medical University of South Carolina, Charleston, SC, USA

Introduction

1

Jaroslava Halper

Abstract

Just like the first edition of this widely successful book the second edition provides latest updates of our understanding of pathophysiology, pathology, clinical presentation and treatment of heritable soft connective tissue diseases. In addition, new knowledge of not only structures but also of functions of basic components of connective tissues (e.g., collagen), and of organs such as tendons has been added as well. Moreover, readers will learn more about new syndromes and new subgroups of previously described syndromes and disorders as well. The authors are not only prominent investigators in their field, but they are also good writers and that should provide an additional incentive for interested readers.

Keywords

Ehlers- Danlos syndrome · Marfan syndrome · Joint hypermobility · Collagen and tendon structure and function · Transforming growth factor β · Cutis laxa · Collagen VI myopathies

J. Halper (✉)
Department of Pathology, College of Veterinary Medicine, and Department of Basic Sciences, AU/UGA Medical Partnership, The University of Georgia, Athens, GA, USA
e-mail: jhalper@uga.edu

This volume represents a second edition of a widely successful and read book Progress in Heritable Soft Connective Tissue Diseases. The content was revised and updated with inclusion of newly identified syndromes and variations of already more or less characterized disorders. Marfan and Ehlers-Danlos syndromes are the best known and studied conditions included in this group of diseases. The first description of Ehlers-Danlos syndrome appeared first in Hippocrates' writing (Airs, Waters and Places) in 400 BC (Parapia and Jackson 2008). It was not until the seventeenth century when sporadic accounts were published in medical literature. Finally, the syndrome was named after Edward Lauritz Ehlers and Henri-Alexander Danlos provided more comprehensive and systematic descriptions of several patients suffering with joint hypermobility and skin fragility more than a century ago (Parapia and Jackson 2008). Marfan syndrome, a more common condition than Ehlers-Danlos syndrome, was named after Antoine Marfan who published the first case report of possible Marfan syndrome in 1896 (Marfan 1896).

The difference in knowledge about the pathogenesis, biochemistry, genetics of these two and related disorders between then and now has widened as it will become obvious upon reading this volume. However, this does not take anything away from the contribution of past physicians as they applied their ingenuity, clinical astuteness

J. Halper (ed.), *Progress in Heritable Soft Connective Tissue Diseases*, Advances in Experimental Medicine and Biology 1348, https://doi.org/10.1007/978-3-030-80614-9_1

and close attention to seemingly unrelated details. On the contrary, they are providing examples how to approach new unknown phenomena. Though the roster of authors has changed to some extent, the quality of chapters has not. The topics have been revised or updated, and some were expanded as necessary.

The first chapters are more general, concentrating more on the physiology, structure and biochemistry of normal soft tissues. That should help in understanding of the pathophysiology of these disorders, many of them related to each other in more than one ways.

Chapters 2 and 3 discuss primarily collagen and tendon structure and function. Though most of tendon disorders and injuries encountered by millions of people on daily basis are non-hereditary in nature, many of the disorders described in this volume (Marfan, Ehlers-Danlos, cutis laxa etc) affect musculoskeletal system (and mainly tendons) to great degree. Chapter 2 informs mostly on collagen structure and structural components, as well as on the interface between structure and biomechanical behavior. Chapter 3 deals with changes and adaptations of structure, not just collagens, but of extracellular matrix with time and development. Moreover, Connizzo et al. interpret the functional changes in tendons through the glasses of bioengineering (Chap. 3). In this sense these two chapters are complementary. The several next chapters (Chaps. 4, 5 and 6) review the role of basic components of extracellular matrix in function of normal connective tissue and their contribution to pathophysiology of soft tissue diseases. Authors of Chap. 7 continue in this vein by concentrating on one growth factor, transforming growth factor β, which plays an outsized role in many disorders of soft connective tissues. Chapter 8 on Marfan syndrome expands this narrative, together with advances in Marfan management.

The Ehlers-Danlos category has grown to 13 bona fide members described in details in Chap. 9 together with related and enlarging group of joint hypermobility disorders. In addition, Chap 10 brings us up to date progress in Ehlers- Danlos syndrome due to abnormal metabolism of glycosaminoglycan chains attached to proteoglycans regulating assembly of collagens.

New entities are included as well, now we have not one but two syndromes associated with professor Bart Loeys (Chaps. 11 and 12). Advances in genetic analysis have greatly expanded our understanding of cutis laxa, and also added many new (and rare subtypes) (Chap 13) – it turns out that the complexity of cutis laxa has greatly expanded over the last several years. Numerous variations of cutis laxa and joint mobility disorders have been discovered taking advantage of recent advancements) in genetic analysis. We have acquired better understanding of pathogenesis and biochemical changes in some other, more established entities, such as Marfan and collagen VI myopathies where better management and treatment is on the horizon (Chaps. 8 and 14, respectively). In particular, antisense agents, cyclosporine A and autophagy stimulating drugs are being investigated for treatment of myopathies due to mutations in in the *COL6A1*, *COL6A2* and *COL6A3* genes (Chap. 14). Even in the case of connective tissues diseases in domestic animals some progress has been made. Warmblood fragile foal syndrome was added to the last chapter, and the section on equine degenerative suspensory ligament desmitis was substantially revised in Chap. 15.

All chapters were contributed by a group of distinguished and preeminent physicians and scientists, all of them not working “just” in the field but making new discoveries described by them. Most authors address clinical, biochemical or/ and genetic similarities among the various entities and conditions, provide guidance how distinguish among them and properly diagnose patients. Readers will notice that seemingly there is an overlap among many of these disorders. And indeed, many of them, if not most are interconnected because of the prominent roles of TGFβ, of fibrillin microfibrils and collagen fibril assembly (and other molecules) playing in connective tissues physiology, and by extension in pathogenesis of many disorders described in the book. What I found particularly helpful that author(s) of each chapter bring their own perspective even when describing closely related mechanism of the disease. These observations from different point of view by each group of authors should help with diagnosis and management of such cases.

Last but not least, the chapters are very readable, more like detective stories than dry description of genetic/biochemical defects. I do hope that basic scientists and clinicians with similar and diverse interests alike will appreciate this volume and will be inspired by it to develop their research in the field.

We hope that you will find this volume not just informative, but also stimulating to ask questions about the basic science behind these syndromes and diseases, and perhaps even inspiring to pursue research on one or more topics elaborated on in this book. This volume should serve as a bridge between basic science and clinical disciplines, and as a reference book not only for established physicians, residents and medical students, but also for scientists whether they are well established investigators or graduate students.

I would like to thank all contributors for their hard work, especially during these unexpectedly difficult times. Special appreciation goes to Dr. Gonzalo Cordova, our editor at Springer Nature., for his kindness, patience and understanding why we had to request additional time to push back on the deadline. Now when all is done, let us say Cheers on work well done!!

References

Marfan AB (1896) Un cas de déformation congénitale des quatre membres plus prononcée aux extrémités caractérisée par l'allongement de os avec un certain degré d'amincissement. Bull Mem Soc Med Hop Paris (Ser 3) 12:220–226

Parapia LA, Jackson C (2008) Ehlers-Danlos syndrome – a historical review. Br J Haematol 141(1):32–35

Basic Structure, Physiology, and Biochemistry of Connective Tissues and Extracellular Matrix Collagens

2

Michael J. Mienaltowski, Nicole L. Gonzales, Jessica M. Beall, and Monica Y. Pechanec

Abstract

The physiology of connective tissues like tendons and ligaments is highly dependent upon the collagens and other such extracellular matrix molecules hierarchically organized within the tissues. By dry weight, connective tissues are mostly composed of fibrillar collagens. However, several other forms of collagens play essential roles in the regulation of fibrillar collagen organization and assembly, in the establishment of basement membrane networks that provide support for vasculature for connective tissues, and in the formation of extensive filamentous networks that allow for cell-extracellular matrix interactions as well as maintain connective tissue integrity. The structures and functions of these collagens are discussed in this chapter. Furthermore, collagen synthesis is a multi-step process that includes gene transcription, translation, post-translational modifications within the cell, triple helix formation, extracellular secretion, extracellular modifications, and then fibril assembly, fibril modifications, and fiber formation. Each step of collagen synthesis and fibril assembly is highly dependent upon the biochemical structure of the collagen molecules created and how they are modified in the cases of development and maturation. Likewise, when the biochemical structures of collagens or are compromised or these molecules are deficient in the tissues – in developmental diseases, degenerative conditions, or injuries – then the ultimate form and function of the connective tissues are impaired. In this chapter, we also review how biochemistry plays a role in each of the processes involved in collagen synthesis and assembly, and we describe differences seen by anatomical location and region within tendons. Moreover, we discuss how the structures of the molecules, fibrils, and fibers contribute to connective tissue physiology in health, and in pathology with injury and repair.

Keywords

Hierarchical structure of tendon · Midsubstance · Enthesis · Ligaments · Supramolecular structures of collagens · Collagens I-XXVIII · Procollagens · Triple helix · Crosslinking · Fibril-associated collagens with interrupted triple helices (FACIT) · Basement membrane collagen ·

M. J. Mienaltowski (✉) · N. L. Gonzales
J. M. Beall · M. Y. Pechanec
Department of Animal Science, University of California Davis, Davis, CA, USA
e-mail: mjmienaltowski@ucdavis.edu; nlgonzales@ucdavis.edu; jbeall@ucdavis.edu; mpechanec@ucdavis.edu

J. Halper (ed.), *Progress in Heritable Soft Connective Tissue Diseases*, Advances in Experimental Medicine and Biology 1348, https://doi.org/10.1007/978-3-030-80614-9_2

Fibril-forming collagens · Beaded filament-forming collagen · Hexagonal network-forming collagen · Transmembrane collagens · Collagen fibril assembly · Protofibrils · Mature tendon fibrils · Small leucine-rich proteoglycans (SLRPs) · Mechanical behavior · Collagen in repair

Abbreviations

ADAMTS	A Disintegrin And Metalloproteinase with Thrombospondin motifs
BM	Bethlem myopathy
BMP	Bone morphogenetic protein
COL	Collagenous
Crtap	Cartilage-associated protein
ECM	Extracellular matrix
EDS	Ehlers-Danlos syndrome
EGR1	Early Growth Response 1
FACIT	Fibril-associated collagens with interrupted triple helices
FGF	Fibroblast growth factor
FMOD	Fibromodulin
GDF5	Growth differentiation factor 5
GPCs	Golgi-to-plasma membrane compartments
HH	Hedgehog
HO	Heterotopic ossification
Hyl	Hydroxylysine
Hyp	Hydroxyproline
IHH	Indian Hedgehog
LHs	Lysyl hydroxylases
LOX	Lysyl oxidase
MACITS	Membrane-associated collagen with triple-helix domains
NC	Non-collagenous
P3Hs	Prolyl 3-hydroxylases
P4Hs	Prolyl 4-hydroxylases
PARP	Proline/arginine- rich protein
RER	Rough endoplasmic reticulum
SLRPs	Small leucine-rich proteoglycans
TGFβ	Transforming growth factor β
UCMP	Ullrich congenital muscular dystrophy

2.1 Introduction

The structure and composition of tendons and ligaments impart distinctive form and function for these connective tissues. Tendons attach muscle to bone, and ligaments attach bone to bone. As connective tissues, both tendons and ligaments are fibrous tissues that are composed of sparse cells dispersed within an extracellular matrix (ECM) rich in collagens, proteoglycans, and water. The specific hierarchical structure of these connective tissues is closely linked to their distinct functions with even positional differences in composition and organization being required for adequate performance within the musculoskeletal system. For example, differences in composition and organization might be necessary by anatomical location (e.g., axial versus limb, flexor versus extensor) or even spatially within a specific tendon or ligament (e.g., midsubstance versus enthesis within a tendon). The composition and organization of these two connective tissues allow them to guide motion, to resist abnormal displacement of bones, guide motion in order to center the actions of several muscles, and to distribute force and share load. Tendons and ligaments both have extensive water content (50–60% for tendons and 60–70% for ligaments) and by dry weight are primarily composed of collagen (70–80% for tendons and more than 80% for ligaments); they also contain elastin, glycoproteins, and proteoglycans like small leucine-rich repeat proteoglycans (SLRPs) (Frank et al. 2007; Rumian et al. 2007). Within musculoskeletal connective tissues like tendons, the hierarchical organization of fibrous collagens and other proteins ultimately contributes significantly to structure and function (Fig. 2.1). Tendons are composed primarily of collagen fibrils, and bundles of these fibrils are organized as fibers. Fibers are grouped together with tenocytes as fascicles. The fascicles are surrounded by the endotenon, which is a cellular, loose connective tissue. Tendons are then organized into groups of fascicles with the contiguous outer edge the collection of fascicles surrounded by an epitenon cover. This hierarchical structure com-

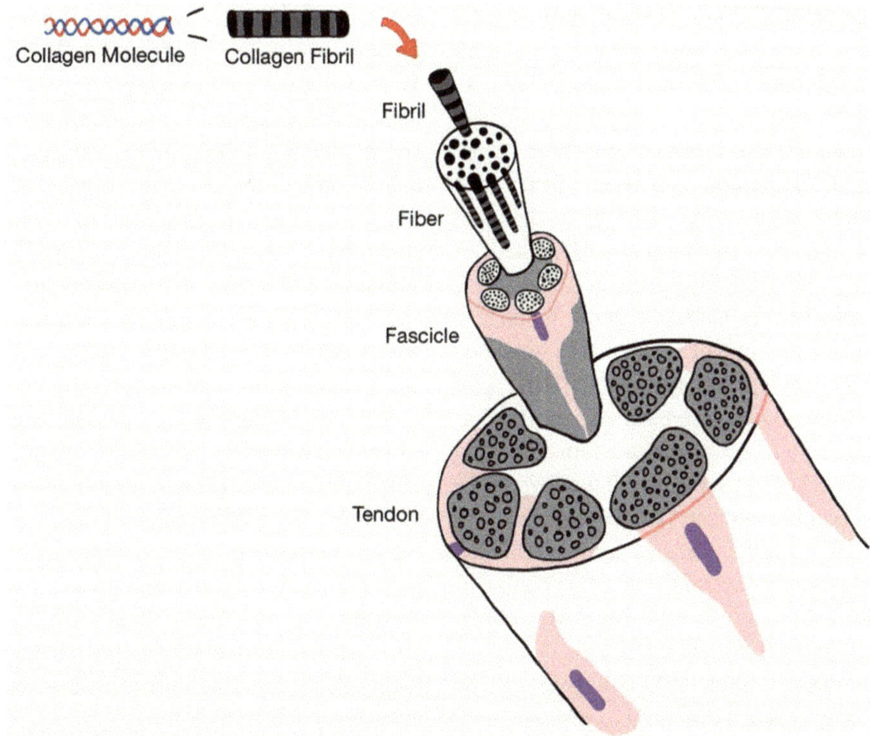

Fig. 2.1 Hierarchical structure of tendon. Tendons are hierarchical structures composed primarily of fibrils. Triple helical collagen molecules assemble to form fibrils, which in turn bundle together to form fibers. Within a mature tendon, fibers are bundled with tenocytes in fascicles; this collection of cells and fibers is surrounded by loose endotenon connective tissue, with the other edge of this ultimate structure being surrounded by epitenon contiguously

posed of assembled collagen fibrils allows for tendons and ligaments to be relatively compliant and flexible at low energy with low loading forces, yet also increase in stiffness with increasing forces and loads.

This chapter provides an overview of the basic structure, physiology, and biochemistry of connective tissues like tendons and ligaments and the ECM of which they are primarily comprised. The chapter begins with a summary of those collagens found in connective tissues with particular focus on collagens found in tendons and ligaments. Afterward, there is an overview of collagen synthesis, assembly, and maturation, which is followed by a characterization of unique protein structural features of the fibrous collagens. Finally, the tissue structure and function of the collagen-rich ECM of connective tissues like tendons and ligaments are discussed in the context of differing roles for the collagens in musculoskeletal physiology, including in tissue repair.

2.2 Collagens

Collagens are proteins that comprise the majority of the ECM of connective tissues. Collagens are typically trimers within which each monomer has at least one collagenous (COL) domain as well as non-collagenous (NC) domains. Each collagen type has its own specific number and structure of COL and NC domains and is classified with categories based upon domain structure and supra-structural organization. Within vertebrate genomes, there are 28 distinct collagen glycoproteins that are encoded by at least 45 genes; a 29th collagen (XXIX) is associated with type VI collagen. Collagens are described with Roman numeral designations (I–XXIX) in chronological order of discovery. Moreover, collagens are composed of combinations of one or more genetically distinct alpha chains, each of which has its own primary (domain) structure that contributes to the classification of the collagen types. Thus, each distinct alpha chain is encoded by a different gene. For example, the human α1(I) chain is encoded by the *COL1A1* gene and the horse α1(III) chain by the *COL3A1* gene. Collagens can be homotrimeric; that is, they are composed of three identical alpha chains, like $[\alpha1(III)]_3$ for collagen III. However, collagens also can be heterotrimeric, comprised of alpha chains encoded by different genes of the same collagen type, like $[\alpha1(I)]\,[\alpha2(I)]_2$ for collagen I from genes *COL1A1* and *COL1A2*. Furthermore, a single collagen type could possibly have multiple chain compositions, like $[\alpha1(V)]\;(\alpha2(V)]_2$, $[\alpha1(V)]\;[\alpha2(V)]\;[\alpha3(V)]$, or $[\alpha1(V)]_3$ for collagen V.

As is shown in Table 2.1, collagens can be grouped by suprastructural organization. Collagens I, II, III, V, and XI are considered fibril-forming and form cross-striated fibrils with a D-periodicity of 67 nm. Fibril-Associated Collagens with Interrupted Triple helices (FACIT) collagens are another category of collagens that associate with collagen fibrils; they interact with collagenous and non-collagenous proteins to regulate collagen fibril organization. Essentially, collagens within each category vary in abundance within tissues because their presence or absence affects the overall higher order of tissue structures as they all assemble together and copolymerize with other non-collagenous macromolecules. As tissues develop and grow or repair and remodel, collagens and non-collagenous proteins and other matrix molecules localize and organize in such a way that tissue structure and function are affected. This is certainly true amongst hierarchically organized tendons and ligaments, which are often described in terms of how form affects function.

In tendons and ligaments, between 80% and 90% of the connective tissue is composed of collagen I by dry weight (Frank et al. 2007). While collagen I is the predominant collagen in these tissues, other fibrillar collagens like collagens III and V certainly play key roles in development, growth, and repair (Birk and Bruckner 2005). Moreover, other collagen types regulate just how fibril and fiber assembly occurs. Compositionally and spatiotemporally specific collagen types ultimately inform the structure, organization, function, or even dysfunction of collagen-rich connective tissues (Franchi et al. 2007). Thus, in tendons, collagens, non-collagenous proteins, and other collagen regulatory molecules found within the ECM generally vary in composition and organization anatomically by tendon and even within tendons (e.g., myotendinous junction vs. tendon proper vs. insertion). Oftentimes cells of the tendon and ligament proper as well as surrounding peritendinous and periligamentous regions respond to niche components and even mechanical cues to promote growth, adaptation, homeostasis, and repair responses. Thus, composite structures of collagens within the ECM are highly dependent upon proper hierarchical organization of collagen molecules, fibrils, and fibers. Compromise of fibril organization affects structure and thus function; likewise in connective tissue engineering, consideration of proper organization of collagen within cell-based therapeutics is essential to recapitulating the function of tendons or ligaments grafts and constructs. In the following sections, suprastructural categories of collagen will be described, starting with fibril-forming collagens.

2.3 Fibril-Forming Collagens

The subfamily of fibril-forming collagens includes collagens I, II, III, V, XI, XXIV and XXVII. Collagens I, II, III, V, and XI have been found in tendons and in ligaments while XXIV and XXVII have been primarily expressed in bone and cartilage, respectively (Birk and Bruckner 2005; Frank et al. 2007; Franchi et al. 2007; Genovese and Karsdal 2016; Nielsen and Karsdal 2016). Fibril-forming collagens each have a long uninterrupted triple helical domain (ca. 300 nm). Moreover, fibril-forming collagen genes cluster into three distinct groups (Boot-Handford and Tuckwell 2003) which carry over

Table 2.1 General classification of collagen types

Classification	Collagen types (genes)	Supramolecular structure
Fibril-forming collagens	I (*COL1A1, COL1A2*)	Striated fibrils
	II (*COL2A1*)	
	III (*COL3A1*)	
	V (*COL5A1, COL5A2, COL5A3*)	Striated fibrils that retain N-regulatory domains
	XI (*COL11A1, COL11A2*)	
	XXIV (*COL24A1*)	Retain N-regulatory domains but supermolecular structure unknown
	XXVII (*COL27A1*)	Unknown
FACIT[a] collagens	IX (*COL9A1, COL9A2, COL9A3*)	Associated with fibrils, other interactions
	XII (*COL12A1*)	
	XIV (*COL14A1*)	
FACIT-like collagens	XVI (*COL16A1*)	Interfacial regions, basement membrane zones
	XIX (*COL19A1*)	
	XXI (*COL21A1*)	
	XXII (*COL22A1*)	
Network-forming collagens		
Basement membrane	IV (*COL4A1, COL4A2, COL4A3, COL4A4, COL4A5, COL4A6*)	Chicken wire network with lateral association
Beaded filament-forming	VI (*COL6A1, COL6A2, COL6A3, Col6A4, COL6A5 or COL29A1, COL6A6*)	Beaded filaments, networks
Anchoring fibrils	VII (*COL7A1*)	Laterally associated antiparallel dimers
Hexagonal networks	VIII (*COL8A1, COL8A2*)	Hexagonal lattices
	X (*COL10A1*)	
Transmembrane collagens	XIII (*COL13A1*)	Transmembrane and shed soluble ecto-domains
	XVII (*COL17A1*)	
	XXIII (*COL23A1*)	
	XXV (*COL25A1*)	
	Gliomedin (*GLDN*)	
	Ectodysplasin A (EDA)	
Multiplexin collagens	XV (*COL15A1*)	Basement membranes, cleaved C-terminal domains influence angiogenesis
	XVIII (*COL18A1*)	
Other molecules with collagenous domains	XXVI (*COL26A1*)	Collagenous domains in primary non-collagenous molecules
	XXVIII (*COL28A1*)	
	Acetylcholinesterase (*COLQ*)	
	Adiponectin & C1q (*ADIPOQ*)	
	Collectins, surfactant protein, others	

[a]Fibril-associated collagen with interrupted triple helix

into functional subclasses. Collagens I, II and III are the most abundant proteins in the vertebrate body and make up the bulk components of all collagen fibrils. Within tendons and ligaments, type III collagen is found in greater abundance during embryonic development (Birk and Mayne 1997). With maturation, levels of collagen III decrease, with the notable exception occurring within chicken tendons where collagen III persists in the endotenon and tendon sheath (Birk and Mayne 1997), in the mature tendons of rabbits within which 5% of all collagen is collagen III (Riechert et al. 2001), and even in human supraspinatus tendon to some extent (Buckley et al. 2013). In ligaments, roughly 10–20% of all collagen has been found to be collagen III in rabbits, humans, and horses (Becker et al. 1991; Keene et al. 1991; Riechert et al. 2001; Frank et al. 2007; Shikh Alsook et al. 2015). Moreover, many studies have reported increased levels of collagen III in injured and healing mature tendons and ligaments across many vertebrate species (Williams et al. 1984; Shikh Alsook et al. 2015; Snedeker and Foolen 2017). Otherwise, collagen I is the predominant collagen of the tendon and ligament midsubstance, However, within the entheses of tendons and ligaments, collagen II is usually present within the fibrocartilaginous zone (Fukuta et al. 1998; Milz et al. 2008). Co-assembled with types I, II, and II are quantitatively minor collagens types V and XI, which have been described on the surfaceome (i.e., plasma membrane and pericellular matrix) of tendon fibroblasts (Smith et al. 2012). Types V and XI are part of a fibril-forming subclass that retains portions of the N-terminal propeptide; they are involved in the regulation of fibril assembly (Birk and Bruckner 2011). Collagens XXIV and XXVII make up the third subclass and have differences relative to the other fibril-forming collagen types including, shorter helical regions that are interrupted (Koch et al. 2003; Plumb et al. 2007). Their presence within tendons and ligaments has not yet been reported.

Fibril-forming collagens are synthesized as procollagens, which contain an N-terminal propeptide and a non-collagenous C-terminal propeptide. The N-propeptide is composed of several non-collagenous domains and a short collagenous domain. Premature assembly of collagen molecules in fibrils is limited because of the presence of the propeptides. Procollagens are secreted from the cell from vacuole-like structures containing these procollagens (Trelsad and Hayashi 1979; Canty and Kadler 2005). The initial assembly of collagen into fibrils is regulated by the processing of the propeptides. Several enzymes with specificity for each different collagen type are involved (Greenspan 2005; Colige et al. 2005). The C-propeptides are processed by bone morphogenetic protein 1(BMP-1)/tolloid proteinases or furin (Peltonen et al. 1985; Canty and Kadler 2005; Greenspan 2005). Certain members of the a-disintegrin-and-metalloproteinase-with thrombospondin-like-motifs family (ADAMTS 2, 3 and 14) as well as BMP-1 are involved in N-propeptide processing (Colige et al. 2005; Canty and Kadler 2005). Generally for collagens I and II, propeptide processing is complete, thus producing a collagen molecule with one large central triple helical domain and terminal, short non-collagenous sequences termed the telopeptides; however, for other fibril-forming collagens propeptide processing can be incomplete. Collagens III, V, and XI can be incompletely processed – retaining a C-telopeptide and a partially processed N-propeptide domain. Retention of propeptides has been implicated in the regulation of fibrillogenesis (Fessler and Fessler 1979; Fessler et al. 1981; Fleischmaier et al. 1981; Moradi-Ameli et al. 1994; Rousseau et al. 1996). Once propeptide processing is complete for collagen molecules, they will self-assemble to form striated fibrils with a D-periodicity of 67 nm. Within each fibril, collagen molecules arrange longitudinally in staggered arrays. Thus, a gap occurs between the ends of neighboring molecules, and this gap overlap structure is present in all collagen fibrils with a 67 nm D-periodic banding pattern (Fig. 2.2).

Collagen fibrils can be heterotypic; thus, they can be assembled from combinations of two or more fibril-forming collagen types. As previously mentioned, collagens I and III are quantitatively the major fibril-forming collagens of tendons and ligaments, and collagen II is often present in the fibrocartilaginous regions particularly near tendon- or ligament-to-bone interfaces. Quantitatively minor amounts of collagens V and XI may also be found in heterotypic collagen fibrils of tendons and ligaments. The addition of collagens V and XI – typically partially processed of the N-propeptide domains – to these heterotypic fibrils signify regulatory activity in fibril assembly. The N-propeptides have a flexible or hinge domain (NC2) between the triple helical domain (COL1) and a short triple helical domain (COL2). The N-terminal domain (NC3) is composed of a variable domain and a proline/arginine- rich protein (PARP) domain. Partial processing removes the PARP yet retains the hinge, COL2, and variable domains (Linsenmayer et al. 1993; Gregory et al. 2000; Hoffman et al. 2010). Thus, quantitatively minor regulatory fibril-forming collagens co-assemble

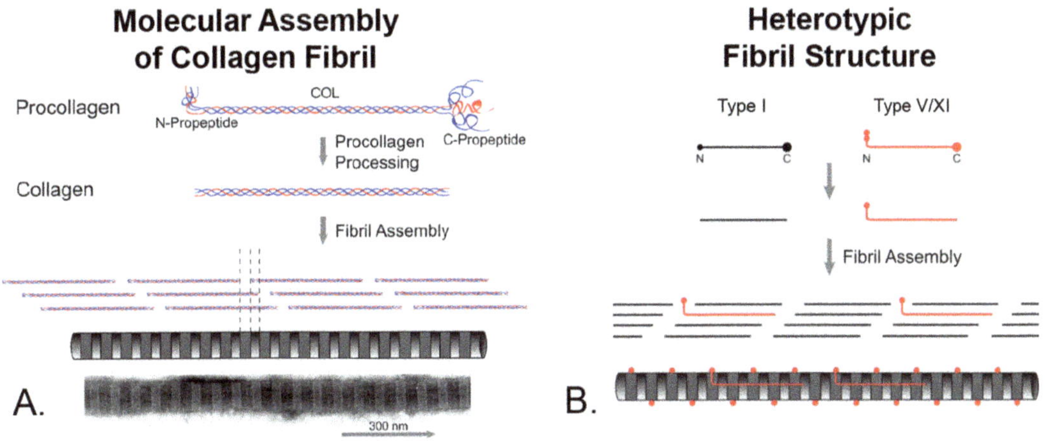

Fig. 2.2 **Fibril-forming collagens and their assembly.** (**a**) Fibril-forming collagens are synthesized as procollagens. Procollagens contain a central COL domain and flanking propeptide N- and C-terminal NC domains. Propeptides are processed and the resulting collagen molecules assemble to form striated fibrils. Each fibrillar collagen molecule is approximately 300 nm (4.4D) in length and 1.5 nm in diameter. Within the fibril, the collagen molecules are staggered N to C in a pattern giving rise to the D-periodic repeat. A D-periodic collagen fibril from tendon is presented at the bottom of the figure panel, with the alternating light and dark pattern demonstrating the respective overlaps and gaps in the fibril. (**b**) Collagen fibers are heterotypic; they are co-assembled from quantitatively major fibril-forming collagens and regulatory fibril-forming collagens. Regulatory fibril-forming collagens have a partially processed N-terminal propeptide, retaining a NC domain that must be in or on the gap region. The heterotypic interaction is involved in nucleation of fibril assembly. (This figure is from Mienaltowski and Birk (2014) and used with permission from the publisher; it was adapted from Birk and Bruckner (2011) with permission from the publisher)

with quantitatively major fibril-forming collagens to form a heterotypic collagen fibril; because the N-terminal domain of the regulatory fibril-forming collagens cannot be integrated into the staggered packing of the helical domains, fibril packing is affected (Fig. 2.2). The rigid COL2 domain of the collagen V or XI molecule can project toward the fibril surface in the gap region of the assembled fibril. Alterations in the spatiotemporal expression of collagens V and/or XI have been shown to affect fibril assembly during embryogenesis to the point of lethality, to impact proper tendon development, to maintain appropriate mechanical properties in tendons and ligaments, and to establish some semblance of structurally and functionally sufficient tendon repair (Hulmes 2002; Wenstrup et al. 2011; Connizzo et al. 2015, 2016; Park et al. 2015; Sun et al. 2015; Johnston et al. 2017). Changes include altered fibril structure, decreased fibril number and abnormal fibril and fiber organization (Wenstrup et al. 2011; Sun et al. 2015; Connizzo et al. 2016). Thus, interactions between the fibrillar collagens affect the organization of collagen fibrils within collagen- rich tissues like tendons and ligaments and are ultimately essential for functional integrity of the tissues (Hulmes 2002; Connizzo et al. 2015; Park et al. 2015; Sun et al. 2015; Johnston et al. 2017).

2.4 Fibril-Associated Collagens with Interrupted Triple Helices

FACIT collagens closely interact with fibril-forming collagens. These molecules affect the surface properties of fibrils, and they contribute to how fibril packing is regulated. Collagens IX, XII, XIV and XX are FACIT collagens. Collagen IX is a FACIT collagen that is also a proteoglycan with covalently attached glycosaminoglycan side chains – either chondroitin sulfate or dermatan sulfate (Noro et al. 1983; Huber et al. 1986). Collagen IX is primarily found interacting with collagen II (van der Rest and Mayne 1988). Likewise, FACIT collagen XII is a proteoglycan (Koch et al. 1992). Collagens XII and XIV have been found throughout musculoskeletal connective tissues, including tendons and ligaments at various times during development (Ansorge et al. 2009; Zhang et al. 2003). Collagen XIV has been found specifically at the bone-ligament interface in bovine entheses and even within the myotendinous junctions of semitendinosus and gracilis muscles following increased levels of training (Niyibizi et al. 1995; Jakobsen et al. 2017). Although it shares similarities to collagen XII and XIV, collagen XX is distributed within many tissues, but its biological role has yet to be elucidated (Willumsen and Karsdal 2016). FACIT collagens can be characterized as having have short COL domains interrupted by NC domains with an N-terminal NC domain projecting into the interfibrillar space (Fig. 2.3). FACIT collagens have two C-terminal domains NC1 and COL1 that are believed to interact with the collagen I fibrils (Olsen 2014). Large globular NC domains at FACIT collagen N-termini protrude from the fibril surface (Birk and Bruckner 2005). FACIT collagens along the surface of fibrils have been shown to affect fiber suprastructures and tendon biomechanics (Zhang et al. 2003; Ansorge et al. 2009).

The FACIT-like collagens – including collagens XVI, XIX, XXI and XXII – have features in common with FACIT collagen, but they are structurally and functionally unique. Their roles in musculoskeletal connective tissues have yet to be elucidated (Myers et al. 1994; Pan et al. 1992; Yoshioka et al. 1992; Chou and Li 2002; Koch et al. 2004). Collagen XXII has been found in the muscle side of myotendinous junctions of rats (Kostrominova and Brooks 2013); it has also been found as a marker along with *Lrg5* in murine committed articular chondrocyte progenitor cells and is enriched in the superficial/surface region of murine juvenile articular cartilage (Feng et al. 2019).

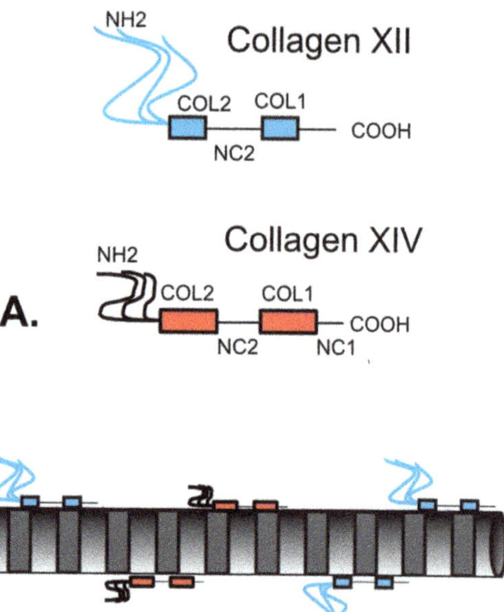

Fig. 2.3 FACIT collagens associate with fibrils. (**a**) The domain structures of FACIT collagens found in tendons and ligaments are illustrated. Note that all FACITs have alternative spliced variants, and collagen XII can have glycosaminoglycan chains attached covalently. The FACIT collagens have 2–3 COL domains and 3–4 NC domains with a large N-terminal NC domain that projects into the inter-fibrillar space. (**b**) The FACIT collagens all associate with the surface of collagen fibrils, including N-truncated isoforms due to alternative splicing in collagen XII. Collagen XII is capable of other non-fibril interactions (not shown). (This figure is from Mienaltowski and Birk (2014) and used with permission from the publisher; it was adapted from Birk and Bruckner (2011) with permission from the publisher)

2.5 Basement Membrane Collagen

Collagen IV is considered a basement membrane collagen. It is the collagenous component of an integrated network of several matrix molecules that form an ECM which defines the interface between tissues (Yurchenco and Patton 2009; Wiradjaja et al. 2010). In a site-dependent manner throughout the body, there are diverse networks of basement membranes composed of varying types of macromolecules. This includes a variety of several subtypes of collagen IV –

which can be composed of different stoichiometries of 6 collagen IV-encoding genes *COL4A1* through *COL4A6* reviewed by Khoshnoodi et al. (2006). Typically basement membranes contain lattices of laminin and collagen IV molecules that each polymerize to form a mesh with its own capacity for withstanding mechanical stress (Morrissey and Sherwood 2015). In musculoskeletal tissues, basement membranes surround muscle, connective tissue, and vasculature (Wiradjaja et al. 2010; Morrissey and Sherwood 2015; Marr et al. 2020). Within tendons, basement membrane exists within the core of the tendon, particularly within the interfascicular matrix; this basement membrane network has been well-visualized using the combination of fluorescence-based confocal microscopy and x-ray micro-computed tomography (Marr et al. 2020). Moreover, basement membranes are found along the sheath, paratenon, and epitenon of tendons in a tendon-specific manner and along the "epiligament," or surrounding surface layer of tissue for ligaments, which are also fed by or adjacent to vasculature (Frank et al. 2007). *COL4A1* encodes a collagen IV molecule that is essential for the formation of a functional basement membrane for tendon sheaths that is intact and retains tendon cells (Taylor et al. 2011). When the *Col4a1* gene is mutated in the mouse, disruptions occur in the organization of the basement membrane, thus leading to adhesions between the tendon and tendon synovium (Taylor et al. 2011).

2.6 Beaded Filament-Forming Collagen

Collagen VI is a beaded filament-forming collagen and is ubiquitous within connective tissue. It is often enriched in pericellular regions and is found as an extensive filamentous network with collagen fibrils. Collagen VI can form an array of different suprastructures, including beaded microfibrils, broad banded structures and hexagonal networks (Furthmayr et al. 1983; von der Mark et al. 1984; Bruns et al. 1986). Moreover, collagen VI interacts with many other extracellular molecules including: collagens I, II, IV, XIV; microfibril-associated glycoprotein (MAGP-1); perlecan; decorin and biglycan; hyaluronan, heparin and fibronectin, as well as integrins and the cell-surface proteoglycan NG2. These interactions are tissue-specific as collagen VI integrates with many different components of the ECM as well as cellular biomolecules (Kielty and Grant 2002). Collagen VI also may influence cell proliferation, apoptosis, migration, and differentiation, as well as ECM homeostasis and tissue mechanical properties (Cescon et al. 2015). Thus, collagen VI plays roles in the development of tissue-specific ECM, in the homeostasis of musculoskeletal and connective tissue structure and mechanics, and in repair to re-establish connective tissue integrity and for cytoprotection (Sardone et al. 2016; Bell et al. 2018; Antoniel et al. 2020). In musculoskeletal tissue, collagen VI has proven to be essential; mutations have been shown to cause various forms of muscular dystrophy as well as proximal joint contractures involving tendons in humans (Lampe and Busby 2005; Voermans et al. 2009; Izu et al. 2011; Pan et al. 2013, 2014). In tendons, when collagen VI is removed via null *Col6a1* mouse model, changes occur in biological processes in the tenocytes (Izu et al. 2011). For example, there is an abnormal increase in matrix metalloproteinase-2 (MMP-2) activity, which is believed to affect compensatory collagen-bound SLRP turnover thus causing dysfunctional fibril assembly and biomechanical deficits (Izu et al. 2011; Sardone et al. 2016).

Collagen VI is commonly formed as a heterotrimer composed of α1(VI), α2(VI) and α3(VI) chains (Chu et al. 1987; Kielty and Grant 2002). Each monomer has a 105 nm triple helical domain with flanking N- and C-terminal globular domains (Birk and Bruckner 2005). The α3(VI) chain of the heterotrimer can be processed extracellularly (Ayad et al. 1998). Structural heterogeneity is introduced into the heterotrimer by alternative splicing of domains, primarily of the α3(VI) N–terminal domain. Three additional α chains of type VI collagen have been described, α4(VI), α5(VI), α6(VI); these chains have high homology with the α3(VI) chain and may form additional isoforms (Fitzgerald et al. 2008; Gara et al. 2008). The supramolecular assembly of collagen VI begins intracellularly (Fig. 2.4). Two collagen VI monomers assemble in a lateral, antiparallel fashion to form a dimer; the monomers are staggered by 30 nm with the C-terminal domains interacting with the helical domains (Engvall et al. 1986). The resulting overlap generates a central 75 nm helical domain flanked by a non-overlapped region with the N– and C-globular domains, each about 30 nm (Furthmayer et al. 1983). The C-terminal domain-helical domain interactions are stabilized by disulfide bonds near the ends of each overlapped region (Ball et al. 2003). The overlapped helices form into a supercoil of the two monomers in the central region (Engvall et al. 1986; Knupp and Squire 2001). Two dimers then align to form, tetramers, also intracellularly. The tetramers are secreted and associate end-to-end to form beaded filaments extracellularly (Furthmayer et al. 1983; Engvall et al. 1986). The newly formed thin, beaded filaments (3–10 nm) have a periodicity of approximately 100 nm (Bruns et al. 1986; Knupp and Squire 2001). Beaded filaments laterally associate to form beaded microfibrils (Bruns

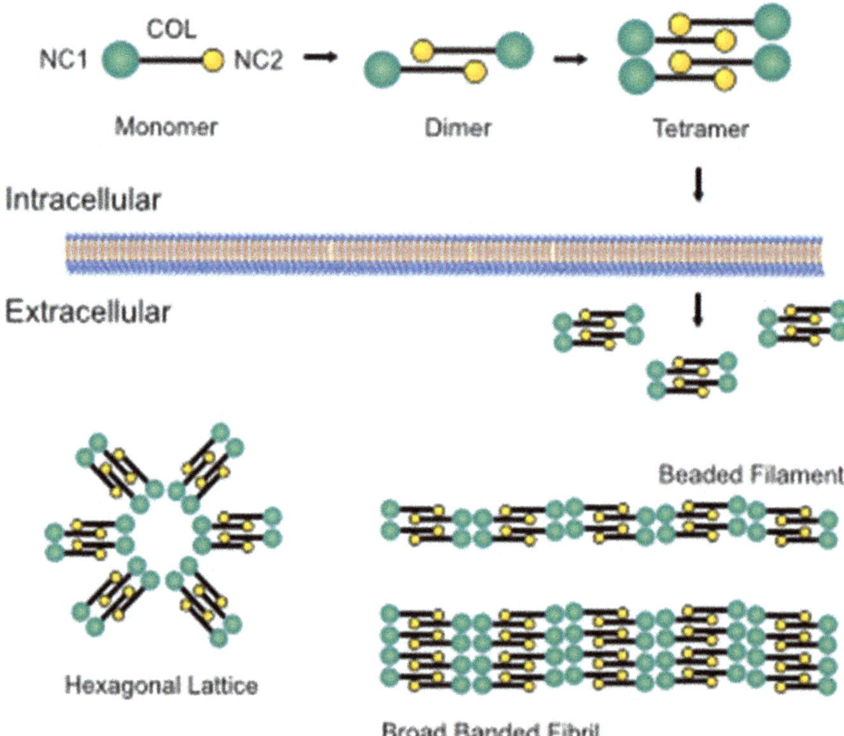

Fig. 2.4 Assembly of collagen VI suprastructures. Collagen VI monomers have a C-terminal NC domain, a central triple helical domain, and an N-terminal NC domain. The monomers assemble N-C to form dimers. Tetramers assemble from two dimers aligned in-register. The tetramers are secreted and form the building blocks of 3 different collagen VI suprastructures: beaded filaments, broad banded fibrils and hexagonal lattices. These suprastructures form via end-to- end interactions of tetramers and varying degrees of lateral association. (This figure is from Mienaltowski and Birk (2014) and used with permission from the publisher; it was adapted from Birk and Bruckner (2011) with permission from the publisher)

et al. 1986). Beaded filaments are capable of forming networks and hexagonal lattices, particularly in the presence of regulatory molecules like biglycan (Wiberg et al. 2002). When making the distinction between beaded filaments and lattices, one would observe sustained lateral growth of beaded filaments possibly with the lateral association of preformed microfibrils; this is contrast to the hexagonal lattices, which form via end-to-end interactions of tetramers occurring non-linearly (Wiberg et al. 2002).

Generally collagen VI-rich composite structures contain hexagonal lattices or beaded filaments which contains many other integrated molecules that all contribute to the ultimate aggregate suprastructure, As mentioned above, this includes the induction of hexagonal lattices by collagen tetramers interacting with biglycan – more effectively causing these lattices then decorin, for example (Wiberg et al. 2002). Moreover, essential interactions between collagen VI and other regulatory molecules like CSPG4/NG2 transmembrane proteoglycan must be maintained, or there can be disruptions to structural integrity or appropriate cellular activity. Tenocytes from patients with Bethlem myopathy (BM) and Ullrich congenital muscular dystrophy (UCMP) – caused by mutations in collagen VI genes – have dramatic changes in fibril morphology, increased incidence of cell death, issues with cellular polarization and cell migration, which could have implication on tendon repair in BM and UCMP patients (Sardone et al. 2016; Antoniel et al. 2020).

2.7 Hexagonal Network-Forming Collagen

Collagens VIII and X are closely related short chain collagens, with comparable gene and protein structures (Yamaguchi et al. 1991; Kielty and Grant 2002). While collagen VIII and X are not generally found within musculoskeletal tissues specifically, they are instead found in the vascular tissue, particularly within, near, or surrounding tendon and cartilage matrix (Kittelberger et al. 1990; MacBeath et al. 1996; Ricard-Blum et al. 2000). A review of collagen VIII describes its structure as found in endothelial cells, particularly in bovine aortic endothelial cells (Sage and Bornstein 1987). One study did localize collagen VIII within the tendon proper and peritenon of bovine Achilles tendon (Sawade and Konomi 1991). Collagen X has a restricted distribution, found in hypertrophic cartilage, cartilaginous tendon-insertion sites, as well as within mineralized zones ligament insertion sites (Fukuta et al. 1998; Shen 2005). Moreover, collagen X has been found in adult osteoarthritic cartilage (Girkontaite et al. 1996). This collagen is a homotrimer composed of $\alpha 1(X)$ chains and the supramolecular form is a hexagonal lattice (Kwan et al. 1991).

2.8 Transmembrane Collagens

Collagens XIII, XVII, XXIII, and XXV are transmembrane collagens and are all homotrimers. Within the structure of these collagens, there is an N-terminal cytoplasmic domain, as well as a large C-terminal domain containing multiple COL domains with NC interruptions providing flexibility (Heikkinen et al. 2012). These collagens are unique in that they all contain a hydrophobic membrane spanning domain (Exposito et al. 2002; Heikkinen et al. 2012). Between the membrane-spanning domain and an adjacent extracellular linker domain is the first COL domain involved in trimerization that is also subject to proteolytic cleavage and is shed extracellularly. Of the transmembrane collagens, so far from the literature only collagen XIII has been found in musculoskeletal tissues, particularly in myotendinous and neuromuscular junctions (Latvanlehto et al. 2010; Heikkinen et al. 2012).

2.9 Procollagen Synthesis, Collagen Fibril Assembly, Growth and Maturation

Collagen synthesis, assembly, and maturation require a sequence of well-controlled intracellular and extracellular events. In the nucleus, collagen genes are transcribed from DNA into mRNA. Then the mRNA transcripts are transported out of the nucleus; they are translated into preprocollagen monomers. The transported monomers undergo N-terminal signaling peptide cleavage. In the RER, post-translational modification occurs prior to the assembly of procollagen triple helices. The extent of these modifications is affected by the rate of triple helix formation, which is in turn affected by the primary structure of the alpha chain propeptides. For example, point mutations in the Gly-X-Y sequence may result in altered molecular properties for that chain as well as dysfunctional regulation of chain selection, helix formation, or post-translational modification. The triple helices are secreted as procollagens to prevent premature assembly of suprastructures within the cells. Once procollagens are secreted extracellularly, pro-peptides are processed by collagen type-specific metalloproteinases. Processing may, however, begin during the transport of newly synthesized procollagens to the cell surface (Canty et al. 2004; Humphries et al. 2008). Processed collagen triple helices are then cross-linked. The relationship between protein structure, post-translational modification, triple helix assembly and collagen fibril formation will be discussed below, as well as the growth and maturation of fibrils in tendons and ligaments in following sections.

2.10 Triple Helix Assembly and the Impact of Primary Structure on Secondary, Tertiary, and Quaternary Structures

Once collagen pre-propeptides are translated, there are many post-translational modifications. It begins with several molecular chaperones recognizing the N-terminal signal peptide found on each pre-propeptide; these chaperones then direct the pre-propeptides into the lumen of the rough endoplasmic reticulum (RER) (Canty and Kadler 2005). The N-terminal signal for transport to the RER is then cleaved so that each pre-propeptide becomes a propeptide, or a pro-alpha chain (Canty and Kadler 2005; Hulmes 2008). Afterward, pro-alpha chains undergo hydroxylation of prolyl and lysyl residues followed by glycosylation of hydroxylysyl residues (Hulmes 2008; Ishikawa and Bachinger 2013). Then three pro-alpha chains will trimerize as homo- or heterotrimers, which involves the selection and alignment of appropriate alpha chains with subsequent assembly into specific trimeric collagen molecules (Khoshnoodi et al. 2006; Ishikawa and Bachinger 2013). Trimerization is initiated through interactions among the non-collagenous trimerization domains of alpha chains at their C-termini in the RER (Hulmes 2008; Ishikawa and Bachinger 2013).

The primary amino acid sequences of collagen alpha chains affect the proteins structures at all other levels. The primary structure of a collagen alpha chain features a COL domain that will coil into a left-handed helix lacking secondary structure-specific intra-chain hydrogen bonds, unlike the right-handed helix of a common motif with secondary structure, the alpha helix (Ramachandran and Kartha 1954; Kramer et al. 1999). Thus, three alpha chains will supercoil to form a triple helix. This parallel, right-handed superhelix is stabilized by inter-chain quaternary hydrogen bonds that are almost perpendicular to the triple helical axis. Due to the imino acid content, collagen polypeptides assume an elongated polyproline II-like helix with a secondary structure of all peptide bonds in the *trans* configuration. The pitch of the polyproline II helix in collagenous polypeptides corresponds to three amino acids, almost exactly. Steric constraints require that only the smallest amino acid glycine can occupy the positions at the center of the triple helix (Ramachandran and Kartha 1954). Thus, the triple helical domains have a repeating (Gly-X-Y)$_n$ structure with the X and Y positions being any of the 21 proteinogenic amino acids of the universal genetic code (Rich and Crick 1961; Kramer et al. 1999). However, the X and Y positions are typically occupied by proline and hydroxyproline, which are necessary for helix formation and stability, respectively (Rich and Crick 1961; Kramer et al. 1999). When glycine residues are replaced with other amino acids, there are interruptions in the triple helix motif which cause the rod-like structures to have rigid kinks or flexible hinges (Vogel et al. 1988; Shoulders and Raines 2009). This change in primary structure thus ultimately changes structural features in the collagen; this change can also provide flexibility by increasing helical domains as seen in collagen V. However, it is also possible that missense mutations in collagen genes result in the substitution of glycine residues in collagen peptides and ultimately whole triple helical polypeptides. These mutations manifest as the underlying etiology for mild or severe, systemic connective tissue diseases (Myllyharju and Kivirikko 2004; Malfait and De Paepe 2009).

Because post-translational modification and triple helix assembly occur concomitantly, once initiated, trimerization must be controlled to allow for alpha chain post-translational events like hydroxylation and glycosylation to occur. During or after collagen alpha chain translation, amino acids within the triple helical domains of each chain can be modified (Ishikawa and Bachinger 2013) (Fig. 2.5). One form of co-translational modification of the unfolded chains is hydroxylation, which is performed by three distinct enzyme families – prolyl 4-hydroxylases (P4Hs), prolyl 3-hydroxylases (P3Hs) and lysyl hydroxylases (LHs) (Ishikawa and Bachinger 2013) – leading to the production of two unique amino acids, hydroxyproline (Hyp) and hydroxylysine (Hyl), which are important downstream for triple helix stability and glycosylation, respectively (Ishikawa and Bachinger 2013). It is important to note that these hydroxylases all have different isoforms in vertebrates with differing tissue-specificity and spatiotemporal expression, substrate preference, and other possible functions particularly when part of complex formation with other proteins (Ishikawa and Bachinger 2013; Gjaltema and Bank 2017). Thus, as the propeptides are synthesized, unique amino acids Hyp and Hyl are introduced by enzymatic hydroxylation of almost all prolyl- and some of the lysyl residues in the Y-positions. Hyp residues are important in triple helix stability. The 4-trans hydroxyl group assists in directing the free Hyp, similar to that of Y-hydroxyproline in collagen triple helices. The forced integration of proline into Y-positions absorbs more ring deformation when compared with physiological hydroxylated collagen (Berg and Prockop 1973; Okuyama et al. 2009). After hydroxylation, more post-translational modifications occur prior to triple helix formation. Some Hyl residues will receive *O*-linked sugar modifications with the addition of carbohydrates b-galactose and a-glucose as catalyzed by galactotransferases and glycosyltransferases, respectively (Ishikawa and Bachinger 2013). There are several theories about the functional role of these *O*-linked modifications; one study speculated that these modifications slightly destabilize collagen triple helices while enhancing their self-assembly (Huang et al. 2012), while other studies point out that these modifications could affect lateral assembly of fibrils (Sricholpech et al. 2012; Pokidysheva et al. 2013). There are more post-translational modifications after secretion and during aggregate assembly.

Besides its impact during triple helix assembly, post-translational glycosylation of collagens affects fibril structure, in particular due to the impact on fibril assembly (Torre-Blanco et al. 1992; Batge et al. 1997; Notbohm et al. 1999). The circumference of collagen triple helical domains is affected by the extent of hydroxylation of lysyl residues and subsequent glucosyl- and galactosylation of hydroxylysyl residues. Moreover, intermolecular center-to-center distances correlate with the extent of glycosylation, particularly if these post-translational modifications occur in those regions of polypeptides situated in overlap regions of the fibril. The extent of glycosylation of Hyl modifications can be manipulated by features such as increased enzyme activity levels (Keller et al. 1985) and disease-causing mutations (Myllyharju and Kivirikko 2004). Thus, rates of procollagen triple helix formation can be substantially reduced in the RER by collagen mutations, and over-modification of collagens via *O*-linked saccharides can result in compromised fibrillar organization. Nevertheless, glycosylation level can serve as a physiological mode of regulation in normal native tissue, par-

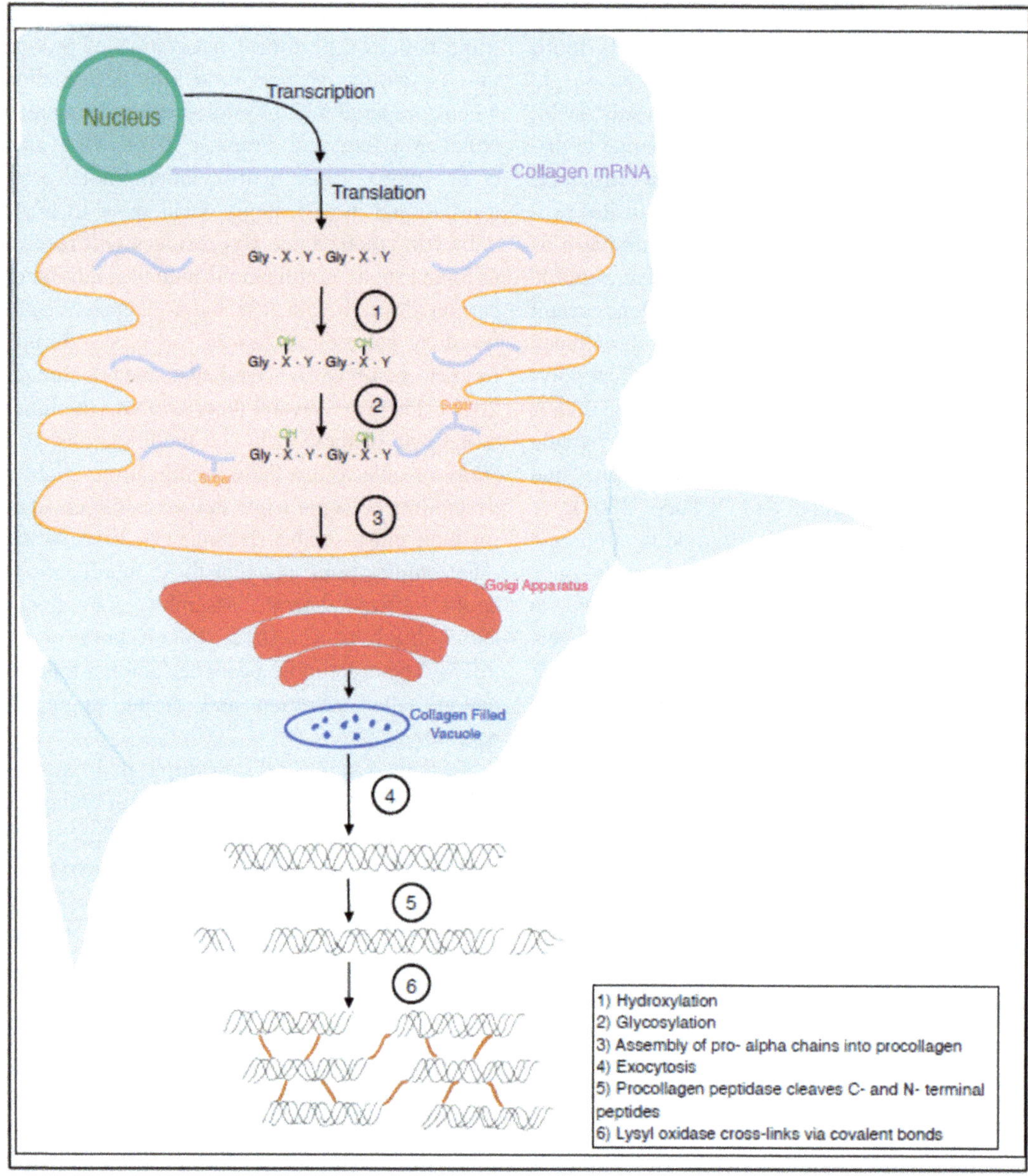

Fig. 2.5 Post-translation modification of collagen molecules and triple helices. A collagen mRNA is translated into preprocollagens. Preprocollagens are transported to the rough endoplasmic reticulum where they undergo hydroxylation, glycosylation, cleavages, and assembly into triple helices that are transferred outside the cell via exocytosis. Outside of the cell, cleavages lead to the formation of tropocollagens that assemble and cross-link to form fibrils

ticularly when levels of post-translational modification are tissue-specific, thus contributing to the structural properties of particular tissues. Intrafibrillar water has also been shown to affect the molecular organization of collagen fibrils. That is, as variable amounts of water are incorporated within the fibrils, intermolecular distances between lateral or longitudinal neighbors have been shown to differ; when collagen fibrils are dried, their D-periodicity shortens and intermolecular lateral distances reduce (Brodsky et al. 1982; Katz et al. 1986).

Collagenous domains of alpha chains are distinctly rich in *cis* -peptide bonds due to their high content of imino acids that favor *cis*-peptide bond formation (Bruckner et al. 1981). Kinetically, a great deal of energy is required to cause *cis*-peptide bonds to comply with triple helix formation, which occurs in a zipper-like, slow-folding isomerization process (Bachinger et al. 1980; Bruckner et al. 1981). At the start of triple helix formation, a variable number of *cis* bonds are distributed throughout the still unfolded procollagen polypeptides; thus, the folding times required for full-length triple helix formation are variable because they are dependent upon kinetic obstacles. It has been shown that cyclophilin B catalyzes the *cis* to *trans* isomerization of Gly-Pro- but not X-Hyp peptide bonds in chick embryo tendon fibroblasts (Steinmann et al. 1991). Cyclophilin B is acting as a peptidyl prolyl *cis/trans*-isomerase; it can be inhibited by the immunosuppressor cyclosporin A (Steinmann et al. 1991). Moreover, cyclophilin B, prolyl-3-hydroxylase, and cartilage-associated protein (Crtap) form a ternary complex with high chaperone activity in the endoplasmic reticulum. For nascent fibrillar procollagens, prolyl-3- hydroxylase catalyzes the introduction of a single 3-Hyp residue at the C-terminal end of the triple helical domain. Ultimately, the ternary complex localizes cyclophilin B-activity to the initiation sites for procollagen folding, allowing for the efficient catalytic isomerization of peptidyl-prolyl *cis* bonds. When this process is perturbed, severe osteogenesis imperfecta (type VII and VIII) has been observed in humans for null mutations in *LEPRE1* and *CRTAP* genes (Marini et al. 2010).

There are more modifications that occur post-translationally after secretion of procollagen and during supramolecular assembly. Extracellular lysyl oxidases exist that convert the amino groups on some of the Hyl and lysine residues in the collagen polypeptide chain to aldehydes (Hulmes 2008). The aldehydes which form aldols or β-ketoamines as they react with aldehydes or amino groups on allysines in other chains to generate quaternary structures that are intra-molecular and inter-molecular covalent crosslinks (Barnard et al. 1987; Bailey 2001; Avery and Bailey 2005). This collagen cross-linking can occur at early stages of aggregation to modulate the suprastructural outcome of fibrillogenesis or the formation of networks. Cross-links in tendon can occur near the N- and C-termini of collagen molecules. These cross-links aid in polymerization to form fibrils and increase with age.

2.11 Collagen Fibril Assembly

In Sect. 2.9, a summary was provided of the steps leading to procollagen triple helix formation. In Sect. 2.10, specific features of collagen primary protein structure were provided with an elucidation of how these features and other post-translational modifications affect secondary, tertiary, and quaternary structures. In this section, collagen assembly will be described from gene to procollagen chains, from triple helix formation to fibril assembly. Collagen fibril assembly is described as a self-assembly process involving a series of events that occur in both intracellular and extracellular compartments of the fibroblast (Kadler et al. 1996). Such compartmentalization allows for many steps in the process from gene to collagen fibril for which regulation can affect collagen assembly. Regulation steps include activating the controls of the stoichiometry of different matrix gene products during synthesis, hydroxylation, glycosylation, folding and triple helix formation, packaging for secretion from the Golgi apparatus, transport via specialized and elongated intracellular compartments with secretion at the cell surface, and the actual formation of extracellular compartments for matrix assembly (Birk et al. 1989). Once these controls are activated, simultaneous regulated interactions of processing enzymes, fibril associated molecules (e.g., proteoglycans and FACITs), and adhesive glycoproteins all allow for the fibroblast to make various collagen types depending on its destination in the body (Fig. 2.6).

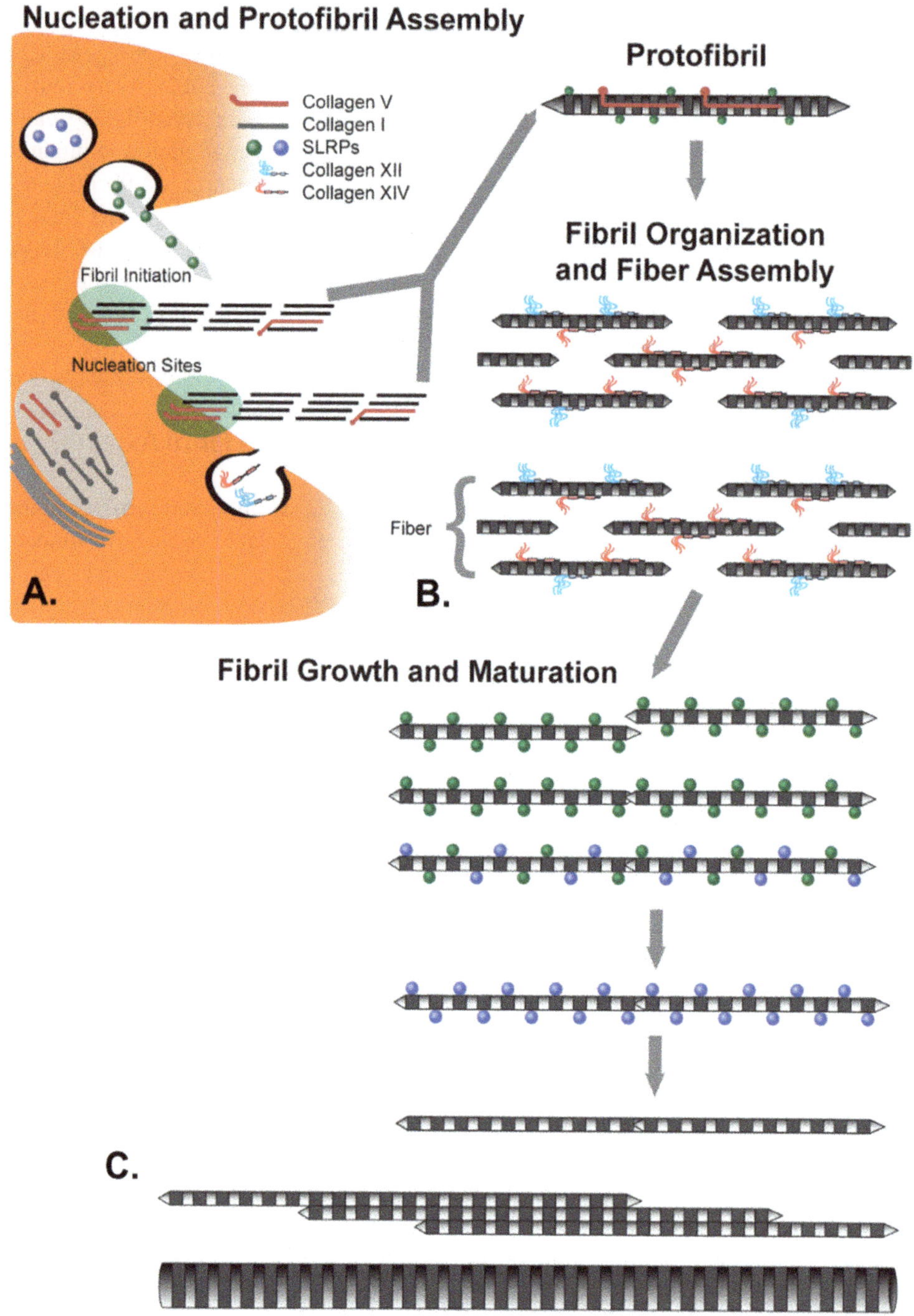

Fig. 2.6 Model of the regulation of fibril assembly. Model of the regulation of fibril assembly. Fibril assembly involves a sequence of events. (**a**) Collagen assembly is initiated at the fibroblast cell surface via nucleators like collagens V and XI; then immature, small diameter, short protofibrils are assembled. The nucleation process is cell directed involving interactions with organizers (e.g., integrins and syndecans) at the cell surface. (**b**) Protofibrils are deposited into the matrix and are stabilized by interactions with matrix regulators like SLRPs and FACITs. (**c**) Changes occur with fibril stabilization resulting from processing, turnover and/or displacement; this affects linear and lateral growth of mature fibrils in a tissue-specific manner. (This figure is from Mienaltowski and Birk (2014) and used with permission from the publisher; it was adapted from Birk and Bruckner (2011) with permission from the publisher)

Assembly of collagen begins in the nucleus which synthesizes collagen mRNAs during transcription. These mRNAs are transported out of the nucleus; there are 45 mRNA variations that make up the 28 collagens seen in vertebrates with Type-I collagen being the most abundant and playing a major role in the body by stabilizing necessary structures (Birk et al. 1989; Kadler 2017). Once transcribed, the collagen mRNA is translated into a preprocollagen, also known as a prepro-α-chain. The collagenous region of the preprocollagen utilizes repeating Gly-X-Y units. The X and Y units are normally imino acids proline and hydroxyproline. Glycine is the only amino acid small enough to fit in the tight spaces in the central position of the triple helix procollagen assembled later in the cell. Due to the size restriction, the glycine concentration is more abundant than any other amino acids in the cell (Kadler et al. 1996).

As mentioned in the previous section, an N-signaling peptide aids in the transport of the preprocollagen to the RER. Once the signaling peptide is cleaved, the propeptides are hydroxylated. In order for hydroxylation to occur, vitamin C must be present as a cofactor. Deficiencies in vitamin C lead to scurvy, which is a disease characterized by inadequate connective tissue renewal due to this vitamin deficiency (Canty and Kadler 2005). Moreover, there have been studies demonstrating that Vitamin C-enriched gelatin supplementation leads to increases in collagen synthesis if taken an hour before exercise (Lis and Baar 2019). Another known defect due to hydroxylation irregularities is Epidermolysis bullosa. This disease is caused by mutations in *PLOD3*, which encodes the lysyl hydroxylase 3 enzyme. This skin disorder manifests due to a lack of type VII collagen. Type VII collagen is characterized as an anchoring fibril – tethering the dermal-epidermal basement membrane to the underlying dermis. Thus, this disease manifests as blisters and erosions leading to ulcers and poor skin healing (Vahidnezhad et al. 2019). Procollagens also undergo glycosylation, as reviewed in the previous section. Glycosylation triggers the assembly of procollagen chains to aggregate into triple helices. If this assembly is defective, this causes osteogenesis imperfecta, which is most often caused by an autosomal dominant genetic disorder of the *COL1A1* and *COL1A2* genes. For a recent review of osteogenesis imperfecta, please see Marini et al. (2017). Collagen fibrils are composites of different matrix molecules, and control of heteropolymeric mixing and trimer type stoichiometry begins within the intracellular compartments. After the procollagens assemble correctly into triple helices, they travel from the RER to the Golgi body to be exocytosed. Secreted triple helices of procollagens are further processed outside of the cell to become tropocollagens.

Along with the procollagens, secretion of different matrix molecules and modifying enzymes occurs with specific spatial, temporal, and circumstantial patterns. Moreover, an important regulator of the expression of key tendon markers for tenocytes and for downstream expression of fibrillar collagens and associated collagen genes is the transcription factor Early Growth Response 1 (EGR1) (Havis and Duprez 2020). EGR1 is involved in the development, homeostasis, and healing process for tendons, bones, and adipose tissue through regulating matrix formation. The loss of *EGR1* causes reduced ECM production, and ultimately compromised tendon morphology (Havis and Duprez 2020). Thus, the character of the assembled fibrillar matrix depends upon not only the collagen types synthesized, but on the regulated interactions with procollagen processing enzymes, fibril-associated molecules (e.g., proteoglycans and FACITs), adhesive glycoproteins, and the involvement of transcription factors. The spatial and temporal regulation of molecules interacting during packaging and transport or at the sites of secretion provides a mechanism whereby limited numbers of matrix molecules can be uniquely assembled to produce the divoersity of structure and function observed across tissues.

Extracellularly in the developing tendon, fibril assembly begins in channels defined by the fibroblast surface (Trelstad and Hayashi 1979; Birk and Trelstad 1986; Canty et al. 2004; Canty and Kadler 2005). Immature protofibrils with uniform diameters and short lengths – relative to mature fibrils – are assembled in these microdomains (Birk and Trelstad 1986; Birk et al. 1989). These extracellular channels have been shown to form at the time of secretion as specialized post-Golgi secretory compartments. They fuse with the fibroblast membrane and are maintained due to slow membrane recycling associated with the presence of the assembled protofibril (Birk and Trelstad 1986; Birk et al. 1989). Others have suggested that intracellular processing of procollagen may occur within elongated Golgi-to-plasma membrane compartments (GPCs), which is then followed by the extrusion of protofibrils through a cellular protrusion where GPCs fuse to fibroblast plasma membranes (Canty et al. 2004; Canty and Kadler 2005). Regardless, protofibrils are ultimately present extracellularly. Once the protofibrils are deposited into the extracellular matrix, more compartmentalization occurs as fibrils form small fibers and then larger structures characteristic of the specific tissue (e.g., large fibers in tendon). Extracellular compartmentalization hierarchy allows for there to be some level of control over the extracellular steps of matrix assembly by the fibroblasts secreting these molecules.

Once secreted, peptidases remove the globular N- and C- propeptides from procollagens forming tropocollagens (Canty and Kadler 2005). Procollagen C-proteinase enhancer-1 is a secreted protein needed in order to accelerate the proteolytic release of the C-propeptide from the fibrillar procollagens (Pulido et al. 2018) while the N-proteinase activity is provided by the members of A Disintegrin And Metalloproteinase with Thrombospondin motifs (ADAMTS) (Canty and Kadler 2005). If there is a defect in ADAMTS expression specifically, heterotopic ossification is seen in the cell (Mead et al. 2018). Heterotopic ossification (HO) causes abnormal bone growth in non-skeletal tissues such as muscles, tendons, or any soft tissue. ADAMTS7 and ADAMTS12 regulate tendon collagen fibril structure while also functioning as an innate inhibitor of HO (Mead et al. 2018). C proteinase, on the other hand, triggers the self-assembly of collagen fibrils seen in the next step (Mead et al. 2018). The final step in collagen synthesis is triggered by lysyl oxidase (LOX). This enzyme cross-links tropocollagens to one another. These cross-links are covalent and help the collagen to aggregate first into long linear subunits as well as aggregates that grow both linearly and laterally to form a subfibril (Trelstad and Hayashi 1979). These cross-links have been shown to be regulated by the SLRP fibromodulin (FMOD). FMOD affects the extent and pattern of lysyl oxidase-mediated collagen cross-linking by inhibiting enzyme access to the telopeptides. Therefore, if collagen is fibromodulin deficient, it causes increased C-telopeptide cross-linking, leading to abnormal fibril formation and decreased mechanical strength (Kalamajski et al. 2014).

Many other factors also have an influence on development and maintenance of collagenous extracellular matrix. Recently it was determined that the circadian clock plays a significant role in tendon development. Researchers have concluded that the circadian rhythm of a given tendon regulates the signaling of bone morphogenetic protein signaling, procollagen I synthesis and secretion, as well as endoplasmic reticulum homeostasis (Yeung and Kadler 2019). Additionally, the roles of decorin and biglycan have also shown to be of massive importance in maintaining collagen fibril structure, realignment, and mechanical properties in mature tendons. Both decorin and biglycan are SLRPs that help maintain tendon morphology, peaking at different times in tendon development, growth, and adaptation. For instance, decorin peaks in early development of tendon structure when the collagen fibrils are undergoing lateral growth, and it continues to be expressed during maturation and aging (Zhang et al. 2005). Although decorin is necessary in the assembly of collagen fibrils in tendon and ligament, it also plays a crucial role in the establishment of normal biomechanical functions of cartilage (Han et al. 2019). On the other hand, biglycan peaks during post-natal development and then rapidly declines. Together, these molecules influence elastic properties, stress relaxation speeds, collagen fibrils diameter, and collagen morphology. Absence of deficiencies of these molecules has shown to have detrimental effects on the proper assembly of collagen fibrils

(Robinson et al. 2017). Other important components of the ECM are lumican and plasma membrane-associated collagen with triple-helix domains (MACITS). Lumican is an extracellular matrix secreted proteoglycan that helps to regulate collagen fibrogenesis as well as maintain ECM structure (Mohammadzadeh et al. 2020) while MACITS stabilize the pericellular ECM network (Izzi et al. 2020). Properly processed collagen aggregates to form an elastic tissue. This elasticity is attributed to the actin cytoskeleton made early on in embryonic development. Chemical disruption in the cytoskeleton causes the tendon to lose its elastic characteristic (Schiele et al. 2015). While procollagen molecules are processed inside the cell, tropocollagen molecules are assembled outside of the cell in the extracellular matrix (ECM). Produced and maintained by tenocytes, the ECM is the compendium of signaling effectors that provides structural and functional support to the cells that it surrounds. This matrix is required to allow for regular cellular maintenance (Iozzo and Gubbiotti 2018). The ECM is also constantly remodeled and maintained. Metalloproteinases allow for the ECM to remodel due to their enzymatic abilities. These include the degradation of matrix components, cell surface receptors, cytokines, and growth factors (Karamanos et al. 2019).

If properly made, the matrix will have three extracellular compartments that are responsible for matrix organization and heavily regulated by fibulin-4. Fibulin-4 is an ECM glycoprotein that is essential in the compartmentalization of the extracellular space (Markova et al. 2016). These compartments are collagen fibrils, bundles, and collagen microaggregates. They allow for the cell's influence to extend into the extracellular space around it mainly through fibropositors (Birk and Trelstad 1986). Plasma membrane protrusions allow collagen fibrils to be deposited in the extracellular channels formed by adjacent fibroblasts (Canty et al. 2004). Fibropositors interact with the extracellular matrix to provide a mechanical interface between the cytoskeleton of the cell and the ECM, allowing for effective communication to the cell for appropriate response to different mechanical loads (Holmes et al. 2018). The ability to form fibropositors comes from the physical linkage of the aggrecan network caused by decorin.

2.12 Assembly and Growth of Mature Tendon Fibrils

In mature tendons, collagen fibrils are long functionally continuous arrangements of molecules with diameters ranging between 20 and 500 nm depending on the tissue and developmental stage (Birk et al. 1995, 1997; Canty and Kadler 2002). During development, the tropocollagen molecules are arranged into a quarter-staggered pattern with enzymatic and non-enzymatic chemical crosslinking (Wess et al. 1998). Enzymatic crosslinking is accomplished through an aldehyde bond formation by lysyl oxidase with non-enzymatic crosslinking by glycation. The collagen fibrils are assembled near the fibroblast surface as uniform and relatively short D periodic protofibrils with diameters in the range of 20–40 nm, lengths of 4–12μm, and tapered ends (Birk et al. 1989, 1995; Kadler et al. 1996; Graham et al. 2000). As a result of this and the staggered molecular organization, the characteristic 67 nm D-periodic banding pattern can be observed. Additionally, the Hodge-Petruska model of staggered tropocollagen has been extended with revealing regions of amorphous and crystalline order in a lateral quasi-hexagonal lattice arrangement when seen in cross-section (Hulmes et al. 1995; Wess et al. 1998; Orgel et al. 2006). The newly assembled protofibrils are deposited and incorporated into the developing tendon extracellular matrix as small bundles of fibrils or immature fibers. The fibrils are stabilized via interactions with macromolecules such as FACITs and SLRPs (Fig. 2.6). Specifically, SLRPs such as decorin interact with the collagen fibrils and regulate fibril growth by binding non-covalently at every D period to an intraperiod site on the surface of the collagen fibrils (Iozzo and Schaefer 2015). Tendon maturation continues with linear fibril growth involving end overlap of the protofibrils, followed by lateral growth in tendons and ligaments. Lateral fibril growth features the association and fusion of fibrils laterally to generate larger diameter fibrils generating a cylindrical fibril structure by molecular rearrangement. To accomplish this, some or all of the collagen stabilizing components are lost or replaced. Throughout this process, the amount of intra- and intermolecular crosslinking plays an important role in regulation of structure turnover and stability. Mechanical stability improves with increased crosslinking, which varies by subregion or zone within tendons and ligaments as well as anatomically throughout the body. At this point, a crimp pattern at the fibril and fascicle level between the collagen fibrils by elastin introduces additional complexity to the tendon or ligament with the amplitude and period of crimp being tissue specific. The detailed roles of collagen accretion and maintenance in fibril assembly, with homeostasis, and in repair or regeneration remain to be elucidated.

2.13 Regulation of Collagen Fibril Assembly and Growth

Regulation of collagen fibrillogenesis is tissue-specific and can have many effectors. Interactions occur amongst many different classes of molecules, such as processing enzymes, heterotypic fibril-forming collagens, FACITs and SLRPs, as well as other glycoproteins (e.g., fibronectin and tenascin X). Many molecules serve as regulators via three main classes of interactions: organizers, nucleators, and regulators (Kadler et al. 2008). During growth and development of tendon, significant increases in collagen fibril diameters occur, which directly affects the mechanical properties of the tendon. (Zhang et al. 2005).

Collagens V and XI have been shown to nucleate collagen fibril formation in self-assembly assays in addition to playing a role in assembly of immature protofibrils (Marchant et al. 1996; Blaschke et al. 2000; Wenstrup et al. 2004a, b). The alpha-chains of collagens V and XI have multiple tissue-specific isoforms of a highly homologous collagen type forming heterotypic fibrils with collagen I and II regulating collagen organization and fibril differences. A mouse model with a targeted deletion within the *Col5a1* gene is lethal in embryos because of a lack of fibril formation in the mesenchyme; though collagen I is synthesized and secreted, protofibrils are not properly assembled (Wenstrup et al. 2004a). Likewise, in two separate mouse models where collagen XI is absent (*cho/cho*, ablated *Col11a1* alleles; *Col2a1*-null mice, from which the α3(XI) chain is derived), animals develop chondrodysplasia (cho) where cartilage is devoid of fibrils. (Seegmiller et al. 1971; Li et al. 1995a, b; Aszodi et al. 1998). Thus, a key role for collagen V/XI in nucleation of assembly of short, small diameter protofibrils is demonstrated.

Under physiological conditions in vitro, collagens I and II can self-assemble after long lag phases in the absence of collagen V. As such, collagen V is critical in providing a fibril-forming site for the fibroblast. During collagen fibril formation, collagen V fibrils produce the necessary site for collagen I allowing for subsequent fibril intermediates, linear fusions, and lateral fusions producing the heterotypic collagen I and V fibril. This feature is tissue specific: in the cornea where smaller fibril diameters are paramount to transparency, collagen V is 10–20% of the fibril- forming collagen content, and the increased nucleation sites contribute to many smaller diameter fibrils (Segev et al. 2006; Birk 2001). In tendons with increased mechanical strength, larger diameter fibrils are produced with decreased collagen V (1–5%) within the fibril-forming collagen content. Perhaps the best example of the impact of collagen I/V interactions on structure is with the classical form of Ehlers-Danlos Syndrome (EDS) – a generalized connective tissue disorder where the majority of patients are heterozygous for mutations in collagen V (Symoens et al. 2012). With this form of EDS, approximately 50% of the normal collagen V amount is present, and affected patients have a dermal phenotype with large, structurally aberrant fibrils. Similarly, in the heterozygous (*Col5a1*$^{+/-}$) mouse model, there was a 50% reduction of collagen V; thus, there were fewer nucleation events than the normal mice (Wenstrup et al. 2004a). A similar situation occurs in the tendon (Wenstrup et al. 2011). The reduction in nucleation sites makes collagen

V a rate-limiting molecule. The regulation of these collagen I/V interactions is coordinated by the sites of assembly and by other molecules that organize and sequester these interactions at the cell surface (Smith et al. 2012).

The nucleation of protofibrils is informed by the localization of cellular structures such as organelles, cytoskeletal components, cytoplasmic membrane domains, and organizing molecules. However, cell-defined extracellular domains are equally important for efficient fibril assembly, and without peri- and extracellular organization procollagens would be freely secreted away from the cell or forming tissues such as in monolayer cultures. Consequently, cell-directed positioning of the deposited matrix is possible as assembled protofibrils undock from the cell surface, are extruded from the cell, and then are incorporated into the extracellular matrix. This then re-primes the nucleation sites for the next round of nucleation. Cell-directed collagen fibril assembly involves organizing molecules such as integrins and fibronectin. Fibronectin mediates cell-collagen interactions when there are assembling collagen fibrils and other ECM molecules. Fibronectin molecules assemble into fibrils via integrin interactions (Zhong et al. 1998; Mao and Schwarzbauer 2005; Kadler et al. 2008). The fibronectin fibril network contains multiple binding sites for collagen fibril assembly. When binding sites are blocked in fibronectin networks, collagen fibril assembly is inhibited (McDonald et al. 1982). Furthermore, modifications to fibronectin-integrin interaction affect collagen fibril assembly, suggesting both that the cytoskeleton is involved in some way with fibril assembly and that other direct interactions of integrins with other surface molecules are essential to fibril assembly.

Once the protofibrils are assembled and deposited into the ECM, further assembly to the mature fibrils involves linear and lateral growth of the preformed intermediates. In tendons and ligaments, two classes of regulatory molecules involved in the regulation of these steps are SLRPs and FACIT collagens. Both classes are fibril-associated, and molecules within each class have their own tissue-specific, temporal, and spatial expression patterns. Differences in expression patterns contribute to the differences in structure and function amongst tissues, including tendons and ligaments.

SLRPs are crucial regulators of linear and lateral fibril growth that are highly expressed during development (Chen and Birk 2013). Of the five classes of SLRPs, two are instrumental in collagen fibril growth, maturation, and homeostasis. Class I (biglycan and decorin) and class II (fibromodulin and lumican) SLRPs are tissue-specific molecules with characteristic components necessary for appropriate function. Disruption of these characteristics (the leucine-rich repeat ear or the N- and C-terminus disulfide caps) leads to disruption of function resulting in detrimental or fatal effects as seen in SLRP single or double deficient mice. As a result, the importance of their regulation of linear and lateral fibril growth is demonstrated (Ameye and Young 2002; Chikravarti 2002; Kalamajski and Oldberg 2010; Chen and Birk 2013; Connizzo et al. 2013; Dourte et al. 2013; Dunkman et al. 2013, 2014). Additionally, SLRPs protect collagen fibrils from proteolysis by sterically hindering access of collagenases to cleavage sites (Geng et al. 2006; Chen and Birk 2013).

In tendons, linear and lateral fibril regulation is controlled largely by decorin and fibromodulin with modulations done by biglycan and lumican (Ezura et al. 2000; Zhang et al. 2005). In this regard, changes to decorin, biglycan, or fibromodulin can result in alterations in fibril growth resulting in larger fibril diameters (Danielson et al. 1997; Ameye and Young 2002; Ameye et al. 2002; Jepsen et al. 2002; Zhang et al. 2005; Dourte et al. 2012). The structural abnormalities to the fibrils alter the mechanical and structural properties of the tendons exacerbated by multiple SLRP class deficiencies. Due to the high level of homology of SLRPs within classes, binding regions of molecules, such as fibromodulin and lumican, to collagen I are similar if not identical (Kalamajski and Oldberg 2009). Though some level of compensation of single SLRP deficiencies are exhibited within a class, deficiencies of two SLRPs in different classes (biglycan and fibromodulin) alter tendon biomechanics, fibril diameters, and promote ectopic ossification within the tendon (Ameye and Young 2002; Ameye et al. 2002; Bi et al. 2007). SLRPs regularly turnover and more easily allow for changes in expression to affect fibrillogenesis and tendon structure throughout development, maturation, and injury. A more detailed description of activities mediated by SLRPs can be found in Chap. 6.

The regulation of linear and lateral fibril growth is also affected by FACIT collagens. As described earlier in the chapter, FACIT collagens are fibril-associated molecules with large non-collagenous domains. FACITs, like SLRPs, demonstrate tissue-specific and chronological expression patterns necessary for proper fibril assembly. These collagens are located on the surface of fibrils and assume differing roles. Collagen IX is involved in regulation of fibril growth in cartilage (van der Rest and Mayne 1988) while collagen XII provides specific bridges between fibrils and other matrix components such as decorin and fibromodulin (Chinquet et al. 2014). It is also believed that Collagen XIV is involved in the regulation of linear tendon fibril growth due to its interactions with decorin (Young et al. 2002; Ansorge et al. 2009). Targeted deletion of collagen XIV in mouse models demonstrated a premature entrance into the fibril growth stage in tendons, resulting in larger diameter fibrils in early developmental stages. Thus, in some tissues, it seems that FACITs may serve as 'gate keepers' regulating the transition from protofibril assembly to fibril growth during development. In the case of collagen XIV, it temporarily stabilizes protofibrils to prevent the initiation of lateral fibril growth. The large non-collagenous domain and its inter-fibrillar location have been long suspected to play an additional role in fibril packing. Control of fibril packing would also influence lateral associations necessary for growth.

Maintenance of the collagen fibrils during aging and after injury is also affected by SLRPs. SLRPs play a role to protect collagen fibrils from proteolysis by sterically hindering access of collagenases to cleavage sites (Geng et al. 2006). During injury, decorin expression increases and inhibits scar formation by decreasing the presence of transforming growth factor β (TGFβ), a potent fibrogenic factor (Jarvinen and Ruoslahti 2010).

2.14 Effects of Composition and Structure on Function for Tendons and Ligaments

Tendons and ligaments are dense connective tissues with similar compositions, though slight differences in content and morphology allow for each tissue type to function in a distinct manner, yet both have large extracellular matrix to cell ratios. The composition of the extracellular matrix in tendon is crucial for transmitting the force generated from muscle to bone. In contrast, ligaments have a more passive role attaching bone to bone by guiding the joints motion and preventing abnormal displacement of bones. Additionally, to a minor extent, ligaments must also withstand loads to resist joint instability. Collagens make up 60–70% and 70–80% of the dry weight for ligaments and tendons, respectively (Frank et al. 2007; Rumian et al. 2007). The predominant fibril-forming collagen of tendons and ligaments is collagen I, though other collagens within tendons and ligaments include fibril-forming collagens III and V, as well as collagens VI, XII, and XIV. Besides collagens, both tendons and ligaments contain many of the macromolecules mentioned throughout this chapter that regulate fibril assembly as well as water content (60–70% in ligaments and 50–60% in tendons) (Rumian et al. 2007). Thus, while the composition of these tissues is quite similar, distinctions can be made in the proportions of water and macromolecules. Moreover, there are several similarities in the morphologies of the two tissues. Both tendons and ligaments have many parallel fibers that run along the axis of tension and these fibers are composed of mainly fibrils containing collagen I with a distinct "crimp" pattern in the collagen fibers of both tissues. There are also several morphological distinctions such as the angle of some fiber subsets. Within ligaments there are many regions where collagen fibers intertwine with adjacent fibers so that subsets of fibers are running obliquely at a 20–30° angle from the other fibers running along one axis (Franchi et al. 2010). These differences in composition and morphology are enough to contribute to the function of the tissues.

The distinct compositions of ligaments and tendons contribute to the nonlinear anisotropic mechanical behavior in these tissues. The mechanical properties of tendon exhibit a nonlinear elasticity that reflects itself in a mechanical stress-strain curve by three distinct regions (Fig. 2.7). A tensile stress-strain curve plots the deformation a material can sustain at certain loads by the force per unit area (stress) and the deformation that a material exhibits at that load (strain). When loading conditions are low, tendons and ligaments are relatively compliant. The first region of the stress-strain curve is the non-linear "toe" region which is an initial load affecting the "crimped" collagen fibers and any viscoelastic properties provided by molecules interacting with the collagens. The "toe" region of the tendon is smaller (only up to 2% strain) because most collagen fibers are oriented parallel to the directions of strain and "uncrimping" occurs. With increasing tensile loads, these tissues – tendons more so than ligaments – become increasingly stiffer. At some point, the stiffness of the material (load/length) will follow a linear slope characterized by the "linear" region. As slippage occurs within collagen fibrils, next between fibrils, and then until the point of tearing as adjacent fibril molecules slip away with tensile failure, the "yield to failure" region is finally reached (termed, ultimate load or point of ultimate tensile stress). As the load proceeds from initial strain to the point of failure, the area under the curve is considered the total energy absorbed. Though both tendons and ligaments are considered to exhibit a nonlinear anisotropic mechanical behavior, differences in their load-elongation curves do exist due to the difference in function. As already mentioned, the morphology of the ligament allows for the "toe" of the curve from the initial loading to be longer because the crimping pattern of the ligament is more greatly affected by fibrils not oriented exactly parallel – sometimes even perpendicular to the direction of the load. The stiffness of tendons and ligaments are also distinct. In humans, the elastic modulus (Young's Modulus) of a tendon (Achilles tendon, 375 ± 102 MPa; biceps tendon, 421 ± 212 MPa; patellar tendon, 660 ± 266 MPa) is generally greater than that of a ligament (meniscofemoral ligament, 355 ± 234 MPa; anterolateral bundle of the posterior cruciate ligament, 294 ± 115 MPa; posterior bundle of the posterior cruciate ligament, 150 ± 69 MPa) (Harner et al. 1995; An et al. 2004). Furthermore, the stiffness for tendons (Achilles tendon 430 N/mm) is generally greater than for ligaments (lateral/medial collateral ligaments, 20 N/mm; anterior cruciate ligament, ACL 182 N/mm) (Fishkin et al. 2002; Barlett 2008; Peltonen et al. 2010). Greater values for elastic modulus and stiffness are indicative of stiffer, less flexible connective tissue that is capable of absorbing and transmitting more energy. In addition, the tensile strength of tendons (50–150 N/m^2) is greater than that of ligaments (26–39 N/m^2) (Martin et al. 1998; Quapp and Weiss 1998; Chandrashekar et al. 2006). Aging also significantly alters the mechanical properties of tendons as the elastic modulus (Young's Modulus) of a patellar tendon in humans aged 29–50 years is around 660 ± 266 MPa while an individual between 64 and 93 is about 504 ± 222 MPa (Johnson et al. 1994).

The compositional differences and distinctions in stiffness and tensile strength between tendons and ligaments all contribute to the understanding of how these two connective tissues function. Tendons and ligaments have different roles. Tendons center the actions of several muscles into one axis of stress or strain. They distribute contractile force of muscles to bones, and they provide the muscle with distance from the insertion that is mechanically beneficial. During locomotion tendons store elastic energy and prevent muscle injury with viscoelasticity. Another key mechanical property of tendons is the viscoelasticity, likely a result of the content of water, collagenous proteins and interactions with

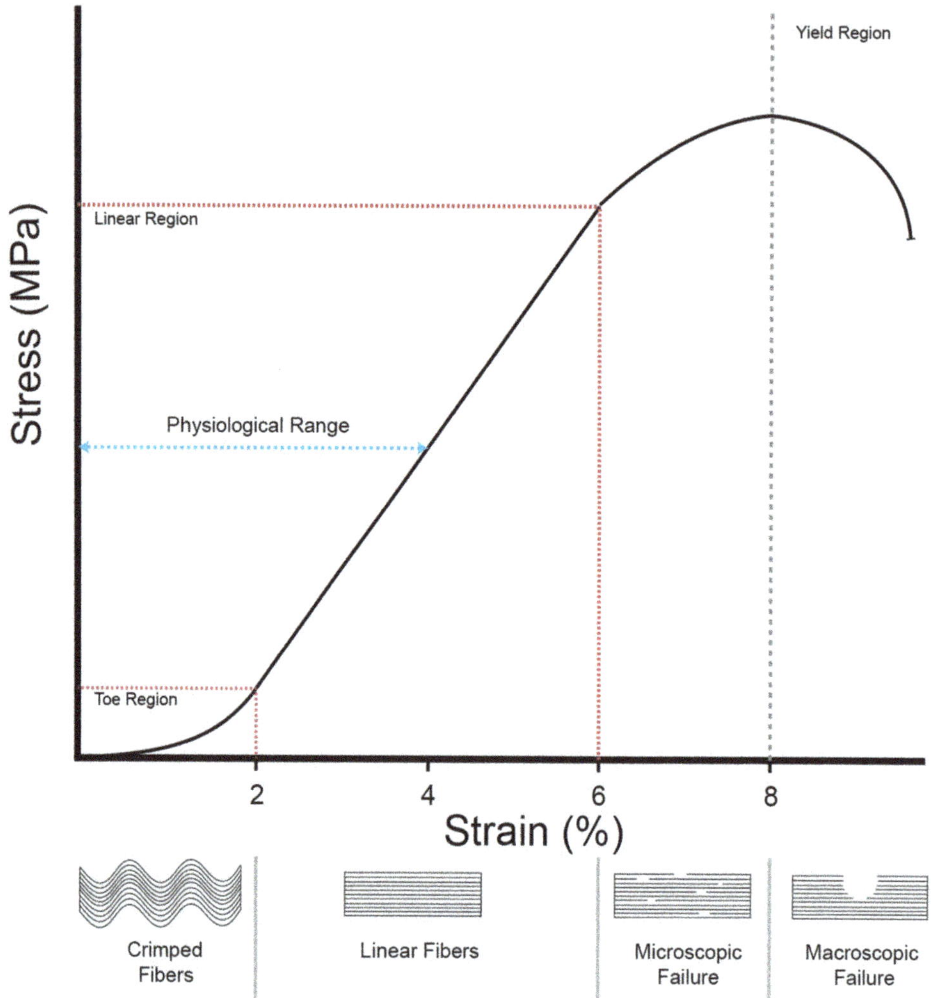

Fig. 2.7 Load-elongation curve for tendons and ligaments. There are three distinct regions within the curve which define the response to tensile loading. In the "Toe Region," initial load is affecting the "crimped" collagen fibers and any viscoelastic properties provided by molecules interacting with the collagens. In the "Linear Region," increasing tensile loads cause lengthening as the tendon or ligament becomes increasingly stiffer. In this region, load/elongation will ultimately follow a linear slope as slippage occurs within and then between collagen fibrils. In the "Yield Region," load continues to increase until the point of tearing as adjacent fibril molecules slip away to tensile failure, complete rupture, or macroscopic failure

proteoglycans that cause the mechanical behavior to be dependent on the rate of mechanical strain (Wang et al. 2012). Therefore, the rate of stress and strain is not constant for a tendon and depends on the time of displacement or load (Robi et al. 2013). Viscoelastic materials are deformable at low strain rates but less so at high strains. As a result, tendons at low strains can absorb more mechanical energy but conversely, become less effective at carrying loads from muscle to bone. But as mentioned previously, the tendon stiffness increases with increased loads therefore the viscoelastic property is overcome at high strain rates. Additionally, three characteristics of tendon viscoelasticity (creep, stress relaxation, and hysteresis) are now considered evident properties of tendons as differences in these properties greatly vary across tendons. Changes to these properties between high- and low-stressed tendons signal the potential for different material composition to compensate for higher fatigue in high-stressed tendons, but much of the micro-structural origin of tendons is still unknown (Pike et al. 2000; Peltonen et al. 2013). Though, recent models have shown that varying fibril length may answer the viscoelastic nature of tendons, this model has yet to be combined with the dissipative theory for the bridges between collagenous and noncollagenous proteins in the matrix to account for both the fibril-matrix interaction and fibril length distributions (Ciarletta and Amar 2009; Shearer et al. 2020).

The relative stiffness and tensile strength of tendons is important for maintaining force transmission. Ligaments, however, have a different function. They guide joint motion by attaching adjacent involved bones, stabilize the joint, and thus control the range of motion when load is applied. Tendons are susceptible to injury from overuse, wear and tear, and abrupt tears or avulsions when great forces are applied. Ligaments, though flexible, have less tensile strength and are prone to shear force injuries. The functions of tendons and ligaments are made evident by thorough consideration of their composition, morphology, and physiology. While tendons and ligaments possess characteristics that might distinguish one connective tissue from another, each of these tissue types also differs by anatomical location. For example, features of an Achilles tendon are not identical to those of a flexor digitorum profundus tendon or a patellar tendon. Likewise, characteristics of an anterior cruciate ligament are not identical to those of a medial collateral ligament. In addition, within each tendon and ligament, there are zones where composition changes. For example, ligaments can be divided into ligament mid- substance, fibrocartilage, mineralized fibrocartilage and bone. Similarly, tendons have musculotendinous junctions, mid-substance, fibrocartilage, and mineralized fibrocartilage to the enthesis. In the following sections, differences in collagen structure and physiology will be described by anatomic location and by zone.

2.15 Effect of Anatomical Location on Tendons and Ligaments

Throughout this chapter, the composition of tendons and ligaments have been described. While tendons and ligaments look very similar throughout their various locations within the body, upon closer inspection several differences may be found that are dependent upon connective tissue anatomy and function. For example, round tendons such as flexor digitorum profundus tendons found in the forearm typically consist of parallel bundles of collagen fibers and are subject to pulling forces, or tensile loads. In contrast, flat tendons like rotator cuff tendons are more complex in their anatomy, consisting of longitudinal, angled and intercrossing collagen fibers; these flat tendons can withstand more complex loading, such as compression, which better allows the tendon to function at its specific anatomical location (Clark and Harryman 1992; Frank et al. 2007). Some tendons contain more proteoglycans than others, particularly those with more fibrocartilaginous tissue, which forms at locations where tendons wrap around bone and are also subjected to transverse compressive loading besides longitudinal tension (Benjamin et al. 1995; Berenson et al. 1996). Flatter rotator cuff tendons contain more proteoglycans throughout than the typical round tendons; additional proteoglycans are believed to be aggrecan and SLRP biglycan (Berenson et al. 1996). Extracellular matrix analysis of ovine tendons and ligaments demonstrated that each tendon (e.g., long digital extensor tendon, superficial digital flexor tendon, patellar tendon) and ligament (e.g., lateral collateral ligament, medial collateral ligament, posterior cruciate ligament, anterior cruciate ligament) had its own unique range for matrix compositions when examining water, glycosaminoglycan and collagen content (Rumian et al. 2007). Likewise, each tendon and ligament demonstrates its own collagen organizational and mechanical features (Rumian et al. 2007).

Several tendons, like those at the extremities of the body, are surrounded by a visceral synovial sheath. The synovial sheath is linked to the outer parietal sheath by the mesotenon, which delivers blood, lymph and nerve supply to the tendon. While some tendons, like those of the ankle, are surrounded by continuous mesotenon, other tendons such as those of the fingers and toes are limited in mesotenon to small sections called vincula. Without continuous mesotenon for blood supply, tendons with vincula rely on surrounding synovial fluid for a majority of their nutrients (Frank et al. 2007). Synovial sheaths also provide lubrication for smooth tendon movement, and assist in the healing following an injury.

Three tendons commonly the subject of study are the patellar tendon, the Achilles or calcaneal tendon, and the supraspinatus tendon. The patella, developed as a bony process at the end of the femur, is superficially embedded in the quadriceps, or patellar tendon (Eyal et al. 2015). When defining tendons versus ligaments, anatomical location alone is sometimes misleading. The patellar tendon, responsible for connecting the quadricep muscle complex to the tibia, is often referred to as the patellar ligament due to the patella being embedded within it, causing an appearance of a bone to bone attachment (Rumian et al. 2007). Further classified as an energy storing tendon, the patellar tendon is similar to the Achilles tendon in that they both have greater elasticity and extensibility, and can withstand more loading cycles prior to fatigue or failure, when compared to positional tendons such as the anterior tibialis tendon. These characteristics of the patellar and Achilles tendons allow them to decrease the energetic cost of movement and play a large role in locomotion efficiency (Thorpe et al. 2016). The Achilles tendon is able to withstand force of up to 6 times the body weight of an individual during exercises such as walking or running. Previous studies with rats have shown that after 4 days of exercise, the Achilles tendon undergoes loading-induced collagen synthesis, showing that the tendon actively responds to mechanical loading (Chen et al. 2020). Similarly, decorin and biglycan, class I SLRPs, have viscoelastic properties that regulate the rupture risk of the high-use Achilles tendons (Gordon et al. 2015). The supraspinatus tendon is a flatter rotator cuff tendon that fans across the shoulder blade to attach the supraspinatus muscle to the humerus. There are increased amounts of glycosaminoglycans in human rotator cuff tendons, with 50% being hyaluronic acid in the supraspinatus tendon. In contrast, hyaluronic acid makes up less than 10% of total glycosaminoglycans of the flexor tendon, and less than 1% in tendon fibrocartilage. With significantly higher levels of glycosaminoglycans being present in rotator cuff tendons, it suggests these tendons have adapted to mechanical loading different from purely tensional loads (Berenson et al. 1996). Many rotator cuff injuries begin with a tear in the supraspinatus tendon, and occur more frequently with age-related degeneration of the rotator cuff tendons (Clark and Harryman 1992). There are two theories behind the reason for this degeneration, the impingement theory and the avascular theory. The impingement theory attributes degeneration of the tendons to the mechanical wear resulting from them being between the coracoacromial arch and the superior humeral head. The avascular theory suggests there is a significant hypovascular area of the supraspinatus tendon, leading to impaired healing resulting in early degeneration and injury (Berenson et al. 1996).

Gross anatomical differences in tendon and ligament size and shape occur as these connective tissues: traverse areas with limited space (e.g., within the wrist), centralize the force of several muscles (e.g., the Achilles tendon), or manage multi-directional forces and movements by intertwining collagen fibers with fibers of nearby connective tissues associated with tension in another axis (e.g., cruciate ligaments and rotator cuff tendons). The precise composition and structural arrangement of each tendon and ligament provides specific mechanical properties that allow connective tissues to function. Thus, to some extent, each tendon and ligament has its own unique features.

2.16 Roles of Collagens in Transition from Midsubstance to Enthesis in Tendons and Ligaments

Much of the chapter up until this point has covered the tendon and ligament midsubstance. Collagen organization also plays a role in the transitional zone that links the tendon to the bone. This transition is the enthesis. Through gradation in structure, composition, and mechanical properties, the enthesis allows for the transfer of stress from the tendon, which is an elastic material, to the bone which is a hard material (Roberts et al. 2019). It is a biological feat for the enthesis to transfer stress from the tendon which has an elastic modulus of 200 MPa to bone which has a modulus on the order of 20 MPa (Derwin et al. 2018). The enthesis is able to accomplish this transfer through its unique compositional gradation which is generally described as four zones: tendon proper, fibrocartilage, mineralized fibrocartilage, and bone (Fig. 2.8). Each zone is made of specific materials in order to accommodate the difference in elasticity between the two ends.

The tendon midsubstance is made from type I collagen fibers that are parallel to one another along the axis of given strain. Additionally, tendon proper is made of small amounts of other matrix proteins like collagen V, VI, XII, XIV, decorin, and versican (Waggett et al. 1998). The midsubstance is followed soon after by fibrocartilage – the section of the enthesis that begins the transition from tendon to bone and is made from type II and III collagen, small amounts of types I, IX, and X collagen, aggrecan, and decorin (Waggett et al. 1998). After fibrocartilage, mineralized fibrocartilage marks the transition towards bony tissue (Thomopoulos et al. 2003). This section is made of type II collagen with significant amounts of type X collagen and aggrecan. Lastly, the enthesis connects to the bone. This is made of type I collagen with a higher mineral content (Thomopoulos et al. 2010). If one were to list these, they would see that each section has one component of the section above it, thus allowing for a smooth transition that enables the sections to have no clear cut off between zones (Genin 2017). This inevitably leads to an efficient way of transferring stress, but an inefficient repair mechanism. This will be talked about later in the section.

Recently, more insight has been gained into the development of entheses, providing knowledge that is crucial to discern the establishments of transition zones. At first, the enthesis begins to form embryonically through a pool of progenitors that express *SOX9*$^{+}$ and *Scleraxis* (*Scx*) (Felsenthal et al. 2018). They quickly differentiate into two different kinds of enthesis: stationary and migrating. Stationary entheses use *Gli1*$^{+}$ progenitors, the descendants of *SOX9*$^{+}$ progenitors, in order to grow linearly. The entire process of stationary enthesis development happens from embryogenesis to maturity. On the other hand, progenitors can become part of a migrating enthesis via the utilization of the compartmentalized population of *Gli1*$^{+}$ progenitors in order to co-populate them alongside *SOX9*$^{+}$ progenitors to eventually replace them during postnatal development as postnatal migrating entheses do not utilize *SOX9*$^{+}$ once they mature. Therefore, migrating entheses can be seen to use one type of progenitor cell from the embryonic template, only to be replaced by another cell lineage that is used in the mature organ (Felsenthal et al. 2018).

Once the fetus matures, cell fate plays a major role in how the enthesis develops. *Scx*$^{+}$/*SOX9*$^{+}$ progenitors have the potential to differentiate into either tenocytes or chondrocytes (Roberts et al. 2019). Bone morphogenetic protein 4 (BMP4) is a known protein that promotes the differentiation of *Scx*$^{+}$/*SOX9*$^{+}$ progenitors into chondrocytes. However, fibroblast growth factor (FGF) signaling also plays a major role in differentiation. In *Fgfr2*-null mice, there is reduced *Scx* expression in the *Scx*$^{+}$/*SOX9*$^{+}$ progenitors and induced biased differentiation into chondrocytes expressing collagen II. FGF signaling performs this regulation through a mechanism that employs upstream notch signaling (Roberts et al. 2019). This differentiation between tenocytes and chondrocytes is what causes them to be present in zones closer to the bone, or closer to the tendon. If they are placed closer to the tendon, progeni-

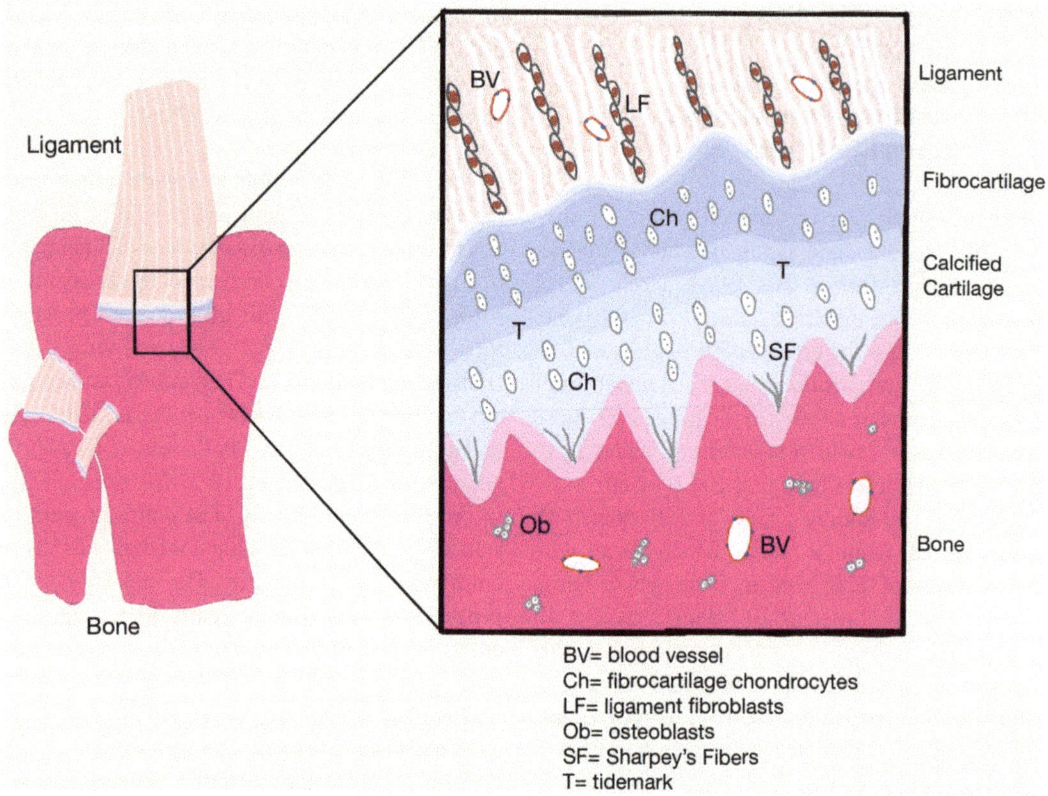

Fig. 2.8 Transition zones. This illustration demonstrates the transition from fibrous tissue to bone at a ligament insertion. The tissue transitions from ligament to fibrocartilage. Increases in the level of calcification are noted in the compositional gradient closer to insertion; this compositional change is demarcated by a tidemark (T) which traces the interface between non-calcified fibrocartilage and calcified fibrocartilage zones. Calcified fibrocartilage interdigitates with the underlying subchondral bone to complete the insertion. Structures of note within the illustration include: ligament fibroblasts (LF), fibrocartilage chondrocytes (Ch), osteoblasts (Ob), Sharpey's fibers (SF), and blood vessels (BV). (This figure has been adapted from Place et al. (2009) and Mienaltowski and Birk (2014) with permission from the publishers)

tors express *Scleraxis*. If they are placed closer to the bone, these chondrocytes are used in the mineralized fibrocartilage section of the enthesis. However, chondrocytes are not immediately mineralized. They must utilize *Hedgehog (HH)* signaling which is required for the mineralization of this section in the fibrocartilage (Roberts et al. 2019). As enthesis cells develop, IHH (Indian Hedgehog Homolog) signaling is involved in making sure the transition from unmineralized to mineralized fibrochondrocytes is maintained despite cell proliferation. *Gli1*$^{+}$ unmineralized fibrochondrocytes will only mineralize in response to activated hedgehog signaling. Any disruption will significantly inhibit their ability to mineralize (Dyment et al. 2015). As the cell matures, growth differentiation factor 5 (GDF5) progenitors contribute to the growth of the enthesis by proliferating in columns between fibrillar bundles in order to distribute their clones throughout the enthesis and not concentrate on certain layers (Dyment et al. 2015). Along with these GDF5 progenitors, the enthesis needs Connexin 43 (Cx43). This gap junction protein is what allows for mineralization of the fibrocartilage, formation and mechanoresponsiveness of the functional tendon, and the shaping of bone microarchitecture (Shen et al. 2020).

The enthesis has structural variations at every level of scale, all of which contribute to its overall tissue mechanics which makes it hard to repair (Deymier et al. 2017). For example, the shape of the tendon-to-bone attachment contributes significantly to the stresses received by the enthesis. Narrowing down to zonal structure and composition, interdigitation, fiber organization and a compliant interface increases the toughness of the transitions. Looking even more molecularly, spatial gradients in mineralization stiffen the matrix (Derwin et al. 2018). This variation informs regulation of enthesis organization. The tendon portion of the insertion is structured to align along the direction of the muscle force, while the bony insertion structure is optimized to transfer the complex multidirectional forces that would develop near the bone. This variance of function is due to the drastic difference in viscoelastic behavior of the insertion along its length. The tendon insertion carries higher tensile loads and a higher level of stress for a longer duration compared to the bony insertion (Waggett et al. 1998). The sharp difference in elasticity, as well as structural capability, is what makes the enthesis difficult to repair. Since these characteristics rely on a gradation that is not repaired after injury, scientists have been struggling to find a way to heal these injuries.

2.17 Collagen in Repair

Movement and physical activities are part of everyday life and require the contraction of skeletal muscles. Upon skeletal muscle contraction, different levels of stress transfer across the structural hierarchy until they reach the tenocytes within native tendons. By transmitting and displacing the forces that act on them, tendons allow for the generation of muscle movement and joint stabilization (Freedman et al. 2018). With these daily movements, there is the potential that overuse or too much force can cause acute rupture or chronic degenerative changes in tendons or ligaments. In tendons, there is a variability of force that acts on them depending on their shape or structure, location, and ultimate use or stresses placed on those tendons. For all tendons, fibrillar collagens including types I, III, V and XI make up the largest proportion of the extracellular matrix. These collagens are involved in tendon growth, adaptation, and homeostasis; following an injury, their activity is often aberrant and produces tissue that does not allow for full recovery (Gaut and Duprez 2016; Mienaltowski et al. 2016; Nguyen et al. 2018). One area of the body, the rotator cuff, is subjected to high levels of daily use and makes up a relatively large proportion of body injuries, both chronic and acute. With 20% of doctor visits being musculoskeletal related, 30% of them are tendon injuries and 17 million people in the United States are affected by a rotator cuff injury, several of which require surgical repair (Abraham et al. 2017). While rotator cuff injuries tend to occur from overuse of the rotation of the shoulder, other tendon based injuries tend to occur from repeated heavy impact, such as patellar injuries. Patellar injuries are commonly referred to as jumper's knee, due to the tears being common in athletes that repeatedly sprint and jump. These injuries are recognized through microinjury to tendon fibers, mucoid degeneration and reduced fibrocartilaginous tissue between the tendon and bone conjunction (Golman et al. 2020). Tendon injuries can be the result of inflammatory degeneration of tissue as in tendinitis or due to non-inflammatory degradation of tissue because of chronic overuse as in tendinosis (Morais et al. 2015).

The structure and function of tendon relies heavily upon the proper organization of its extracellular matrix, followed by adequate homeostasis to keep pace with activity demands placed on connective tissue. In tendon repair, many of the same growth factor signaling pathways (TGFβ/SMAD2/3 and FGF/ERK MAPK) that play a role in development also play a role in healing (Docheva et al. 2014; Gaut and Duprez 2016). While it would be optimal for such signaling pathways to recapitulate development in repair, this is not necessarily the case, particularly in mature tendon, likely because of factors like inflammation, as well as changes to stresses to the affected tissue or even stress shielding at the injury site(s) where surrounding tissue receives stress and thus essential mechanical cues are not received in the healing tissue (Dunkman et al. 2014; Morais et al. 2015; Freedman et al. 2018; Baar 2019). Moreover, the cells involved in mature tendon healing might not be capable of mounting a matrix assembly response identical to that of developing tendon because the programming of their gene expression may differ from developing tendon via genetic and epigenetic determinants of cell fate (Ingraham et al. 2003; Dyment et al. 2013; Mienaltowski et al. 2014, 2016, 2019; Sakabe et al. 2018; Moser et al. 2018). Finally, age has been demonstrated to be related to higher levels of injury from wear and tear but also from a decreased ability of tendons and ligaments to fully repair following an injury (Dunkman et al. 2014; Mienaltowski et al. 2016). Collagen types I, II and III make up a significant amount of the extracellular matrix as fibril-forming collagens that form supra-molecular support structures; however, they are metabolized relatively slowly. The covalent cross-links within collagen fibrils play a large role in the stabilization and support of the fibrils, yet also contribute to the long half-life of collagen which in turn predisposes collagen to age related modifications. It is shown in tendons that there is an accumulation of partially degraded collagen in the ECM with increasing age, which ultimately weakens the tendons, leaving them more susceptible to injuries (Birch 2018). As tendons age, it has been determined that collagens become less aligned, have more elongated and fewer cells, and have larger diameter fibrils when compared to less-aged tendons. Some studies have also shown that as tendons age, the amount of decorin increases, showing correlation to the age effects that tendons have on injury susceptibility (Dunkman et al. 2014). Overall, as tendons age they become less stiff due to changes in their ECM and fibril structure, allowing less force to result in injury (Tuite et al. 1997). In tendon healing, the quality of the ECM assembled ultimately is affected by many factors.

2.18 Summary

Proper tendon and ligament function is highly dependent upon its tissue structure, more specifically their extracellular matrices and in particular the collagen that comprises 70–80% of their dry weight. Tendon and ligament collagens belong to several subfamilies that can be grouped by their general suprastructural forms with predominant categories including: fibril-forming, FACIT, network-forming, basement membrane, beaded filament- forming, and transmembrane-spanning collagens. The many collagens provide considerable diversity for the functional extracellular matrix suprastructures of connective tissues like tendons and ligaments. This diversity is further compounded by the numbers of different alpha chains formed; by alternative splicing; by the many types of post-translational modifications both within the cell and once secreted, as well as the heterotypic combinations of interacting collagens. Diversity is further compounded by the myriad of possible suprastructures formed, all macromolecular heteropolymers containing different collagens and other fibril-associated molecules. Within tendons and ligaments, the assembly of these collagen suprastructures relies upon the primary structures of these molecules that contain domains that will affect downstream secondary, tertiary, and quaternary structures. Clearly, the content and organization of these connective tissues affect how they function in absorbing and transmitting forces as well as in maintaining stability. While generalizations can be made for both connective tissues, much has been reported regarding the differences amongst tendons and ligaments throughout the body. Furthermore, even within any one tendon or ligament, zonal differences in composition and structure were demonstrated. Likewise, changes in composition inform structural alterations seen with development, growth and exercise adaptation, as well as aberrances associated with age, overuse, and even in healing across the tendon or ligament or within specific regions (e.g., mid-substance, enthesis, etc.). Thus, while an understanding of how collagen is assembled and organized is critical for tendon and ligament repair and regeneration, it is also essential to focus on how these general mechanisms generate unique structures that will determine each tissue's distinct features and functional properties given the existing physiological state of the connective tissue.

References

Abraham AC, Shah SA, Thomopoulos S (2017) Targeting inflammation in rotator cuff tendon degeneration and repair. Tech Should Elbow Surg 18(3):84–90

Ameye L, Young MF (2002) Mice deficient in small leucine-rich proteoglycans: novel in vivo models for osteoporosis, osteoarthritis, Ehlers-Danlos syndrome, muscular dystrophy, and corneal diseases. Glycobiology 12(9):107R–116R

Ameye L, Aria D, Jepsen K, Oldberg A, Xu T, Young MF (2002) Abnormal collagen fibrils in tendons of biglycan/fibromodulin-deficient mice lead to gait impairment, ectopic ossification, and osteoarthritis. FASEB J 16(7):673–680

An KN, Sun YL, Luo ZP (2004) Flexibility of type I collagen and mechanical property of connective tissue. Biorheology 41:239–246

Ansorge HL, Meng X, Zhang G, Veit G, Sun M, Klement JF, Beason DP, Soslowsky LJ, Koch M, Birk DE (2009) Type XIV collagen regulates fibrillogenesis: premature collagen fibril growth and tissue dysfunction in null mice. J Biol Chem 284(13):8427–8438

Antoniel M, Traina F, Merlini L, Andrenacci D, Tigani D, Santi S, Cenni V, Sabatelli P, Faldini C, Squarzoni S (2020) Tendon extracellular matrix remodeling and defective cell polarization in the presence of collagen VI mutations. Cell 9(2):409

Aszodi A, Chan D, Hunziker E, Bateman JF, Fassler R (1998) Collagen II is essential for the removal of the notochord and the formation of intervertebral discs. J Cell Biol 143(5):1399–1412

Avery NC, Bailey AJ (2005) Enzymic and nonenzymic cross-linking mechanisms in relation to turnover of collagen: relevance to aging and exercise. Scand J Med Sci Sports 15(4):231–240

Ayad S, Boot-Handford RP, Humphries MJ, Kadler KE, Shuttleworth CA (1998) The extracellular matrix facts book, 2nd edn. Academic, San Diego, pp 69–76

Baar K (2019) Stress relaxation and targeted nutrition to treat patellar tendinopathy. Int J Sport Nutr Exerc Metab 29(4):453–457

Bachinger HP, Bruckner P, Timpl R, Prockop DJ, Engel J (1980) Folding mechanism of the triple helix in type-III collagen and type-III pN-collagen. Role of disulfide bridges and peptide bond isomerization. Eur J Biochem 106(2):619–632

Bailey AJ (2001) Molecular mechanisms of ageing in connective tissues. Mech Ageing Dev 122(7):735–755

Ball S, Bella J, Kielty C, Shuttleworth A (2003) Structural basis of type VI collagen dimer formation. J Biol Chem 278(17):15326–15332

Barnard K, Light ND, Sims TJ, Bailey AJ (1987) Chemistry of the collagen cross-links. Origin and partial characterization of a putative mature crosslink of collagen. Biochem J 244(2):303–309

Bartlett J (2008) Anterior cruciate ligament graft fixation: is the issue mechanical or biological? ANZ J Surg 78(3):118

Batge B, Winter C, Notbohm H, Acil Y, Brinckmann J, Muller PK (1997) Glycosylation of human bone collagen I in relation to lysyl hydroxylation and fibril diameter. J Biochem 122(1):109–115

Becker J, Schuppan D, Rabanus JP, Rauch R, Niechoy U, Gelderblom HR (1991) Immunoelectron microscopic localization of collagens type I, V, VI and of procollagen type III in human periodontal ligament and cementum. J Histochem Cytochem 39(1):103–110

Bell R, Remi Gendron N, Anderson M, Flatow EL, Andrarawis-Puri N (2018) A potential new role for myofibroblasts in remodeling of sub-rupture fatigue tendon injuries by exercise. Sci Rep 8(1):8933

Benjamin M, Qin S, Ralphs JR (1995) Fibrocartilage associated with human tendons and their pulleys. J Anat 187:625–633

Berenson MC, Blevins FT, Plaas AH, Vogel KG (1996) Proteoglycans of human rotator cuff tendons. J Orthop Res 14(4):518–525

Berg RA, Prockop DJ (1973) The thermal transition of a non-hydroxylated form of collagen. Evidence for a role for hydroxyproline in stabilizing the triple-helix of collagen. Biochem Biophys Res Commun 52(1):115–120

Bi Y, Ehirchiou D, Kilts TM, Inkson CA, Embree MC, Sonoyama W, Li L, Leet AI, Seo BM, Zhang L, Shi S, Young MF (2007) Identification of tendon stem/progenitor cells and the role of the extracellular matrix in their niche. Nat Med 13(10):1219–1227

Birch HL (2018) Extracellular matrix and ageing. In: Harris JR, Korolchuk VI (eds) Biochemistry and cell biology of ageing: part I biomedical science. Springer Nature, Singapore, pp 169–190

Birk DE (2001) Type V, collagen: heterotypic type I/V collagen interactions in the regulation of fibril assembly. Micron 32(3):223–237

Birk DE, Bruckner P (2005) Collagen suprastructures. Curr Top Chem 247:185–205

Birk DE, Bruckner P (2011) Collagens, suprastructures, and collagen fibril assembly. In: Mecham RP (ed) The extracellular matrix: an overview. Springer, Berlin, pp 77–116

Birk DE, Mayne R (1997) Localization of collagen types I, III and V during tendon development. Changes in collagen types I and III are correlated with changes in fibril diameter. Eur J Cell Biol 72(4):352–361

Birk DE, Trelstad RL (1986) Extracellular compartments in tendon morphogenesis: collagen fibril, bundle, and macroaggregate formation. J Cell Biol 103(1):231–240
Birk DE, Zycband EI, Winkelmann DA, Trelstad RL (1989) Collagen fibrillogenesis in situ: fibril segments are intermediates in matrix assembly. Proc Natl Acad Sci U S A 86(12):4549–4553
Birk DE, Nurminskaya MV, Zycband EI (1995) Collagen fibrillogenesis in situ: fibril segments undergo post-depositional modifications resulting in linear and lateral growth during matrix development. Dev Dyn 202(3):229–243
Birk DE, Zycband EI, Woodruff S, Winkelmann DA, Trelstad RL (1997) Collagen fibrillogenesis in situ: fibril segments become long fibrils as the developing tendon matures. Dev Dyn 208(3):291–298
Blaschke UK, Eikenberry EF, Hulmes DJ, Galla HJ, Bruckner P (2000) Collagen XI nucleates self- assembly and limits lateral growth of cartilage fibrils. J Biol Chem 275(14):10370–10378
Boot-Handford RP, Tuckwell DS (2003) Fibrillar collagen: the key to vertebrate evolution? A tale of molecular incest. BioEssays 25(2):142–151
Brodsky B, Eikenberry EF, Belbruno KC, Sterling K (1982) Variations in collagen fibril structure in tendons. Biopolymers 21(5):935–951
Bruckner P, Eikenberry EF, Prockop DJ (1981) Formation of the triple helix of type I procollagen in cellulo. A kinetic model based on cis-trans isomerization of peptide bonds. Eur J Biochem 118(3):607–613
Bruns RR, Press W, Engvall E, Timpl R, Gross J (1986) Type VI collagen in extracellular, 100-nm periodic fi laments and fibrils: identification by immunoelectron microscopy. J Cell Biol 103(2):393–404
Buckley MR, Evans E, Satchel LN, Matuszewski PE, Chen Y-L, Elliott DM, Soslowsky LJ, Dodge GR (2013) Distributions of types I, II, and III collagen by region in the human supraspinatus tendon. Connect Tissue Res 54(6):374–379
Canty EG, Kadler KE (2002) Collagen fibril biosynthesis in tendon: a review and recent insights. Comp Biochem Physiol A Mol Integr Physiol 133(4):979–985
Canty EG, Kadler KE (2005) Procollagen trafficking, processing and fibrillogenesis. J Cell Sci 118:1341–1353
Canty EG, Lu Y, Meadows RS, Shaw MK, Holmes DF, Kadler KE (2004) Coalignment of plasma membrane channels and protrusions (fibripositors) specifies the parallelism of tendon. J Cell Biol 165(4):553–563
Cescon M, Gattazzo F, Chen P, Bonaldo P (2015) Collagen VI at a glance. J Cell Sci 128(19):3525–3531
Chakravarti S (2002) Functions of lumican and fibromodulin: lessons from knockout mice. Glycoconj J 19(4–5):287–293
Chandrashekar N, Mansouri H, Slauterbeck J, Hashemi J (2006) Sex-based differences in the tensile properties of the human anterior cruciate ligament. J Biomech 39(16):2943–2950
Chen S, Birk DE (2013) The regulatory roles of small leucine-rich proteoglycans in extracellular matrix assembly. FEBS J 280(10):2120–2137
Chen M, Shetye S, Rooney SI, Soslowski LJ (2020) Short and long-term exercise results in a differential Achilles tendon mechanical response. J Biomech Eng 142(8):081011
Chiquet M, Birk DE, Bönnemann CG, Koch M (2014) Collagen XII: protecting bone and muscle integrity by organizing collagen fibrils. Int J Biochem Cell Biol 53:51–54
Chou MY, Li HC (2002) Genomic organization and characterization of the human type XXI collagen (COL21A1) gene. Genomics 79(3):395–401
Chu ML, Mann K, Deutzmann R, Pribula-Conway D, Hsu-Chen CC, Bernard MP, Timpl R (1987) Characterization of three constituent chains of collagen type VI by peptide sequences and cDNA clones. Eur J Biochem 168(2):309–317
Ciarletta P, Amar BM (2009) A finite dissipative theory of temporary interfibrillar bridges in the extracellular matrix of ligaments and tendons. J R Soc Interface 9:909–924
Clark JM, Harryman DT 2nd (1992) Tendons, ligaments, and capsule of the rotator cuff. Gross and microscopic anatomy. J Bone Joint Surg Am 74(5):713–725
Colige A, Ruggiero F, Vandenberghe I, Dubail J, Kesteloot F, Van Beeumen J, Beschin A, Brys L, Lapiere CM, Nusgens B (2005) Domains and maturation processes that regulate the activity of ADAMTS-2, a metalloproteinase cleaving the aminopropeptide of fibrillar procollagens types I-III and V. J Biol Chem 280(41):34397–34408
Connizzo BK, Sarver JJ, Birk DE, Soslowsky LJ (2013) Effect of age and proteoglycan deficiency on collagen fiber re-alignment and mechanical properties in mouse supraspinatus tendon. J Biomech Eng 135(2):021019
Connizzo BK, Freedman BR, Fried JH, Sun M, Birk DE, Soslowsky LJ (2015) Regulatory role of collagen V in establishing mechanical properties of tendons and ligaments is tissue dependent. J Orthop Res 33(6):882–888
Connizzo BK, Adams SM, Adams TH, Birk DE, Soslowsky LJ (2016) Collagen V expression is crucial in regional development of the supraspinatus tendon. J Orthop Res 34(12):2154–2161
Danielson KG, Baribault H, Holmes DF, Graham H, Kadler KE, Iozzo RV (1997) Targeted disruption of decorin leads to abnormal collagen fibril morphology and skin fragility. J Cell Biol 136(3):729–743
Derwin KA, Galatz LM, Ratcliffe A, Thomopoulos S (2018) Enthesis repair: challenges and opportunities of effective tendon-to-bone healing. J Bone Joint Surg Am 100(109):1–7
Deymier AC, An Y, Boyle JJ, Schwartz AG, Birman V, Genin GM, Thomopoulos S, Barber AH (2017) Micro-mechanical properties of the tendon-to-bone attachment. Acta Biomater 56:25–35
Docheva D, Müller SA, Majewski M, Evans CH (2014) Biologics for tendon repair. Adv Drug Deliv Rev 84:222–239

Dourte LM, Pathmanathan L, Jawad AF, Iozzo RV, Mienaltowski MJ, Birk DE, Soslowsky LJ (2012) Influence of decorin on the mechanical, compositional, and structural properties of the mouse patellar tendon. J Biomech Eng 134(3):031005

Dourte LM, Pathmanathan L, Mienaltowski MJ, Jawad AF, Birk DE, Soslowsky LJ (2013) Mechanical, compositional, and structural properties of the mouse patellar tendon with changes in biglycan gene expression. J Orthop Res 31(9):1430–1437

Dunkman AA, Buckley MR, Mienaltowski MJ, Adams SM, Thomas SJ, Satchell L, Kumar A, Pathmanathan L, Beason DP, Iozzo RV, Birk DE, Soslowsky LJ (2013) Decorin expression is important for age-related changes in tendon structure and mechanical properties. Matrix Biol 32(1):3–13

Dunkman AA, Buckley MR, Mienaltowski MJ, Adams SM, Thomas SJ, Kumara A, Beason DP, Iozzo RV, Birk DE, Soslowsky LJ (2014) The injury response of aged tendons in the absence of biglycan and decorin. Matrix Biol 35:232–238

Dyment NA, Liu C-F, Kazemi N, Aschbacher-Smith LE, Kenter K, Breidenbach AP, Shearn JT, Wylie C, Rowe DW, Butler DL (2013) The paratenon contributes to scleraxis-expressing cells during patellar tendon healing. PLoS One 8(3):e59944

Dyment NA, Breidenbach AP, Schwartz AG, Russell RP, Ashbacher-Smith L, Liu H, Hagiwara Y, Jiang R, Thomopoulos S, Butler DL, Rowe DW (2015) GDF5 progenitors give rise to fibrocartilage cells that mineralize via hedgehog signaling to form the zonal enthesis. Dev Biol 405(1):96–107

Engvall E, Hessle H, Klier G (1986) Molecular assembly, secretion, and matrix deposition of type VI collagen. J Cell Biol 102(3):703–710

Exposito JY, Cluzel C, Garrone R, Lethias C (2002) Evolution of collagens. Anat Rec 268:302–316

Eyal S, Blitz E, Shwartz Y, Akiyama H, Schweitzer R, Zelzer E (2015) On the development of the Patella. Development 142:1831–1839

Ezura Y, Chakravarti S, Oldberg A, Chervoneva I, Birk DE (2000) Differential expression of lumican and fibromodulin regulate collagen fibrillogenesis in developing mouse tendons. J Cell Biol 151(4):779–788

Felsenthal N, Rubin S, Stern T, Krief S, Pal D, Pryce BA, Schwitzer R, Zelzer E (2018) Development of migrating tendon-to-bone attachments involves replacement of progenitor populations. Development 145:1–10

Feng C, Chan WCW, Lam Y, Wang X, Chen P, Niu B, Ng VCW, Yeo JC, Stricker S, Cheah KSE, Koch M, Mundlos S, Ng HH, Chan D (2019) Lgr5 and Col22a1 mark progenitor cells in the lineage toward juvenile articular chondrocytes. Stem Cell Rep 13(4):713–729

Fessler LI, Fessler JH (1979) Characterization of type III procollagen from chick embryo blood vessels. J Biol Chem 254(1):233–239

Fessler LI, Timpl R, Fessler JH (1981) Assembly and processing of procollagen type III in chick embryo blood vessels. J Biol Chem 256(5):2531–2537

Fishkin Z, Miller D, Ritter C, Ziv I (2002) Changes in human knee ligament stiffness secondary to osteoarthritis. J Orthop Res 20(2):204–207

Fitzgerald J, Rich C, Zhou FH, Hansen U (2008) Three novel collagen VI chains, alpha4(VI), alpha5(VI), and alpha6(VI). J Biol Chem 283(29):20170–20180

Fleischmajer R, Timpl R, Tuderman L, Raisher L, Wiestner M, Perlish JS, Graves PN (1981) Ultrastructural identification of extension aminopropeptides of type I and III collagens in human skin. Proc Natl Acad Sci U S A 78(12):7360–7364

Franchi M, Trire A, Quaranta M, Orsini E, Ottani V (2007) Collagen structure of tendon relates to function. Sci World J 7:404–420

Franchi M, Quaranta M, Macciocca M, Leonardi L, Ottani V, Bianchini P, Diaspro A, Ruggeri A (2010) Collagen fibre arrangement and functional crimping pattern of the medial collateral ligament in the rat knee. Knee Surg Sports Traumatol Arthrosc 18(12):1671–1678

Frank CBS, Shrive CB, Frank IKY, Hart DA (2007) Form and function of tendon and ligament. In: Einhorn TA, O'Keefe RJ, Buckwalter JA (eds) Orthopaedic basic science, 3rd edn. American Academy of Orthopaedic Surgeons, Rosemont, pp 199–222

Freedman BR, Rodriguez AB, Leiphart RJ, Newton JB, Ban E, Sarver JJ, Mauck RL, Shenoy VB, Soslowsky LJ (2018) Dynamic loading and tendon healing affect multiscale tendon properties and ECM stress transmission. Sci Rep 8(1):10854

Fukuta S, Oyama M, Kavalkovich K, Fu FH, Niyibizi C (1998) Identification of types II, IX and X collagens at the insertion site of the bovine Achilles tendon. Matrix Biol 17(1):65–73

Furthmayr H, Wiedemann H, Timpl R, Odermatt E, Engel J (1983) Electron-microscopical approach to a structural model of intima collagen. Biochem J 211(2):303–311

Gara SK, Grumati P, Urciuolo A, Bonaldo P, Kobbe B, Koch M, Paulsson M, Wagener R (2008) Three novel collagen VI chains with high homology to the alpha3 chain. J Biol Chem 283(16):10658–10670

Gaut L, Duprez D (2016) Tendon development and diseases. Wiley Interdiscip Rev Dev Biol 5(1):5–23

Geng Y, McQuillan D, Roughley PJ (2006) SLRP interaction can protect collagen fibrils from cleavage by collagenases. Matrix Biol 25(8):484–491

Genin GM (2017) Unification through disarray. Nat Mater 16(6):607–608

Genovese F, Karsdal MA (2016) Chapter 27 – type XXVII collagen. In: Karsdal MA (ed) Biochemistry of collagens, laminins and elastin: structure, function and biomarkers. Elsevier, Amsterdam, pp 155–158

Girkontaite I, Frischholz S, Lammi P, Wagner K, Swoboda B, Aigner T, von der Mark K (1996) Immunolocalization of type X collagen in normal fetal and adult osteoarthritic cartilage with monoclonal antibodies. Matrix Biol 15(4):231–238

Gjaltema RAF, Bank RA (2017) Molecular insights into prolyl and lysyl hydroxylation of fibrillar collagens

in health and disease. Crit Rev Biochem Mol Biol 52(1):74–95
Golman M, Wright ML, Wong TT, Lynch TS, Ahmad CS, Thomopoulos S, Popkin CA (2020) Rethinking patellar tendinopathy and partial patellar tendon tears. Am J Sports Med 48(2):359–369
Gordon JA, Freedman BR, Zuskov A, Iozzo RV, Birk DE, Soslowsky LJ (2015) Achilles tendons from decorin- and biglycan-null mouse models have inferior mechanical and structural properties predicted by an image-based empirical damage model. J Biomech 48(10):2110–2115
Graham HK, Holmes DF, Watson RB, Kadler KE (2000) Identification of collagen fibril fusion during vertebrate tendon morphogenesis. The process relies on unipolar fibrils and is regulated by collagen-proteoglycan interaction. J Mol Biol 295(4):891–902
Greenspan DS (2005) Biosynthetic processing of collagen molecules. Top Curr Chem 247:149–183
Gregory KE, Oxford JT, Chen Y, Gambee JE, Gygi SP, Aebersold R, Neame PJ, Mechling DE, Bachinger HP, Morris NP (2000) Structural organization of distinct domains within the non- collagenous N-terminal region of collagen type XI. J Biol Chem 275(15):11498–11506
Han B, Li Q, Wang C, Patel P, Adams SM, Doyran B, Nia HT, Oftadeh R, Zhou S, Li CY, Liu XS, Lu L, Enomoto-Iwamoto M, Quin L, Mauck RL, Iozzo RV, Birk DE, Han L (2019) Decorin regulates the aggrecan network integrity and biomechanical functions of cartilage extracellular matrix. ACS Nano 13:11320–11333
Harner CD, Xerogeanes JW, Livesay GA, Carlin GJ, Smith BA, Kusayama T, Kashiwaguchi S, Woo SL (1995) The human posterior cruciate ligament complex: an interdisciplinary study. Ligament morphology and biomechanical evaluation. Am J Sports Med 23(6):736–745
Havis E, Duprez D (2020) EGR1 transcription factor is a multifaceted regulator of matrix production in tendons and other connective tissues. Int J Mol Sci 21(1664):1–25
Heikkinen A, Tu H, XIII Collagen PT (2012) A type II transmembrane protein with relevance to musculoskeletal tissues, microvessels and inflammation. Int J Biochem Cell Biol 44(5):714–717
Hoffman GG, Branam AM, Huang G, Pelegri F, Cole WG, Wenstrup RM, Greenspan DS (2010) Characterization of the six zebrafish clade B fibrillary procollagen genes, with evidence for evolutionarily conserved alternative splicing within the proalpha1(V) C-propeptide. Matrix Biol 29(4):261–275
Holmes DF, Yeung CYC, Garva R, Zindy E, Taylor SH, Watson S, Kalson NS, Kadler KE (2018) Synchronized mechanical oscillations at the cell-matrix interface in the formation of tensile tissue. PNAS 115(40):E9288–E9297
Huang P-W, Chang J-M, Horng J-C (2012) Effects of glycosylated (2S,4R)-hydroxyproline on the stability and assembly of collagen triple helices. Amino Acids 48(12):2765–2772
Huber S, van der Rest M, Bruckner P, Rodriguez E, Winterhalter KH, Vaughan L (1986) Identification of the type IX collagen polypeptide chains. The alpha 2(IX) polypeptide carries the chondroitin sulfate chain(s). J Biol Chem 261(13):5965–5968
Hulmes DJ (2002) Building collagen molecules, fibrils, and suprafibrillar structures. J Struct Biol 137(1–2):2–10
Hulmes DJ (2008) Collagen diversity, synthesis and assembly. In: Fratzl P (ed) Collagen: structure and mechanics. Springer, New York, pp 15–48
Hulmes DJ, Wess TJ, Prockop DJ, Fratzl P (1995) Radial packing, order, and disorder in collagen fibrils. Biophys J 68:1661–1670
Humphries SM, Lu Y, Canty EG, Kadler KE (2008) Active negative control of collagen fibrillogenesis in vivo, intracellular cleavage of the type I procollagen propeptides in tendon fibroblasts without intracellular fibrils. J Biol Chem 283(18):12129–12135
Ingraham J, Hauck R, Ehrlich H (2003) Is the tendon embryogenesis process resurrected during tendon healing? Plast Reconstr Surg 112(3):844–854
Iozzo RV, Gubbiotti MA (2018) Extracellular matrix: the driving force of mammalian diseases. Matrix Biol 71–72:1–9
Iozzo RV, Schaefer L (2015) Proteoglycan form and function: a comprehensive nomenclature of proteoglycans. Matrix Biol 42:11–55
Ishikawa Y, Bachinger HP (2013) A molecular ensemble in the rER for procollagen maturation. Biochim Biophys Acta, Mol Cell Res 1833(11):2749–2491
Izu Y, Ansorge HL, Zhang G, Soslowsky LJ, Bonaldo P, Chu ML, Birk DE (2011) Dysfunctional tendon collagen fibrillogenesis in collagen VI null mice. Matrix Biol 30(1):53–61
Izzi V, Heljasvaara R, Heikkinen A, Karppinen SM, Koivunen J, Pihlajaniemi T (2020) Exploring the roles of MACIT and multiplexin collagens in stem cells and cancer. Semin Cancer Biol 62:134–148
Jakobsen JR, Mackey AL, Knudsen AB, Koch M, Kjaer M, Krogsgaard MR (2017) Composition and adaptation of human myotendinous junction and neighboring muscle fibers to heavy resistance training. Scand J Med Sci Sports 27(12):1547–1559
Järvinen TA, Ruoslahti E (2010) Target-seeking antifibrotic compound enhances wound healing and suppresses scar formation in mice. Proc Natl Acad Sci U S A 107(50):21671–21676
Jepsen KJ, Wu F, Peragallo JH, Paul J, Roberts L, Ezura Y, Oldberg A, Birk DE, Chakravarti S (2002) A syndrome of joint laxity and impaired tendon integrity in lumican- and fibromodulin-deficient mice. J Biol Chem 277(38):35532–35540
Johnson GA, Tramaglini DM, Levine RE, Ohno K, Choi NY, Woo SL (1994) Tensile and viscoelastic properties of human patellar tendon. J Orthop Res 12(6):796–803
Johnston JM, Connizzo BK, Shetye SS, Robinson KA, Huegel J, Rodriguez AB, Sun M, Adams SM, Birk DE, Soslowsky LJ (2017) Collagen V haploinsufficiency in a murine model of classic Ehlers-Danlos syndrome

is associated with deficient structural and mechanical healing in tendons. J Orthop Res 35(12):2707–2715
Kadler KE (2017) Fell Muir lecture: collagen fibril formation in vitro and in vivo. Int J Exp Pathol 98:4–16
Kadler KE, Holmes DF, Trotter JA, Chapman JA (1996) Collagen fibril formation. Biochem J 316(Pt1):1–11
Kadler KE, Hill A, Canty-Laird EG (2008) Collagen fibrillogenesis: fibronectin, integrins, and minor collagens as organizers and nucleators. Curr Opin Cell Biol 20(5):495–501
Kalamajski S, Oldberg A (2009) Homologous sequence in lumican and fibromodulin leucine-rich repeat 5-7 competes for collagen binding. J Biol Chem 284(1):534–539
Kalamajski S, Oldberg A (2010) The role of small leucine-rich proteoglycans in collagen fibrillogenesis. Matrix Biol 29(4):248–253
Kalamajski S, Cuiping L, Tillgren V, Rubin K, Oldberg A, Rai J, Weis MA, Eyre DR (2014) Increased C-telopeptide cross-linking of tendon type I collagen in fibromodulin-deficient mice. J Biol Chem 289(27):18873–18879
Karamanos NK, Theocharis AD, Neill T, Iozzo RV (2019) Matrix modeling and remodeling: a biological interplay regulating tissue homeostasis and diseases. Matrix Biol 75–76:1–11
Katz EP, Wachtel EJ, Maroudas A (1986) Extrafibrillar proteoglycans osmotically regulate the molecular packing of collagen in cartilage. Biochim Biophys Acta 882(1):136–139
Keene DR, Sakai LY, Burgeson RE (1991) Human bone contains type III collagen, type VI collagen, and fibrillin: type III collagen is present on specific fibers that may mediate attachment of tendons, ligaments, and periosteum to calcified bone cortex. J Histochem Cytochem 39(1):59–69
Keller H, Eikenberry EF, Winterhalter KH, Bruckner P (1985) High post-translational modification levels in type II procollagen are not a consequence of slow triple helix formation. Coll Relat Res 5(3):245–251
Khoshnoodi J, Cartailler JP, Alvares K, Veis A, Hudson BG (2006) Molecular recognition in the assembly of collagens: terminal noncollagenous domains are key recognition modules in the formation of triple helical protomers. J Biol Chem 281(50):38117–38121
Kielty C, Grant ME (2002) The collagen family: structure, assembly, and organization in the extracellular matrix. In: Royce PM, Steinmann B (eds) Connective tissue and its heritable disorders. Wiley-Liss, New York, pp 159–222
Kittleberger R, Davis PF, Flynn DW, Greenhill NS (1990) Distribution of type VIII collagen in tissues: an immunohistochemical study. Connect Tissue Res 24(3–4):303–318
Knupp C, Squire JM (2001) A new twist in the collagen story–the type VI segmented supercoil. EMBO J 20(3):372–376
Koch M, Bernasconi C, Chiquet M (1992) A major oligomeric fibroblast proteoglycan identified as a novel large form of type-XII collagen. Eur J Biochem 207(3):847–856
Koch M, Laub F, Zhou P, Hahn RA, Tanaka S, Burgeson RE, Gerecke DR, Ramirez F, Gordon MK (2003) Collagen XXIV, a vertebrate fibrillar collagen with structural features of invertebrate collagens: selective expression in developing cornea and bone. J Biol Chem 278(44):43236–43244
Koch M, Schulze J, Hansen U, Ashwodt T, Keene DR, Brunken WJ, Burgeson RE, Bruckner P, Bruckner-Tuderman L (2004) A novel marker of tissue junctions, collagen XXII. J Biol Chem 279(21):22514–22521
Kostrominova K, Brooks SV (2013) Age-related changes in structure and extracellular matrix protein expression levels in rat tendons. Age (Dordr) 35(6):2203–2214
Kramer RZ, Bella J, Mayville P, Brodsky B, Berman HM (1999) Sequence dependent conformational variations of collagen triple-helical structure. Nat Struct Biol 6:454–457
Kwan AP, Cummings CE, Chapman JA, Grant ME (1991) Macromolecular organization of chicken type X collagen in vitro. J Cell Biol 114(3):597–604
Lampe AK, Bushby KM (2005) Collagen VI related muscle disorders. J Med Genet 42(9):673–685
Latvanlehto A, Fox MA, Sormunen R, Tu H, Oikarainen T, Koski A, Naumenko N, Shakirzyanova A, Kallio M, Ilves M, Giniatullin R, Sanes JR, Pihlajaniemi T (2010) Muscle-derived collagen XIII regulates maturation of the skeletal neuromuscular junction. J Neurosci 30(37):12230–12241
Li SW, Prockop DJ, Helminen H, Fassler R, Lapvetelainen T, Kiraly K, Peltarri A, Arokoski J, Lui H, Arita M et al (1995a) Transgenic mice with targeted inactivation of the Col2 alpha 1 gene for collagen II develop a skeleton with membranous and periosteal bone but no endochondral bone. Genes Dev 9(22):2821–2830
Li Y, Lacerda DA, Warman ML, Beier DR, Yoshioka H, Ninomiya Y, Oxford JT, Morris NP, Andrikopoulos K, Ramirez F et al (1995b) A fibrillar collagen gene, Col11a1, is essential for skeletal morphogenesis. Cell 80(3):423–430
Linsenmayer TF, Gibney E, Igoe F, Gordon MK, Fitch JM, Fessler LI, Birk DE (1993) Type V collagen: molecular structure and fibrillar organization of the chicken alpha 1(V) NH2-terminal domain, a putative regulator of corneal fibrillogenesis. J Cell Biol 121(5):1181–1189
Lis DM, Baar K (2019) Effects of different vitamin C-enriched collagen derivatives on collagen synthesis. Int J Sport Nut Exer Metab 29:526–531
MacBeath JRE, Kielty CM, Shuttleworth A (1996) Type VIII collagen is a product of vascular smooth-muscle cells in development and disease. Biochem J 319(3):993–998
Malfait F, De Paepe A (2009) Bleeding in the heritable connective tissue disorders: mechanisms, diagnosis and treatment. Blood Rev 23(5):191–197
Mao Y, Schwarzbauer JE (2005) Fibronectin fibrillogenesis, a cell-mediated matrix assembly process. Matrix Biol 24(6):389–399

Marchant JK, Hahn RA, Linsenmayer TF, Birk DE (1996) Reduction of type V collagen using a dominant-negative strategy alters the regulation of fibrillogenesis and results in the loss of corneal- specific fibril morphology. J Cell Biol 135(5):1415–1426
Marini JC, Cabral WA, Barnes AM (2010) Null mutations in LEPRE1 and CRTAP cause severe recessive osteogenesis imperfecta. Cell Tissue Res 339(1):59–70
Marini JC, Forlino A, Bächinger HP, Bishop NJ, Byers PH, De Paepe A, Fassier F, Fratzl-Zelman N, Kozloff KM, Krakow D, Montpetit K, Semler O (2017) Osteogenesis imperfecta. Nat Rev Dis Primers 3:17052
Markova DZ, Pan TC, Zhang RZ, Zhang G, Sasaki T, Arita M, Birk DE, Chu ML (2016) Forelimb contractures and abnormal tendon collagen fibrillogenesis in fibulin-4 null mice. Cell Tissue Res 364(3):637–646
Marr N, Hopkinson M, Hibbert AP, Pitsillides AA, Thorpe CT (2020) Biomodal whole-mount imaging of tendon using confocal microscopy and x-ray microputed tomography. Biol Procedures Online 22:13
Martin RB, Burr DB, Sharkey NA (1998) Mechanical properties of ligament and tendon. Skeletal tissue mechanics. Springer, New York, pp 309–346
McDonald JA, Kelley DG, Broekelmann TJ (1982) Role of fibronectin in collagen deposition: fab' to the gelatin-binding domain of fibronectin inhibits both fibronectin and collagen organization in fibroblast extracellular matrix. J Cell Biol 92(2):485–492
Mead TJ, McCulloch DR, Ho JC, Du Y, Adams SM, Birk DE, Apte SS (2018) The metalloproteinase-proteoglycans ADAMTS7 and ADAMTS12 provide an innate, tendon-specific protective mechanism against heterotopic ossification. JCI Insight 3(7):e92941
Mienaltowski MJ, Birk DE (2014) Structure, physiology, and biochemistry of collagens. In: Halper J (ed) Progress in heritable soft connective tissue diseases, Advances in experimental medicine and biology, vol 802. Springer, Dordrecht, pp 5–29
Mienaltowski MJ, Adams SM, Birk DE (2014) Tendon proper- and peritenon-derived progenitor cells have unique tenogenic properties. Stem Cell Res Ther 5(4):86
Mienaltowski MJ, Dunkman AA, Buckley MR, Beason DP, Adams SM, Birk DE, Soslowsky LJ (2016) Injury response of geriatric mouse patellar tendons. J Orthop Res 34(7):1256–1263
Mienaltowski MJ, Canovas A, Fates VA, Hampton AR, Pechanec MY, Islas-Trejo A, Medrano JF (2019) Transcriptome profiles of isolated murine Achilles tendon proper- and peritenon-derived progenitor cells. J Orthop Res 36(6):1409–1418
Milz S, Jakob J, Buttner A, Tischer T, Putz R, Benjamin M (2008) The structure of the coracoacromial ligament: fibrocartilage differentiation does not necessarily mean pathology. Scand J Med Sci Sports 18(1):16–22
Mohammadzadeh N, Melleby AO, Palmero S, Sjaastad I, Chakravarti S, Engebretsen KVT, Christensen G, Lunde IG, Tønnessen T (2020) Moderate loss of the extracellular matrix proteoglycan Lumican attenuates cardiac fibrosis in mice subjected to pressure overload. Cardiology 145:187–198
Moradi-Ameli M, Rousseau JC, Kleman JP, Champliaud MF, Boutillon MM, Bernillon J, Wallach J, Van der Rest M (1994) Diversity in the processing events at the N-terminus of type-V collagen. Eur J Biochem 221(3):987–995
Morais DS, Torres J, Guedes RM, Lopes MA (2015) Current approaches and future trends to promote tendon repair. Ann Biomed Eng 43(9):2025–2035
Morrissey MA, Sherwood DR (2015) An active role for basement membrane assembly and modification in tissue sculpting. J Cell Sci 128:1661–1668
Moser HL, Doe AP, Meier K, Garnier S, Laudier D, Akiyama H, Zumstein MA, Galatz LM, Huang AH (2018) Genetic lineage tracing of targeted cell populations during enthesis healing. J Orthop Res 36(12):3275–3284
Myers JC, Yang H, D'Ippolito JA, Presente A, Miller MK, Dion AS (1994) The triple-helical region of human type XIX collagen consists of multiple collagenous subdomains and exhibits limited sequence homology to alpha 1(XVI). J Biol Chem 269(28):18549–18557
Myllyharju J, Kivirikko KI (2004) Collagens, modifying enzymes and their mutations in humans, flies and worms. Trends Genet 20(1):33–43
Nguyen PK, Pan XS, Li J, Kuo CK (2018) Roadmap of molecular, compositional, and functional markers during embryonic tendon development. Connect Tissue Res 59(5):495–508
Nielsen SH, Karsdal MA (2016) Chapter 24 – type XXIV collagen. In: Karsdal MA (ed) Biochemistry of collagens, laminins and elastin: structure, function and biomarkers. Elsevier, Amsterdam, pp 143–145
Niyibizi C, Visconti CS, Kavalkovich K, Woo SL (1995) Collagens in an adult bovine medial collateral ligament: immunofluorescence localization by confocal microscopy reveals that type XIV collagen predominates at the ligament-bone junction. Matrix Biol 14(9):743–751
Noro A, Kimata K, Oike Y, Shinomura T, Maeda N, Yano S, Takahashi N, Suzuki S (1983) Isolation and characterization of a third proteoglycan (PG-Lt) from chick embryo cartilage which contains disulfide-bonded collagenous polypeptide. J Biol Chem 258(15):9323–9331
Notbohm H, Nokelainen M, Myllyharju J, Fietzek PP, Muller PK, Kivirikko KI (1999) Recombinant human type II collagens with low and high levels of hydroxylysine and its glycosylated forms show marked differences in fibrillogenesis in vitro. J Biol Chem 274(13):8988–8992
Okuyama K, Hongo C, Wu G, Mizuno K, Noguchi K, Ebisuzaki S, Tanaka Y, Nishino N, Bachinger HP (2009) High-resolution structures of collagen-like peptides [(Pro-Pro-Gly)$_4$-X$_{aa}$-Y$_{aa}$-Gly-(Pro-Pro-Gly)$_4$]: implications for triple-helix hydration and Hyp(X) puckering. Biopolymers 91(5):361–372

Olsen BR (2014) Chapter 10 – matrix molecules and their ligands. In: Principles of tissue engineering, 4th edn. Academic, London, pp 189–208

Orgel JP, Irving TC, Miller A, Wess TJ (2006) Microfibrillar structure of type I collagen in situ. Proc Natl Acad Sci U S A 103:9001–9005

Pan TC, Zhang RZ, Mattei MG, Timpl R, Chu ML (1992) Cloning and chromosomal location of human alpha 1(XVI) collagen. Proc Natl Acad Sci U S A 89(14):6565–6569

Pan TC, Zhang RZ, Markova D, Arita M, Zhang Y, Bogdanovich S, Khurana TS, Bonnemann CG, Birk DE, Chu ML (2013) COL6A3 protein deficiency in mice leads to muscle and tendon defects similar to human collagen VI congenital muscular dystrophy. J Biol Chem 288(20):14320–14331. PMC#3656288

Pan TC, Zhang RZ, Arita M, Bogdanovich S, Adams SM, Kumar Gara S, Wagener R, Khurana TS, Birk DE, Chu ML (2014) A mouse model for dominant collagen VI disorders: heterozygous deletion of Col6a3 Exon 16. J Biol Chem 289(15):10293–10307

Park AC, Phillips CL, Pfeiffer FM, Roenneburg DA, Kernien JF, Adams SM, Davidson JM, Birk DE, Greenspan DS (2015) Homozygosity and heterozygosity for null Col5a2 alleles produce embryonic lethality and a novel classic Ehlers-Danlos Syndrome-related phenotype. Am J Pathol 185(7):2000–2011

Peltonen L, Halila R, Ryhanen L (1985) Enzymes converting procollagens to collagens. J Cell Biochem 28(1):15–21

Peltonen J, Cronin NJ, Avela J, Finni T (2010) In vivo mechanical response of human Achilles tendon to a single bout of hopping exercise. J Exp Biol 213(Pt 8):1259–1265

Peltonen J, Cronin NJ, Stenroth L, Finni T, Avela J (2013) Viscoelastic properties of the Achilles tendon in vivo. Springerplus 2(1):212

Pike AV, Ker RF, Alexander RM (2000) The development of fatigue quality in high and low-stressed tendons of sheep (Ovis aries). J Exp Biol 203(Pt 14):2187–2193

Place ES, Evans ND, Stevens MM (2009) Complexity in biomaterials for tissue engineering. Nat Mater 8(6):457–470

Plumb DA, Dhir V, Mironov A, Poulsom R, Kadler KE, Thornton DJ, Briggs MD, Boot-Handford RP (2007) Collagen XXVII is developmentally-regulated and forms thin fibrillar structures distinct from those of classical vertebrate fibrillar collagens. J Biol Chem 282(17):12791–12795

Pokidysheva E, Zientek KD, Ishikawa Y, Mizuno K, Vranka JA, Montgomery NT, Keene DR, Kawaguchi T, Okuyama K, Bächinger HP (2013) Posttranslational modifications in type I collagen from different tissues extracted from wild type and prolyl 3-hydroxylase 1 null mice. J Biol Chem 288(34):24742–24752

Pulido D, Sharma U, Sandrine VLG, Hussain SA, Cordes S, Mariana N, Bettler E, Moali C, Aghajari N, Hohenester E, Hulmes DJS (2018) Structural basis for the acceleration of procollagen processing by procollagen C-proteinase enhancer-1. Structure 26(10):1384–1392

Quapp KM, Weiss JA (1998) Material characterization of human medial collateral ligament. J Biomech Eng 120(6):757–763

Ramachandran GN, Kartha G (1954) The structure of collagen. Nature 174:269–270

Ricard-Blum S, Dublet B, van der Rest M (2000) Unconventional collagens. Types VI, VII, VIII, IX, X, XII, XIV, XVI, and XIX. Oxford University Press, North Carolina

Rich A, Crick FHC (1961) The molecular structure of collagen. J Mol Biol 3(5):483–506

Riechert K, Labs K, Lindenhayn K, Sinha P (2001) Semiquantitative analysis of types I and III collagen from tendons and ligaments in a rabbit model. J Orthop Sci 6(1):68–74

Roberts RR, Bobzin L, Teng CS, Pal D, Tuzon CT, Schweitzer R, Merrill AE (2019) FGF signaling patterns cell fate at the interface between tendon and bone. Development 146:1–15

Robi K, Jakob N, Matevz K, Matjaz V (2013) Chapter 2: the physiology of sports injuries and repair processes. Curr Iss Sports Exer Med. https://doi.org/10.5772/54234

Robinson KA, Sun M, Barnum CE, Weiss SN, Huegel J, Shetye SS, Lin L, Saez D, Adams SM, Iozzo RV, Soslowsky LJ, Birk DE (2017) Decorin and biglycan are necessary for maintaining collagen fibril structure, fiber realignment, and mechanical properties of mature tendons. Matrix Biol 64:81–93

Rousseau JC, Farjanel J, Boutillon MM, Hartmann DJ, van der Rest M, Moradi-Ameli M (1996) Processing of type XI collagen. Determination of the matrix forms of the alpha1(XI) chain. J Biol Chem 271(39):23743–23748

Rumian AP, Wallace AL, Birch HL (2007) Tendons and ligaments are anatomically distinct but overlap in molecular and morphological features – a comparative study in an ovine model. J Orthop Res 25:458–464

Sage H, Bornstein P (1987) Type VIII Collagen. In: Mayne R, Burgeson RE (eds) Structure and function of collagen types. Academic, Orlando, pp 173–194

Sakabe T, Sakai K, Maeda T, Sunaga A, Furuta N, Schweitzer R, Sasaki T, Sakai T (2018) Transcription factor scleraxis vitally contributes to progenitor lineage direction in wound healing of adult tendon in mice. J Biol Chem 293(16):5766–5780

Sardone F, Santi S, Tagliavini F, Traina F, Merlini L, Squarzoni S, Cescon M, Wagener R, Maraldi NM, Bonaldo P, Faldini C, Sabatelli P (2016) Collagen VI-NG2 axis in human tendon fibroblasts under conditions mimicking injury response. Matrix Biol 55:90–105

Sawade H, Konomi H (1991) The α1 chain of type VIII collagen is associated with many but not all microfibrils of elastic fiber system. Cell Struct Funct 16:455–466

Schiele NR, Flotow FV, Tochka ZL, Hockaday LA, Marturano JE, Thibodeau JJ, Kuo CK (2015)

Actin cytoskeleton contributes to the elastic modulus of embryonic tendon during early development. J Orthop Res 33(6):874–881
Seegmiller R, Fraser FC, Sheldon H (1971) A new chondrodystrophic mutant in mice. Electron microscopy of normal and abnormal chondrogenesis. J Cell Biol 48(3):580–593
Segev F, Heon E, Cole WG, Wenstrup RJ, Young F, Slomovic AR, Rootman DS, Whitaker-Menezes D, Chervoneva I, Birk DE (2006) Structural abnormalities of the cornea and lid resulting from collagen V mutations. Invest Ophthalmol Vis Sci 47(2):565–573
Shearer T, Parnell WJ, Lynch B, Screen HRC, Abrahams DI (2020) A recruitment model of tendon viscoelasticity that incorporates fibril creep and explains strain-dependent relaxation. ASME. J Biomech Eng 142(7):071003
Shen G (2005) The role of type X collagen in facilitating and regulating endochondral ossification of articular cartilage. Orthod Craniofacial Res 8(1):11–17
Shen H, Schwartz AG, Civitelli R, Thomopoulos S (2020) Connexin 43 is necessary for murine tendon enthesis formation and response to loading. J Bone Miner Res 35(8):1494–1503
Shikh Alsook MK, Gabriel A, Salouci M, Piret J, Alzamel N, Moula N, Denoix J-M, Antoine N, Baise E (2015) Characterization of collagen fibrils after equine suspensory ligament injury: An ultrastructural and biochemical approach. Vet J 204(1):117–122
Shoulders MD, Raines RT (2009) Collagen structure and stability. Annu Rev Biochem 78:929–958
Smith SM, Thomas CE, Birk DE (2012) Pericellular proteins of the developing mouse tendon: a proteomic analysis. Connect Tissue Res 53(1):2–13
Snedeker JG, Foolen J (2017) Tendon injury and repair – a perspective on the basic mechanisms of tendon disease and future clinical therapy. Acta Biomater 63:18–36
Sricholpech M, Perdivara I, Yokoyama M, Nagaoka H, Terajima M, Tomer KB, Yamauchi M (2012) Lysyl hydroxylase 3-mediated glucosylation in type I collagen: molecular loci and biological significance. J Biol Chem 287(27):22998–23009
Steinmann B, Bruckner P, Superti-Furga A (1991) Cyclosporin A slows collagen triple-helix formation in vivo: indirect evidence for a physiologic role of peptidyl-prolyl cis-trans-isomerase. J Biol Chem 266(2):1299–1303
Sun M, Connizzo BK, Adams SM, Freedman BR, Wenstrup RJ, Soslowsky LJ, Birk DE (2015) Targeted deletion of collagen V in tendons and ligaments results in a classic Ehlers-Danlos syndrome joint phenotype. Am J Pathol 185(5):1436–1447
Symoens S, Syx D, Malfait F, Callewaert B, De Backer J, Vanakker O, Coucke P, De Paepe A (2012) Comprehensive molecular analysis demonstrates type V collagen mutations in over 90% of patients with classic EDS and allows to refi ne diagnostic criteria. Hum Mutat 33(10):1485–1493
Taylor SH, Al-Youha S, Van Agtmael T, Lu Y, Wong J, mcGrouther DA, Kadler KA (2011) Tendon is covered by a basement membrane epithelium that is required for cell retention and the prevention of adhesion formation. PLoS One 6(1):e16337
Thomopoulos S, Williams GR, Gimbel JA, Favata M, Soslowsky LJ (2003) Variation of biomechanical, structural, and compositional properties along the tendon to bone insertion. J Orthop Res 21:413–419
Thomopoulos S, Genin GM, Galatz LM (2010) The development and morphogenesis of the tendon-to-bone insertion what development can teach us about healing. J Musculoskelet Neuronal Interact 10(1):35–45
Thorpe CT, Riley GP, Birch HL, Clegg PD, Screen HRC (2016) Fascicles and the interfascicular matrix show adaptation for fatigue resistance in energy storing tendons. Acta Biomater 42:308–315
Torre-Blanco A, Adachi E, Hojima Y, Wootton JA, Minor RR, Prockop DJ (1992) Temperature-induced posttranslational over-modification of type I procollagen. Effects of over-modification of the protein on the rate of cleavage by procollagen N-proteinase and on self-assembly of collagen into fibrils. J Biol Chem 267(4):2650–2655
Trelsad RL, Hayashi K (1979) Tendon collagen fibrillogenesis: intracellular subassemblies and cell surface changes associated with fibril growth. Dev Biol 71:228–242
Tuite DJ, Renstrom PAFH, O'Brien M (1997) The aging tendon. Scand J Med Sci Sports 7:72–77
Vahidnezhad H, Youssefian L, Saeidian AH, Touati A, Pajouhanfar S, Baghdadi T, Shadmehri AA, Giunta C, Kraenzlin M, Syx D, Malfait F, Has C, Lwin SM, Karamzadeh R, Liu L, Guy A, Hamid M, Kariminejad A, Zeinali S, McGrath JA, Uitto J (2019) Mutations in PLOD3, encoding lysyl hydroxylase 3, causes a complex connective tissue disorder including recessive dystrophic epidermolysis bullosa-like blistering phenotype with abnormal anchoring fibrils and type VII collagen deficiency. Matrix Biol 81:91–106
van der Rest M, Mayne R (1988) Type IX collagen proteoglycan from cartilage is covalently cross- linked to type II collagen. J Biol Chem 263(4):1615–1618
Voermans NC, Bonnemann CG, Hamel BC, Jungbluth H, van Engelen BG (2009) Joint hypermobility as a distinctive feature in the differential diagnosis of myopathies. J Neurol 256(1):13–27
Vogel BE, Doelz R, Kadler KE, Hojima Y, Engel J, Prockop DJ (1988) A substitution of cysteine for glycine 748 of the alpha 1 chain produces a kink at this site in the procollagen I molecule and an altered N-proteinase cleavage site over 225 nm away. J Biol Chem 263(35):19249–19255
von der Mark H, Aumailley M, Wick G, Fleischmajer R, Timpl R (1984) Immunochemistry, genuine size and tissue localization of collagen VI. Eur J Biochem 142(3):493–502

Waggett AD, Ralphs JR, Kwan APL, Woodnutt D, Benjamin M (1998) Characterization of collagens and proteoglycans at the insertion of the human Achilles tendon. Matrix Biol 16:457–470

Wang JHC, Guo Q, Li B (2012) Tendon biomechanics and mechanobiology – a minireview of basic concepts and recent advancements. J Hand Ther 25(2):133–141

Wenstrup RJ, Florer JB, Brunskill EW, Bell SM, Chervoneva I, Birk DE (2004a) Type V collagen controls the initiation of collagen fibril assembly. J Biol Chem 279(51):53331–53337

Wenstrup RJ, Florer JB, Cole WG, Willing MC, Birk DE (2004b) Reduced type I collagen utilization: a pathogenic mechanism in COL5A1 haploinsufficient Ehlers-Danlos syndrome. J Cell Biochem 92(1):113–124

Wenstrup RJ, Smith SM, Florer JB, Zhang G, Beason DP, Seegmiller RE, Soslowsky LJ, Birk DE (2011) Regulation of collagen fibril nucleation and initial fibril assembly involves coordinate interactions with collagens V and XI in developing tendon. J Biol Chem 286(23):20455–20465

Wess TJ, Hammersley AP, Wess L, Miller A (1998) Molecular packing of type I collagen in tendon. J Mol Biol 275:255–267

Wiberg C, Heinegard D, Wenglen C, Timpl R, Morgelin M (2002) Biglycan organizes collagen VI into hexagonal-like networks resembling tissue structures. J Biol Chem 277(51):49120–49126

Williams IF, McCullagh KG, Silver IA (1984) The distribution of Types I and III collagen and fibronectin in the healing equine tendon. Connect Tissue Res 12(3–4):221–227

Willumsen N, Karsdal MA (2016) Chapter 20 – type XX collagen. In: Karsdal MA (ed) Biochemistry of collagens, laminins and elastin: structure, function and biomarkers. Elsevier, Amsterdam, pp 127–129

Wiradjaja F, DiTommaso T, Smyth I (2010) Basement membranes in development and disease. Birth Defects Res C Embryo Today 90(1):8–31

Yamaguchi N, Mayne R, Ninomiya Y (1991) The alpha 1 (VIII) collagen gene is homologous to the alpha 1 (X) collagen gene and contains a large exon encoding the entire triple helical and carboxylterminal non-triple helical domains of the alpha 1 (VIII) polypeptide. J Biol Chem 266(7):4508–4513

Yeung CYC, Kadler KE (2019) Importance of the circadian clock in tendon development. Curr Top Dev Biol 133:310–332

Yoshioka H, Zhang H, Ramirez F, Mattei MG, Moradi-Ameli M, van der Rest M, Gordon MK (1992) Synteny between the loci for a novel FACIT-like collagen locus (D6S228E) and alpha 1 (IX) collagen (COL9A1) on 6q12-q14 in humans. Genomics 13(3):884–886

Young BB, Zhang G, Koch M, Birk DE (2002) The roles of types XII and XIV collagen in fibrillogenesis and matrix assembly in the developing cornea. J Cell Biochem 87(2):208–220

Yurchenco PD, Patton BL (2009) Developmental and pathogenic mechanisms of basement membrane assembly. Curr Pharm Des 15(12):1277–1294

Zhang G, Young BB, Birk DE (2003) Differential expression of type XII collagen in developing chicken metatarsal tendons. J Anat 202(5):411–420

Zhang G, Young BB, Ezura BB, Favata M, Soslowsky LJ, Chakravarti S, Birk DE (2005) Development of tendon structure and function: regulation of collagen fibrillogenesis. J Musculoskelet Neuronal Interact 5(1):5–21

Zhong C, Chrzanowska-Wodnicka M, Brown J, Shaub A, Belkin AM, Burridge K (1998) Rho- mediated contractility exposes a cryptic site in fibronectin and induces fibronectin matrix assembly. J Cell Biol 141(2):539–551

3 Tendon Extracellular Matrix Assembly, Maintenance and Dysregulation Throughout Life

Seyed Mohammad Siadat, Danae E. Zamboulis, Chavaunne T. Thorpe, Jeffrey W. Ruberti, and Brianne K. Connizzo

Abstract

In his Lissner Award medal lecture in 2000, Stephen Cowin asked the question: "How is a tissue built?" It is not a new question, but it remains as relevant today as it did when it was asked 20 years ago. In fact, research on the organization and development of tissue structure has been a primary focus of tendon and ligament research for over two centuries. The tendon extracellular matrix (ECM) is critical to overall tissue function; it gives the tissue its unique mechanical properties, exhibiting complex non-linear responses, viscoelasticity and flow mechanisms, excellent energy storage and fatigue resistance. This matrix also creates a unique microenvironment for resident cells, allowing cells to maintain their phenotype and translate mechanical and chemical signals into biological responses. Importantly, this architecture is constantly remodeled by local cell populations in response to changing biochemical (systemic and local disease or injury) and mechanical (exercise, disuse, and overuse) stimuli. Here, we review the current understanding of matrix remodeling throughout life, focusing on formation and assembly during the postnatal period, maintenance and homeostasis during adulthood, and changes to homeostasis in natural aging. We also discuss advances in model systems and novel tools for studying collagen and non-collagenous matrix remodeling throughout life, and finally conclude by identifying key questions that have yet to be answered.

Keywords

Tendon · Collagen remodeling · Non-collageneous matrix · Homeostasis · Aging

S. M. Siadat · J. W. Ruberti
Department of Bioengineering, Northeastern University, Boston, MA, USA
e-mail: siadat.s@northeastern.edu; j.ruberti@northeastern.edu

D. E. Zamboulis
Institute of Life Course and Medical Sciences, Faculty of Health and Life Sciences, University of Liverpool, Liverpool, UK
e-mail: D.E.Zamboulis@liverpool.ac.uk

C. T. Thorpe
Comparative Biomedical Sciences, The Royal Veterinary College, University of London, London, UK
e-mail: cthorpe@rvc.ac.uk

B. K. Connizzo (✉)
Department of Biomedical Engineering, Boston University, Boston, MA, USA
e-mail: connizzo@bu.edu

J. Halper (ed.), *Progress in Heritable Soft Connective Tissue Diseases*, Advances in Experimental Medicine and Biology 1348, https://doi.org/10.1007/978-3-030-80614-9_3

Abbreviations

AGEs	Advanced glycation end-products
Aha	Azidohomoalanine
BATs	Bioartificial tendons
CHP	Collagen hybridizing peptide
CMP	Collagen mimetic peptide
COMP	Cartilage oligomeric matrix protein
CSA	Cross sectional area
DAMPs	Damage-associated molecular patterns
DIBO	Dibenzooctyne
DIC	Differential interference contrast
DTAF	Dichlorotriazinyl aminofluorescein
ECM	Extracellular matrix
EDS	Ehlers-Danlos Syndrome
EM	Electron microscopy
FACIT	Fibril-associated collagens with interrupted triple helices
FIC	Flow-induced crystallization
FN	Fibronectin
FRET	Förster resonance energy transfer
GAG	Glycosaminoglycan
GFP	Green fluorescent protein
GPC	Golgi to plasma membrane carrier
IL	Interleukin
LEs	Ligament equivalents
Met	Methionine
MMP	Matrix metalloproteinase
N_3-Pro	Azido-proline
PGE_2	Prostaglandin E_2
ROS	Reactive oxygen species
SASP	Senescence-associated secretory phenotype
SHG	Second harmonic generation
SLRP	Small leucine rich proteoglycan
TECs	Tissue engineered constructs
TEM	Transmission electron microscopy
TSCs	Tendon stem cells
TSP	Thrombospondin

3.1 Introduction

In his Lissner Award medal lecture in 2000, Stephen Cowin asked the question: "How is a tissue built?" It is not a new question, but it remains as relevant today as it did when it was asked 20 years ago (Cowin 2000). In fact, research on the organization and development of tissue structure has been a primary focus of tendon and ligament research for over two centuries. The tendon extracellular matrix (ECM) is critical to overall tissue function; it gives the tissue its unique mechanical function, exhibiting complex nonlinear responses, viscoelasticity and flow mechanisms, excellent energy storage and fatigue resistance (Butler et al. 1997; Connizzo et al. 2013a; Franchi et al. 2007; Thorpe and Screen 2016; Thompson et al. 2017). This matrix also creates a unique microenvironment for resident cells, allowing cells to maintain their phenotype and translate mechanical and chemical signals into biological responses (Thompson et al. 2017; Wall et al. 2018; Wang et al. 2013a; Dyment et al. 2020). Importantly, this architecture is constantly remodeled by local cell populations in response to functional changes such as exercise, as well as in response to tissue damage or injury. Here, we review our current understanding of matrix remodeling throughout life, focusing on formation and assembly during the postnatal period, maintenance and homeostasis during adulthood, and changes to homeostasis in natural aging.

3.1.1 Tendon Composition, Structure, and Function

The dry weight of the tendon ECM can be dissected into two main components: the collagenous structural hierarchy, and the non-collagenous matrix (Fig. 3.1). Both components are essential to tendon function and biology, although the collagenous structure has been studied far more extensively. Type I collagen is the primary protein in tendon, accounting for 65–80% of the dry mass of the tendon (Brinckmann and Bachinger 2005; Kannus 2000). The asymmetric triple-helix collagen molecules coil to form the triple helix of a collagen molecule (Mienaltowski and Birk 2014). Collagen molecules then link in a quarter staggered orientation to form fibrils. Collagen fibrils, now considered to be the basic unit of tendon, are bundled together within a collagen fiber. Collagen fibers are then bundled together and bound via a fine sheath of tissue; this structure is now called a fascicle. Fascicles then bundle to

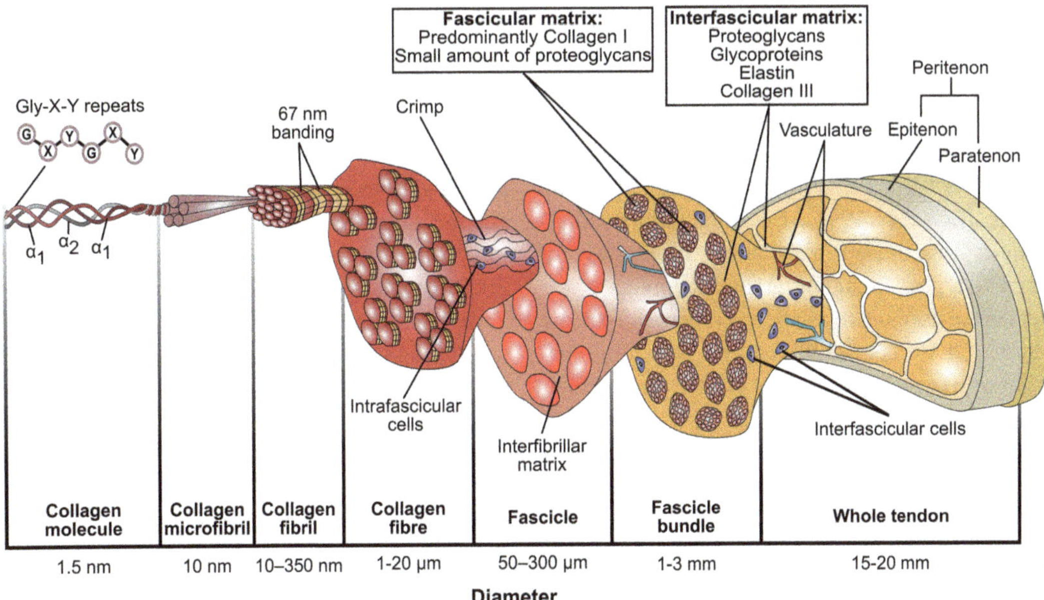

Fig. 3.1 Hierarchical organization of the equine superficial digital flexor tendon with specific detail related to the interfascicular and interfibrillar matrix composition. (Reproduced from O'Brien et al. 2020)

form whole tendon, which is surrounded by the epitenon sheath. The non-collagenous matrix in tendon is found interspersed between collagen fibrils, fibers, and fascicles in the interfibrillar, interfiber, and interfascicular region of the tendon, respectively, and is mainly composed of proteoglycans, glycoproteins, and minor collagens (Kannus et al. 1998; Taye et al. 2020; Thorpe et al. 2016a).

The structural organization of the tendon ECM is a major contributor to overall tissue function. During mechanical loading, collagen fascicles, fibers, and fibrils exhibit a number of dynamic responses that allow for reduction of stress concentrations and prevent structural damage (Connizzo et al. 2013a; Franchi et al. 2007). This includes uncrimping (Lavagnino et al. 2017; Patterson-Kane et al. 1997; Miller et al. 2012a), or the reduction in the wavy formation of the collagen fibers, and fiber/fibril re-alignment (Miller et al. 2012b; Connizzo et al. 2013b; Lake et al. 2010), when these structures re-orient towards the axis of loading and consolidate to a single fiber direction. In addition, collagen fascicles, fibers, and fibrils have all demonstrated the capacity to slide against one another, although this ability is more often attributed to the properties of the non-collagenous compartment rather than the collagen structure itself (Connizzo et al. 2014a; Rigozzi et al. 2013; Thorpe et al. 2015a; Szczesny and Elliott 2014). In addition, proteoglycans and their glycosaminoglycan (GAG) chains present in the non-collagenous compartment attract and trap water molecules allowing for complex fluid flow and viscoporoelasticity (Butler et al. 1997; Rigozzi et al. 2013; Legerlotz et al. 2013a; Connizzo and Grodzinsky 2017; Buckley et al. 2013). It is crucial to note however that both the structure and function of tendons and ligaments varies significantly based on tissue site, and more specifically based on the functional demands of the tissue.

3.1.2 Function-Based Variations in Tendon Composition and Structure

All tendons within the appendicular skeleton transfer muscle-generated force to the bony skeleton, positioning the limbs during locomotion. In addition to a positional function, specific tendons also store and release energy as they stretch and recoil with each stride, reducing the energetic cost of locomotion (McNeill 2002). The major energy storing tendons in the human are the Achilles and hamstring tendons, whereas in large quadrupeds, such as the horse, the digital flexor tendons are the predominant energy storing tissues (Shepherd et al. 2014; Lichtwark and Wilson 2005; Biewener 1998). Energy storing tendons require specialised mechanical properties for their function, including greater compliance and enhanced fatigue resistance, properties that are conferred by compositional and structural specialisations at different levels of the tendon hierarchy (Thorpe and Screen 2016; Thorpe et al. 2013a). Here, we specify research performed in energy storing or positional tendons for clarity wherever relevant.

Tendon structure and composition are also dramatically different at the junction with muscle and bone compared to the midsubstance. The enthesis, or insertion site, has unique compositional and structural properties that allow it to minimize stress concentrations at the junction of dissimilar materials (Deymier-Black et al. 2015; Thomopoulos et al. 2003; Saadat et al. 2016). Tissue function at these sites is also altered, demonstrating more complex multi-scale mechanical responses (Connizzo et al. 2016a). In addition, some tendons exhibit unique anatomical positions that alter function. Tendons that wrap around bony structures exhibit cartilaginous-like tissue regions with higher levels of the large proteoglycan aggrecan and enhanced mechanical function in compression (Connizzo and Grodzinsky 2018a; Wren et al. 2000; Koob and Vogel 1987; Fang et al. 2014). For the purposes of this discussion, we focus on general changes across multiple species in the collagen structure and non-collagenous matrix at the midsubstance of the tendon and not in specialized regions.

3.1.3 Tendon Cell Populations

Remodeling of the extracellular matrix is cell-mediated, and therefore an understanding of cell populations within tendon is necessary for discussion of this highly complex process. Early in development, tendon is highly cellular, with proliferative cells appearing homogenous with more rounded cell nuclei. Following deposition of the extracellular matrix, tendon becomes hypocellular with limited mitotic activity and a heterogeneous cell population with cells with long and spindle shaped nuclei in the fascicles and the more rounded, densely packed cells in the interfascicular matrix (Oryan and Shoushtari 2008; Russo et al. 2015; Grinstein et al. 2019; Zamboulis et al. 2020). Until recently the main cell types that had been described in tendon were tenocytes and tendon progenitor/stem cells (TSCs) as well as tissue-resident immune cells, vascular cells, neuronal cells, and chondrocyte-like cells at the tendon insertion (Kannus 2000; Ackermann et al. 2016; Thomopoulos et al. 2010; Bi et al. 2007; Lee et al. 2018; Mienaltowski et al. 2018). With the advent of single-cell sequencing, the investigation of cell heterogeneity within tissues has been made possible and its recent use in tendon research has unveiled several tendon cell subtypes (Paolillo et al. 2019; Harvey et al. 2019; Kendal et al. 2020; De Micheli et al. 2020; Yin et al. 2016), but the role of the identified clusters in the development, maintenance, and aging of tendon still remains to be elucidated.

3.2 Postnatal Development

3.2.1 Collagen Fibril Formation

The highly dynamic nature of fibrillogenesis and growth of fibrils in the complex extracellular environment has made it challenging to precisely separate the events that cause conversion of soluble collagen to an insoluble fibril. *In vitro* polymerization of tissue-extracted collagen molecules in solution has shed light on fibrillogenesis kinetics and thermodynamics. Collagen molecules polymerize spontaneously at physiological pH, temperature, and ionic strength (Gross and Kirk 1958; Wood 1964; Williams et al. 1978; Vanamee and Porter 1951) demonstrating the same detailed fine structure of native fibrils (Vanamee and Porter 1951; Bahr 1950; Noda and Wyckoff 1951; Schmitt et al. 1942). Slight deviations from physiological conditions lead to formation of abnormal fibrils (Gross 1956). Thermodynamically, type I collagen fibrillogenesis *in vitro* is an entropy-driven and endothermic self-assembly process (Kadler et al. 1987) which is driven by the loss of solvent molecules from the collagen surface. *In vitro* self-assembly, however, cannot explain the formation of highly organized native collagenous tissues such as tendon with a multi-hierarchical structure comprising molecules, fibrils, fibers, and fascicles all parallel to the long axis of the tendon (Franchi et al. 2007). Formation of unorganized networks of fibrils varying in diameter and direction *in vitro* (Wood and Keech 1960; Bard and Chapman 1973), points to the critical role of cellular environment *in vivo*. It is clear that collagen production and fibrillogenesis is under the direct control of fibroblasts (Wolbach and Howe 1926; Maximow 1928; Stearns 1940a, b; Wassermann 1954; Porter and Pappas 1959; Chapman 1961; Peach et al. 1961; Ross and Benditt 1961; Goldberg and Green 1964). What is not exactly clear, is the site and mechanism of initial fibril formation which has been the subject of studies for almost two centuries (Schwann 1839, 1847). The literature contains contradictory explanations regarding whether the collagen fibrils of the connective tissues arise within the cytoplasm (Ferguson 1912; Bradbury and Meek 1958; Godman and Porter 1960), on the surface (Porter and Pappas 1959; Mall 1902), or in the intercellular spaces (Stearns 1940a, b; Ross and Benditt 1961; Mallory 1903; Hertzler 1910; Baitsell 1915, 1916, 1921, 1925; Isaacs 1916, 1919; Gross et al. 1955; Ross and Benditt 1962) of collagen-secreting cells.

After the advent of electron microscopy, several studies demonstrated vesicular components containing small fibrils just below the cell surface (Bradbury and Meek 1958; Godman and Porter 1960; Sheldon and Kimball 1962; Voelz 1964; Welsh 1966; Trelstad 1971). High voltage electron microscopy revealed collagen fibrils within small surface recesses in chick embryo cornea (Birk and Trelstad 1984), tendon (Birk and Trelstad 1986; Yang and Birk 1986), and dermis (Ploetz et al. 1991) fibroblasts. It was suggested that cells directly produce fibrils within these deep and narrow recesses and place them into the ECM (Fig. 3.2a). However, it was previously shown that fibrils can be produced by any action that causes shrinkage of the intercellular substance (Isaacs 1919), increasing the possibility of formation of artificial fibrils due to fixation or dehydration in prepared samples for electron microscopy. Canty et al. (2004) using serial section and 3-D reconstructions of chick embryonic tendon fibroblasts revealed fibrils within closed intracellular Golgi to plasma membrane carriers (GPCs). Further, using pulse-chase experiments, procollagen fragments were detected within the GPCs (Canty et al. 2004). It was proposed that the GPCs were on their way to plasma membrane protrusions, which were named fibril depositors or fibripositors. It has been widely accepted now that fibripositors are the site of fibril assembly *in vivo* (Holmes et al. 2018); fibril segments are formed intracellularly and then discharged into extracellular space by the non-muscle myosin II mechanism (Fig. 3.2b) (Kalson et al. 2013; Canty et al. 2004).

However, fibripositors are absent during postnatal development (Humphries et al. 2008) and therefore cannot explain the persistent production of *de novo* fibrils in postnatal tendon and throughout life (Chang et al. 2020) when cells lose their ability to directly access damaged or developing fibrils in the dense and mature ECM (Isaacs 1919; Kalson et al. 2015). The fibripositor theory is also unclear regarding intracellular processing of procollagen. It has been shown that removal of the carboxyl propeptides lowers the solubility of procollagen (Kadler and Watson 1995) and is an essential step for the assembly of collagen fibrils (Prockop et al. 1979a, b). While procollagen processing has been reported within intracellular compartments of postnatal murine (Humphries et al. 2008) and chick embryonic (Canty et al. 2004) tendon fibroblasts, the enzymes for procollagen cleavage have been detected primarily within the extracellular culture medium (Hojima et al. 1985; Kessler and Goldberg 1978; Duksin et al. 1978; Leung et al. 1979; Jimenez et al. 1971) and not extracts of the cells (Goldberg et al. 1975). The required ionic

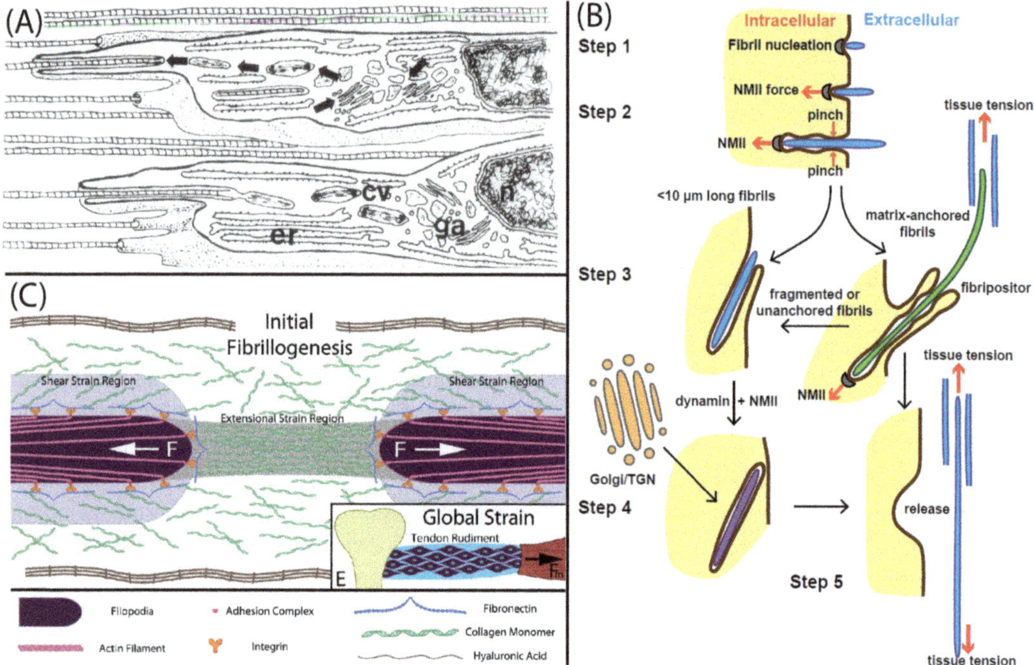

Fig. 3.2 Possible mechanisms of fibril formation. (A) Collagen fibrillogenesis model proposed by Trelstad and Hayashi (1979). Collagen is synthesized in the endoplasmic reticulum (er), packaged in the Golgi apparatus (ga), and transferred in condensation vacuoles (cv) to deep cytoplasmic recesses (site of fibril assembly). **(B)** The processes of collagen fibril nucleation and movement in the fibripositor model proposed by Kalson et al. (2013). The initial collagen fibril nucleation occurs at the plasma membrane by accretion of collagen molecules or collagen aggregates. NMII powers the transport of newly formed fibrils in fibripositors. **(C)** Flow-induced crystallization model by Paten et al. (2016) elucidating the early stage of tendon morphogenesis *in vivo*: (1) cell recruitment, (2) cell migration and organization, (3) ECM molecular synthesis e.g., collagen monomers, fibronectin, elastin, proteoglycans and hyaluronic acid, (4) initial fibrillogenesis by filopodia on the fibroblasts via exerting a contractile force on collagen-binding complexes, and (5) tissue strains cause formation of additional fibrils precisely where they are required for tissue connectivity

calcium concentration for enzyme activity (Hojima et al. 1985) is also orders of magnitude larger than intracellular calcium concentration (Bronner 2001). Furthermore, the procollagen proteinases are neutral metalloproteinases (Kessler and Goldberg 1978; Duksin et al. 1978; Leung et al. 1979; Goldberg et al. 1975; Njieha et al. 1982; Bornstein et al. 1972) and have negligible activity at pH 6 or below (Hojima et al. 1985, 1994). The acidic pH of Golgi network transport carriers and secretory vacuoles (Demaurex et al. 1998) is incompatible with the neutral pH condition required for procollagen processing and fibrillogenesis of collagen molecules. N'Diaye et al. recently showed that the extracellular space is the main action site of bone morphogenetic protein 1, which is required for type I procollagen C-terminal processing in postnatal lung fibroblasts (N'Diaye et al. 2020). It is possible that the detected intracellular collagen fragments in other studies (Canty et al. 2004; Humphries et al. 2008) are processed extracellularly and then rapidly endocytosed.

Several studies suggest that intracytoplasmic fibrils are evidence for the ability of fibroblasts to phagocytose extracellular collagen fibrils in rapidly remodeling (Ten Cate 1972; Ten Cate and Deporter 1974, 1975; Ten Cate and Freeman 1974; Listgarten 1973) or developing (Dyer and Peppler 1977) tissues. Intracellular mature fibrils have been reported with loss of banding (Ten Cate 1972), coiled in membrane-bound structures (Ten Cate 1972), and with poorly-visualized structures (Listgarten 1973). Some fibrils were observed situated partly within the fibroblast and partly outside of it while demonstrating the presence of enzyme activity (Deporter and Ten Cate 1973). All of this suggests that the observed intracellular fibrils were once extracellular and on their way to be degraded intracellularly. It has been shown that intracellular cross-banded collagen fibrils appear even when collagen synthesis is blocked (Everts et al. 1985; Everts and Beertsen 1987; Beertsen et al. 1984) and that cytoplasmic actin filament systems are involved in the phagocytosis of collagen (Everts et al. 1985, 1989). Furthermore, quantitative radio-autography after injection of ^{3}H-proline revealed that collagen precursors (procollagen) were released outside of the cell fibroblasts (Marchi and Leblond 1983, 1984). The observed intracytoplasmic collagen fibrils did not contain the new labeled proline, but were instead associated with lysosomes and digestive vacuoles, had lost their banding and were at various stages of degeneration.

Several studies suggest that fibril formation could operate independently of the cell surface or at some nominal distance from it, guided by long-range spatial cues provided by cell traction (Stopak et al. 1985) or mechanical forces (Gross et al. 1955; Paten et al. 2016; Lewis 1917). Wolbach followed histologic sequences in the development of connective tissue of guinea pigs under a scorbutic condition (Wolbach and Howe 1926; Wolbach 1933). It was suggested that rapid appearance and large volume of intercellular collagen fibrils is due to presence of a liquid precursor of collagen in the extracellular space, and that the collagen fibril formation is influenced by forces acting on this homogeneous collagen. Another study followed the progress of a healing wound in the connective tissue of a living rabbit's ear, demonstrating that intercellular connective tissue fibrils formed extracellularly as a result of fibroblastic activity (Stearns 1940a, b). The fibroblasts participated directly in the process by the projection of cytoplasmic material from their surface. Since this cytoplasmic material disappeared as the fibrils formed, it was suggested that the secreted material was utilized in the production of fibrils guided by applied tension and orientation of fibroblast cells. Emerging evidence suggests the presence of a newly synthesized precursor – tropocollagen – that is free in the ECM (Gross et al. 1955) and diffuses away from the secretory cells (Revel and Hay 1963), and that individual collagen fibrils can form from precursor molecules/microfibrils produced by more than one cell (Lu et al. 2018).

Paten et al. demonstrated *in vitro* how tension can directly drive initial fibrillogenesis (Paten et al. 2016). It was shown that organized fibrils can be formed by slowly drawing a microneedle from the slightly concentrated surface of a collagen solution droplet. They then proposed a model for early connective tissue development in which extensional strain triggers fibril formation extracellularly directly in the path of force. Paten et al. further expanded the concept to address the establishment of continuity in collagenous tissue, suggesting that the amplification of the extensional strain rate between the ends of early fibrils can rapidly fuse them by flow-induced crystallization (FIC) (Fig. 3.2c). They further estimated that the required collagen concentration and contraction rates necessary for FIC is achievable by the local cell population. While it has not yet been demonstrated experimentally, the FIC model has the potential to explain (1) the abundance of short fibril segments during initial tendon morphogenesis and their end-to-end growth (Birk et al. 1995, 1997), (2) the synchronized alignment of collagen fibrils far from the main cell body (Young et al. 2014), and (3) the role of hyaluronic acid (Goldberg and Green 1964; Green and Hemerman 1964), fibronectin (Sottile and Hocking 2002; McDonald et al. 1982; Paten et al. 2019), actin filaments (Johnson and Galis 2003), and integrins (Li et al. 2003) which have been all shown previously to be necessary for collagen fibrillogenesis. While the precise manner in which collagen molecules are manipulated to drive the formation, growth, and remodeling of collagen fibrils has not been agreed upon, it is likely guided by a common physical and regulated by multiple factors to establish long-range connectivity and growth of collagenous structures into the path of force, where it is needed.

3.2.2 Post-formation Assembly

Embryonic growth occurs by an increase in both fibril number and diameter (Parry and Craig 1977, 1978; Scott et al. 1981; Scott and Hughes 1986). In the postnatal period, tendon growth continues by increases in fibril diameter and length (Parry and Craig 1977, 1978; Parry et al. 1978a; Eikenberry et al. 1982; Michna 1984) in a multi-stage growth/stabilization process (Nurminskaya and Birk 1998). The manner in which molecules or fibril segments add to the growing fibril *in vivo* is not completely understood. Fibril growth involves both an intrinsic self-assembly process (diffusion-controlled) and extrinsic regulation (interface-controlled) by other fibril-associated molecules, and the local environment of collagen fibrils (Hoffmann et al. 2019). The data from growing native fibrils have provided evidence for models of fibril fusion (Graham et al. 2000; Kadler et al. 2000), molecular accretion (Kalson et al. 2015; Holmes et al. 2010), and possibly a combination of both (Birk et al. 1997; Ezura et al. 2000).

Interfibrillar fusion can potentially involve tip-to-tip, tip-to-shaft, and shaft-to-shaft fusion (Birk et al. 1995). However, bipolar fibrils with two C-ends or fibrils with multiple switch regions have not been found, either *in vivo* or *in vitro* (Fig. 3.3) (Kadler et al. 1996). End-to-end fusion of unipolar and bipolar fibrils will decrease the unipolar fibril population. Therefore, an enriched bipolar fibril population, unable to fuse further, could determine the limit of fibril growth in length. Fibril fusion can also be regulated by fibril-associated proteoglycans or some other macromolecule through maintaining interfibrillar spacing and inhibition of lateral segment fusion (Scott et al. 1981). It has been shown that mature rat tail tendon comprises several fibrils in the process of fusion or separation with some intrafibrillar proteoglycans inside large collagen fibrils (Scott 1990). Furthermore, fibrils' tips in embryonic chick metatarsal leg tendons have less surface bound proteoglycans compared to the fibril shaft allowing for tip-to-tip fusion and longitudinal fibril growth (Graham et al. 2000).

Direct evidence for molecular accretion *in vivo* is scarce due to the difficulty of visualizing and tracking of single collagen molecules (see Sect. 3.5.3). It has been shown that slow stretching of a cell culture tendon-like construct increases fibril diameter and volume fraction (Kalson et al. 2011). However, interfibrillar fusion alone could not explain the increase in fibril volume fraction. *In vitro* studies have also shown direct evidence for growing fibrils from acid-soluble collagen (Holmes and Chapman 1979). Fractured ends of isolated fibrils from avian embryonic tendon can further grow in the opposite axial direction by molecular accretion (Holmes et al. 2010). Kalson et al. (2015) presented a growth model based on 3D-electron microscopy of mouse tail tendon (Kalson et al. 2015). During the embryonic growth stage, fibril number, diameter, and length increase by fibril nucleation and axial growth. During postnatal growth, fibril number remains constant but fibril diameter and length continue to grow likely by molecular accretion. Birk et al. (1997) proposed a model in which thin fibril intermediates are formed by molecular accretion in chicken embryo metatarsal tendon (Birk et al. 1997). Then, longer and larger diameter fibrils are produced by lateral associations of preformed segments. The longer fibrils would have multiple polarity changes which would determine the regions able to associate. Growth would follow by molecular rearrangement to reconstitute cylindrical fibrils. Enzymatic intervention is also considered in this model to degrade poorly cross-linked fibrils in regions of polarity reversal and generate short polar units that could participate in further growth. Ezura et al. (2000) also suggested a fibril growth model in the developing mouse flexor tendons where fibril intermediates form by molecular accretion and are stabilized through their interactions with small leucine-rich repeat proteoglycans (Ezura et al. 2000). The change in composition of the matrix proteoglycans leads to a multi-step fusion/growth process. More tissue specific models are needed to fully explain the

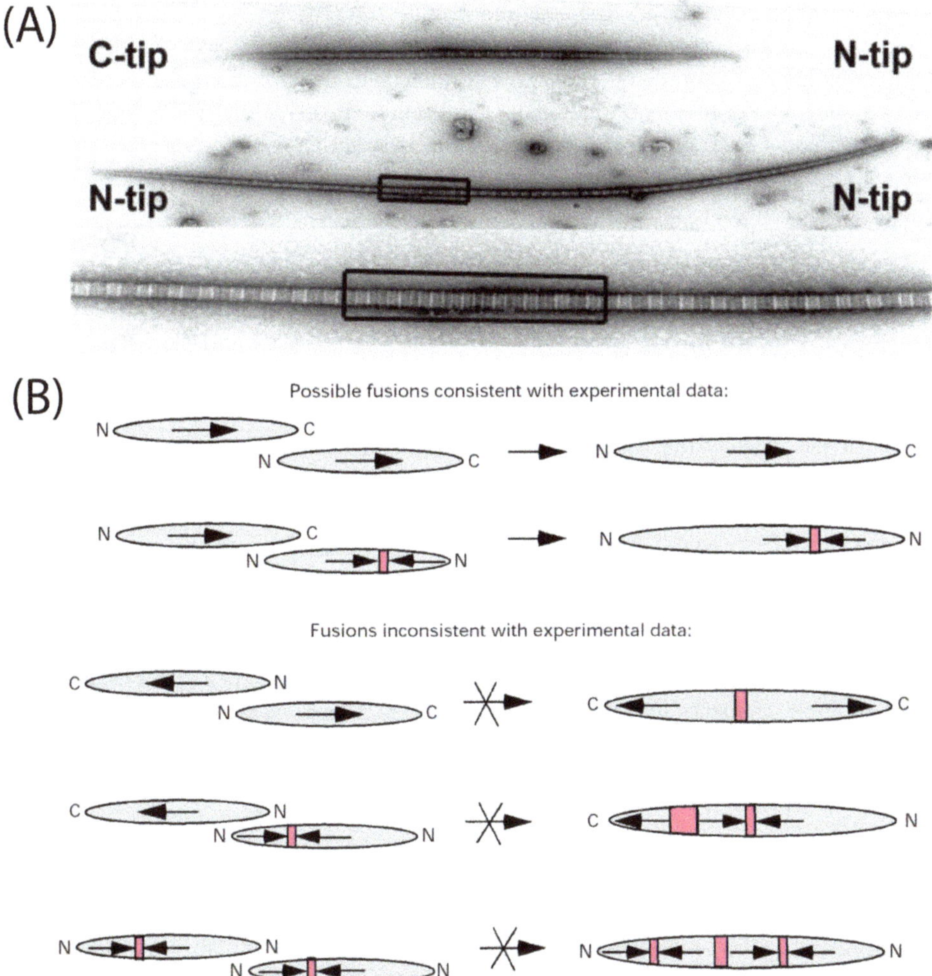

Fig. 3.3 Collagen fibril polarity and fusion. (**a**) Unipolar and bipolar collagen fibrils from embryonic chick tendon. Reproduced with permission from Kadler et al. (2000). The molecular switch region of a bipolar fibril is shown in magnification. (**b**) Possible models of fibril end-to-end fusion based on fibril's polarity. Arrows indicate molecular polarity within a fibril and pink boxes indicate regions of polarity reversal. Reproduced from Kadler et al. (1996)

combination of fibril associated molecules in every stage of fibril growth and stabilization which establishes the biological and mechanical functionality of tendons (Robinson et al. 2005).

Fibril growth mechanisms might be different in tissues with different mechanical and biological functions. For example, fibrils from sea cucumber dermis (Trotter et al. 1998) and sea urchin spine ligament (Trotter et al. 2000) display symmetrical mass distributions with a single transition zone in the center, making fibril fusion an unlikely growth mechanism (Trotter et al. 1998). Most likely, fibril growth throughout life in tendon is maintained by molecular accretion as well as linear and lateral association of fibril segments. In the early stages of development, tissue architecture is defined by fibril growth in number and length possibly through flow-induced crystallization (Paten et al. 2016) and/or spontaneous end-to-end fusion of small fibril segments

(Graham et al. 2000). Later in development and upon removal of lateral growth inhibitors, fibrils rapidly grow by lateral fusion (Scott et al. 1981) followed by molecular accretion to maintain a uniform (Parry and Craig 1984), energetically-stable shape. Cross-linked, adult fibrils may grow and remodel further by molecular accretion upon mechanical loading or injury of tendon.

3.2.3 Regulators of Matrix Growth and Development

Regardless of the mechanism, fibril growth in tendon and ligaments is highly regulated (Parry et al. 1978b). Fibrils *in vivo* are cylindrical with uniform diameter (Parry and Craig 1984), but reconstituted fibrils *in vitro* have a broad diameter distribution (Bard and Chapman 1973). Presence of an upper limit for fibril diameter may be due to the difficulty of the addition of new molecules or fibril segments and points to the participation of several regulatory processes, detailed below.

3.2.3.1 Water Structures

Collagen structure and stability is driven by molecular interaction with water molecules (Finch and Ledward 1972; Luescher et al. 1974; Kopp et al. 1990; Bigi et al. 1987; Miles and Ghelashvili 1999; Na 1989; Tiktopulo and Kajava 1998; Burjanadze 1982). Initial fibril formation is an endothermic, but entropy driven process (Kadler et al. 1987; Cassel 1966) arising from release of water molecules (Streeter and de Leeuw 2011; Kauzmann 1959). Post formation assembly can also be regulated by stabilization of water molecules (Cooper 1970), where breakers of water structure promote fibril formation, and makers of water structure are inhibitory (Hayashi and Nagai 1972). Mature fibrils *in vivo* are cross-linked by covalent bonds between neighboring molecules. However, the young and growing fibrils are stabilized by non-covalent hydrogen bonds (Bailey et al. 1998) and have the potential to bind more water molecules (Kopp et al. 1990). In fact, proteoglycans (Birk et al. 1996) or hyaluronate (Scott et al. 1981; Scott 1984) can stabilize the water layer associated with the collagen molecules. Release of these trapped water molecules could provide the increase of entropy required to drive the association of molecules into the fibrils.

Collagen structural models (Ramachandran and Chandrasekharan 1968; Ramachandran et al. 1973; Berg and Prockop 1973; Yee et al. 1974; Privalov et al. 1979) suggest that there are two types of intermolecular and intramolecular hydrogen bonds in fibrils: (I) a direct interchain hydrogen bond forms between the glycine residue and the residue in the second position of the neighboring chain, and (II) an additional hydrogen bond which links two adjacent tropocollagens using a bridging water molecule. This water-mediated hydrogen bonding makes two thirds of hydrogen bonds that connect neighboring peptides (Cameron et al. 2007) and therefore is a dominant interaction in stabilizing the fibrillar structure (Leikin et al. 1995; Kuznetsova et al. 1998). These water bridges are dynamically linked with freely exchangeable hydrogen atoms (Tourell and Momot 2016). Furthermore, water molecules can be confined by hydrophobic groups of neighboring tropocollagens (Hulmes et al. 1973) to maximize the number of water-water hydrogen bonds (Southall et al. 2002; Dill 1990). Since molecular assembly is driven by decreasing the number of unfulfilled hydrogen-binding opportunities at the protein-water interface (Fernández 2016), the trapped water molecules and the water bridges may have an important role in the collagen molecular assembly during fibril growth and remodeling (Martin et al. 2020).

3.2.3.2 Surface-Associated Proteoglycans

Proteoglycans are a superfamily of molecules distinguished by the covalent attachment of one or more highly negatively charged glycosaminoglycan chains to their core proteins (Comper and Laurent 1978), and they play a significant regulatory role during fibrillogenesis. Surface-associated proteoglycans and their glycosaminoglycan chains extend around the fibril and through steric effects limit lateral fibril growth (Scott et al. 1981; Scott 1980, 1984; Scott and Orford 1981). A three phase model of fibrillogenesis and fiber maturation in rat tail tendon was proposed by Scott et al. (1981) In phase 1 (up to day 40 after conception), tropocollagen interacts with dermatan sulphate-rich proteoglycan during or immediately after formation of microfibrils. The hyaluronate and proteoglycan-rich environment and collagen synthesis increase the number of thin fibrils, rather than growth in diameter of established fibrils. In phase 2 (from day 40 to approximately day 120 after conception), concentrations of chondroitin sulphate-rich proteoglycan and hyaluronate decrease, promoting the addition of collagen to extant fibrils rather than formation of new fibrils, resulting in rapid increase of fibril diameter without axial periodicity change. In phase 3 (day 120 after conception onwards), fibril growth slows down and reaches its final structure.

Direct *in vivo* evidence for the role of proteoglycans in the regulation of collagen assembly and growth has been achieved by development of animals deficient in small leucine rich proteoglycans (SLRPs). The principal SLRPs found in tendon are decorin, biglycan, fibromodulin, and lumican. Both decorin and biglycan are expressed in the interfibrillar matrix and interfascicular matrix in postnatal development but they present distinct temporal patterns (Zamboulis et al. 2020; Zhang et al. 2006; Ansorge et al. 2012). Interfibrillar biglycan abundance in the mouse is highest early in development whereas decorin abundance peaks later during development; both are low in abundance at maturity (Zhang et al. 2006; Ansorge et al. 2012). Equine tendon shares the same temporal expression for decorin but biglycan abundance peaks later (Zamboulis et al. 2020). Both proteoglycans have a regulatory role in collagen fibril assembly during tendon development. Biglycan is believed to promote fibril diameter growth, whereas decorin is believed to control lateral fusion of the fibrils and increase fibril stability (Zhang et al. 2005). Decorin and biglycan-deficient mice show abnormal fibril structure and lateral fusion during development resulting in an increased number of small fibrils with a simultaneous presence of collagen fibrils with unusually larger diameter and decreased failure strength and stiffness once in maturity (Zhang et al. 2006; Ameye et al. 2002; Corsi et al. 2002). Decorin and biglycan also share a binding site for collagen type I (Schönherr et al. 1995) and an increase in biglycan abundance in decorin-deficient mice was observed, alluding to compensation between the two proteins (Zhang et al. 2006).

Both fibromodulin and lumican are found in the interfibrillar matrix of mouse tendon, with lumican expression peaking during early postnatal development and fibromodulin abundance peaking in the later stages (Ezura et al. 2000). In contrast, the temporal expression in the equine interfascicular matrix was reversed, with fibromodulin abundance early and lumican peaking towards the end (Zamboulis et al. 2020). Fibromodulin and lumican share a binding site on collagen type I implying that they are likely to have functional overlap (Svensson et al. 2000). Fibromodulin and lumican deficient and double deficient mice showed abnormal fibril structure, with lumican deficient mice displaying an increase in larger diameter fibrils and fibromodulin deficient mice an increase in smaller diameter fibrils at maturity. In the fibromodulin deficient mice, increased cross-linking of collagen was also observed (Kalamajski et al. 2014) and lumican expression was increased, suggesting compensation (Ezura et al. 2000). In the lumican deficient mice, the phenotype was less severe and tendon mechanical properties were not affected. Interestingly, the mechanical properties of double knockout mice were dependent on the number of functioning alleles pointing toward a regulatory role for fibromodulin and a modulatory role for lumican (Ezura et al. 2000; Jepsen et al. 2002).

Asporin and lubricin (PRG4) are also expressed in tendon interfibrillar and interfascicular matrix, but have received much less attention than the principal SLRPs. In developing equine tendon, asporin demonstrates a temporal pattern in the interfascicular matrix where it is increased in early development and subsequently decreases but remains present in mature tendon (Zamboulis et al. 2020; Henry et al. 2001; Peffers et al. 2015).

The role of asporin in tendon fibrillogenesis and mechanical properties has not been documented yet, but the skin of asporin deficient mice had increased expression of collagen type I and III, increased toughness, as well as a two-fold increase in decorin and biglycan levels (Maccarana et al. 2017). Lubricin, a large proteoglycan important for matrix lubrication (Rees et al. 2002; Kohrs et al. 2011; Sun et al. 2015a; Funakoshi et al. 2008; Nugent et al. 2006), is also found in the interfascicular matrix of equine tendon, with increasing abundance with development and in low abundance pericellularly in the interfibrillar matrix (Zamboulis et al. 2020). In lubricin deficient mice the gliding resistance of fascicles against each other was increased compared to null mice, confirming lubricin may play an important role in interfascicular lubrication (Kohrs et al. 2011). However, the role of lubricin in fibrillogenesis has not yet been elucidated in tendon.

3.2.3.3 pN-Collagen

There are several observations suggesting that N-propeptides are confined to the fibril surface (Watson et al. 1992; Holmes et al. 1991) where they block accretion of further molecules (Fleischmajer et al. 1981, 1983, 1985, 1987a, b; Nowack et al. 1976; Veis et al. 1973; Lapiere and Nusgens 1974; Timpl et al. 1975; Lenaers and Lapiere 1975). As a result, further lateral growth would be regulated by enzymic cleavage of the propeptides. The important role of N-propeptide has been observed in the studies of dermatosparaxis and Ehlers-Danlos syndrome (EDS) type VIIB. Dermatosparaxis is caused by partial loss of procollagen N-proteinase activity (Lapiere et al. 1971; Lenaers et al. 1971; Becker and Timpl 1976). Presence of N-propeptide on the surface of these fibrils results in a non-circular cross sections (Watson et al. 1998). Remarkably, it has been shown that dermatosparactic collagen fibrils will gain a normal appearance after implantation in normal animals (Shoshan et al. 1974), suggesting the existence of a dynamic mechanism for fibril growth and degradation. Also, Ehlers-Danlos syndrome type VIIB fibrils in which pN-collagen is only partially cleaved have rough-bordered and non-circular cross sections (Watson et al. 1992; Holmes et al. 1993).

Growth models (Hulmes 1983; Chapman 1989) have been proposed for collagen fibrils in which accretion of collagen molecules is inhibited by N-propeptides on the fibril surface. Growth of pN-collagen fibrils is inhibited due to the steric blocking of interaction sites by the N-propeptides. The growth inhibitor part of the molecules (the N-terminus) is confined to the fibril surface and the C-ends are buried inside the interior of the fibril. Since the growth inhibitors cannot act as a site for further accretion, their surface density increases with lateral growth. Growth of fibril diameter continues until fluidity in intermolecular contacts is restricted due to steric hindrance. This first critical diameter depends on the lateral width of the inhibitor segment, allowing for growth of fibrils with preferred diameters in different tissues (the inhibitor might vary in different tissues and stages of development, but the same mechanism still applies). When a fibril reaches uniformity at this critical diameter, accretion is limited to the fibril ends and growth is only in axial direction. Lateral growth can proceed to a second critical diameter after enzymatic removal of the growth inhibitor.

Romanic et al. (1991) in an *in vitro* study demonstrated that pN-collagen III can co-polymerize with collagen I, but cannot be deposited on previously assembled collagen I fibrils (Romanic et al. 1991). It was shown that the presence of pN-collagen III can (1) inhibit the rate of collagen I assembly, (2) decrease the amount of collagen I incorporated into fibrils, and (3) decrease the diameter of fibrils in comparison with fibrils generated under the same conditions from collagen I alone. Fibril diameter progressively decreased with increasing the initial molar ratio of pN-collagen III to collagen I. Therefore, it was concluded that pN-collagen III coats the surface of collagen I fibrils early in the process of fibril assembly and hinders lateral growth of the fibrils. But it does not bind to the growing tips of fibrils, resulting in formation of thin fibrils.

3.2.3.4 Minor Collagens

Other types of collagens are synthesized simultaneously with type I collagen (Gay et al. 1976; Burke et al. 1977; Foidart et al. 1980, 1983). The structural similarities of fibril forming collagens allow them to polymerize within the same "heterotypic" fibrils (Henkel and Glanville 1982; Fleischmajer et al. 1990). In tendon, approximately 95% of collagen is type I, with the remaining being mostly type III (Birch et al. 1999; Makisalo et al. 1989; Riley et al. 1994a; Amiel et al. 1984). Collagen type III is found both in the interfibrillar and interfascicular matrix of the developing tendon. In equine tendon, collagen type III expression increases throughout development in both the interfibrillar and interfascicular matrix reaching peak abundance towards the end of maturation (Zamboulis et al. 2020). Collagen type III distribution in the avian tendon is observed throughout the interfibrillar and interfascicular matrix early in development but solely in the interfascicular matrix later (Birk and Mayne 1997; Kuo et al. 2008). The decrease in collagen type III expression in avian tendon is also associated with the appearance of collagen fibrils with larger diameters implying participation of collagen type III in the regulation of collagen fibrillogenesis (Birk and Mayne 1997). Furthermore, collagen type III deficient mice demonstrated disrupted collagen fibrillogenesis and larger diameter fibrils, confirming the involvement of collagen type III in fibrillogenesis (Liu et al. 1997).

Collagen type V has also demonstrated a growth regulatory effect on collagen fibrillogenesis (Wenstrup et al. 2004; Birk et al. 1990a) and its mutations have been identified in patients with classic EDS (Malfait and De Paepe 2014; Symoens et al. 2012). Collagen type V is found in the interfibrillar and interfascicular matrix of the developing equine tendon and in the interfibrillar matrix of mouse tendon in association with the tenocyte surface (Zamboulis et al. 2020; Wenstrup et al. 2011; Smith et al. 2012, 2014; Sun et al. 2015b). Reduction of collagen V expression during development also results in formation of fibrils with larger diameters in other tissues such as the dermis (Wenstrup et al. 2006) and cornea (Segev et al. 2006). Corneal stroma, which contains collagen fibrils of uniformly small diameter (Hay and Revel 1969), is relatively rich in type V collagen with 20% type V to 80% type I (McLaughlin et al. 1989). Studies of type I/V interactions in the mature corneal stroma have shown that type I and type V collagen co-assemble into fibrils (Fitch et al. 1984; Birk et al. 1986, 1988; Linsenmayer et al. 1985, 1990) and decreasing the levels of type V collagen secreted by corneal fibroblasts *in situ* results in assembly of large-diameter fibrils with a broad size distribution (Marchant et al. 1996). *In vitro* fibrillogenesis studies (Birk et al. 1990a; Adachi and Hayashi 1986) also showed that fibrils produced from only type I collagen were thicker than hybrid fibrils of type I and type V collagen. In addition, collagen V-null mice tendons are smaller than their wild type counterparts and exhibit reduced mechanical function (Connizzo et al. 2015). However, the effect of collagen V deficiency on mechanical function is much more dramatic in joint stabilizing tendons and ligaments, suggesting a relationship between mechanical loading and collagen V mediated fibril development (Connizzo et al. 2015). Collagen type XI is found to be present early in development both in the mouse and equine interfibrillar matrix, and thought to play synergistic roles with collagen type V (Zamboulis et al. 2020; Wenstrup et al. 2011). Col11a1-null mouse models (Sun et al. 2020) show decreased body weights and their flexor digitorum longus tendon has abnormal collagenous matrix structure with a significant decrease in biomechanical properties. Absence of collagen type XI disrupts the parallel alignment of fibrils and increases fibril diameter, similar to collagen type V.

Collagen type XII and XIV are closely related members of the fibril-associated collagens with interrupted triple helices (FACIT) collagen class and have been identified in the interfibrillar matrix in mouse (Izu et al. 2020; Ansorge et al. 2009), and the interfibrillar and interfascicular matrix in the developing avian (Young et al. 2000; Zhang et al. 2003) and equine tendon (Zamboulis et al. 2020). Collagen type XIV levels are high in early development and decrease thereafter to barely detectable levels in mature tendon whereas collagen type XII is more abundant in early development but also present throughout development, maturation, and aging (Zamboulis et al. 2020; Izu et al. 2020; Ansorge et al. 2009; Young et al. 2000; Zhang et al. 2003). Collagen type XII regulates lateral network formation and fiber domain compartmentalisation, as well as collagen type I secretion. Collagen type XIV plays a role in the early stages of tendon fibrillogenesis and entry into lateral growth, in accordance with its temporal expression. Absence of collagen type XII in Col12a1-null mouse model results in larger tendons with abnormal collagen fibril packing, increased stiffness, and decreased overall type I collagen (Izu et al. 2020). Also, type XIV collagen deficient mouse tendons demonstrate premature fibril growth and larger fibril diameters, but no deficiency in biomechanical properties at maturity (Ansorge et al. 2009). Despite being closely related, there does not appear to be a compensatory relationship in expression patterns (Izu et al. 2020; Ansorge et al. 2009).

Finally, collagen type VI has also been identified both in the interfibrillar matrix of developing mouse tendon, especially in the pericellular region, and in the interfibrillar and interfascicular matrix in equine developing tendon (Zamboulis et al. 2020; Smith et al. 2012; Izu et al. 2011). During development, collagen type VI was found to be implicated in maintaining the cell shape, microdomain structure and fiber organisation. Collagen VI deficient mice displayed abnormal fibril assembly in the pericellular region with more dense fibrils of smaller diameter and frequent very large or twisted fibrils (Izu et al. 2011). Other collagens such as collagen type IV and XXI show temporal expression in the development of the equine interfascicular matrix but they have received less attention and their role is not currently known.

3.2.3.5 Elastin, Fibrillins, and Fibulins

Elastin is found at the core of elastic fibers surrounded by a fibrillin-rich microfibril scaffold (Kielty et al. 2002). In tendon, its abundance is function-dependent, with a greater abundance of elastin found in energy storing tendons (Thorpe and Screen 2016; Godinho et al. 2017). Elastin is present during embryonic development and increases in response to mechanical loading (Oryan and Shoushtari 2008; Zamboulis et al. 2020; Wagenseil et al. 2010; Luo et al. 2018). Spatially, elastin is localized sparsely in the interfibrillar matrix parallel to the tendon axis and more densely in the interfascicular matrix, with both a parallel and perpendicular organization in relation to the tendon axis. Elastin haploinsufficiency in mice resulted in alterations to collagen fibril structure, favoring an increase in large diameter fibrils and reduced interfibrillar matrix, but these changes were site-specific (Eekhoff et al. 2017). The effect of elastin depletion on tissue function has also been debated, with some studies showing significant mechanical disruption and others demonstrating no effect (Eekhoff et al. 2017; Grant et al. 2015; Fang and Lake 2016). When fascicle and interfascicular matrix were interrogated separately following elastase treatment in equine tendons, fascicles did not show any changes in their mechanical properties. However, the interfascicular matrix was significantly compromised, suggesting a different role for interfibrillar and interfascicular elastin (Godinho et al. 2020).

Fibrillin-1 and 2 are known to be involved in elastogenesis and regulate activation and bioavailability of TGF-β superfamily members (Chaudhry et al. 2007; Boregowda et al. 2008). Fibrillin-1 and 2 are present in the interfibrillar and interfascicular matrix in mature tendon, co-localizing with elastin and also pericellularly on their own (Ritty et al. 2002; Kharaz et al. 2018). In developing equine tendon, fibrillin-1 and 2 were identified in the interfibrillar and interfascicular matrix with fibrillin-1 showing an increase in abundance during development in the interfascicular matrix only (Zamboulis et al. 2020). Fibrillin-1 deficiency in mice did not disrupt the tendon structure apart from generating smaller tendons (Tran et al. 2019) and fibrillin-2 deficiency resulted in a decrease in collagen cross-linking but did not affect tendon structure (Boregowda et al. 2008). It is possible that similar to elastin deficiency, the interfascicular matrix is more profoundly affected by fibrillin-1 and 2 deficiencies than the fascicles.

Fibulin-4 and 5 are indispensable for elastogenesis (McLaughlin et al. 2006; Nakamura et al. 2002; Yanagisawa et al. 2002) and fibulin-4 is found in the tendon interfibrillar matrix co-localized with fibrillins (Markova et al. 2016). In fibulin-4 deficient mice, forelimb contractures were noted and collagen fibrillogenesis was disrupted in tendons (Markova et al. 2016). Fibulin-5 is found in the interfibrillar matrix but also the interfascicular matrix where its abundance in equine tendon peaks early in development (Zamboulis et al. 2020). In fibulin-5 deficient mice, malformed elastic fibers were found in tendon with no other changes to the composition or structure of the tendon. In addition, the linear modulus of the Achilles tendon was increased in the fibulin-5 deficient mice whereas the positional tibialis anterior tendon did not show any changes in mechanical properties (Eekhoff et al. 2021). Taken together, this supports a role for elastic fibers in the mechanical properties of functionally distinct tendons or tendon compartments beyond regulation of collagen fibrillogenesis.

3.2.3.6 Thrombospondins

Thrombospondins, specifically TSP-1, TSP-4, and COMP (TSP-5), have also recently been identified in the interfibrillar and interfascicular matrix of tendons (Kannus et al. 1998; Zamboulis et al. 2020; DiCesare et al. 1994; Hauser et al. 1995; Smith et al. 1997; Fang et al. 2000; Södersten et al. 2006; Havis et al. 2014; Schulz et al. 2016). COMP levels in the developing interfibrillar and interfascicular matrix increase with development and have been shown to be associated with loading (Zamboulis et al. 2020; Smith et al. 1997). In COMP deficient mice, the tendon structure exhibited larger fibril diameters with an increase in irregular shape, suggesting a role in collagen fibrillogenesis. In addition, collagen accumulation in the endoplasmic reticulum was detected in isolated dermal fibroblasts *in vitro*, alluding to its intracellular role in the secretion of collagen, which is dependent on the formation of a COMP-collagen complex (Schulz et al. 2016). TSP-4 has been reported to have a similar spatiotemporal expression as COMP, a function associated with loading, and also to be increased in COMP deficient mice (Schulz et al. 2016; Cingolani et al. 2011; Frolova et al. 2014). Similarly to COMP deficient mice, in TSP-4 deficient mice, tendons exhibited larger fibril diameters (Frolova et al. 2014). TSP-2 and TSP-3 have also been reported in the interfibrillar matrix of mouse tendon (Havis et al. 2014; Frolova et al. 2014; Kyriakides et al. 1998) and TSP-2 deficiency (Kyriakides et al. 1998) resulted in a similar collagen fibril phenotype noted in TSP-4 and COMP deficient mice (Schulz et al. 2016; Frolova et al. 2014).

3.3 Maintenance of the Matrix During Adulthood

3.3.1 Matrix Turnover

The pioneering studies of Schoenheimer and his collaborators in the 1930s changed the perception of proteins from a static collection of material to a material existing in a state of dynamic flux, where the balance of synthesis and degradation is critical to homeostatic maintenance of structure (Cohn 2002; Wilkinson 2018). The study of matrix turnover in maintaining adult tissue homeostasis, and the regulation of this process, has been the focus of much research over the past century since then and could be the key to preventing injury.

3.3.1.1 Collagenous Matrix

It is well established that collagen is one of the longest lived proteins in many tissues within the body, with a relatively low rate of turnover in skin, tendon and cartilage compared to other ECM proteins (Thorpe et al. 2010; Maroudas et al. 1998; Sivan et al. 2006, 2008; Verzijl et al. 2000). However, the specific rate of collagen turnover within tendon is still a matter of controversy, with conflicting data reported in the literature. Several studies have reported negligible turnover of tendon collagen within an individual's lifetime, with a half-life of 198 years in the energy storing equine superficial digital flexor tendon determined by measuring the rate of aspartic acid racemization, and no collagen turnover detected in the healthy adult human Achilles tendon using ^{14}C bomb pulse data (Thorpe et al. 2010; Heinemeier et al. 2018, 2013a). However, other studies have reported relatively rapid collagen synthesis in tendon, with fractional synthesis rates of 0.04–0.06% $hour^{-1}$ calculated in the human patellar tendon, which equates to half-lives ranging from 48 to 64 days (Miller et al. 2005; Babraj et al. 2005; Smeets et al. 2019). There are several potential explanations for these large discrepancies.

The studies in which high fractional synthesis rates were reported used stable isotope labelling over a very short time period, and it is unlikely that all newly synthesized collagen would be incorporated into the matrix, such that fractional synthesis rates would be overestimated. Indeed, using the tracer *cis*-[^{18}F]fluoro-proline combined with positron emission tomography and measuring protein incorporation in the rat Achilles tendon, it has been demonstrated that only approximately 20% of the proline taken up in the tissue was incorporated into the tendon matrix (Skovgaard et al. 2011).

The studies in which extremely long half-lives have been reported may also be affected by several factors. In these studies, the collagenous fraction of the matrix is purified using enzymatic digestion or protein extraction techniques (Thorpe et al. 2010; Heinemeier et al. 2018). Such purification techniques may result in some collagen loss; indeed approximately 13% of collagen was lost during purification by guanidine hydrochloride extraction (Thorpe et al. 2010). This is likely to represent more recently synthesized collagen that is less tightly cross-linked into the matrix, and therefore the half-life calculated based on the remaining collagen would be overestimated. There are also limitations associated with the methods used to estimate half-life; calculation of protein turnover rates using racemization of aspartic acid relies on assumptions made during calculations, as accumulation of D-Aspartic acid is affected by several factors, including temperature, pH, and protein structure (Thorpe et al. 2010). Precision of ^{14}C measurements is limited by variability in tissue radiocarbon levels within the population, which has progressively decreased over the past 50 years (Hodgins and U. S. Department of Justice 2009).

More recent studies also help to explain these previous contradictory findings, suggesting there may be pools of collagen within tendon that have differential turnover rates. Indeed, more collagen neopeptides, which are a marker of turnover, were identified within the interfascicular matrix compared to the fascicular matrix in the equine superficial digital flexor tendon (Thorpe et al. 2016a). These findings are supported by a recent study using *in vivo* isotope labelling combined with laser capture microdissection and mass spectrometry to measure the turnover rates of individual proteins within the fascicular and interfascicular matrices in the rat Achilles tendon (Choi et al. 2020). Results revealed significantly faster turnover of collagen in the interfascicular matrix compared to the fascicles, with a half-lives of 1.6 and 2.7 years for type I collagen in interfascicular matrix and fascicles respectively. While no studies have directly determined differences in turnover rates of extracellular matrix proteins between small and large animals, it is likely that protein turnover is more rapid in rodent models compared to humans, as previous studies have demonstrated a negative correlation between median protein turnover rate constants and lifespan (Swovick et al. 2018), and the half-life of serum albumin is approximately tenfold greater in the human compared to the rat (Chaudhury et al. 2003; Jeffay 1960).

Emerging evidence also suggests the presence of a sacrificial collagen matrix within tendon fascicles, with a recent study in murine tendon identifying the presence of thin collagen fibrils that are interspersed between thicker fibrils, and are synthesized and removed from the tendon within a 24 h period, while the bulk of the collagen remains unchanged (Chang et al. 2020). This rapidly turned over collagen may act to protect the long-lived collagen from mechanical damage, and also helps to explain previous studies which have measured both a high rate of synthesis, but very low rates of bulk turnover.

There is also evidence to suggest that collagen half-life varies between tendons with different functions, with a half-life of 198 years in the energy storing equine superficial digital flexor tendon compared to 34 years in the positional common digital extensor tendon (Thorpe 2010). While a lower rate of collagen turnover in high strain energy storing tendons may seem counter-intuitive, slower turnover may protect the tendon from remodeling which would weaken its structure, with the trade-off that when damage does occur it is more difficult to repair.

3.3.1.2 Non-collagenous Matrix

While only a small number of studies have measured rates of collagen turnover in tendon, even fewer have assessed turnover of non-collagenous proteins. It is, however, well established that non-collagenous protein turnover occurs at a more rapid rate than collagen turnover, with the exception of elastin, which is known to have very low turnover rate. While elastin half-life in tendon has not been measured, in other connective tissues there is compelling evidence that following development elastic fibers are not replaced throughout an individual's lifetime (Shapiro et al. 1991; Sherratt 2009). Aspartic acid racemization has been used to estimate turnover of the non-collagenous fraction of the extracellular matrix in functionally distinct equine tendons. However, this study was unable to provide turnover rates of individual proteins and a small amount of soluble collagen was detected in the fraction analysed, which is likely to affect the results (Thorpe et al. 2010). Despite these limitations, this study did show that turnover of non-collagenous proteins differed in tendons with different functions, with more rapid turnover in energy storing tendons compared to positional tendons (2.2 years vs. 3.5 years), which may allow for greater reparative capacity in injury-prone energy storing tendons (Thorpe et al. 2010).

Metabolism of different proteoglycan classes has been studied in tendon explants using radio-labelling, with results demonstrating relatively rapid turnover of newly synthesised large proteoglycans (half-life approx. 2 days) compared to small leucine rich proteoglycans (half-life approx. 20 days) and showing that different pathways are involved in the degradation of large and small proteoglycans (Samiric et al. 2004). However, this approach is only able to measure the turnover of newly synthesised proteoglycans rather than those already present within the matrix, which may be metabolized at a slower rate. More recent approaches using isotope labelling *in vivo* have measured turnover rates of a range of tendon proteoglycans, with half-lives ranging from 21 days for decorin to 72 days for lumican (Choi et al. 2020). There is also evidence to suggest that turnover rates of non-collagenous proteins may vary according to their location within the tendon matrix. Turnover of interfascicular decorin occurs at a faster rate than that of interfibrillar decorin (Choi et al. 2020). The reasons for this are unclear but indicate that proteoglycans may have distinct roles in different tendon regions.

3.3.2 Mechanical Stimulation for Matrix Remodeling

It is well established that mechanical stimulation drives the natural remodeling of the tendon ECM, and specifically the collagen structure (Zamboulis et al. 2020; Smith et al. 2002; Screen et al. 2005a; Batson et al. 2003; Bohm et al. 2015; Pan et al. 2018; Quigley et al. 2018; Theodossiou et al. 2019). Tenocytes can sense changes in their mechanical environment through cell-cell and cell-matrix interactions and transduce mechanical signals, which then trigger adaptive responses, a process called mechanohomeostasis (Fig. 3.4) (Maeda et al. 2012; Lavagnino et al. 2015; Heinemeier et al. 2003; Maeda et al. 2011; Havis et al. 2016). Since mechanotransduction pathways are comprehensively reviewed elsewhere (Wall et al. 2016, 2018; Humphrey et al. 2014), we report here on downstream changes in ECM structure in response to changes in mechanical stimuli. In addition, we focus on adaptations to normal loading and sub-failure damage rather than massive tissue injury/repair processes which

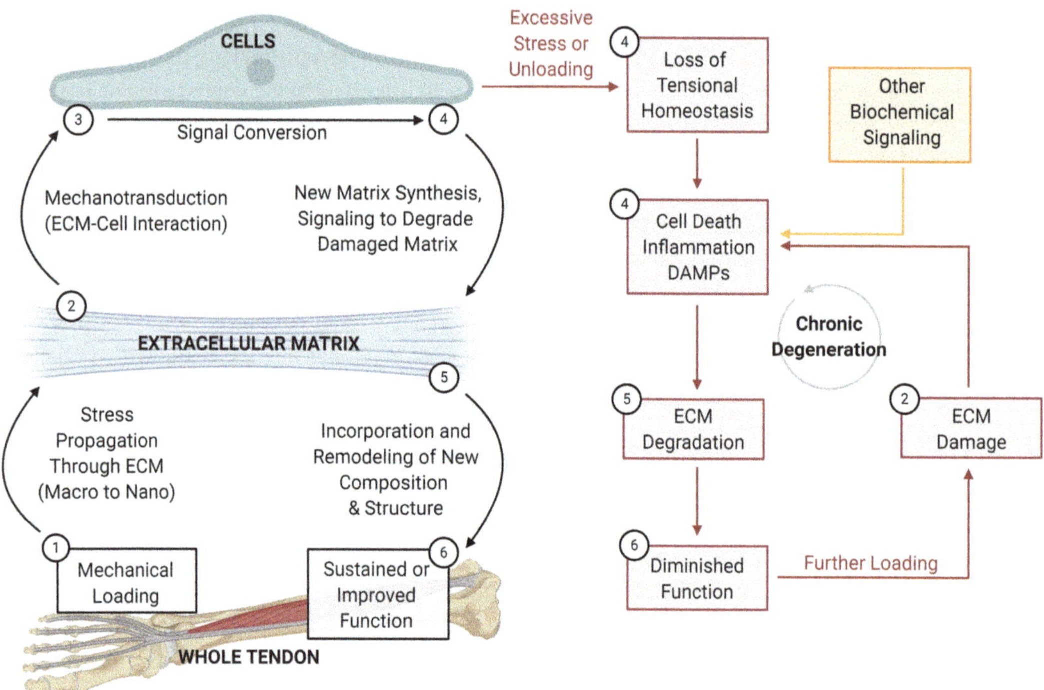

Fig. 3.4 Schematic of adult matrix mechanohomeostasis. (1) Multiaxial and multimodal mechanical loading on the tendon applies stress macroscopically to the tissue, which then (2) propagates through the multiscale hierarchy of the tendon matrix via interactions between the collagenous and non-collagenous matrix. Stress is then transduced from physical to biochemical signals in the cell via mechanotransduction (3), and these signals then trigger (4) catabolic or anabolic responses. In the case of normal loading or positive adaptation due to exercise (left), (5) new matrix is synthesized and incorporated into the existing structure while damaged matrix is removed resulting in (6) sustained or improved tissue function. In the case of excessive loading (overuse) or stress deprivation (disuse), there is a loss of tensional homeostasis at the cellular level which leads to the production of inflammatory markers and damage-associated molecular patterns (DAMPs) as well as increased matrix degradation and cell death (right). These signals can be spread to other cells through paracrine signaling, and can also be caused by other biochemical signaling or cellular changes (e.g., cell aging) in the absence of changes to mechanical loading (see Sect. 3.3.3). This process can lead to diminished function, and enter the tissue into a chronic degenerative cycle whereby further loading causes more matrix damage, eventually leading to tissue rupture and/or tendinopathy. (Created with BioRender.com)

are well described elsewhere (Thomopoulos et al. 2015; Andarawis-Puri et al. 2015; Andarawis-Puri and Flatow 2018).

3.3.2.1 Exercise

Alterations in mechanical stimuli can influence ECM turnover of adult tendons, with exercise and disuse both reported to result in a range of adaptations. However, the response seen in tendon is far less pronounced than that seen in muscle, and results are contradictory. In humans, there is evidence of tendon hypertrophy in response to exercise, with increases in patellar tendon cross sectional area (CSA) (Couppé et al. 2008; Farup et al. 2014). Studies have also reported increased markers of collagen synthesis and breakdown in peritendinous tissue both as a result of acute exercise and longer-term training in human Achilles and patellar tendons (Langberg et al. 1999, 2001; Astill et al. 2017). As collagen turnover rate in the tendon core is very low it has been suggested that additional newly synthesized collagen may be deposited around the edge of the tendon, resulting in increased CSA (Magnusson and Kjaer 2019). However, other studies which have taken tendon biopsies to assess collagen synthesis post-exercise in the patellar tendon report conflicting results, with some observing increased collagen synthesis (Miller et al. 2005) and others reporting no change (Dideriksen et al. 2013; Hansen et al. 2009). This limited responsiveness is supported by studies which have either detected no, or very limited changes, in collagen and growth factor gene expression in response to exercise (Dideriksen et al. 2013; Heinemeier et al. 2013b; Sullivan et al. 2009).

These findings are in contrast to those reported in a variety of small animal models, which have demonstrated upregulation of tendon associated genes and increases in mechanical properties as a result of exercise or increased loading (Heinemeier et al. 2007, 2012; Olesen et al. 2006). However, the majority of small animal studies have been performed in animals that are not yet fully mature, such that they may have more capacity for adaptation to loading than skeletally mature human tendon. In addition, the type and duration of exercise performed is likely to influence results, with studies of the rat supraspinatus tendon demonstrating that a single bout of exercise tends to decrease mechanical properties, whereas chronic exercise results in improved mechanical properties (Rooney et al. 2017). This is accompanied by more matrix-related gene changes in chronic compared to acute exercise groups (Rooney et al. 2015). *Ex vivo* studies have also been performed to uncover the effects of loading on tendon metabolism, with mechanical loading of artificial tendon constructs *in vitro* resulting in little change in tendon related genes at physiological levels of loading, but upregulation of genes associated with tendon development as a result of overloading (Herchenhan et al. 2020). By contrast, exposing fascicles from rat tail tendons to moderate degrees of loading increased collagen synthesis without generating mechanical or structural changes (Screen et al. 2005a; Legerlotz et al. 2013b).

3.3.2.2 Disuse or Stress Deprivation

Disuse has been shown to result in a marked decline in tendon mechanical properties, both in humans and animal models (Magnusson and Kjaer 2019; Rumian et al. 2009; Almeida-Silveira et al. 2000; Matsumoto et al. 2003; Couppé et al. 2012). However, the mechanisms by which these alterations occur are unclear, as the majority of studies do not report any alterations in tendon dimensions or mass as a result of unloading (Kinugasa et al. 2010; de Boer et al. 2007; Heinemeier et al. 2009). Some studies have reported decreased patellar tendon collagen synthesis as a result of lower limb suspension in the human, even after relatively short periods of disuse (de Boer et al. 2007; Dideriksen et al. 2017), accompanied by increased matrix metalloproteinase 2 (MMP-2) expression (Boesen et al. 2013). By contrast, results from animal studies are variable and sometimes contradictory; hind limb suspension in the rat resulted in very few alterations in the Achilles tendon (Heinemeier et al. 2009), whereas denervation-induced unloading of the mouse patellar tendon caused decreased expression of type I collagen, increased expression of MMP-13 and a decrease in collagen fibril diameter (Mori et al. 2007). Explant models have been used to further investigate the effect of unloading on tendon metabolism, with stress deprivation of murine tail tendon fascicles resulting in increased levels of matrix degrading enzymes and reduced mechanical properties (Abreu et al. 2008; Lavagnino et al. 2003, 2005; Wunderli et al. 2018). More recent studies demonstrate decreased expression of genes associated with both matrix synthesis and degradation in stress deprived murine flexor tendons (Connizzo et al. 2019).

The contradictory findings from animal studies may be due to differences in species and ages in studies, the particular model of mechanical stimulation employed, and also whether the experiments have been performed *in vivo* or *ex vivo*. In addition, it has been reported that functionally distinct tendons also display a differential response to unloading *ex vivo*, with more rapid and extensive changes seen in positional compared to energy storing tendons (Choi et al. 2019). Further, stress deprivation may preferentially affect the interfascicular matrix, with greater deterioration in this region compared to the fascicles in unloaded rat tail tendon (Rowson et al. 2016). Different types of mechanical stimulation can also generate different responses and, *in vitro*, tenocytes are mostly stimulated using tension, which likely mirrors the mechanical stimulation interfibrillar tenocytes experience *in vivo*. However, *in vitro* shear stress stimulation of adult tenocytes, which is likely experienced by interfascicular tenocytes, generated an "anti-fibrotic" expression pattern with decreased transcription of collagen type I and III (Fong et al. 2005). In addition, the responses to mechanical stimulation may also be influenced by the age of the cells or tissues *in vitro* (Zamboulis et al. 2020; Fong et al. 2005) and the magnitude of loading (Zhang and Wang 2013).

3.3.2.3 Sub-failure Microdamage

In addition to normal exercise, studies have sought to understand the capacity for intrinsic repair of microdamage that occurs due to tendon overload. Several *in vivo* models have been developed to induce tendon fatigue damage, including treadmill running and repetitive reaching activities (Glazebrook et al. 2008; Carpenter et al. 1998; Gao et al. 2013). A model developed by Fung et al. (2010) in which the rat patellar tendon is clamped and loaded directly while the animal is under anaesthesia allows precise loads to be applied to the tendon, while the number of cycles applied can be varied to induce different degrees of damage. This model has been used to extensively characterise the structural, mechanical and molecular changes within tendon to varying levels of fatigue damage at different time points. Results show that structural alterations become more pronounced as severity of fatigue loading progresses, with isolated collagen fiber kinking in response to low-level fatigue loading which becomes more widespread in moderate fatigue loading and is accompanied by fiber separation. Severely fatigue loaded tendons exhibit widespread matrix disruption and fiber thinning (Fung et al. 2010). These structural changes are associated with alterations in mechanical properties, with a single bout of moderate fatigue loading being sufficient to induce accumulation of structural damage associated with non-recoverable loss of stiffness (Bell et al. 2018). These studies indicate a limited ability for intrinsic repair of damage above a certain threshold, even when no further loading is applied.

Considering the molecular changes as a result of fatigue loading, expression of genes associated with matrix remodeling, including collagens and MMPs, were negatively correlated with the degree of damage (Andarawis-Puri et al. 2012), suggesting an impaired ability to repair microdamage as the damage worsens. In addition, apoptosis within the tendon increased with damage (Andarawis-Puri et al. 2014), likely due to alterations in cell microenvironment. Increased apoptosis will likely decrease the capacity for matrix remodeling, leading to further damage accumulation. The authors of these studies suggest that restoration of cell microenvironment may be key to improving the capacity of resident tendon cells to successfully remodel regions of microdamage (Andarawis-Puri and Flatow 2018). Exercise performed post-fatigue loading provides a method of influencing cell microenvironment and subsequently matrix synthesis, and, depending on timing, can either worsen or improve repair. Exercise initiated 2 weeks after fatigue loading resulted in increased levels of procollagen-I, indicative of matrix remodeling, whereas exercise that commenced immediately after fatigue loading caused further damage to the tendon, accompanied by increased levels of aggrecan and collagen type III, proteins that are both associated with a failed healing response (Bell et al. 2015). It is likely that post-fatigue loading exercise also influences matrix degradation, however this is yet to be directly determined.

There is also emerging evidence to suggest that initial overload damage within the tendon may occur within the interfascicular matrix. In bovine and equine flexor tendon explants exposed to cyclic loading *in vitro*, initial damage occurred preferentially to the interfascicular matrix, with upregulation of inflammatory mediators observed in this region (Spiesz et al. 2015; Thorpe et al. 2015b). The high shear environment within the interfascicular matrix of energy storing tendons, caused by interfascicular sliding as the tendons stretch, is likely to expose the resident cells to a complex strain environment incorporating tension, shear and compression (Cook and Screen 2018). Overload may therefore induce cell-mediated degradation, and subsequent loss of interfascicular matrix structure is likely to alter cell microenvironment within the fascicles, leading to propagation of damage throughout the tissue. However, the majority of rodent tendons lack an interfascicular matrix structure (Liu et al. 2016; Lee and Elliott 2019), and therefore the response of the interfascicular matrix to microdamage cannot be studied using these models, limiting our knowledge in this area.

3.3.3 Biochemical Disruption of Matrix Homeostasis

There are a variety of biochemical stimulators that can influence tendon homeostasis. While inflammation occurs in the initial response to tendon injury, inflammatory mediators, including prostaglandins and cytokines, are also upregulated in tendon in response to exercise (Langberg et al. 1999, 2002). Blocking prostaglandin E_2 (PGE_2) by administration of non-steroidal anti-inflammatories resulted in decreased peritendinous collagen synthesis in response to exercise in the human patellar tendon, and collagenase upregulation in rat tendon cells (Christensen et al. 2011; Tsai et al. 2010). In addition, peritendinous infusion of (interleukin-6) IL-6 elevates collagen synthesis in a similar manner to exercise (Andersen et al. 2011). Inflammatory mediators also influence proteolytic activity, with IL-1β acting in synergy with mechanical stretch to increase levels of matrix degrading enzymes in rabbit tendon fibroblasts and human patellar tendon derived cells (Archambault et al. 2002; Yang et al. 2005). Recent studies also show that IL-1 and IL-6 can directly lead to matrix degeneration using an *in vitro* model system (Connizzo and Grodzinsky 2018b, 2020). Collectively, these results suggest that inflammatory mediators are important stimulators of collagen turnover in tendon that can act independently of loading. However, it is likely that only a very small proportion of newly synthesized collagen is incorporated into the matrix, and therefore upregulation of matrix metalloproteases does not necessarily alter tendon mechanical properties or collagen content (Marsolais et al. 2007). Interestingly, it seems that regular mechanical loading is required to protect rat tail tendons cultured in the presence of inflammatory cells from degradation and loss of mechanical properties (Marsolais et al. 2007), highlighting the importance of mechanical stimuli for maintenance of tendon homeostasis.

Systemic diseases can also affect tendon metabolism and increase the risk of tendon injury. Diabetes is associated with increased prevalence of tendinopathy and disorganization of the collagen fibers within human tendon (Abate et al. 2013). Tendons from diabetic mice have smaller cross-sectional areas, reduced mechanical properties and altered collagen fiber alignment, and these alterations vary between tendon types (Connizzo et al. 2014b). It is hypothesized that these changes are caused by the accumulation of advanced glycation end-products (AGEs) due to the increased availability of glucose, causing loss of both biological and mechanical function (Abate et al. 2013). These AGEs are also known to accumulate naturally during the aging process. Studies have also shown that treating rat tendon-derived cells with high glucose results in downregulation of ECM-associated genes (Wu et al. 2017), indicating alterations in tendon homeostasis.

Obesity is another recognized risk factor for tendon injury, initially postulated to be caused by the increased mechanical strain due to weight. However, it has recently been established that adipose tissue is a potent releaser of signaling molecules, with raised serum levels of inflammatory markers present in obese individuals suggesting the presence of low grade inflammation, which could disrupt tendon homeostasis (Abate et al. 2013; Cilli et al. 2004). Indeed, in diabetic and obese mice, collagen and MMP expression is elevated during tendon healing, with increased macrophages and delayed remodelling (Ackerman et al. 2017a). Other metabolic disorders are also associated with tendon pathologies, including hypercholesterolemia, which results in cholesterol deposits in tendon, accompanied by alterations in tenocyte gene and protein expression, matrix turnover, tissue vascularity, and cytokine production (Soslowsky and Fryhofer

2016). It appears these disorders all affect tendon homeostasis via a variety of mechanisms which often involve inflammatory mediators, resulting in altered turnover and disruptions to the tendon matrix leading to increased risk of pathology.

3.3.4 Circadian Regulation

Recent studies have also unveiled the importance of the circadian clock in regulating tendon protein turnover, with rhythmic expression of several clock-associated genes resulting in nocturnal procollagen synthesis and diurnal fibril assembly in mice. This pool of newly synthesized collagen is then rhythmically degraded. This could be a primary mechanism for repairing microdamage that accumulates over a single day of use, but the incorporation of these newly synthesized collagen fragments into the existing matrix has not yet been confirmed. Disabling the circadian clock results in formation of abnormal collagen fibrils and collagen accumulation, indicating that protein homeostasis in tendon is maintained by circadian regulation of a sacrificial collagen matrix (Chang et al. 2020). While endogenous circadian rhythms have been observed in human tendon cells, studies have not yet been able to detect any alterations in expression of clock-associated genes within tendon as a result of exercise or immobilization (Yeung and Kadler 2019; Yeung et al. 2014). However, expression levels of clock genes in these studies were very low, and there were high levels of variability between individuals. Therefore, more studies are needed to determine if the alterations in tendon turnover as a result of changes to loading environment occur via circadian regulation.

3.4 Dysregulation of ECM Structure and Function During Aging

Aging is one of the primary risk factors for degenerative tendon injuries, particularly in the Achilles tendon and the rotator cuff tendons (Wertz et al. 2013; Strocchi et al. 1991; Minagawa et al. 2013; Longo et al. 2011; May and Garmel 2020). These injuries cause significant pain, frailty and a loss of independence, leading to a general reduction in quality of life (Kjær et al. 2020). Age-related disorders are associated with a degenerative tendon state, rather than an acute tendon rupture, which is thought to be a result of repetitive damage to the extracellular matrix (Fig. 3.4). However, the ability to study tendon aging in a controlled and repeatable fashion is quite challenging. Results are heavily dependent on the tendon being studied, the methodology used and on the ages defined as 'young' and 'old'. Furthermore, aging is a complex and multifactorial process, involving natural changes in structure and function as well as alterations to the biological processes that regulate tissue architecture.

3.4.1 Changes to Matrix Structure and Function with Age

Historically, aging studies in tendon have focused on alterations in tendon structure and function in aged individuals, and in particular on detecting changes in the collagenous structure. However, findings of age-related changes in collagen morphology appear to be species- and tendon-dependent. Collagen content has been shown to increase (Stammers et al. 2020), decrease (Couppé et al. 2009; Sugiyama et al. 2019), or remain unchanged (Birch et al. 1999; Thorpe et al. 2010; Kostrominova and Brooks 2013) with increasing age in a variety of model systems, despite also reporting downregulation of collagen mRNA expression (Kostrominova and Brooks 2013). Equine research shows a decrease in tendon fibril diameters with increasing age (Parry et al. 1978a), hypothesized to lead to increased fibrillar interaction and reduced interfibrillar sliding (Ribitsch et al. 2020). Alterations in collagen cross-linking are also debated in the literature, with increases in mature cross-links observed in old human subjects (Couppé et al. 2009) while overall cross-linking levels decreased with age in mouse tail tendon fascicles (Stammers et al. 2020). However, non-enzymatic crosslinking associated with advanced glycation end-products was increased in both studies.

Studies in age-related alterations in the structural organization of the collagenous and non-collagenous tissue compartments are similarly inconclusive. Studies in rat tail fascicles using polarized Raman spectroscopy demonstrate changes in collagen fiber orientation with aging, specifically indicating a more homogeneous tissue structure (Van Gulick et al. 2019), yet histological studies report disruption of collagen fiber organization in aged mouse tendons (Sugiyama et al. 2019). Other studies have also demonstrated altered crimp morphology in the flexor tendon of older horses (Patterson-Kane et al. 1997). Crimp frequency and amplitude in the murine flexor and patellar tendons were no different with age, but the change in crimp amplitude in response to mechanical loading was larger in older flexor tendons (Zuskov et al. 2020). Interestingly, the number of collagen fascicles was observed to decrease with age, suggesting a shift towards a greater proportion of interfascicular matrix in older tendons (Ali et al. 2018; Gillis et al. 1997).

Similar to age-associated changes in collagen content, inconsistent differences in glycosaminoglycan (GAG) levels have been observed. GAG content is decreased with age in the human supraspinatus tendon, but not in the biceps tendon (Riley et al. 1994b). However, GAG content was no different in male or female murine flexor tendons (Connizzo et al. 2019). Research in the equine model showed tendon-specific changes in GAG content, with age-associated decreases in positional tendons but no difference in energy storing tendons (Thorpe et al. 2010). This alludes that changes in GAG content with aging may be specific not only to the tendon studied but also perhaps to regional differences within the tendon. For example, accumulation of GAGs has been reported in tendinopathy samples, which is highly associated with aging, and tendons with regions that wrap around bone such as the rotator cuff and the insertion of the Achilles tendon (Thornton and Hart 2011; Archambault et al. 2007; Attia et al. 2012; Majima et al. 2000).

With respect to the other non-collagenous components of tendon, there are few studies investigating age-related changes. Measures of DNA content, and therefore tissue cellularity, do not change in aged equine tendons (Birch et al. 1999). One recent study in aged murine flexor tendons demonstrated a significant reduction in cell density in aged murine flexor tendons, but this change appeared to be sex-dependent with no differences found in age-matched female tendons (Connizzo et al. 2019). Cell density has also been shown to decrease in both rabbits and rats (Magnusson and Kjaer 2019; Nakagawa et al. 1994). In addition to cell number, tenocyte shape has also been reported to be altered in aging, with a shift towards a higher nucleus to cytoplasm ratio and a reduction of other organelles (Ippolito et al. 1980). Elastic fibers, typically found between collagen fibers and fascicles, have been reported to decrease and become more disorganized during aging (Godinho et al. 2017; Eekhoff et al. 2017; Ippolito et al. 1980), potentially altering sliding and stretch mechanisms at the microscale. Lubricin, which acts as a lubricant to enable gliding function (Funakoshi et al. 2008; Sun et al. 2006; Taguchi et al. 2009), has been reported to increase with age in rabbit tendons (Thornton et al. 2015) but remain unchanged in human Achilles tendon (Peffers et al. 2015). Finally, aged mice have been reported to have increased calcification, reduced vascularization, and increased adipose tissue (Zhang and Wang 2015; Marqueti et al. 2017).

Changes in tissue structure do not appear to translate into clear deficits in macroscopic tissue function. In fact, age-related changes in quasi-static mechanical properties appear to vary based on the specific tendon studied, the protocol used to assess changes, and the boundary conditions (gripping, testing environment, etc.) for experimentation (Ackerman et al. 2017b; Vogel 1980; Shadwick 1990; Haut et al. 1992). Tendon mechanical properties have shown to both decrease (Vogel 1980) and increase with age in rat tail tendons (Shadwick 1990; Nielsen et al. 1998). In rat patellar tendon, mechanical properties were weakly positively correlated with age (Haut et al. 1992) or decreased with age (Dressler et al. 2002). Achilles tendon function is decreased in older humans (Lindemann et al. 2020), and either decreased (Pardes et al. 2017) or no different (Gordon et al. 2015) in aged rodents compared to mature counterparts. Rotator cuff tendons do not appear to have altered macroscale function with aging (Connizzo et al. 2013b; Lin et al. 2020). Interestingly, measures of dynamic tissue function through fatigue loading (Zuskov et al. 2020; Thorpe et al. 2017), dynamic macroscopic testing (Pardes et al. 2017; Dunkman et al. 2013), and measures of dynamic responses at the fiber (Connizzo et al. 2013b; Li et al. 2013) and fibril (Thorpe et al. 2013b) levels all suggest a diminished mechanical function in the aging population. In addition, nanomechanical testing revealed increased fluid flow and poroelasticity in aged supraspinatus tendons but decreased compressive function (Connizzo and Grodzinsky 2018a), alluding to deficits in dynamic mechanical function. These dynamic and nanoscale evaluations are indicative of changes present in the extracellular matrix, but could be more associated with changes in the interfibrillar or interfascicular matrix (Thorpe et al. 2013b, 2015a, 2017) rather than the collagenous matrix.

3.4.2 Matrix Turnover in Aged Tendons

Like many other tissues, it has been well established that the matrix repair response in aged tendons is impaired (Ackerman et al. 2017b; Mienaltowski et al. 2016). Recent studies have focused primarily on massive injury responses as a result of partial or full-thickness tendon tears. However, we focus here on the ability of aged cells to regulate everyday tissue homeostasis. Although tendons typically are thought to have very low matrix turnover at maturity, tenocytes do become metabolically active, begin to proliferate and actively remodel the matrix in response to changes in mechanical stimulus (Heinemeier et al. 2012; Rooney et al. 2014, 2015; Magnusson and Kjaer 2003; Kjaer et al. 2005). As reported here and in studies before, only a small fraction of the collagen present in tendons, hypothesized to be associated with small diameter collagen fibrils, is homeostatically regulated for daily remodeling to comply with functional demands (Chang et al. 2020; Thorpe et al. 2010; Yeung and Kadler 2019; Birch et al. 2016). However, the turnover rate of this small fraction has not been studied extensively in aged tendons to date (Birch et al. 2016). One study in equine tendons suggested that there is a decline in collagen turnover in aged tendons, while other studies of diseased tendon show increased collagen turnover rate (de Mos et al. 2007). Recent investigations of collagen synthetic activity in horse tendons reported no differences though (Thorpe et al. 2015b), suggesting no difference in the capacity to remodel the matrix. However, recent studies in mouse tendon explants demonstrated that although there were no differences between young and aged mice synthetic activity at baseline, age-related declines were evident when subjected to stress deprivation (Connizzo et al. 2019). Perhaps an injurious stimulus is necessary to illuminate larger deficits in matrix synthesis due to the generally low metabolic activity of tendon *in vivo*, and this highlights potential deficits that could be present in homeostatic remodeling and tissue repair but are not yet explored.

The interfascicular matrix has recently been shown to contain more proteins and more protein fragments than the collagenous compartment, indicating greater matrix degradation and turnover (Thorpe et al. 2016a). Since dynamic reorganizations such as collagen sliding and re-alignment are responsible for much of the daily function of tendons, the interfascicular compartment is likely more often damaged and remodeled accordingly. In fact, the interfascicular matrix, and not the fibrous matrix, was recently shown to be the primary location of adaptation to mechanical loading during development, highlighting the importance of this compartment in understanding overall tissue turnover (Zamboulis et al. 2020). While protein quantity does not change with aging in the interfascicular matrix, the number of protein fragments decreased indicating decreased matrix turnover and accumulation of tissue damage, potentially leading to chronic disease (Thorpe et al. 2016a).

3.4.3 Aging-Associated Changes in Cell Function Affecting Matrix Homeostasis

One difficulty in identifying mechanisms for age-related tissue degeneration is the inability to disentangle changes in the ECM and changes in cell behavior. Since resident cell populations are critical to maintaining and repairing the extracellular matrix in mechano-homeostasis (Fig. 3.4), it is likely that changes with age in the extracellular matrix are preceded by cellular adaptations. Age-related cellular changes have been characterized extensively in other organ systems, defined as nine primary hallmarks of aging (López-Otín et al. 2013; Hernandez-Segura et al. 2018). This includes genomic instability, telomere attrition, epigenetic alterations, loss of proteostasis, deregulated nutrient sensing, mitochondrial dysfunction, cellular senescence, stem cell exhaustion, and altered intercellular communication. The first four hallmarks represent primary causes of cellular damage, while the other five are either responses to that damage, initially attempting to mitigate the damage but eventually becoming damaging themselves, or consequences of that damage. Since there are several other reviews that describe these hallmarks and their effects on cell behavior extensively (López-Otín et al. 2013; Folgueras 2018; Guerville et al. 2020; Rebelo-Marques et al. 2018), we focus here on those changes that may be relevant to understanding age-related changes in matrix turnover, based on literature in the tendon and ligament field as well as studies performed in other fibrous tissues.

3.4.3.1 DNA Damage and Matrix Turnover

Although we have DNA repair mechanisms, damage naturally accumulates over time via exposure to environmental toxins, simple DNA replication errors, or damage molecules. Telomeres protect the terminal ends of chromosomes from deterioration, but since cells are not able to copy the ends of DNA efficiently the telomere region shortens with each cell division. After some time, this can lead to cell growth arrest, limiting the ability of tissues to grow and regenerate with aging. Stem cells harvested from the periodontal ligament were reported to have significantly shorter telomere length with increasing donor age, and this corresponded with reduced regenerative properties (Ng et al. 2020; Trivanović et al. 2015). In contrast, relative telomere length was not decreased in aged equine tendons (Thorpe et al. 2016b). Given differences in collagen turnover rate between the two tissues, with periodontal ligament being associated with much faster turnover, DNA damage accumulation due to telomere shortening may be dependent on specific tendon function. Since tenocytes more generally have a fairly low proliferation rate at maturity (Grinstein et al. 2019), it is unclear what role replication-based damage would play in tenocyte behavior; it is more likely that these mechanisms would impact tendon-derived stem or progenitor cells (Kohler et al. 2013).

Proteins are constantly being synthesized and degraded throughout our lifetime to maintain an efficient and effective functional tissue. The regulation of protein assembly inside the cell, an array of quality control mechanisms, is called protein homeostasis or proteostasis (López-Otín et al. 2013; Klaips et al. 2018). These mechanisms become less efficient in aged organisms, which can result in protein aggregation as well as the production of damaged or misfolded proteins which can cause cell and tissue dysfunction. Decreased expression of genes encoding molecular chaperones facilitating protein folding and proteostasis has been reported in fibroblasts harvested from skin in patients with classic EDS (Chiarelli et al. 2019a, b). Given the tendon and ligament phenotype in this disease, this work provides evidence that loss or inefficient proteostasis could be a mechanism for disrupted matrix production in fibrous tissues. Furthermore, another recent study reported interplay between collagen synthesis and endoplasmic reticulum stress via the circadian clock, whereby targeting protein misfolding in disease could restore collagen homeostasis (Pickard et al. 2019). Future work is necessary to determine if loss of proteostasis is an age-related phenomenon in tenocytes or tendon stem cells and what direct effects this might have on matrix homeostasis.

3.4.3.2 Mitochondrial Dysfunction and Oxidative Stress

A byproduct of mitochondrial energy production is the presence of free radicals or reactive oxygen species (ROS), which can be potentially damaging to the cell. For many years, ROS were thought to be the major culprit behind aging but recent studies showing that lowering ROS does not impact health have challenged this idea (López-Otín et al. 2013; Hekimi et al. 2011; Van Remmen et al. 2003). Production of ROS is important for signaling cell stress, but this process is a delicate balance. Over time, increasing production of ROS results in dysfunction of the mitochondria which in turn can lead to cells becoming less efficient at producing energy and causing damage to other cellular components (López-Otín et al. 2013). One recent study reported increases in the expression of peroxiredoxin, an antioxidant, in degenerated tendon, suggesting that oxidative stress may be a factor in the etiology or progression of age-related tendon disease (Wang et al. 2001). In addition, a reduction in catalase and heat shock proteins discovered through proteomic analysis suggests that aged tendons may be prone to ROS-based damage (Peffers et al. 2014). However, other studies have found no changes in oxidative-stress related genes (Peffers et al. 2015). Several studies have suggested that DNA damage in tendon cells can be induced through the production of ROS via mechanical overload or underload (Yudoh et al. 2005; Zapp et al. 2020). In fact, repression of oxidative stress through drug therapies diminishes the aberrant differentiation of tendon-derived cells subjected to excessive mechanical overload (Hsiao et al. 2019; Morikawa et al. 2014). Physiological loading was found to reduce the production of oxidative products such as ROS, protecting cells from premature senescence and matrix degeneration (Zhang and Wang 2015).

Besides the study of reactive oxygen species produced in the mitochondria, there have not been many studies on the role of mitochondrial function in tendon aging more generally. One recent study found that treating rat-derived tendon fibroblasts with advanced glycation end-products caused alterations in mitochondrial DNA content as well as a shift in matrix remodeling towards degradation rather than synthesis (Patel et al. 2019). Mitochondrial biomarkers are upregulated in the early phases of tendon healing, and therefore dysfunction in this organelle may also play a role in impaired tissue healing found in aged individuals (Thankam et al. 2018). Furthermore, the export of mitochondrial calcium is a key process in the process of matrix calcification during tendon calcification, suggesting a link with matrix production (Yue et al. 2016). Given these links between mitochondrial function and matrix production, clearly more research into the role of mitochondrial function in normal tenocyte or tendon stem cell homeostasis is warranted.

3.4.3.3 Cellular Senescence and SASP in Matrix Degradation

Cellular senescence is a natural repair response to damage, which can arise due to overexpression of certain oncogenes, by excessive cell replication, or by the presence of certain DNA damage-causing molecules (Acosta et al. 2013; Blagosklonny 2011; Blokland et al. 2020). Senescence is critical for wound repair and tumor suppression in young and mature individuals, preventing damaged cells from continuing to proliferate and propagate throughout the tissue. However, in old tissues, clearance of these cells is deficient, likely due to deteriorating immune function and thus, senescent cells accumulate within the matrix. One major concern with the presence of senescent cells is their ability to produce pro-inflammatory cytokines, called the senescence-associated secretory phenotype (SASP) (Miller et al. 2012b; Connizzo et al. 2013b). Inflammatory signaling produces many deleterious effects on matrix metabolism, as discussed above in Sect. 3.3.3 (Acosta et al. 2013; Zhang et al. 2015; Tsuzaki et al. 2003; Fedorczyk et al. 2010). Therefore, exposure to high levels of inflammatory cytokines may tip the scales towards matrix degeneration over adaptation

(Connizzo and Grodzinsky 2018b, 2020). Importantly, the SASP reinforces senescence through paracrine signaling (Acosta et al. 2013), thus a small population of senescent cells in aged tissues can lead to significant declines in tissue maintenance (Campisi 1998). This inflammatory signaling can also be further stimulated by damage-associated molecules produced in the extracellular matrix regularly, such as soluble decorin, tenascin-c and fibrinogen (Blokland et al. 2020).

Both tenocytes and tendon stem cells (TSCs) have been induced to replicative senescence *in vitro*, and cells harvested from aged subjects have been shown to favor senescence induction earlier than young counterparts (Kohler et al. 2013; Arnesen and Lawson 2006). Cells prematurely induced to senescence in the laboratory have been critical at understanding the age-related process, but cell culture alone does not replicate native cell-cell and cell-matrix connections for studying ECM remodeling. Therefore, the link between cellular senescence and dysregulation of ECM maintenance has not yet been fully elucidated. There does appear to be a connection between matrix synthesis and cellular senescence. One study demonstrated that collagen I is upregulated in senescent fibroblasts harvested from human subjects and in cells subjected to hydrogen peroxide to induce senescence (Murano et al. 1991; Dumont et al. 2000), indicating a role for collagen production in senescence. Interestingly, research using senescence-accelerated mouse models demonstrated that senescence-prone cells respond to collagenase-injection with altered expression profiles favoring matrix degradation over synthesis (Ueda et al. 2019). In fact, increased expression of MMPs has been reported before generally with aging and also specifically in aging tendon and senescent cells (Dudhia et al. 2007; Jones et al. 2006; Yu et al. 2013; Millis et al. 1992). In the absence of mechanical signals (as in the case of disuse or injury), aged mouse flexor explants exhibited increased expression of MMPs and cellular senescence markers (p16/p19/p53) (Connizzo et al. 2019). Therefore, there does appear to be a relationship between senescence and collagen turnover although it is unclear whether collagen is typically increased or decreased due to discrepancies between studies. Senescence has been implicated in fibrosis of the lung and in cutaneous wounds (Waters et al. 2018; Jun and Lau 2017), but further work is needed to clarify this link in tendon and ligament tissues.

3.4.3.4 Tendon Stem Cell Exhaustion and Matrix Repair

Like other cells, stem cells are also subject to age-related changes such as DNA damage accumulation, telomere shortening and cellular senescence. Over time, these lead to changes in the behavior of the stem cells present as well as a reduction in the pool of stem cells available. Age-related changes in TSCs is one of the more commonly studied mechanisms of aging in the tendon and ligament literature (Lui and Wong 2019; Zhou et al. 2010; Dai et al. 2019), thought to be a primary mechanism for age-related declines in tendon healing. TSCs are present in lower numbers in aged rabbit (Zhang and Wang 2010), rat (Zhou et al. 2010), and human tendons (Kohler et al. 2013; Ruzzini et al. 2014). Since these cells are often recruited to injury sites to aid in tissue repair, this reduction in cell number is hypothesized to be a primary determinant of diminished healing capacity.

While multiple studies have also shown that the self-renewal capacity of TSCs is not altered with aging (Kohler et al. 2013; Zhou et al. 2010; Ruzzini et al. 2014), the functional capacity of these cells to perform duties necessary for matrix remodeling and repair is indeed altered. Aged TSCs exhibit lower proliferative capacity and reduced migration (Zhang and Wang 2015; Kohler et al. 2013; Zhou et al. 2010), suggesting insufficiencies in recruitment of TSCs to repair sites in aged tendons. However, the recruitment of TSCs to an injury site *in vivo* has not yet been explored in detail, and studies to date have primarily been performed in cell culture. Structural differences to the tendon ECM with aging as discussed above may further alter the ability of TSCs to migrate to wounds *in vivo*.

Only a few studies have investigated the ability of aged or senescent TSCs to perform their duties with regards to matrix synthesis. One recent study revealed significant deficits in the ability to form three-dimensional tissue organoids, citing poor ability to produce and organize collagen matrix and reduced expression of matrix-related genes, including collagen I and key regulators of fibrillogenesis (Yan et al. 2020). In addition, organoids formed from aged TSCs also exhibited significant apoptosis and senescence. Expression of ECM and ECM-remodeling genes was found to be significantly reduced in other studies of aged mouse and human tendons, specifically reporting reduced collagen expression and reduced collagen production in aged TSCs (Klatte-Schulz et al. 2012; Han et al. 2017; Gehwolf et al. 2016). This could suggest that aged TSCs, and specifically senescent TSCs, may respond to injury via fibrotic mechanisms.

3.4.3.5 Altered Intercellular Communication and Mechanosensing

Tissues are able to grow and function normally due to the ability of cells to communicate with each other, constantly transferring information locally to nearby cells through direct cell-cell junctions or through the interstitial matrix via secretion of soluble factors (López-Otín et al. 2013; Rebelo-Marques et al. 2018). Aging can alter the ability of cells to perform this function and in the case of stem cells, impact cell fate and function. Signaling in tenocytes during development, homeostasis and injury has been extensively studied as it is critical to transduction of mechanical signals in order to facilitate tissue adaptation (Wall et al. 2016; Wall and Banes 2005). Dysregulated cell-cell communication was reported in aged TSCs recently (Popov et al. 2017), but interestingly this has not been explored in aging tenocytes yet. This avenue of investigation may be critical to understanding the dysfunction of matrix maintenance that occurs with age and we strongly encourage more research in this area.

3.5 Novel Systems and Tools to Study ECM Maintenance and Regulation

At the heart of the research discussed above is the dynamic addition and removal of material from critical structures within the tissue. The net flux of molecular components to developing and extant structures is positive during matrix assembly/growth, zero during maintenance and negative during degradation. For collagenous tissue assembly, maintenance and dysregulation it is critical to track (1) the production/export of new ECM molecules, (2) the degradation of existing ECM molecules, (3) the trafficking of ECM molecules from the cells to the matrix, and (4) the fate of the ECM molecules as they incorporate into matrix structures. This work heavily relies on novel tools and model systems used to track ECM molecules. Here we focus on those that can be used extracellularly, where they can help illuminate the dynamics of component exchange in the compartment that resides between the cells and the structural matrix.

3.5.1 In Vitro Model Systems

One major hurdle to studying ECM maintenance throughout life is the difficulty in measuring matrix production and breakdown in real-time without disruption of the intricate tissue structure. A number of simpler *in vitro* model systems have been designed to address this concern. Generally, *in vitro* culture allows for complete control and accurate measurement of applied mechanical and biological stimuli through the use of novel bioreactors (Wang et al. 2013a, b; Dyment et al. 2020; Janvier et al. 2020; Tohidnezhad et al. 2020; Chen et al. 2016; Butler et al. 2009), allowing for simple and straightforward experiments. Recent developments in tissue engineering strategies have allowed researchers to produce three-dimensional tissue engineered constructs (TECs), bioartificial tendons (BATs), and ligament equivalents (LEs) (Chen et al. 2016;

Deng et al. 2009; Butler et al. 2008; Garvin et al. 2003; Huang et al. 1993). Typical cell sources for engineered neo-tendons include mesenchymal stem cells, fibroblasts, embryonic tendon cells, and tendon progenitor or stem cells (TSCs), which are harvested and expanded using traditional culture methods. Cells are then supplied with appropriate growth factors and mechanical cues to stimulate production of tendon-like matrix. Mechanical cues include the use of custom bioreactors to stimulate tenogenic differentiation through static and cyclic tensile loading as well as the use of spatial or organizational cues, such as high aspect ratio channels and aligned substrate morphology in order to stimulate cells to form aligned tendon-like collagenous tissue.

Through this work, researchers have established that mechanical stimulation is critical for formation of appropriate collagen fibril morphology *in vitro* (Kalson et al. 2011; Mubyana and Corr 2018; Schiele et al. 2013; Kapacee et al. 2008). The arrangement of geometric constraints (posts, channels, etc.) and the topographical surface in these systems can dictate both matrix alignment and cell phenotype, opening the door for studying links between substrate-specific mechanotransduction and matrix assembly (Schiele et al. 2013; Nirmalanandhan et al. 2007; Bayer et al. 2010). Furthermore, these systems have been critical in identifying which cell types can be induced to a tenogenic lineage and the necessary conditions to do so (Chen et al. 2016; Rajpar and Barrett 2019; Angelidis et al. 2010; Harris et al. 2004). These studies have paved the path for *in vivo* studies using larger and more complex tissue engineered constructs for tendon repair and also aided in the establishment of metrics to define repair capacity for tendon-derived cell populations, all while revealing the sophistication and complexity of tendon and ligament cell biology. However, these neo-tendons and more sophisticated TECs have not been able to faithfully recapitulate the mature tendon matrix, lacking hierarchical fibrillar structure and mechanical integrity, and thus can only be used for studying the initial stages of ECM production and not adult maintenance. Furthermore, studies to date have only focused on the collagenous matrix and have not investigated the interfibrillar and interfascicular matrix development. Finally, the study of age-related dysfunction would be difficult in systems requiring cell expansion due replicative senescence-prone aged cell populations.

Though the technique has been used since the late 1980s (Dyment et al. 2020; Wunderli et al. 2020), explant culture models have gained popularity again recently to study matrix turnover *in vitro* without major disruption of the hierarchical ECM structure. Explants can be harvested either as whole tendon with adjacent muscle and bone intact (murine rotator cuff (Connizzo and Grodzinsky 2018b)), intact tendon midsubstance (canine (Hannafin et al. 1995; Ikeda et al. 2010), rabbit (Abrahamsson et al. 1991), equine (Murphy and Nixon 1997), avian (Flick et al. 2006), and murine flexor tendon (Connizzo et al. 2019)), functional tendon sub-units (rat tail tendon fascicle (Lavagnino et al. 2016; Wunderli et al. 2017; Leigh et al. 2008; Screen et al. 2005b)), or cut pieces of tendon (human (Wong et al. 2009; Costa-Almeida et al. 2018) and bovine tendon explants (Koob and Vogel 1987; Samiric et al. 2006)). Historically, these explant models have been used primarily to understand the role of mechanical stimulus in preventing degeneration of tissue ECM either by studying stress deprivation or applying mechanical stress or strain to explants via custom-built bioreactors (Koob and Vogel 1987; Lavagnino et al. 2003; Connizzo et al. 2019; Hannafin et al. 1995; Flick et al. 2006; Gardner et al. 2012). They have also been used more recently to study inflammation, disease, and injury through the use of medium additives and other chemicals to simulate various biological environments (Connizzo and Grodzinsky 2018b, 2020; Abrahamsson et al. 1991; Wong et al. 2009; Fessel et al. 2014). Combining these model systems with the use of traditional labeling pulse-chase experiments allow for the measurement of matrix (proteoglycan, collagen) synthesis (Connizzo and Grodzinsky 2018b; Koob et al. 1992; Robbins et al. 1997). Recent studies have identified sex- and age-related differences in matrix synthesis and overall tissue metabolism despite no initial differences at baseline (Connizzo et al. 2019), highlighting the power of these model systems in studying cell-mediated ECM remodeling in real-time.

A major benefit of explant culture models is the preservation of the intact hierarchical fibrillar matrix and the internal cell population, allowing for study of natural cell-matrix interactions and more minor changes to the ECM. Furthermore, it is possible that explants could be a viable model system to study the interfibrillar matrix, as this tissue is also kept intact during harvest. Explant tissues can be harvested from transgenic and aged animals, allowing us to pinpoint the roles of certain regulatory proteins in ECM homeostasis directly as well as to capitalize on the use of previously established injury and disease models. However, explants are inherently separated from other cell types and tissues that may be relevant for ECM maintenance, such as systemic innervation, lymphatics, and vasculature. Furthermore, tissue harvest induces injury and appropriate culture conditions for tissue maintenance is still an ongoing avenue of investigation by multiple laboratories (Abrahamsson et al. 1991; Wunderli et al. 2017; van Vijven et al. 2020; Vogel and Hernandez 1992). Nevertheless, with the increase in novel tools for measuring real-time ECM regulation, there is untapped potential for explant culture systems in studying ECM turnover.

3.5.2 In Vivo Model Systems

In vivo animal models, and in particular transgenic rodent models, have long been used in tendon research to study tendon matrix development, maturation, and aging (Delgado Caceres et al. 2018; Hast et al. 2014; Carpenter et al. 1999; Robinson et al. 2017). However, traditional transgenic animal models are limited by an inability to separate temporal regulation and compensatory effects due to the involvement of many regulatory proteins during tendon development (Connizzo et al. 2013a; Theodossiou and Schiele 2019). However, precise control of expression via the establishment of novel inducible mouse lines have since allowed for the study of temporal expression patterns during healing and growth (Gumucio et al. 2020; Disser et al. 2019; Ackerman et al. 2017c), as well as the ability to label cell populations for lineage tracing and local expression pattern studies (Yoshida et al. 2016; Soeda et al. 2010; Dyment et al. 2014, 2015). Using tamoxifen-inducible scleraxis-cre mouse models, researchers recently established that decorin and biglycan contribute critically to normal tendon homeostasis and aging, despite their low expression relative to developmental time points (Robinson et al. 2017; Leiphart et al. 2020). However, one study demonstrated deleterious effects of tamoxifen injection on tendon homeostasis and healing (Best et al. 2020) alluding to pro-fibrotic mechanisms, and another demonstrated altered rotator cuff healing (Cho et al. 2015), warranting further exploration. Using a doxycycline-induced green fluorescent protein (GFP) reporter model, researchers were able to pinpoint the transition from development to homeostasis, and correlate this to tissue growth (Grinstein et al. 2019).

In addition to these inducible models, the generation of tendon-specific knockout mice and cell lines via targeting of scleraxis-lineage cells has revolutionized the study of tendon development, homeostasis, and aging (Gumucio et al. 2020; Yoshida et al. 2016; Pryce et al. 2007; Schweitzer et al. 2001; Killian and Thomopoulos 2016). First, the study of key ECM regulatory proteins that were previously unexplored due to embryonic or perinatal lethality is now possible. Through these efforts an essential role for collagen XI in tendon development was discovered where the lack of collagen XI resulted in altered fibrillar structure and organization, as well as reduced tissue function (Sun et al. 2020). In addition, these models have established a critical role for MMP-14 during development in the formation of collagen fibrils, in sharp contrast to more well-known function in facilitating matrix breakdown (Taylor et al. 2015). Furthermore, recent work has identified the critical role of collagen V in the regulation of regionally-dependent (Connizzo et al. 2016b, c) and site-specific tendon structure and function (Sun et al. 2015b; Connizzo et al. 2015). Along with novel inducible models, these technologies are primed to study the role of matrix regulators in tendon homeostasis without disruption from developmental processes or systemic changes associated with genetic knockdown.

Despite these major advances, there are still a number of hurdles in studying tendon matrix homeostasis and regulation. One difficult area of study is the dysfunction of ECM regulation during aging. Mice considered for the study of aging should ideally be between 18 and 24 months of age, after most biomarkers of aging are present and before survivorship drops significantly (Flurkey et al. 2007). Maintaining aging rodent colonies is both expensive and time-consuming, especially when considering novel transgenic lines for which breeding must be performed in house. One solution is to consider mouse models of accelerated aging, such as models of progeroid syndromes, models of mitochondrial mutations, senescence-prone mice or models of 'inflammageing' (Folgueras 2018; Kõks et al. 2016; Butterfield and Poon 2005). We are not yet aware of any studies investigating tendon aging with these models nor is there much evidence of tendon disease, and therefore this presents an interesting future avenue of exploration.

3.5.3 Tools for Labelling Collagen Turnover

The advent of electron microscopy (EM), its application to living systems and the recent extension of its capacity to produce highly-detailed 3-D serial reconstructions of tendon nanoscale structure has advanced our understanding of tendon morphology and development tremendously (Fig. 3.5a, b) (Birk et al. 1990b; Starborg et al. 2013; Trelstad et al. 1982). EM has sufficient resolution to observe the details of cell/matrix interaction with nearly molecular resolution. However, EM requires dehydration and fixation of tissue, it thus cannot address the critical question of directionality or magnitude of the flux of molecules, leaving matrix assembly and degradation dynamics an indirect and speculative endeavor. While it is possible to image "single" collagen fibrils and collagen matrix remodeling with label-free methods such as second harmonic generation (SHG, Fig. 3.5c) (Campagnola et al. 2002; Cox et al. 2003; Theodossiou et al. 2006), differential interference contrast (DIC) (Petroll and Ma 2003; Bhole et al. 2009) and confocal reflection (Brightman et al. 2000; Kim et al. 2006), these methods are also limited by a number of constraints: SHG reportedly has nanoscale resolution [recently claimed at 30 nm (Bancelin et al. 2014)] but only captures fibrils with non-centrosymmetric organization because it relies on a lack of inversion symmetry (Campagnola et al. 2002). DIC cannot resolve fibrils in dense tissue and is subject to orientation angle contrast dependency (Siadat et al. 2021a) and confocal reflection microscopy has resolution limitations and density/contrast difficulties as well. In addition, all of them are unable to track the fate of single molecules during their transit to and from the matrix.

The ultimate goal of labelling is to determine the spatial and temporal fates of target molecules from their translation to the site of action to removal from service, all in real time in a living animal. It would be even better if their exact locations and orientation with structures could be determined as well (Alzola et al. 2021). Fortunately, the labelling and tracking of matrix molecules has been proceeding apace for years secondary to advances in labelling techniques and microscopy methods. As far as we can tell, no combination of molecular probe and imaging method has met this lofty standard to study tendon extracellular matrix. However, there are a number of probe/microscopy combinations that can reasonably be used to ask particular, circumscribed questions with excellent results.

3.5.3.1 Collagen-Binding Protein Labels

Collagen labels based on a bacterial adhesion protein with specificity for collagen (CNA35) and on an integrin (GST-α_1I) were recently demonstrated in Krahn et al. (2006) The labels were shown to be more specific than dichlorotriazinyl aminofluorescein (DTAF) which has been the standard for tracking collagen formation. CNA5

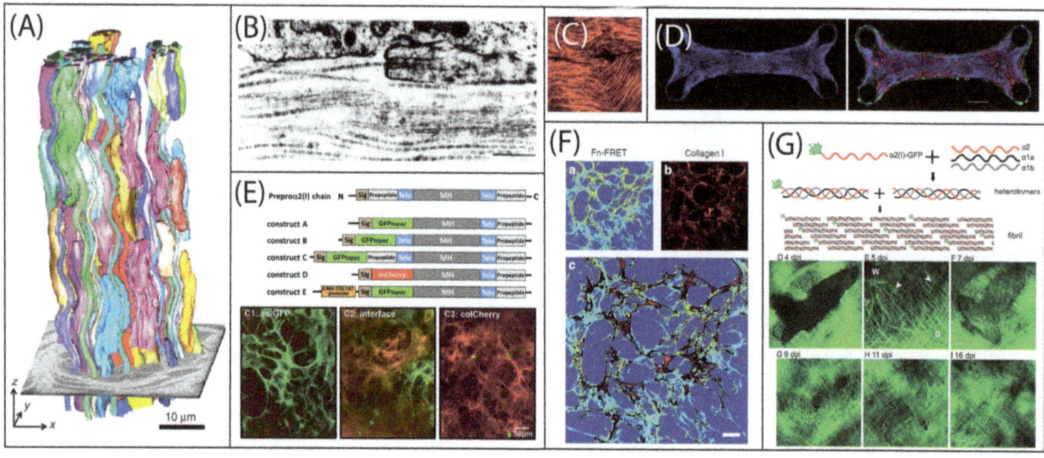

Fig. 3.5 **Tools for labelling collagen synthesis, remodeling, and incorporation.** (**A**) 3View® analysis of resin embedded sample of a newborn mouse tendon. Colors represent different cells/bundles of fibrils. From Starborg et al. (2013) with permission. (**B**) Conventional transmission electron micrograph of collagen formation in a developing chick tendon showing the details of the cell-fibril interface. From Trelstad et al. (1982) with permission. Scale bar is 300 nm. (**C**) Label free, confocal SHG images of collagen in frozen rat foot flexor tendon in transmission mode; 880 nm pumping frequency. From Theodossiou et al. (2006) with permission. (**D**) Labelling of engineered collagen-rich cardiac tissue. Collagen is stained with reversible collagen binding dye produced in bacteria: CNA35-m Turquoise2. Cells: green; mitochondria: red. Modified from Aper et al. (2014) with permission. Scale bar is 100µm. (**E**) Endogenous label incorporation The GFPtopaz and mCherry labels indicate that individual collagen molecules are incorporated into the same network of forming fibers. From Lu et al. (2018) with permission. (**F**) Fret labelled Fibronectin (Fn-FRET) and type 1 Collagen mechanochemical interaction probed from Kubow et al. (2015) with permission. Collagen was shown to colocalize principally with unloaded FN (yellow). Scale bar is 20µm. (**G**, top) GFP labelled collagen I zebrafish line generation. The N-terminal region of the collagen I $\alpha 2$ chains were selected for placement of the label. GFP-tagged alpha chain trimerises with unlabeled "**a**" and "**b**" chains. A mix of heterotrimers (labeled and unlabeled) are capable of forming fibrils with the label residing in the intratrimer gaps. (**G**, bottom) Progression of the closing of incision wound in the flank skin of transgenic zebrafish. Wound gape due to tension release shown at 4 dpi is closed with loose network of deposited collagen fibrils showing poor organization at 5 dpi. The collagen network is repaired over the next 11 days and reorganized into an orthogonal pattern by 16 dpi. Scale bars are 15µm. (Modified from Morris et al. (2018) with permission)

had better affinity for collagen than GST-α_1I and did not show substantial cross-reactivity with NCPs in the matrix. The binding of the probe is reversible which makes it "unlikely" to affect matrix production and permits time course investigations of matrix development. However, the probe is not specific for type I and also binds collagen III and IV. An interesting application for the probe was one in which the probe was bound to collagen and made "activatable" via MMP-2 proteolysis (Xia et al. 2011). More recently, the same group added six genetically encoded collagen probes produced in bacteria that fuse CNA35 to fluorescent proteins across the visible spectrum in an engineered, collagen rich tissue (Fig. 3.5d) (Aper et al. 2014). There have been multiple collagen binding proteins discovered which can be used to label collagen (Chilakamarthi et al. 2014), which would all operate in a manner similar to CNA35. While the primary utility of these probes is the real-time, multi-color imaging of live tissue, *in situ*, the multiple color probes could make it possible to perform sequential collagen deposition tracking experiments provided the reversibility of previously bound probe does not permit exchange with newly added probes. However, the size of the probe could inhibit proper assembly of matrix given that the molecular weight is the combination of the CNA35 (35 kDa) and the fluorophore (e.g. 93 kDa for tdTomato). Furthermore, there is a troubling lack of investigations, demonstrating the effect of collagen binding proteins on collagen assembly kinetics.

3.5.3.2 Bio-orthogonal Labels

Advances in bio-orthogonal chemistry have led to the development of a series of functionalized metabolites that act as chemical reporters (Grammel and Hang 2013; Dieterich et al. 2006). These can be viewed as analogous to the early radio tracer experiments with the exception that they are non-toxic, easily incorporated with little regulation, have low off target effects and can be readily illuminated fluorescently with high resolution, *in vitro* and *in vivo*. The process involves separating the incorporation of the reporter from its detection. This prevents the addition of bulky fluorophores until a readout is desired, which also presents opportunities for pulse chase experiments. Proteome tagging using non-canonical amino acids with reactive handles has the potential to revolutionize live cell imaging and tracking of molecular moieties. Non-canonical amino acids are incorporated into the target molecule using the cell's own machinery. Amgarten et al. labelled collagen with azido-proline (N_3-Pro) in fetal ovine osteoblast culture via supplementation of the growth medium with cis-4-azido-L-proline (Amgarten et al. 2015). The incorporation of the N_3-Pro minimally affects collagen formation as expected, and provides a substrate for dibenzooctyne (DIBO) fluorescent probe. While they show that the N_3-Pro did not affect cell viability, the DIBO reacted with some intracellular components including actin increasing background fluorescence which required additional treatment to reduce. Nonetheless, bio-orthogonal collagen labelling has enormous potential as a live cell and *in vivo* imaging technique. In 2014, Mirigian et al. performed bio-orthogonal pulse chase experiments in dermal human fibroblasts with and without a type I collagen chain mutation (Mirigian et al. 2014). They incorporated the non-canonical amino acid azidohomoalanine (Aha), a methionine (Met) analog, into cell secreted collagen by supplementing Met and Cys-free DMEM with Aha. They reported quiet incorporation of the Aha into the collagen with no discernible effect on post-translational modification, stability or structure of the triple helix. The utility of the tracing was demonstrated by successful measurement of pro-collagen folding kinetics in a normal and osteogenesis imperfecta patient's cells, which is a highly challenging pulse-chase experiment due to the short pulse window.

3.5.3.3 ECM Proteins Conjugated to Labels

Rather than add proteins or peptides that target and bind to ECM components already in the tissue, it is sometimes possible to add labelled ECM proteins themselves to the system as participating tracking molecules. The theory behind this approach is that ECM proteins will behave as they would whether they are secreted by the cell or added to the system. Collagen has a long history of being directly labelled and added exogenously to living systems where it has shown an ability to "home" to its proper morphological position. Stopak et al. injected covalently labelled (FITC) collagen type I into chick limb buds to track its incorporation into tissue rudiments including tendon (Stopak et al. 1985).

In an excellent demonstration of the utility of conjugated ECM protein labels, Sivakumar et al. added fibronectin (FN) conjugated to AlexaFluor 488 or 555 (Sivakumar 2006). These FN labels were dynamically tracked throughout the construction of matrix by osteoblast cells in a culture system showing a remarkable view of matrix assembly dynamics (Kadler et al. 2008). In an extension of this concept, exogenous FN labelling can be adapted in conjunction with Förster resonance energy transfer (FRET) to produce mechano-sensitive imaging (Kubow et al. 2015). In a seminal report, Kubow et al. added FRET labels to plasma FN such that mechanical unfolding of the molecule displaced the FRET labels and produced a detectable signal in a live culture system (Fig. 3.5f). The co-localization of the fibronectin FRET signal with collagen (immunolabelled) permitted the observation of collagen and FN interaction principally when the FN was relaxed and not under load. The collagen-FN mechanochemical reciprocal relationship was also recently probed in a cell-free system whereby collagen fibril nucleation was catalyzed by FN under conditions of extensional strain (Paten et al. 2019). Because labels can interfere with functionality of the protein, efforts have been made to reduce the size and degree of labelling of the probe (Siadat et al. 2021b). An interesting alternative approach was recently described by Doyle which attempts to preserve the intermolecular lysines for association in fibrils rather than labelling sites (Doyle 2018). To do this, Doyle labels the collagen (atto-488 NHS-ester dye) as a formed gel, then reverts the gel back to the molecular state and dilutes the labelled monomers with unlabeled collagen (~2:98%). Reformation of the mixed collagen produces a bright collagen network suitable for cell culture.

3.5.3.4 Endogenous Labels

Some of the most impressive work has been done with endogenous labels in live cultures and in living animals. While *in vitro* systems have substantial and well-known limitations relative to *in vivo* systems, there are a number of advantages which permit excellent observational fidelity. One of the more striking examples of *in vitro* imaging of labelled collagen assembly dynamics was performed using the osteoblast-like cell line MLO-A5 (Lu et al. 2018). The cells were transfected with GFPtpz and mCherry-collagen expression plasmids with careful attention paid to the placement of the label (Fig. 3.5e). The dual collagen labels permitted the dynamic observation of the interface that developed between differentially-labelled cell systems. Co-cultures of two different colored collagen expressing cells, produced a collagenous ECM that fused both colors, indicating a mixing of collagen molecules from each construct to form new fibrils (Fig. 3.5e). While labelling in cell culture is quite informative, it is always striking to see labelling performed well in a living system. In a recent and elegant paper, Morris et al. label type I collagen in a living zebrafish and dynamically track the progress of repair of a wound in skin (Morris et al. 2018). In their experiment, they drove expression of col1a2-GFP using a krtt1c19e promoter known to express in the basal epidermis which produces skin collagen type I in early development Tg(krt19:col1a2-GFP). The placement of the GFP label at the N-terminal region of the collagen molecule ostensibly minimizes the effect of the label on the assembly kinetics and morphology of collagen fibrils formed from them. This work stunningly demonstrates the progression of collagen disruption, organizational control and deposition during repair of a skin wound in the zebrafish (Fig. 3.5g).

3.5.3.5 Collagen Hybridizing Peptide

While labelling intact and functional collagen is informative, it is also quite important to develop labels which can identify collagen that is damaged. The principal role of collagen as a load bearing material makes understanding its failure mechanisms and subsequent repair critical to the development and timely application of clinical treatments for a broad range of injuries. Collagen molecular damage has been evaluated by a number of different methods including increased digestion susceptibility (Willett et al. 2007) and changes in denaturation endotherms (Willett et al. 2008). However, in 2012 Li et al. presented a paper on a caged collagen mimetic peptide (CMP) or collagen hybridizing peptide (CHP) which could be photo-triggered to fold into a triple helix capable of binding heat-denatured or MMP-digested collagen (Li et al. 2012). Zitnay et al. convincingly demonstrated that the CHP would bind preferentially to damaged collagen in 12% strain-overloaded rat tail tendon fascicles using transmission electron microscopy (TEM) and gold nanoparticle labelled CHP (Zitnay et al. 2017). The intensity of CHP staining of cyclically-loaded tendon increased with the frequency and number of the load cycles. More recently, the authors used this technique to measure the molecular damage to rat tail tendon fascicle collagen during cyclic fatigue loading (Zitnay et al. 2020), which has significant implications for our understanding of overuse injury.

3.6 Conclusions and Avenues for Future Work

We review here the large body of work investigating the formation, assembly, and maintenance of the tendon extracellular matrix. It is clear that a vast majority of this work has historically focused on embryonic and postnatal development, and despite nearly a century of research, there are still knowledge gaps and debates among the experts regarding how collagen fibrils form and assemble into the intricate tendon hierarchical structure. The exact growth mechanisms in tendon are still currently unknown. We still do not understand how cells and matrix work together to establish initial continuity in the mechanical structure of developing animals. It remains unclear if mechanical force drives fibril assembly at the molecular level or if fibrils are synthesized first and then organized. While it has been established that traction forces applied by resident cells are necessary for fibril formation, the precise mechanism and location that cells use these forces to convert soluble collagen monomers into fibrils are still to be determined. Furthermore, the question of how fibrils lengthen in a growing tendon under load while preserving mechanical integrity remains unresolved. We are also still understanding how collagen molecules within a matrix that endures high mechanical forces and a large number of cycles have such a long half-life.

There are also still a number of open questions regarding the mechanisms of adult matrix turnover or adaptation. If we ultimately want to understand how chronic matrix degeneration occurs, as in the case of tendinopathy, we want to identify the initiators of matrix remodeling and what events would make this process go awry. One of the missing gaps in this field is a lack of understanding in the repair of sub-failure damage or microdamage and how these mechanisms are different from a massive injury response. In addition, it would be beneficial to know where and how microdamage is initiated and to develop methods to track this damage. Since mechanical function and tissue structure are highly dependent on functional needs, it's possible that the turnover of individual matrix proteins is also functionally specific. Protein turnover in functionally distinct tendons varies with protein type but relative turnover rates for individual proteins between tendon types remain to be determined. Moreover, it is possible that turnover at the junction of tendon with another dissimilar tissue, such as at the enthesis or the myotendinous junction, is more rapid than in the midsubstance. Answers to these questions would dramatically improve our understanding of adult tissue maintenance and potentially provide clues to chronic degeneration.

Age-related cellular mechanisms are likely to blame for the dysfunction of normal tissue homeostasis that could lead to chronic degeneration, but the mechanisms behind these deficits have not been fully established. It is uncertain whether there are changes in mechanosensing or mechanotransduction, preventing cells from sensing and converting appropriate mechanical signals to elicit remodeling, or whether the dysfunction is in the processes of matrix remodeling itself, limiting the synthesis, assembly or incorporation of new ECM. More work is needed to identify what these cellular changes are and how they influence the ability to maintain tissue architecture. In addition, while cellular changes have been studied extensively in tendon stem cells and particularly in relation to the injury response, fewer studies have investigated age-related changes in mature tenocytes which we expect to be responsible for local tissue repair in the absence of inflammatory cell recruitment. Finally, it is important to note that many of the 'hallmarks of aging' are extremely interconnected and most of them have not yet been directly investigated in tendon; therefore, there are likely aging mechanisms that influence matrix homeostasis that have yet to be uncovered.

There are also still major deficits in our basic knowledge of the tendon composition and structure, specifically in the non-collagenous matrix and the cell populations present. Much of the research presented has focused on regulation of the collagen structure, with considerably less attention placed on the regulation of the non-collagenous compartment, specifically the interfibrillar and interfascicular matrix as well as paratenon and epitenon. Studying the non-collagenous matrix is quite challenging due to low abundance and difficulty in precise extraction, as well as absence of *in vitro* systems focusing on it. In addition, *in vivo* models permitting genetic modification (rodents) lack an interfascicular compartment posing another hurdle in the study of the non-collagenous matrix, whilst larger animal models (horse) have an interfascicular compartment but do not lend themselves to genetic modification and longitudinal studies due to time and cost constraints.

Furthermore, this chapter focuses on the regulation and dysregulation of the tendon ECM throughout life, all of which is cell-mediated. However, we still do not have a complete understanding of the specific cell populations that are present in whole tendon and their localization. Recent studies have focused on identifying and characterizing cell populations, highlighting the vast heterogeneity and complexity of the population within tendon compartments. With the advent of single-cell sequencing, investigation of cell heterogeneity within tissues has been made possible and its recent use in tendon research has unveiled several tendon cell subtypes that could be responsible for matrix remodeling (Paolillo et al. 2019; Harvey et al. 2019; Kendal et al. 2020; De Micheli et al. 2020; Yin et al. 2016). Therefore, there appear to be many different subpopulations of cells responsible for producing ECM but the role of the identified clusters in the development, maintenance, and aging of tendon still remains to be elucidated.

Many of these questions will still require years of research to answer, but the development of novel models and tools to study ECM remodeling provide substantial promise for future investigation. With the ability to label and track collagen, and hopefully someday non-collagenous proteins, mechanisms of matrix incorporation and linear growth that have evaded detection in previous years may now be uncovered. Increased knowledge of the processes controlling matrix growth and incorporation could provide guidance for tissue engineering approaches. Furthermore, if key regulators of matrix homeostasis during adulthood and into aging are identified, it may become possible to identify the tipping point between positive adaptation and degeneration leading to progressive tendinopathy. Not only will this allow us to understand the process of degeneration, it will also put research one step closer to developing therapeutics and/or preventative interventions for tendon injury and disease.

References

Abate M, Schiavone C, Salini V, Andia I (2013) Occurrence of tendon pathologies in metabolic disorders. Rheumatology 52:599–608. https://doi.org/10.1093/rheumatology/kes395

Abrahamsson SO, Lundborg G, Lohmander LS (1991) Long-term explant culture of rabbit flexor tendon: effects of recombinant human insulin-like growth factor-I and serum on matrix metabolism. J Orthop Res 9:503–515. https://doi.org/10.1002/jor.1100090406

Abreu EL, Leigh D, Derwin KA (2008) Effect of altered mechanical load conditions on the structure and function of cultured tendon fascicles. J Orthop Res 26:364–373. https://doi.org/10.1002/jor.20520

Ackerman JE, Geary MB, Orner CA, Bawany F, Loiselle AE (2017a) Obesity/type II diabetes alters macrophage polarization resulting in a fibrotic tendon healing response. PLoS One 12:e0181127. https://doi.org/10.1371/journal.pone.0181127

Ackerman JE, Bah I, Jonason JH, Buckley MR, Loiselle AE (2017b) Aging does not alter tendon mechanical properties during homeostasis, but does impair flexor tendon healing. J Orthop Res 35:2716–2724. https://doi.org/10.1002/jor.23580

Ackerman JE, Best KT, O'Keefe RJ, Loiselle AE (2017c) Deletion of EP4 in S100a4-lineage cells reduces scar tissue formation during early but not later stages of tendon healing. Sci Rep 7:8658. https://doi.org/10.1038/s41598-017-09407-7

Ackermann PW, Salo P, Hart DA (2016) Tendon innervation. Adv Exp Med Biol 920:35–51. https://doi.org/10.1007/978-3-319-33943-6_4

Acosta JC, Banito A, Wuestefeld T, Georgilis A, Janich P, Morton JP et al (2013) A complex secretory program orchestrated by the inflammasome controls paracrine senescence. Nat Cell Biol 15:978–990. https://doi.org/10.1038/ncb2784

Adachi E, Hayashi T (1986) In vitro formation of hybrid fibrils of type V collagen and type I collagen. Limited growth of type I collagen into thick fibrils by type V collagen. Connect Tissue Res 14:257–266

Ali OJ, Comerford EJ, Clegg PD, Canty-Laird EG (2018) Variations during ageing in the three-dimensional anatomical arrangement of fascicles within the equine superficial digital flexor tendon. Eur Cell Mater 35:87–102. https://doi.org/10.22203/eCM.v035a07

Almeida-Silveira MI, Lambertz D, Pérot C, Goubel F (2000) Changes in stiffness induced by hindlimb suspension in rat Achilles tendon. Eur J Appl Physiol 81:252–257. https://doi.org/10.1007/s004210050039

Alzola RP et al. (2021) Method for measurement of collagen monomer orientation in fluorescence microscopy.

J Biom Opt 26(7):076501. https://doi.org/10.1117/1.JBO.26.7.076501
Ameye L, Aria D, Jepsen K, Oldberg A, Xu T, Young MF (2002) Abnormal collagen fibrils in tendons of biglycan/fibromodulin-deficient mice lead to gait impairment, ectopic ossification, and osteoarthritis. FASEB J 16:673–680. https://doi.org/10.1096/fj.01-0848com
Amgarten B, Rajan R, Martínez-Sáez N, Oliveira BL, Albuquerque IS, Brooks RA et al (2015) Collagen labelling with an azide-proline chemical reporter in live cells. Chem Commun 51:5250–5252. https://doi.org/10.1039/c4cc07974d
Amiel D, Frank C, Harwood F, Fronek J, Akeson W (1984) Tendons and ligaments: a morphological and biochemical comparison. J Orthop Res 1:257–265. https://doi.org/10.1002/jor.1100010305
Andarawis-Puri N, Flatow EL (2018) Promoting effective tendon healing and remodeling. J Orthop Res 36:3115–3124. https://doi.org/10.1002/jor.24133
Andarawis-Puri N, Sereysky JB, Sun HB, Jepsen KJ, Flatow EL (2012) Molecular response of the patellar tendon to fatigue loading explained in the context of the initial induced damage and number of fatigue loading cycles. J Orthop Res 30:1327–1334. https://doi.org/10.1002/jor.22059
Andarawis-Puri N, Philip A, Laudier D, Schaffler MB, Flatow EL (2014) Temporal effect of in vivo tendon fatigue loading on the apoptotic response explained in the context of number of fatigue loading cycles and initial damage parameters. J Orthop Res 32:1097–1103. https://doi.org/10.1002/jor.22639
Andarawis-Puri N, Flatow EL, Soslowsky LJ (2015) Tendon basic science: development, repair, regeneration, and healing. J Orthop Res 33:780–784. https://doi.org/10.1002/jor.22869
Andersen MB, Pingel J, Kjaer M, Langberg H (2011) Interleukin-6: a growth factor stimulating collagen synthesis in human tendon. J Appl Physiol 110:1549–1554. https://doi.org/10.1152/japplphysiol.00037.2010
Angelidis IK, Thorfinn J, Connolly ID, Lindsey D, Pham HM, Chang J (2010) Tissue engineering of flexor tendons: the effect of a tissue bioreactor on adipoderived stem cell–seeded and fibroblast-seeded tendon constructs. J Hand Surg 35:1466–1472. https://doi.org/10.1016/j.jhsa.2010.06.020
Ansorge HL, Meng X, Zhang G, Veit G, Sun M, Klement JF et al (2009) Type XIV collagen regulates fibrillogenesis: premature collagen fibril growth and tissue dysfunction in null mice. J Biol Chem 284:8427–8438. https://doi.org/10.1074/jbc.M805582200
Ansorge HL, Adams S, Jawad AF, Birk DE, Soslowsky LJ (2012) Mechanical property changes during neonatal development and healing using a multiple regression model. J Biomech 45:1288–1292. https://doi.org/10.1016/j.jbiomech.2012.01.030
Aper SJA, Van Spreeuwel ACC, Van Turnhout MC, Van Der Linden AJ, Pieters PA, Van Der Zon NLL et al (2014) Colorful protein-based fluorescent probes for collagen imaging. PLoS One 9:e114983. https://doi.org/10.1371/journal.pone.0114983
Archambault J, Tsuzaki M, Herzog W, Banes AJ (2002) Stretch and interleukin-1β induce matrix metalloproteinases in rabbit tendon cells in vitro. J Orthop Res 20:36–39. https://doi.org/10.1016/s0736-0266(01)00075-4
Archambault JM, Jelinsky SA, Lake SP, Hill AA, Glaser DL, Soslowsky LJ (2007) Rat supraspinatus tendon expresses cartilage markers with overuse. J Orthop Res 25:617–624. https://doi.org/10.1002/jor.20347
Arnesen SM, Lawson MA (2006) Age-related changes in focal adhesions lead to altered cell behavior in tendon fibroblasts. Mech Ageing Dev 127:726–732. https://doi.org/10.1016/j.mad.2006.05.003
Astill BD, Katsma MS, Cauthon DJ, Greenlee J, Murphy M, Curtis D et al (2017) Sex-based difference in Achilles peritendinous levels of matrix metalloproteinases and growth factors after acute resistance exercise. J Appl Physiol 1985 122:361–367. https://doi.org/10.1152/japplphysiol.00878.2016
Attia M, Scott A, Duchesnay A, Carpentier G, Soslowsky LJ, Huynh MB et al (2012) Alterations of overused supraspinatus tendon: a possible role of glycosaminoglycans and HARP/pleiotrophin in early tendon pathology. J Orthop Res 30:61–71. https://doi.org/10.1002/jor.21479
Babraj JA, Cuthbertson DJR, Smith K, Langberg H, Miller B, Krogsgaard MR et al (2005) Collagen synthesis in human musculoskeletal tissues and skin. Am J Physiol Endocrinol Metab 289:E864–E869. https://doi.org/10.1152/ajpendo.00243.2005
Bahr G (1950) The reconstitution of collagen fibrils as revealed by electron microscopy. Exp Cell Res 1:603–606. https://doi.org/10.1016/0014-4827(50)90010-3
Bailey AJ, Paul RG, Knott L (1998) Mechanisms of maturation and ageing of collagen. Mech Ageing Dev 106:1–56. https://doi.org/10.1016/s0047-6374(98)00119-5
Baitsell GA (1915) The origin and structure of a fibrous tissue which appears in living cultures of adult frog tissues. J Exp Med 21:455–478
Baitsell GA (1916) The origin and structure of a fibrous tissue formed in wound healing. J Exp Med 23:739–756
Baitsell GA (1921) A study of the development of connective tissue in the amphibia. Am J Anat 28:447–475
Baitsell GA (1925) Memoirs: on the origin of the connective-tissue ground-substance in the chick embryo. J Cell Sci 2:571–589
Bancelin S, Aimé C, Gusachenko I, Kowalczuk L, Latour G, Coradin T et al (2014) Determination of collagen fibril size via absolute measurements of second-harmonic generation signals. Nat Commun 5:4920. https://doi.org/10.1038/ncomms5920
Bard JB, Chapman JA (1973) Diameters of collagen fibrils grown in vitro. Nat New Biol 246:83–84. https://doi.org/10.1038/newbio246083a0
Batson EL, Paramour RJ, Smith TJ, Birch HL, Patterson-Kane JC, Goodship AE (2003) Are the material properties and matrix composition of equine

flexor and extensor tendons determined by their functions? Equine Vet J 35:314–318. https://doi.org/10.2746/042516403776148327

Bayer ML, Yeung C-YC, Kadler KE, Qvortrup K, Baar K, Svensson RB et al (2010) The initiation of embryonic-like collagen fibrillogenesis by adult human tendon fibroblasts when cultured under tension. Biomaterials 31:4889–4897. https://doi.org/10.1016/j.biomaterials.2010.02.062

Becker U, Timpl R (1976) NH2-terminal extensions on skin collagen from sheep with a genetic defect in conversion of procollagen into collagen. Biochemistry 15:2853–2862. https://doi.org/10.1021/bi00658a024

Beertsen W, Everts V, Hoeben K, Niehof A (1984) Microtubules in periodontal ligament cells in relation to tooth eruption and collagen degradation. J Periodontal Res 19:489–500. https://doi.org/10.1111/j.1600-0765.1984.tb01304.x

Bell R, Boniello MR, Gendron NR, Flatow EL, Andarawis-Puri N (2015) Delayed exercise promotes remodeling in sub-rupture fatigue damaged tendons. J Orthop Res 33:919–925. https://doi.org/10.1002/jor.22856

Bell R, Gendron NR, Anderson M, Flatow EL, Andarawis-Puri N (2018) A potential new role for myofibroblasts in remodeling of sub-rupture fatigue tendon injuries by exercise. Sci Rep 8:8933. https://doi.org/10.1038/s41598-018-27196-5

Berg RA, Prockop DJ (1973) The thermal transition of a non-hydroxylated form of collagen. Evidence for a role for hydroxyproline in stabilizing the triple-helix of collagen. Biochem Biophys Res Commun 52:115–120

Best KT, Studentsova V, Ackerman JE, Nichols AEC, Myers M, Cobb J et al (2020) Effects of tamoxifen on tendon homeostasis and healing: considerations for the use of tamoxifen-inducible mouse models. J Orthop Res. https://doi.org/10.1002/jor.24767

Bhole AP, Flynn BP, Liles M, Saeidi N, Dimarzio CA, Ruberti JW (2009) Mechanical strain enhances survivability of collagen micronetworks in the presence of collagenase: implications for load-bearing matrix growth and stability. Philos Trans R Soc A Math Phys Eng Sci 367:3339–3362. https://doi.org/10.1098/rsta.2009.0093

Bi Y, Ehirchiou D, Kilts TM, Inkson CA, Embree MC, Sonoyama W et al (2007) Identification of tendon stem/progenitor cells and the role of the extracellular matrix in their niche. Nat Med 13:1219–1227. https://doi.org/10.1038/nm1630

Biewener AA (1998) Muscle-tendon stresses and elastic energy storage during locomotion in the horse. Comp Biochem Physiol B Biochem Mol Biol 120:73–87. https://doi.org/10.1016/s0305-0491(98)00024-8

Bigi A, Cojazzi G, Roveri N, Koch MHJ (1987) Differential scanning calorimetry and X-ray-diffraction study of tendon collagen thermal-denaturation. Int J Biol Macromol 9:363–367. https://doi.org/10.1016/0141-8130(87)90010-9

Birch HL, Bailey JV, Bailey AJ, Goodship AE (1999) Age-related changes to the molecular and cellular components of equine flexor tendons. Equine Vet J 31:391–396. https://doi.org/10.1111/j.2042-3306.1999.tb03838.x

Birch HL, Peffers MJ, Clegg PD (2016) Influence of ageing on tendon homeostasis. Adv Exp Med Biol 920:247–260. https://doi.org/10.1007/978-3-319-33943-6_24

Birk DE, Mayne R (1997) Localization of collagen types I, III and V during tendon development. Changes in collagen types I and III are correlated with changes in fibril diameter. Eur J Cell Biol 72:352–361

Birk DE, Trelstad RL (1984) Extracellular compartments in matrix morphogenesis: collagen fibril, bundle, and lamellar formation by corneal fibroblasts. J Cell Biol 99:2024–2033. https://doi.org/10.1083/jcb.99.6.2024

Birk DE, Trelstad RL (1986) Extracellular compartments in tendon morphogenesis: collagen fibril, bundle, and macroaggregate formation. J Cell Biol 103:231–240. https://doi.org/10.1083/jcb.103.1.231

Birk DE, Fitch JM, Linsenmayer TF (1986) Organization of collagen types I and V in the embryonic chicken cornea. Invest Ophthalmol Vis Sci 27:1470–1477

Birk DE, Fitch JM, Babiarz JP, Linsenmayer TF (1988) Collagen type I and type V are present in the same fibril in the avian corneal stroma. J Cell Biol 106:999–1008

Birk DE, Fitch JM, Babiarz JP, Doane KJ, Linsenmayer TF (1990a) Collagen fibrillogenesis in vitro: interaction of types I and V collagen regulates fibril diameter. J Cell Sci 95(Pt 4):649–657

Birk DE, Zycband EI, Winkelmann DA, Trelstad RL (1990b) Collagen fibrillogenesis in situ: discontinuous segmental assembly in extracellular compartments. Ann N Y Acad Sci 580:176–194

Birk DE, Nurminskaya MV, Zycband EI (1995) Collagen fibrillogenesis in situ: fibril segments undergo post-depositional modifications resulting in linear and lateral growth during matrix development. Dev Dyn 202:229–243. https://doi.org/10.1002/aja.1002020303

Birk DE, Hahn RA, Linsenmayer CY, Zycband EI (1996) Characterization of collagen fibril segments from chicken embryo cornea, dermis and tendon. Matrix Biol 15:111–118

Birk DE, Zycband EI, Woodruff S, Winkelmann DA, Trelstad RL (1997) Collagen fibrillogenesis in situ: fibril segments become long fibrils as the developing tendon matures. Dev Dyn 208:291–298. https://doi.org/10.1002/(SICI)1097--0177(199703)208:3<291::AID-AJA1>3.0.CO;2-D

Blagosklonny MV (2011) Cell cycle arrest is not senescence. Aging 3:94–101. https://doi.org/10.18632/aging.100281

Blokland KEC, Pouwels SD, Schuliga M, Knight DA, Burgess JK (2020) Regulation of cellular senescence by extracellular matrix during chronic fibrotic diseases. Clin Sci Lond Engl 1979 134:2681–2706. https://doi.org/10.1042/CS20190893

Boesen AP, Dideriksen K, Couppé C, Magnusson SP, Schjerling P, Boesen M et al (2013) Tendon and skeletal muscle matrix gene expression and functional responses to immobilisation and rehabilitation in young males: effect of growth hormone administration. J Physiol 591:6039–6052. https://doi.org/10.1113/jphysiol.2013.261263

Bohm S, Mersmann F, Arampatzis A (2015) Human tendon adaptation in response to mechanical loading: a systematic review and meta-analysis of exercise intervention studies on healthy adults. Sports Med – Open 1. https://doi.org/10.1186/s40798-015-0009-9

Boregowda R, Paul E, White J, Ritty TM (2008) Bone and soft connective tissue alterations result from loss of fibrillin-2 expression. Matrix Biol 27:661–666. https://doi.org/10.1016/j.matbio.2008.09.579

Bornstein P, Ehrlich HP, Wyke AW (1972) Procollagen: conversion of the precursor to collagen by a neutral protease. Science 175:544–546. https://doi.org/10.1126/science.175.4021.544

Bradbury S, Meek GA (1958) The fine structure of the adipose cell of the leech Glossiphonia complanata. J Biophys Biochem Cytol 4:603–607. https://doi.org/10.1083/jcb.4.5.603

Brightman AO, Rajwa BP, Sturgis JE, McCallister ME, Robinson JP, Voytik-Harbin SL (2000) Time-lapse confocal reflection microscopy of collagen fibrillogenesis and extracellular matrix assembly in vitro. Biopolymers 54:222–234. https://doi.org/10.1002/1097-0282(200009)54:3<222::aid--bip80>3.0.co;2-k

Brinckmann J, Bachinger HP (2005) Collagen: primer in structure, processing a. assembly. Springer, Berlin Heidelberg

Bronner F (2001) Extracellular and intracellular regulation of calcium homeostasis. Sci World J 1:919–925

Buckley MR, Sarver JJ, Freedman BR, Soslowsky LJ (2013) The dynamics of collagen uncrimping and lateral contraction in tendon and the effect of ionic concentration. J Biomech 46:2242–2249. https://doi.org/10.1016/j.jbiomech.2013.06.029

Burjanadze TV (1982) Evidence for the role of 4-hydroxyproline localized in the 3rd position of the triplet (Gly-X-Y) in adaptational changes of thermostability of a collagen molecule and collagen fibrils. Biopolymers 21:1489–1501. https://doi.org/10.1002/bip.360210803

Burke JM, Balian G, Ross R, Bornstein P (1977) Synthesis of types I and III procollagen and collagen by monkey aortic smooth muscle cells in vitro. Biochemistry 16:3243–3249. https://doi.org/10.1021/bi00633a031

Butler SL, Kohles SS, Thielke RJ, Chen C, Vanderby R (1997) Interstitial fluid flow in tendons or ligaments: a porous medium finite element simulation. Med Biol Eng Comput 35:742–746

Butler DL, Juncosa-Melvin N, Boivin GP, Galloway MT, Shearn JT, Gooch C et al (2008) Functional tissue engineering for tendon repair: a multidisciplinary strategy using mesenchymal stem cells, bioscaffolds, and mechanical stimulation. J Orthop Res 26:1–9. https://doi.org/10.1002/jor.20456

Butler DL, Hunter SA, Chokalingam K, Cordray MJ, Shearn J, Juncosa-Melvin N et al (2009) Using functional tissue engineering and bioreactors to mechanically stimulate tissue-engineered constructs. Tissue Eng Part A 15:741–749. https://doi.org/10.1089/ten.tea.2008.0292

Butterfield DA, Poon HF (2005) The senescence-accelerated prone mouse (SAMP8): a model of age-related cognitive decline with relevance to alterations of the gene expression and protein abnormalities in Alzheimer's disease. Exp Gerontol 40:774–783. https://doi.org/10.1016/j.exger.2005.05.007

Cameron IL, Short NJ, Fullerton GD (2007) Verification of simple hydration/dehydration methods to characterize multiple water compartments on tendon type 1 collagen. Cell Biol Int 31:531–539. https://doi.org/10.1016/j.cellbi.2006.11.020

Campagnola PJ, Millard AC, Terasaki M, Hoppe PE, Malone CJ, Mohler WA (2002) Three-dimensional high-resolution second-harmonic generation imaging of endogenous structural proteins in biological tissues. Biophys J 82:493–508. https://doi.org/10.1016/s0006-3495(02)75414-3

Campisi J (1998) The role of cellular senescence in skin aging. J Investig Dermatol Symp Proc 3:1–5

Canty EG, Lu Y, Meadows RS, Shaw MK, Holmes DF, Kadler KE (2004) Coalignment of plasma membrane channels and protrusions (fibripositors) specifies the parallelism of tendon. J Cell Biol 165:553–563. https://doi.org/10.1083/jcb.200312071

Carpenter JE, Flanagan CL, Thomopoulos S, Yian EH, Soslowsky LJ (1998) The effects of overuse combined with intrinsic or extrinsic alterations in an animal model of rotator cuff tendinosis. Am J Sports Med 26:801–807. https://doi.org/10.1177/03635465980260061101

Carpenter JE, Thomopoulos S, Soslowsky LJ (1999) Animal models of tendon and ligament injuries for tissue engineering applications. Clin Orthop 367:S296–S311

Cassel JM (1966) Collagen aggregation phenomena. Biopolym Orig Res Biomol 4:989–997

Chang J, Garva R, Pickard A, Yeung CC, Mallikarjun V, Swift J et al (2020) Circadian control of the secretory pathway maintains collagen homeostasis. Nat Cell Biol 22:74–86. https://doi.org/10.1038/s41556-019-0441-z

Chapman JA (1961) Morphological and chemical studies of collagen formation. I. The fine structure of guinea pig granulomata. J Biophys Biochem Cytol 9:639–651. https://doi.org/10.1083/jcb.9.3.639

Chapman JA (1989) The regulation of size and form in the assembly of collagen fibrils in vivo. Biopolymers 28:1367–1382. https://doi.org/10.1002/bip.360280803

Chaudhry SS, Cain SA, Morgan A, Dallas SL, Shuttleworth CA, Kielty CM (2007) Fibrillin-1 regulates the bioavailability of TGFβ1. J Cell Biol 176:355–367. https://doi.org/10.1083/jcb.200608167

Chaudhury C, Mehnaz S, Robinson JM, Hayton WL, Pearl DK, Roopenian DC et al (2003) The major histocompatibility complex–related Fc Receptor for IgG (FcRn) binds albumin and prolongs its lifespan. J Exp Med 197:315–322. https://doi.org/10.1084/jem.20021829

Chen B, Ding J, Zhang W, Zhou G, Cao Y, Liu W et al (2016) Tissue engineering of tendons: a comparison of muscle-derived cells, tenocytes, and dermal fibroblasts as cell sources. Plast Reconstr Surg 137:536e–544e. https://doi.org/10.1097/01.prs.0000479980.83169.31

Chiarelli N, Carini G, Zoppi N, Ritelli M, Colombi M (2019a) Molecular insights in the pathogenesis of classical Ehlers-Danlos syndrome from transcriptome-wide expression profiling of patients' skin fibroblasts. PLoS One 14. https://doi.org/10.1371/journal.pone.0211647

Chiarelli N, Ritelli M, Zoppi N, Colombi M (2019b) Cellular and molecular mechanisms in the pathogenesis of classical, vascular, and hypermobile Ehlers–Danlos syndromes. Gene 10. https://doi.org/10.3390/genes10080609

Chilakamarthi U, Kandhadi J, Gunda S, Thatipalli AR, Kumar Jerald M, Lingamallu G et al (2014) Synthesis and functional characterization of a fluorescent peptide probe for non invasive imaging of collagen in live tissues. Exp Cell Res 327:91–101. https://doi.org/10.1016/j.yexcr.2014.05.005

Cho E, Zhang Y, Pruznak A, Kim HM (2015) Effect of tamoxifen on fatty degeneration and atrophy of rotator cuff muscles in chronic rotator cuff tear: an animal model study. J Orthop Res 33:1846–1853. https://doi.org/10.1002/jor.22964

Choi RK, Smith MM, Smith S, Little CB, Clarke EC (2019) Functionally distinct tendons have different biomechanical, biochemical and histological responses to in vitro unloading. J Biomech 95:109321. https://doi.org/10.1016/j.jbiomech.2019.109321

Choi H, Simpson D, Wang D, Prescott M, Pitsillides AA, Dudhia J et al (2020) Heterogeneity of proteome dynamics between connective tissue phases of adult tendon. elife 9:e55262. https://doi.org/10.7554/eLife.55262

Christensen B, Dandanell S, Kjaer M, Langberg H (2011) Effect of anti-inflammatory medication on the running-induced rise in patella tendon collagen synthesis in humans. J Appl Physiol 110:137–141. https://doi.org/10.1152/japplphysiol.00942.2010

Cilli F, Khan M, Fu F, Wang JH (2004) Prostaglandin E2 affects proliferation and collagen synthesis by human patellar tendon fibroblasts. Clin J Sport Med 14:232–236. https://doi.org/10.1097/00042752-200407000-00006

Cingolani OH, Kirk JA, Seo K, Koitabashi N, Lee D-I, Ramirez-Correa G et al (2011) Thrombospondin-4 is required for stretch-mediated contractility augmentation in cardiac muscle. Circ Res 109:1410–1414. https://doi.org/10.1161/circresaha.111.256743

Cohn M (2002) Mini-series: significant contributions to biological chemistry over the past 125 years: biochemistry in the United States in the first half of the twentieth century. Biochem Mol Biol Educ 30:77–85. https://doi.org/10.1002/bmb.2002.494030020034

Comper WD, Laurent TC (1978) Physiological function of connective tissue polysaccharides. Physiol Rev 58:255–315. https://doi.org/10.1152/physrev.1978.58.1.255

Connizzo BK, Grodzinsky AJ (2017) Tendon exhibits complex poroelastic behavior at the nanoscale as revealed by high-frequency AFM-based rheology. J Biomech 54:11–18. https://doi.org/10.1016/j.jbiomech.2017.01.029

Connizzo BK, Grodzinsky AJ (2018a) Multiscale poroviscoelastic compressive properties of mouse supraspinatus tendons are altered in young and aged mice. J Biomech Eng 140. https://doi.org/10.1115/1.4038745

Connizzo BK, Grodzinsky AJ (2018b) Release of pro-inflammatory cytokines from muscle and bone causes tenocyte death in a novel rotator cuff in vitro explant culture model. Connect Tissue Res 59:423–436. https://doi.org/10.1080/03008207.2018.1439486

Connizzo BK, Grodzinsky AJ (2020) Lose-dose administration of dexamethasone is beneficial in preventing secondary tendon damage in a stress-deprived joint injury explant model. J Orthop Res. https://doi.org/10.1002/jor.24451

Connizzo BK, Yannascoli SM, Soslowsky LJ (2013a) Structure-function relationships of postnatal tendon development: a parallel to healing. Matrix Biol 32:106–116. https://doi.org/10.1016/j.matbio.2013.01.007

Connizzo BK, Sarver JJ, Birk DE, Soslowsky LJ, Iozzo RV (2013b) Effect of age and proteoglycan deficiency on collagen fiber re-alignment and mechanical properties in mouse supraspinatus tendon. J Biomech Eng 135:021019. https://doi.org/10.1115/1.4023234

Connizzo BK, Sarver JJ, Han L, Soslowsky LJ (2014a) In situ fibril stretch and sliding is location-dependent in mouse supraspinatus tendons. J Biomech 47:3794–3798. https://doi.org/10.1016/j.jbiomech.2014.10.029

Connizzo BK, Bhatt PR, Liechty KW, Soslowsky LJ (2014b) Diabetes alters mechanical properties and collagen fiber re-alignment in multiple mouse tendons. Ann Biomed Eng 42:1880–1888. https://doi.org/10.1007/s10439-014-1031-7

Connizzo BK, Freedman BR, Fried JH, Sun M, Birk DE, Soslowsky LJ (2015) Regulatory role of collagen V in establishing mechanical properties of tendons and ligaments is tissue dependent. J Orthop Res 33:882–888. https://doi.org/10.1002/jor.22893

Connizzo BK, Adams SM, Adams TH, Jawad AF, Birk DE, Soslowsky LJ (2016a) Multiscale regression modeling in mouse supraspinatus tendons reveals that dynamic processes act as mediators in structure-function relationships. J Biomech 49:1649–1657. https://doi.org/10.1016/j.jbiomech.2016.03.053

Connizzo BK, Adams SM, Adams TH, Birk DE, Soslowsky LJ (2016b) Collagen V expression is crucial in regional development of the supraspinatus tendon. J Orthop Res 34:2154–2161. https://doi.org/10.1002/jor.23246

Connizzo BK, Han L, Birk DE, Soslowsky LJ (2016c) Collagen V-heterozygous and -null supraspinatus tendons exhibit altered dynamic mechanical behaviour at multiple hierarchical scales. Interface Focus 6:20150043. https://doi.org/10.1098/rsfs.2015.0043

Connizzo BK, Piet JM, Shefelbine SJ, Grodzinsky AJ (2019) Age-associated changes in the response of tendon explants to stress deprivation is sex-dependent. Connect Tissue Res:1–15. https://doi.org/10.1080/03008207.2019.1648444

Cook JL, Screen HRC (2018) Tendon pathology: have we missed the first step in the development of pathology? J Appl Physiol 125:1349–1350. https://doi.org/10.1152/japplphysiol.00002.2018

Cooper A (1970) Thermodynamic studies of the assembly in vitro of native collagen fibrils. Biochem J 118:355–365

Corsi A, Xu T, Chen XD, Boyde A, Liang J, Mankani M et al (2002) Phenotypic effects of biglycan deficiency are linked to collagen fibril abnormalities, are synergized by decorin deficiency, and mimic Ehlers-Danlos-like changes in bone and other connective tissues. J Bone Miner Res 17:1180–1189. https://doi.org/10.1359/jbmr.2002.17.7.1180

Costa-Almeida R, Berdecka D, Rodrigues MT, Reis RL, Gomes ME (2018) Tendon explant cultures to study the communication between adipose stem cells and native tendon niche. J Cell Biochem 119:3653–3662. https://doi.org/10.1002/jcb.26573

Couppé C, Kongsgaard M, Aagaard P, Hansen P, Bojsen-Moller J, Kjaer M et al (2008) Habitual loading results in tendon hypertrophy and increased stiffness of the human patellar tendon. J Appl Physiol 105:805–810. https://doi.org/10.1152/japplphysiol.90361.2008

Couppé C, Hansen P, Kongsgaard M, Kovanen V, Suetta C, Aagaard P et al (2009) Mechanical properties and collagen cross-linking of the patellar tendon in old and young men. J Appl Physiol 107:880–886. https://doi.org/10.1152/japplphysiol.00291.2009

Couppé C, Suetta C, Kongsgaard M, Justesen L, Hvid LG, Aagaard P et al (2012) The effects of immobilization on the mechanical properties of the patellar tendon in younger and older men. Clin Biomech 27:949–954. https://doi.org/10.1016/j.clinbiomech.2012.06.003

Cowin SC (2000) How is a tissue built? J Biomech Eng 122:553–569. https://doi.org/10.1115/1.1324665

Cox G, Kable E, Jones A, Fraser I, Manconi F, Gorrell MD (2003) 3-dimensional imaging of collagen using second harmonic generation. J Struct Biol 141:53–62. https://doi.org/10.1016/s1047-8477(02)00576-2

Dai G-C, Li Y-J, Chen M-H, Lu P-P, Rui Y-F (2019) Tendon stem/progenitor cell ageing: modulation and rejuvenation. World J Stem Cells 11:677–692. https://doi.org/10.4252/wjsc.v11.i9.677

de Boer MD, Selby A, Atherton P, Smith K, Seynnes OR, Maganaris CN et al (2007) The temporal responses of protein synthesis, gene expression and cell signalling in human quadriceps muscle and patellar tendon to disuse. J Physiol 585:241–251. https://doi.org/10.1113/jphysiol.2007.142828

De Micheli AJ, Swanson JB, Disser NP, Martinez LM, Walker NR, Oliver DJ et al (2020) Single-cell transcriptomic analysis identifies extensive heterogeneity in the cellular composition of mouse Achilles tendons. Am J Phys Cell Physiol 319:C885–C894. https://doi.org/10.1152/ajpcell.00372.2020

de Mos M, van El B, DeGroot J, Jahr H, van Schie HTM, van Arkel ER et al (2007) Achilles tendinosis: changes in biochemical composition and collagen turnover rate. Am J Sports Med 35:1549–1556. https://doi.org/10.1177/0363546507301885

Delgado Caceres M, Pfeifer CG, Docheva D (2018) Understanding tendons: lessons from transgenic mouse models. Stem Cells Dev 27:1161–1174. https://doi.org/10.1089/scd.2018.0121

Demaurex N, Furuya W, D'Souza S, Bonifacino JS, Grinstein S (1998) Mechanism of acidification of the trans-Golgi network (TGN). In situ measurements of pH using retrieval of TGN38 and furin from the cell surface. J Biol Chem 273:2044–2051. https://doi.org/10.1074/jbc.273.4.2044

Deng D, Liu W, Xu F, Yang Y, Zhou G, Zhang WJ et al (2009) Engineering human neo-tendon tissue in vitro with human dermal fibroblasts under static mechanical strain. Biomaterials 30:6724–6730. https://doi.org/10.1016/j.biomaterials.2009.08.054

Deporter DA, Ten Cate AR (1973) Fine structural localization of acid and alkaline phosphatase in collagen-containing vesicles of fibroblasts. J Anat 114:457–461

Deymier-Black AC, Pasteris JD, Genin GM, Thomopoulos S (2015) Allometry of the Tendon enthesis: mechanisms of load transfer between tendon and bone. J Biomech Eng 137:111005. https://doi.org/10.1115/1.4031571

DiCesare P, Hauser N, Lehman D, Pasumarti S, Paulsson M (1994) Cartilage oligomeric matrix protein (COMP) is an abundant component of tendon. FEBS Lett 354:237–240. https://doi.org/10.1016/0014-5793

Dideriksen K, Sindby AKR, Krogsgaard M, Schjerling P, Holm L, Langberg H (2013) Effect of acute exercise on patella tendon protein synthesis and gene expression. Springerplus 2:109–109. https://doi.org/10.1186/2193-1801-2-109

Dideriksen K, Boesen AP, Reitelseder S, Couppé C, Svensson R, Schjerling P et al (2017) Tendon collagen synthesis declines with immobilization in elderly humans: no effect of anti-inflammatory medication. J Appl Physiol 1985 122:273–282. https://doi.org/10.1152/japplphysiol.00809.2015

Dieterich DC, Link AJ, Graumann J, Tirrell DA, Schuman EM (2006) Selective identification of newly synthesized proteins in mammalian cells using bioorthogonal noncanonical amino acid tagging (BONCAT). Proc Natl Acad Sci 103:9482–9487. https://doi.org/10.1073/pnas.0601637103

Dill KA (1990) Dominant forces in protein folding. Biochemistry 29:7133–7155. https://doi.org/10.1021/bi00483a001

Disser NP, Sugg KB, Talarek JR, Sarver DC, Rourke BJ, Mendias CL (2019) Insulin-like growth factor 1 signaling in tenocytes is required for adult tendon growth. FASEB J 33:12680–12695. https://doi.org/10.1096/fj.201901503R

Doyle AD (2018) Fluorescent labeling of rat-tail collagen for 3D fluorescence imaging. BIO-Protoc 8. https://doi.org/10.21769/bioprotoc.2919

Dressler MR, Butler DL, Wenstrup R, Awad HA, Smith F, Boivin GP (2002) A potential mechanism for age-related declines in patellar tendon biomechanics. J Orthop Res 20:1315–1322. https://doi.org/10.1016/S0736-0266(02)00052-9

Dudhia J, Scott CM, Draper ERC, Heinegård D, Pitsillides AA, Smith RK (2007) Aging enhances a mechanically-induced reduction in tendon strength by an active process involving matrix metalloproteinase activity. Aging Cell 6:547–556. https://doi.org/10.1111/j.1474-9726.2007.00307.x

Duksin D, Davidson JM, Bornstein P (1978) The role of glycosylation in the enzymatic conversion of procollagen to collagen: studies using tunicamycin and concanavalin A. Arch Biochem Biophys 185:326–332. https://doi.org/10.1016/0003-9861(78)90174-1

Dumont P, Burton M, Chen QM, Gonos ES, Frippiat C, Mazarati JB et al (2000) Induction of replicative senescence biomarkers by sublethal oxidative stresses in normal human fibroblast. Free Radic Biol Med 28:361–373. https://doi.org/10.1016/s0891-5849(99)00249-x

Dunkman AA, Buckley MR, Mienaltowski MJ, Adams SM, Thomas SJ, Satchell L et al (2013) Decorin expression is important for age-related changes in tendon structure and mechanical properties. Matrix Biol 32:3–13. https://doi.org/10.1016/j.matbio.2012.11.005

Dyer RF, Peppler RD (1977) Intracellular collagen in the nonpregnant and IUD-containing rat uterus. Anat Rec 187:241–247. https://doi.org/10.1002/ar.1091870209

Dyment NA, Hagiwara Y, Matthews BG, Li Y, Kalajzic I, Rowe DW (2014) Lineage tracing of resident tendon progenitor cells during growth and natural healing. PLoS One 9:e96113. https://doi.org/10.1371/journal.pone.0096113

Dyment NA, Breidenbach AP, Schwartz AG, Russell RP, Aschbacher-Smith L, Liu H et al (2015) Gdf5 progenitors give rise to fibrocartilage cells that mineralize via hedgehog signaling to form the zonal enthesis. Dev Biol 405:96–107. https://doi.org/10.1016/j.ydbio.2015.06.020

Dyment NA, Barrett JG, Awad HA, Bautista CA, Banes AJ, Butler DL (2020) A brief history of tendon and ligament bioreactors: impact and future prospects. J Orthop Res. https://doi.org/10.1002/jor.24784

Eekhoff JD, Fang F, Kahan LG, Espinosa G, Cocciolone AJ, Wagenseil JE et al (2017) Functionally distinct tendons from elastin haploinsufficient mice exhibit mild stiffening and tendon-specific structural alteration. J Biomech Eng 139. https://doi.org/10.1115/1.4037932

Eekhoff JD, Steenbock H, Berke IM, Brinckmann J, Yanagisawa H, Wagenseil JE et al (2021) Dysregulated assembly of elastic fibers in fibulin-5 knockout mice results in a tendon-specific increase in elastic modulus. J Mech Behav Biomed Mater 113:104134. https://doi.org/10.1016/j.jmbbm.2020.104134

Eikenberry EF, Brodsky BB, Craig AS, Parry DAD (1982) Collagen fibril morphology in developing chick metatarsal tendon: 2. Electron microscope studies. Int J Biol Macromol 4:393–398

Everts V, Beertsen W (1987) The role of microtubules in the phagocytosis of collagen by fibroblasts. Coll Relat Res 7:1–15. https://doi.org/10.1016/s0174-173x(87)80017-1

Everts V, Beertsen W, Tigchelaar-Gutter W (1985) The digestion of phagocytosed collagen is inhibited by the proteinase inhibitors leupeptin and E-64. Coll Relat Res 5:315–336. https://doi.org/10.1016/s0174-173x(85)80021-2

Everts V, Hembry RM, Reynolds JJ, Beertsen W (1989) Metalloproteinases are not involved in the phagocytosis of collagen fibrils by fibroblasts. Matrix 9:266–276. https://doi.org/10.1016/s0934-8832(89)80002-2

Ezura Y, Chakravarti S, Oldberg A, Chervoneva I, Birk DE (2000) Differential expression of lumican and fibromodulin regulate collagen fibrillogenesis in developing mouse tendons. J Cell Biol 151(4):779–788

Fang F, Lake SP (2016) Multiscale mechanical integrity of human supraspinatus tendon in shear after elastin depletion. J Mech Behav Biomed Mater 63:443–455. https://doi.org/10.1016/j.jmbbm.2016.06.032

Fang C, Carlson CS, Leslie MP, Tulli H, Stolerman E, Perris R et al (2000) Molecular cloning, sequencing, and tissue and developmental expression of mouse cartilage oligomeric matrix protein (COMP). J Orthop Res 18:593–603. https://doi.org/10.1002/jor.1100180412

Fang F, Sawhney AS, Lake SP (2014) Different regions of bovine deep digital flexor tendon exhibit distinct elastic, but not viscous, mechanical properties under both compression and shear loading. J Biomech 47:2869–2877. https://doi.org/10.1016/j.jbiomech.2014.07.026

Farup J, Rahbek SK, Vendelbo MH, Matzon A, Hindhede J, Bejder A et al (2014) Whey protein hydrolysate augments tendon and muscle hypertrophy independent of resistance exercise contraction mode. Scand J Med Sci Sports 24:788–798. https://doi.org/10.1111/sms.12083

Fedorczyk JM, Barr AE, Rani S, Gao HG, Amin M, Amin S et al (2010) Exposure-dependent increases in IL-1beta, substance P, CTGF, and tendinosis in flexor digitorum tendons with upper extremity repetitive strain injury. J Orthop Res 28:298–307. https://doi.org/10.1002/jor.20984

Ferguson JS (1912) The behavior and relations of living con-nective tissue cells in the fins of fish embryos with special reference to the histogenesis of the collaginous or white fibers. Am J Anat 13:129

Fernández A (2016) Non-Debye frustrated hydration steers biomolecular association: interfacial tension for the drug designer. FEBS Lett 590:3481–3491

Fessel G, Cadby J, Wunderli S, van Weeren R, Snedeker JG (2014) Dose- and time-dependent effects of genipin crosslinking on cell viability and tissue mechanics – toward clinical application for tendon repair. Acta

Biomater 10:1897–1906. https://doi.org/10.1016/j.actbio.2013.12.048
Finch A, Ledward DA (1972) Shrinkage of collagen fibres: a differential scanning calorimetric study. Biochim Biophys Acta 278:433–439
Fitch JM, Gross J, Mayne R, Johnson-Wint B, Linsenmayer TF (1984) Organization of collagen types I and V in the embryonic chicken cornea: monoclonal antibody studies. Proc Natl Acad Sci USA 81:2791–2795. https://doi.org/10.1073/pnas.81.9.2791
Fleischmajer R, Timpl R, Tuderman L, Raisher L, Wiestner M, Perlish JS et al (1981) Ultrastructural identification of extension aminopropeptides of type I and III collagens in human skin. Proc Natl Acad Sci USA 78:7360–7364. https://doi.org/10.1073/pnas.78.12.7360
Fleischmajer R, Olsen BR, Timpl R, Perlish JS, Lovelace O (1983) Collagen fibril formation during embryogenesis. Proc Natl Acad Sci USA 80:3354–3358. https://doi.org/10.1073/pnas.80.11.3354
Fleischmajer R, Perlish JS, Timpl R (1985) Collagen fibrillogenesis in human skin. Ann NY Acad Sci 460:246–257. https://doi.org/10.1111/j.1749-6632.1985.tb51172.x
Fleischmajer R, Perlish JS, Olsen BR (1987a) Amino and carboxyl propeptides in bone collagen fibrils during embryogenesis. Cell Tissue Res 247:105–109. https://doi.org/10.1007/BF00216552
Fleischmajer R, Perlish JS, Olsen BR (1987b) The carboxylpropeptide of type I procollagen in skin fibrillogenesis. J Invest Dermatol 89:212–215. https://doi.org/10.1111/1523-1747.ep12470949
Fleischmajer R, Perlish JS, Burgeson RE, Shaikh-Bahai F, Timpl R (1990) Type I and type III collagen interactions during fibrillogenesis. Ann NY Acad Sci 580:161–175. https://doi.org/10.1111/j.1749-6632.1990.tb17927.x
Flick J, Devkota A, Tsuzaki M, Almekinders L, Weinhold P (2006) Cyclic loading alters biomechanical properties and secretion of PGE2 and NO from tendon explants. Clin Biomech Bristol Avon 21:99–106. https://doi.org/10.1016/j.clinbiomech.2005.08.008
Flurkey KM, Currer J, Harrison DE (2007) Chapter 20 – mouse models in aging research. In: Fox JG, Davisson MT, Quimby FW, Barthold SW, Newcomer CE, Smith AL (eds) Mouse biomed. res, 2nd edn. Academic, Burlington, pp 637–672
Foidart JM, Berman JJ, Paglia L, Rennard S, Abe S, Perantoni A et al (1980) Synthesis of fibronectin, laminin, and several collagens by a liver-derived epithelial line. Lab Investig 42:525–532
Foidart JB, Foidart JM, Hassell J, Mahieu P (1983) Localization by immunofluorescent microscopy of several collagen types and of a basement membrane proteoglycan in rat glomerular epithelial and mesangial cell cultures. Ren Physiol 6:163–170. https://doi.org/10.1159/000172897
Folgueras AR (2018) Freitas-Rodríguez Sandra, Velasco Gloria, López-Otín Carlos. Mouse models to disentangle the hallmarks of human aging. Circ Res 123:905–924. https://doi.org/10.1161/CIRCRESAHA.118.312204
Fong KD, Trindade MC, Wang Z, Nacamuli RP, Pham H, Fang TD et al (2005) Microarray analysis of mechanical shear effects on flexor tendon cells. Plast Reconstr Surg 116:1393–1404.; discussion 1405-6. https://doi.org/10.1097/01.prs.0000182345.86453.4f
Franchi M, Trire A, Quaranta M, Orsini E, Ottani V (2007) Collagen structure of tendon relates to function. ScientificWorldJournal 7:404–420. https://doi.org/10.1100/tsw.2007.92
Frolova EG, Drazba J, Krukovets I, Kostenko V, Blech L, Harry C et al (2014) Control of organization and function of muscle and tendon by thrombospondin-4. Matrix Biol 37:35–48. https://doi.org/10.1016/j.matbio.2014.02.003
Funakoshi T, Schmid T, Hsu HP, Spector M (2008) Lubricin distribution in the goat infraspinatus tendon: a basis for interfascicular lubrication. J Bone Joint Surg Am 90:803–814. https://doi.org/10.2106/JBJS.G.00627
Fung DT, Wang VM, Andarawis-Puri N, Basta-Pljakic J, Li Y, Laudier DM et al (2010) Early response to tendon fatigue damage accumulation in a novel in vivo model. J Biomech 43:274–279. https://doi.org/10.1016/j.jbiomech.2009.08.039
Gao HGL, Fisher PW, Lambi AG, Wade CK, Barr-Gillespie AE, Popoff SN et al (2013) Increased serum and musculotendinous fibrogenic proteins following persistent low-grade inflammation in a rat model of long-term upper extremity overuse. PLoS One 8:e71875. https://doi.org/10.1371/journal.pone.0071875
Gardner K, Lavagnino M, Egerbacher M, Arnoczky SP (2012) Re-establishment of cytoskeletal tensional homeostasis in lax tendons occurs through an actin-mediated cellular contraction of the extracellular matrix. J Orthop Res 30:1695–1701. https://doi.org/10.1002/jor.22131
Garvin J, Qi J, Maloney M, Banes AJ (2003) Novel system for engineering bioartificial tendons and application of mechanical load. Tissue Eng 9:967–979. https://doi.org/10.1089/107632703322495619
Gay S, Martin GR, Muller PK, Timpl R, Kuhn K (1976) Simultaneous synthesis of types I and III collagen by fibroblasts in culture. Proc Natl Acad Sci USA 73:4037–4040. https://doi.org/10.1073/pnas.73.11.4037
Gehwolf R, Wagner A, Lehner C, Bradshaw AD, Scharler C, Niestrawska JA et al (2016) Pleiotropic roles of the matricellular protein Sparc in tendon maturation and ageing. Sci Rep 6:32635. https://doi.org/10.1038/srep32635
Gillis C, Pool RR, Meagher DM, Stover SM, Reiser K, Willits N (1997) Effect of maturation and aging on the histomorphometric and biochemical characteristics of equine superficial digital flexor tendon. Am J Vet Res 58:425–430
Glazebrook MA, Wright JR Jr, Langman M, Stanish WD, Lee JM (2008) Histological analysis of achilles ten-

dons in an overuse rat model. J Orthop Res 26:840–846. https://doi.org/10.1002/jor.20546
Godinho MSC, Thorpe CT, Greenwald SE, Screen HRC (2017) Elastin is localised to the interfascicular matrix of energy storing tendons and becomes increasingly disorganised with ageing. Sci Rep 7. https://doi.org/10.1038/s41598-017-09995-4
Godinho MS, Thorpe CT, Greenwald SE, Screen HRC (2020) Elastase treatment of tendon specifically impacts the mechanical properties of the interfascicular matrix. BioRxiv 2020(09):18.303081. https://doi.org/10.1101/2020.09.18.303081
Godman GC, Porter KR (1960) Chondrogenesis, studied with the electron microscope. J Biophys Biochem Cytol 8:719–760. https://doi.org/10.1083/jcb.8.3.719
Goldberg B, Green H (1964) An analysis of collagen secretion by established mouse fibroblast lines. J Cell Biol 22:227–258
Goldberg B, Taubman MB, Radin A (1975) Procollagen peptidase: its mode of action on the native substrate. Cell 4:45–50. https://doi.org/10.1016/0092-8674(75)90132-4
Gordon JA, Freedman BR, Zuskov A, Iozzo RV, Birk DE, Soslowsky LJ (2015) Achilles tendons from decorin- and biglycan-null mouse models have inferior mechanical and structural properties predicted by an image-based empirical damage model. J Biomech 48:2110–2115. https://doi.org/10.1016/j.jbiomech.2015.02.058
Graham HK, Holmes DF, Watson RB, Kadler KE (2000) Identification of collagen fibril fusion during vertebrate tendon morphogenesis. The process relies on unipolar fibrils and is regulated by collagen-proteoglycan interaction. J Mol Biol 295:891–902. https://doi.org/10.1006/jmbi.1999.3384
Grammel M, Hang HC (2013) Chemical reporters for biological discovery. Nat Chem Biol 9:475–484. https://doi.org/10.1038/nchembio.1296
Grant TM, Yapp C, Chen Q, Czernuszka JT, Thompson MS (2015) The mechanical, structural, and compositional changes of tendon exposed to elastase. Ann Biomed Eng 43:2477–2486. https://doi.org/10.1007/s10439-015-1308-5
Green H, Hemerman D (1964) Production of hyaluronate and collagen by fibroblast clones in culture. Nature 201:710. https://doi.org/10.1038/201710a0
Grinstein M, Dingwall HL, O'Connor LD, Zou K, Capellini TD, Galloway JL (2019) A distinct transition from cell growth to physiological homeostasis in the tendon. elife 8. https://doi.org/10.7554/elife.48689
Gross J (1956) The behavior of collagen units as a model in morphogenesis. J Biophys Biochem Cytol 2:261–274
Gross J, Kirk D (1958) The heat precipitation of collagen from neutral salt solutions: some rate-regulating factors. J Biol Chem 233:355–360
Gross J, Highberger JH, Schmitt FO (1955) Extraction of collagen from connective tissue by neutral salt solutions. Proc Natl Acad Sci USA 41:1–7
Guerville F, De Souto Barreto P, Ader I, Andrieu S, Casteilla L, Dray C et al (2020) Revisiting the hallmarks of aging to identify markers of biological age. J Prev Alzheimers Dis 7:56–64. https://doi.org/10.14283/jpad.2019.50
Gumucio JP, Schonk MM, Kharaz YA, Comerford E, Mendias CL (2020) Scleraxis is required for the growth of adult tendons in response to mechanical loading. JCI Insight 5. https://doi.org/10.1172/jci.insight.138295
Han W, Wang B, Liu J, Chen L (2017) The p16/miR-217/EGR1 pathway modulates age-related tenogenic differentiation in tendon stem/progenitor cells. Acta Biochim Biophys Sin 49:1015–1021. https://doi.org/10.1093/abbs/gmx104
Hannafin JA, Arnoczky SP, Hoonjan A, Torzilli PA (1995) Effect of stress deprivation and cyclic tensile loading on the material and morphologic properties of canine flexor digitorum profundus tendon: an in vitro study. J Orthop Res 13:907–914. https://doi.org/10.1002/jor.1100130615
Hansen M, Miller BF, Holm L, Doessing S, Petersen SG, Skovgaard D et al (2009) Effect of administration of oral contraceptives in vivo on collagen synthesis in tendon and muscle connective tissue in young women. J Appl Physiol 1985 106:1435–1443. https://doi.org/10.1152/japplphysiol.90933.2008
Harris MT, Butler DL, Boivin GP, Florer JB, Schantz EJ, Wenstrup RJ (2004) Mesenchymal stem cells used for rabbit tendon repair can form ectopic bone and express alkaline phosphatase activity in constructs. J Orthop Res 22:998–1003. https://doi.org/10.1016/j.orthres.2004.02.012
Harvey T, Flamenco S, Fan C-M (2019) A Tppp3+Pdgfra+ tendon stem cell population contributes to regeneration and reveals a shared role for PDGF signalling in regeneration and fibrosis. Nat Cell Biol 21:1490–1503. https://doi.org/10.1038/s41556-019-0417-z
Hast MW, Zuskov A, Soslowsky LJ (2014) The role of animal models in tendon research. Bone Joint Res 3:193–202. https://doi.org/10.1302/2046-3758.36.2000281
Hauser N, Paulsson M, Kale AA, Dicesare PE (1995) Tendon extracellular matrix contains pentameric thrombospondin-4 (TSP-4). FEBS Lett 368:307–310. https://doi.org/10.1016/0014-5793(95)00675-y
Haut RC, Lancaster RL, DeCamp CE (1992) Mechanical properties of the canine patellar tendon: some correlations with age and the content of collagen. J Biomech 25:163–173
Havis E, Bonnin MA, Olivera-Martinez I, Nazaret N, Ruggiu M, Weibel J et al (2014) Transcriptomic analysis of mouse limb tendon cells during development. Development 141:3683–3696. https://doi.org/10.1242/dev.108654
Havis E, Bonnin M-A, Esteves De Lima J, Charvet B, Milet C, Duprez D (2016) TGFβ and FGF promote tendon progenitor fate and act downstream of muscle contraction to regulate tendon differentiation during chick limb development. Development 143:3839–3851. https://doi.org/10.1242/dev.136242

Hay ED, Revel JP (1969) Fine structure of the developing avian cornea. Monogr Dev Biol 1:1–144

Hayashi T, Nagai Y (1972) Factors affecting the interactions of collagen molecules as observed by in vitro fibril formation. I. Effects of small molecules, especially saccharides. J Biochem 72:749–758

Heinemeier K, Langberg H, Olesen JL, Kjaer M (2003) Role of TGF-b1 in relation to exercise-induced type I collagen synthesis in human tendinous tissue. J Appl Physiol 95:2390–2397

Heinemeier KM, Olesen JL, Haddad F, Langberg H, Kjaer M, Baldwin KM et al (2007) Expression of collagen and related growth factors in rat tendon and skeletal muscle in response to specific contraction types. J Physiol 582:1303–1316. https://doi.org/10.1113/jphysiol.2007.127639

Heinemeier KM, Olesen JL, Haddad F, Schjerling P, Baldwin KM, Kjaer M (2009) Effect of unloading followed by reloading on expression of collagen and related growth factors in rat tendon and muscle. J Appl Physiol 1985 106:178–186. https://doi.org/10.1152/japplphysiol.91092.2008

Heinemeier KM, Skovgaard D, Bayer ML, Qvortrup K, Kjaer A, Kjaer M et al (2012) Uphill running improves rat Achilles tendon tissue mechanical properties and alters gene expression without inducing pathological changes. J Appl Physiol 1985 113:827–836. https://doi.org/10.1152/japplphysiol.00401.2012

Heinemeier KM, Schjerling P, Heinemeier J, Magnusson SP, Kjaer M (2013a) Lack of tissue renewal in human adult Achilles tendon is revealed by nuclear bomb 14C. FASEB J 27:2074–2079. https://doi.org/10.1096/fj.12-225599

Heinemeier KM, Bjerrum SS, Schjerling P, Kjaer M (2013b) Expression of extracellular matrix components and related growth factors in human tendon and muscle after acute exercise. Scand J Med Sci Sports 23:e150–e161. https://doi.org/10.1111/j.1600-0838.2011.01414.x

Heinemeier KM, Schjerling P, Øhlenschlæger TF, Eismark C, Olsen J, Kjær M (2018) Carbon-14 bomb pulse dating shows that tendinopathy is preceded by years of abnormally high collagen turnover. FASEB J 32:4763–4775. https://doi.org/10.1096/fj.201701569R

Hekimi S, Lapointe J, Wen Y (2011) Taking a 'good' look at free radicals in the aging process. Trends Cell Biol 21:569–576. https://doi.org/10.1016/j.tcb.2011.06.008

Henkel W, Glanville RW (1982) Covalent crosslinking between molecules of type I and type III collagen. The involvement of the N-terminal, nonhelical regions of the alpha 1 (I) and alpha 1 (III) chains in the formation of intermolecular crosslinks. Eur J Biochem 122:205–213. https://doi.org/10.1111/j.1432-1033.1982.tb05868.x

Henry SP, Takanosu M, Boyd TC, Mayne PM, Eberspaecher H, Zhou W et al (2001) Expression pattern and gene characterization of asporin. J Biol Chem 276:12212–12221. https://doi.org/10.1074/jbc.m011290200

Herchenhan A, Dietrich-Zagonel F, Schjerling P, Kjaer M, Eliasson P (2020) Early growth response genes increases rapidly after mechanical overloading and unloading in tendon constructs. J Orthop Res 38:173–181. https://doi.org/10.1002/jor.24513

Hernandez-Segura A, Nehme J, Demaria M (2018) Hallmarks of cellular senescence. Trends Cell Biol 28:436–453. https://doi.org/10.1016/j.tcb.2018.02.001

Hertzler AE (1910) Pseudoperitoneum, varicosity of the peritoneum and sclerosis of the mesentery: with a preliminary note on development of fibrous tissue. J Am Med Assoc 54:351–356

Hodgins GW, U. S. Department of Justice. Measuring atomic bomb-derived 14C levels in human remains to determine Year of Birth and/or Year of Death. 2009

Hoffmann GA, Wong JY, Smith ML (2019) On force and form: mechano-biochemical regulation of extracellular matrix. Biochemistry 58:4710–4720. https://doi.org/10.1021/acs.biochem.9b00219

Hojima Y, van der Rest M, Prockop DJ (1985) Type I procollagen carboxyl-terminal proteinase from chick embryo tendons. Purification and characterization. J Biol Chem 260:15996–16003

Hojima Y, Morgelin MM, Engel J, Boutillon MM, van der Rest M, McKenzie J et al (1994) Characterization of type I procollagen N-proteinase from fetal bovine tendon and skin. Purification of the 500-kilodalton form of the enzyme from bovine tendon. J Biol Chem 269:11381–11390

Holmes DF, Chapman JA (1979) Axial mass distributions of collagen fibrils grown in vitro: results for the end regions of early fibrils. Biochem Biophys Res Commun 87:993–999. https://doi.org/10.1016/s0006-291x(79)80005-4

Holmes DF, Watson RB, Kadler KE (1991) On the regulation of collagen-fibril shape and form. Biochem Soc Trans 19:808–811. https://doi.org/10.1042/bst0190808

Holmes DF, Watson RB, Steinmann B, Kadler KE (1993) Ehlers-Danlos syndrome type VIIB. Morphology of type I collagen fibrils formed in vivo and in vitro is determined by the conformation of the retained N-propeptide. J Biol Chem 268:15758–15765

Holmes DF, Tait A, Hodson NW, Sherratt MJ, Kadler KE (2010) Growth of collagen fibril seeds from embryonic tendon: fractured fibril ends nucleate new tip growth. J Mol Biol 399:9–16. https://doi.org/10.1016/j.jmb.2010.04.008

Holmes DF, Lu Y, Starborg T, Kadler KE (2018) Collagen fibril assembly and function. Curr Top Dev Biol 130:107–142. https://doi.org/10.1016/bs.ctdb.2018.02.004

Hsiao M-Y, Lin P-C, Liao W-H, Chen W-S, Hsu C-H, He C-K et al (2019) The effect of the repression of oxidative stress on tenocyte differentiation: a preliminary study of a rat cell model using a novel differential tensile strain bioreactor. Int J Mol Sci 20. https://doi.org/10.3390/ijms20143437

Huang D, Chang T, Aggarwal A, Lee R, Ehrlich HP (1993) Mechanisms and dynamics of mechanical strengthening in ligament-equivalent fibroblast-populated collagen matrices. Ann Biomed Eng 21:289–305. https://doi.org/10.1007/BF02368184

Hulmes DJ (1983) A possible mechanism for the regulation of collagen fibril diameter in vivo. Coll Relat Res 3:317–321. https://doi.org/10.1016/s0174-173x(83)80013-2

Hulmes DJ, Miller A, Parry DA, Piez KA, Woodhead-Galloway J (1973) Analysis of the primary structure of collagen for the origins of molecular packing. J Mol Biol 79:137–148

Humphrey JD, Dufresne ER, Schwartz MA (2014) Mechanotransduction and extracellular matrix homeostasis. Nat Rev Mol Cell Biol 15:802–812. https://doi.org/10.1038/nrm3896

Humphries SM, Lu Y, Canty EG, Kadler KE (2008) Active negative control of collagen fibrillogenesis in vivo. Intracellular cleavage of the type I procollagen propeptides in tendon fibroblasts without intracellular fibrils. J Biol Chem 283:12129–12135. https://doi.org/10.1074/jbc.M708198200

Ikeda J, Zhao C, Moran SL, An K-N, Amadio PC (2010) Effects of synovial interposition on healing in a canine tendon explant culture model. J Hand Surg 35:1153–1159. https://doi.org/10.1016/j.jhsa.2010.03.023

Ippolito E, Natali PG, Postacchini F, Accinni L, De Martino C (1980) Morphological, immunochemical, and biochemical study of rabbit achilles tendon at various ages. J Bone Joint Surg Am 62:583–598

Isaacs R (1916) An interpretation of connective tissue and neurogliar fibrillae. Anat Rec Phila X 206

Isaacs R (1919) The structure and mechanics of developing connective tissue. Anat Rec 17:242–270

Izu Y, Ansorge HL, Zhang G, Soslowsky LJ, Bonaldo P, Chu M-L et al (2011) Dysfunctional tendon collagen fibrillogenesis in collagen VI null mice. Matrix Biol 30:53–61. https://doi.org/10.1016/j.matbio.2010.10.001

Izu Y, Adams SM, Connizzo BK, Beason DP, Soslowsky LJ, Koch M et al (2020) Collagen XII mediated cellular and extracellular mechanisms regulate establishment of tendon structure and function. Matrix Biol. https://doi.org/10.1016/j.matbio.2020.10.004

Janvier AJ, Canty-Laird E, Henstock JR (2020) A universal multi-platform 3D printed bioreactor chamber for tendon tissue engineering. J Tissue Eng 11:2041731420942462. https://doi.org/10.1177/2041731420942462

Jeffay H (1960) The metabolism of serum proteins: III. Kinetics of serum protein metabolism during growth. J Biol Chem 235:2352–2356

Jepsen KJ, Wu F, Peragallo JH, Paul J, Roberts L, Ezura Y et al (2002) A syndrome of joint laxity and impaired tendon integrity in lumican- and fibromodulin-deficient mice. J Biol Chem 277:35532–35540. https://doi.org/10.1074/jbc.M205398200

Jimenez SA, Dehm P, Prockop DJ (1971) Further evidence for a transport form of collagen. Its extrusion and extracellular conversion to tropocollagen in embryonic tendon. FEBS Lett 17:245–248. https://doi.org/10.1016/0014-5793(71)80156-4

Johnson C, Galis ZS (2003) Quantitative assessment of collagen assembly by live cells. J Biomed Mater Res A 67:775–784. https://doi.org/10.1002/jbm.a.10136

Jones GC, Corps AN, Pennington CJ, Clark IM, Edwards DR, Bradley MM et al (2006) Expression profiling of metalloproteinases and tissue inhibitors of metalloproteinases in normal and degenerate human achilles tendon. Arthritis Rheum 54:832–842. https://doi.org/10.1002/art.21672

Jun J-I, Lau LF (2017) CCN2 induces cellular senescence in fibroblasts. J Cell Commun Signal 11:15–23. https://doi.org/10.1007/s12079-016-0359-1

Kadler KE, Watson RB (1995) Procollagen C-peptidase: procollagen C-proteinase. Methods Enzymol 248:771–781. https://doi.org/10.1016/0076-6879(95)48052-8

Kadler KE, Hojima Y, Prockop DJ (1987) Assembly of collagen fibrils de novo by cleavage of the type I pC-collagen with procollagen C-proteinase. Assay of critical concentration demonstrates that collagen self-assembly is a classical example of an entropy-driven process. J Biol Chem 262:15696–15701

Kadler KE, Holmes DF, Trotter JA, Chapman JA (1996) Collagen fibril formation. Biochem J 316(Pt 1):1–11. https://doi.org/10.1042/bj3160001

Kadler KE, Holmes DF, Graham H, Starborg T (2000) Tip-mediated fusion involving unipolar collagen fibrils accounts for rapid fibril elongation, the occurrence of fibrillar branched networks in skin and the paucity of collagen fibril ends in vertebrates. Matrix Biol 19:359–365. https://doi.org/10.1016/s0945-053x(00)00082-2

Kadler KE, Hill A, Canty-Laird EG (2008) Collagen fibrillogenesis: fibronectin, integrins, and minor collagens as organizers and nucleators. Curr Opin Cell Biol 20:495–501. https://doi.org/10.1016/j.ceb.2008.06.008

Kalamajski S, Liu C, Tillgren V, Rubin K, Oldberg A, Rai J et al (2014) Increased C-telopeptide cross-linking of tendon type I collagen in fibromodulin-deficient mice. J Biol Chem 289:18873–18879. https://doi.org/10.1074/jbc.m114.572941

Kalson NS, Holmes DF, Herchenhan A, Lu Y, Starborg T, Kadler KE (2011) Slow stretching that mimics embryonic growth rate stimulates structural and mechanical development of tendon-like tissue in vitro. Dev Dyn 240:2520–2528. https://doi.org/10.1002/dvdy.22760

Kalson NS, Starborg T, Lu Y, Mironov A, Humphries SM, Holmes DF et al (2013) Nonmuscle myosin II powered transport of newly formed collagen fibrils at the plasma membrane. Proc Natl Acad Sci USA 110:E4743–E4752. https://doi.org/10.1073/pnas.1314348110

Kalson NS, Lu Y, Taylor SH, Starborg T, Holmes DF, Kadler KE (2015) A structure-based extracellular matrix expansion mechanism of fibrous tissue growth. elife 4. https://doi.org/10.7554/eLife.05958

Kannus P (2000) Structure of the tendon connective tissue. Scand J Med Sci Sports 10:312–320. https://doi.org/10.1034/j.1600-0838.2000.010006312.x

Kannus P, Jozsa L, Järvinen TAH, Järvinen TLN, Kvist M, Natri A et al (1998) Location and distribution of non-collagenous matrix proteins in musculoskeletal tissues of rat. Histochem J 30:799–810. https://doi.org/10.1023/a:1003448106673

Kapacee Z, Richardson SH, Lu Y, Starborg T, Holmes DF, Baar K et al (2008) Tension is required for fibripositor formation. Matrix Biol 27:371–375. https://doi.org/10.1016/j.matbio.2007.11.006

Kauzmann W (1959) Some factors in the interpretation of protein denaturation. Adv Protein Chem 14:1–63

Kendal AR, Layton T, Al-Mossawi H, Appleton L, Dakin S, Brown R et al (2020) Multi-omic single cell analysis resolves novel stromal cell populations in healthy and diseased human tendon. Sci Rep 10. https://doi.org/10.1038/s41598-020-70786-5

Kessler E, Goldberg B (1978) A method for assaying the activity of the endopeptidase which excises the nonhelical carboxyterminal extensions from type I procollagen. Anal Biochem 86:463–469. https://doi.org/10.1016/0003-2697(78)90770-4

Kharaz YA, Canty-Laird EG, Tew SR, Comerford EJ (2018) Variations in internal structure, composition and protein distribution between intra- and extra-articular knee ligaments and tendons. J Anat 232:943–955. https://doi.org/10.1111/joa.12802

Kielty CM, Sherratt MJ, Shuttleworth CA (2002) Elastic fibres. J Cell Sci 115:2817–2828

Killian ML, Thomopoulos S (2016) Scleraxis is required for the development of a functional tendon enthesis. FASEB J 30:301–311. https://doi.org/10.1096/fj.14-258236

Kim A, Lakshman N, Petroll WM (2006) Quantitative assessment of local collagen matrix remodeling in 3-D culture: the role of rho kinase. Exp Cell Res 312:3683–3692. https://doi.org/10.1016/j.yexcr.2006.08.009

Kinugasa R, Hodgson JA, Edgerton VR, Shin DD, Sinha S (2010) Reduction in tendon elasticity from unloading is unrelated to its hypertrophy. J Appl Physiol 109:870–877. https://doi.org/10.1152/japplphysiol.00384.2010

Kjaer M, Langberg H, Miller BF, Boushel R, Crameri R, Koskinen S et al (2005) Metabolic activity and collagen turnover in human tendon in response to physical activity. J Musculoskelet Neuronal Interact 5:41–52

Kjær BH, Juul-Kristensen B, Warming S, Magnusson SP, Krogsgaard MR, Boyle E et al (2020) Associations between shoulder symptoms and concomitant pathology in patients with traumatic supraspinatus tears. JSES Int 4:85–90. https://doi.org/10.1016/j.jses.2019.11.001

Klaips CL, Jayaraj GG, Hartl FU (2018) Pathways of cellular proteostasis in aging and disease. J Cell Biol 217:51–63. https://doi.org/10.1083/jcb.201709072

Klatte-Schulz F, Pauly S, Scheibel M, Greiner S, Gerhardt C, Schmidmaier G et al (2012) Influence of age on the cell biological characteristics and the stimulation potential of male human tenocyte-like cells. Eur Cell Mater 24:74–89. https://doi.org/10.22203/ecm.v024a06

Kohler J, Popov C, Klotz B, Alberton P, Prall WC, Haasters F et al (2013) Uncovering the cellular and molecular changes in tendon stem/progenitor cells attributed to tendon aging and degeneration. Aging Cell 12:988–999. https://doi.org/10.1111/acel.12124

Kohrs RT, Zhao C, Sun Y-L, Jay GD, Zhang L, Warman ML et al (2011) Tendon fascicle gliding in wild type, heterozygous, and lubricin knockout mice. J Orthop Res 29:384–389. https://doi.org/10.1002/jor.21247

Kõks S, Dogan S, Tuna BG, González-Navarro H, Potter P, Vandenbroucke RE (2016) Mouse models of ageing and their relevance to disease. Mech Ageing Dev 160:41–53. https://doi.org/10.1016/j.mad.2016.10.001

Koob TJ, Vogel KG (1987) Proteoglycan synthesis in organ cultures from regions of bovine tendon subjected to different mechanical forces. Biochem J 246:589–598

Koob TJ, Clark PE, Hernandez DJ, Thurmond FA, Vogel KG (1992) Compression loading in vitro regulates proteoglycan synthesis by tendon fibrocartilage. Arch Biochem Biophys 298:303–312

Kopp J, Bonnet M, Renou JP (1990) Effect of collagen crosslinking on collagen-water interactions (a Dsc investigation). Matrix 9:443–450. https://doi.org/10.1016/S0934-8832(11)80013-2

Kostrominova TY, Brooks SV (2013) Age-related changes in structure and extracellular matrix protein expression levels in rat tendons. Age (Dordr) 35:2203–2214. https://doi.org/10.1007/s11357-013-9514-2

Krahn KN, Bouten CVC, Van Tuijl S, Van Zandvoort MAMJ, Merkx M (2006) Fluorescently labeled collagen binding proteins allow specific visualization of collagen in tissues and live cell culture. Anal Biochem 350:177–185. https://doi.org/10.1016/j.ab.2006.01.013

Kubow KE, Vukmirovic R, Zhe L, Klotzsch E, Smith ML, Gourdon D et al (2015) Mechanical forces regulate the interactions of fibronectin and collagen I in extracellular matrix. Nat Commun 6:8026. https://doi.org/10.1038/ncomms9026

Kuo CK, Petersen BC, Tuan RS (2008) Spatiotemporal protein distribution of TGF-βs, their receptors, and extracellular matrix molecules during embryonic tendon development. Dev Dyn 237:1477–1489. https://doi.org/10.1002/dvdy.21547

Kuznetsova N, Chi SL, Leikin S (1998) Sugars and polyols inhibit fibrillogenesis of type I collagen by disrupting hydrogen-bonded water bridges between the helices. Biochemistry 37:11888–11895. https://doi.org/10.1021/bi980089+

Kyriakides TR, Zhu Y-H, Smith LT, Bain SD, Yang Z, Lin MT et al (1998) Mice that lack thrombospondin 2 display connective tissue abnormalities that are associated with disordered collagen fibrillogenesis, an increased vascular density, and a bleeding diathesis. J Cell Biol 140:419–430. https://doi.org/10.1083/jcb.140.2.419

Lake SP, Miller KS, Elliott DM, Soslowsky LJ (2010) Tensile properties and fiber alignment of human supraspinatus tendon in the transverse direction demonstrate inhomogeneity, nonlinearity, and regional isotropy. J Biomech 43:727–732. https://doi.org/10.1016/j.jbiomech.2009.10.017

Langberg H, Skovgaard D, Karamouzis M, Bülow J, Kjær M (1999) Metabolism and inflammatory mediators in the peritendinous space measured by microdialysis during intermittent isometric exercise in humans. J Physiol 515:919–927. https://doi.org/10.1111/j.1469-7793.1999.919ab.x

Langberg H, Rosendal L, Kjær M (2001) Training-induced changes in peritendinous type I collagen turnover determined by microdialysis in humans. J Physiol 534:297–302. https://doi.org/10.1111/j.1469-7793.2001.00297.x

Langberg H, Olesen JL, Gemmer C, Kjær M (2002) Substantial elevation of interleukin-6 concentration in peritendinous tissue, in contrast to muscle, following prolonged exercise in humans. J Physiol 542:985–990. https://doi.org/10.1113/jphysiol.2002.019141

Lapiere CM, Nusgens B (1974) Polymerization of procollagen in vitro. Biochim Biophys Acta 342:237–246. https://doi.org/10.1016/0005-2795(74)90078-6

Lapiere CM, Lenaers A, Kohn LD (1971) Procollagen peptidase: an enzyme excising the coordination peptides of procollagen. Proc Natl Acad Sci USA 68:3054–3058. https://doi.org/10.1073/pnas.68.12.3054

Lavagnino M, Arnoczky SP, Tian T, Vaupel Z (2003) Effect of amplitude and frequency of cyclic tensile strain on the inhibition of MMP-1 mRNA expression in tendon cells: an in vitro study. Connect Tissue Res 44:181–187. https://doi.org/10.1080/03008200390215881

Lavagnino M, Arnoczky SPSP, Frank K, Tian T (2005) Collagen fibril diameter distribution does not reflect changes in the mechanical properties of in vitro stress-deprived tendons. J Biomech 38:69–75. https://doi.org/10.1016/j.jbiomech.2004.03.035

Lavagnino M, Wall ME, Little D, Banes AJ, Guilak F, Arnoczky SP (2015) Tendon mechanobiology: current knowledge and future research opportunities. J Orthop Res 33:813–822. https://doi.org/10.1002/jor.22871

Lavagnino M, Oslapas AN, Gardner KL, Arnoczky SP (2016) Hypoxia inhibits primary cilia formation and reduces cell-mediated contraction in stress-deprived rat tail tendon fascicles. Muscles Ligaments Tendons J 6:193–197. https://doi.org/10.11138/mltj/2016.6.2.193

Lavagnino M, Brooks AE, Oslapas AN, Gardner KL, Arnoczky SP (2017) Crimp length decreases in lax tendons due to cytoskeletal tension, but is restored with tensional homeostasis. J Orthop Res 35:573–579. https://doi.org/10.1002/jor.23489

Lee AH, Elliott DM (2019) Comparative multi-scale hierarchical structure of the tail, plantaris, and Achilles tendons in the rat. J Anat 234:252–262. https://doi.org/10.1111/joa.12913

Lee KJ, Clegg PD, Comerford EJ, Canty-Laird EG (2018) A comparison of the stem cell characteristics of murine tenocytes and tendon-derived stem cells. BMC Musculoskelet Disord 19. https://doi.org/10.1186/s12891-018-2038-2

Legerlotz K, Riley GP, Screen HRC (2013a) GAG depletion increases the stress-relaxation response of tendon fascicles, but does not influence recovery. Acta Biomater 9:6860–6866. https://doi.org/10.1016/j.actbio.2013.02.028

Legerlotz K, Jones GC, Screen HR, Riley GP (2013b) Cyclic loading of tendon fascicles using a novel fatigue loading system increases interleukin-6 expression by tenocytes. Scand J Med Sci Sports 23:31–37. https://doi.org/10.1111/j.1600-0838.2011.01410.x

Leigh DR, Abreu EL, Derwin KA (2008) Changes in gene expression of individual matrix metalloproteinases differ in response to mechanical unloading of tendon fascicles in explant culture. J Orthop Res 26:1306–1312. https://doi.org/10.1002/jor.20650

Leikin S, Rau DC, Parsegian VA (1995) Temperature-favoured assembly of collagen is driven by hydrophilic not hydrophobic interactions. Nat Struct Biol 2:205–210

Leiphart R, Shetye S, Weiss S, Dyment N, Soslowsky LJ (2020) Induced knockdown of decorin, alone and in tandem with biglycan knockdown, directly increases aged tendon viscoelasticity. J Biomech Eng. https://doi.org/10.1115/1.4048030

Lenaers A, Lapiere CM (1975) Type III procollagen and collagen in skin. Biochim Biophys Acta 400:121–131. https://doi.org/10.1016/0005-2795(75)90132-4

Lenaers A, Ansay M, Nusgens BV, Lapiere CM (1971) Collagen made of extended -chains, procollagen, in genetically-defective dermatosparaxic calves. Eur J Biochem 23:533–543. https://doi.org/10.1111/j.1432-1033.1971.tb01651.x

Leung MK, Fessler LI, Greenberg DB, Fessler JH (1979) Separate amino and carboxyl procollagen peptidases in chick embryo tendon. J Biol Chem 254:224–232

Lewis MR (1917) Development of connective-tissue fibers in tissue cultures of chick embryos. Carnegie Institution of Washington, Washington

Li S, Van Den Diepstraten C, D'Souza SJ, Chan BM, Pickering JG (2003) Vascular smooth muscle cells orchestrate the assembly of type I collagen via alpha-2beta1 integrin, RhoA, and fibronectin polymerization. Am J Pathol 163:1045–1056

Li Y, Foss CA, Summerfield DD, Doyle JJ, Torok CM, Dietz HC et al (2012) Targeting collagen strands by photo-triggered triple-helix hybridization. Proc Natl Acad Sci 109:14767–14772. https://doi.org/10.1073/pnas.1209721109

Li Y, Fessel G, Georgiadis M, Snedeker JG (2013) Advanced glycation end-products diminish tendon collagen fiber sliding. Matrix Biol 32:169–177. https://doi.org/10.1016/j.matbio.2013.01.003

Lichtwark GA, Wilson AM (2005) In vivo mechanical properties of the human Achilles tendon during one-legged hopping. J Exp Biol 208:4715–4725. https://doi.org/10.1242/jeb.01950

Lin DJ, Burke CJ, Abiri B, Babb JS, Adler RS (2020) Supraspinatus muscle shear wave elastography (SWE): detection of biomechanical differences with varying tendon quality prior to gray-scale morphologic changes. Skelet Radiol 49:731–738. https://doi.org/10.1007/s00256-019-03334-6

Lindemann I, Coombes BK, Tucker K, Hug F, Dick TJM (2020) Age-related differences in gastrocnemii muscles and Achilles tendon mechanical properties in vivo. J Biomech 112:110067. https://doi.org/10.1016/j.jbiomech.2020.110067

Linsenmayer TF, Fitch JM, Gross J, Mayne R (1985) Are collagen fibrils in the developing avian cornea composed of two different collagen types? Evidence from monoclonal antibody studies. Ann NY Acad Sci 460:232–245. https://doi.org/10.1111/j.1749-6632.1985.tb51171.x

Linsenmayer TF, Fitch JM, Birk DE (1990) Heterotypic collagen fibrils and stabilizing collagens. Controlling elements in corneal morphogenesis? Ann NY Acad Sci 580:143–160

Listgarten MA (1973) Intracellular collagen fibrils in the periodontal ligament of the mouse, rat, hamster, guinea pig and rabbit. J Periodontal Res 8:335–342. https://doi.org/10.1111/j.1600-0765.1973.tb00767.x

Liu X, Wu H, Byrne M, Krane S, Jaenisch R (1997) Type III collagen is crucial for collagen I fibrillogenesis and for normal cardiovascular development. Proc Natl Acad Sci USA 94:1852–1856. https://doi.org/10.1073/pnas.94.5.1852

Liu Y, Andarawis-Puri N, Eppell SJ (2016) Method to extract minimally damaged collagen fibrils from tendon. J Biol Methods 3(4):e54. https://doi.org/10.14440/jbm.2016.121

Longo UG, Berton A, Khan WS, Maffulli N, Denaro V (2011) Histopathology of rotator cuff tears. Sports Med Arthrosc Rev 19:227–236. https://doi.org/10.1097/JSA.0b013e318213bccb

López-Otín C, Blasco MA, Partridge L, Serrano M, Kroemer G (2013) The hallmarks of aging. Cell 153:1194–1217. https://doi.org/10.1016/j.cell.2013.05.039

Lu Y, Kamel-El Sayed SA, Wang K, Tiede-Lewis LM, Grillo MA, Veno PA et al (2018) Live imaging of type I collagen assembly dynamics in osteoblasts stably expressing GFP and mCherry-tagged collagen constructs. J Bone Miner Res 33:1166–1182. https://doi.org/10.1002/jbmr.3409

Luescher M, Ruegg M, Schindler P (1974) Effect of hydration upon the thermal stability of tropocollagen and its dependence on the presence of neutral salts. Biopolymers 13:2489–2503. https://doi.org/10.1002/bip.1974.360131208

Lui PPY, Wong CM (2019) Biology of tendon stem cells and tendon in aging. Front Genet 10:1338. https://doi.org/10.3389/fgene.2019.01338

Luo Y, Li N, Chen H, Fernandez GE, Warburton D, Moats R et al (2018) Spatial and temporal changes in extracellular elastin and laminin distribution during lung alveolar development. Sci Rep 8. https://doi.org/10.1038/s41598-018-26673-1

Maccarana M, Svensson RB, Knutsson A, Giannopoulos A, Pelkonen M, Weis M et al (2017) Asporin-deficient mice have tougher skin and altered skin glycosaminoglycan content and structure. PLoS One 12:e0184028. https://doi.org/10.1371/journal.pone.0184028

Maeda T, Sakabe T, Sunaga A, Sakai K, Rivera AL, Keene DR et al (2011) Conversion of mechanical force into TGF-beta-mediated biochemical signals. Curr Biol 21:933–941. https://doi.org/10.1016/j.cub.2011.04.007

Maeda E, Ye S, Wang W, Bader DL, Knight MM, Lee DA (2012) Gap junction permeability between tenocytes within tendon fascicles is suppressed by tensile loading. Biomech Model Mechanobiol 11:439–447. https://doi.org/10.1007/s10237-011-0323-1

Magnusson SP, Kjaer M (2003) Region-specific differences in Achilles tendon cross-sectional area in runners and non-runners. Eur J Appl Physiol 90:549–553. https://doi.org/10.1007/s00421-003-0865-8

Magnusson SP, Kjaer M (2019) The impact of loading, unloading, ageing and injury on the human tendon. J Physiol 597:1283–1298. https://doi.org/10.1113/JP275450

Majima T, Marchuk LL, Sciore P, Shrive NG, Frank CB, Hart DA (2000) Compressive compared with tensile loading of medial collateral ligament scar in vitro uniquely influences mRNA levels for aggrecan, collagen type II, and collagenase. J Orthop Res 18:524–531. https://doi.org/10.1002/jor.1100180403

Makisalo SE, Paavolainen PP, Lehto M, Skutnabb K, Slatis P (1989) Collagen types I and III and fibronectin in healing anterior cruciate ligament after reconstruction with carbon fibre. Injury 20:72–76. https://doi.org/10.1016/0020-1383(89)90143-5

Malfait F, De Paepe A (2014) The Ehlers-Danlos syndrome. Adv Exp Med Biol 802:129–143. https://doi.org/10.1007/978-94-007-7893-1_9

Mall FP (1902) On the development of the connective tissues from the connective-tissue syncytium. Am J Anat 1:329–365

Mallory FB (1903) A hitherto undescribed fibrillar substance produced by connective-tissue cells. J Med Res 10:334–341

Marchant JK, Hahn RA, Linsenmayer TF, Birk DE (1996) Reduction of type V collagen using a dominant-negative strategy alters the regulation of fibrillogenesis and results in the loss of corneal-specific fibril morphology. J Cell Biol 135:1415–1426. https://doi.org/10.1083/jcb.135.5.1415

Marchi F, Leblond CP (1983) Collagen biogenesis and assembly into fibrils as shown by ultrastructural and 3H-proline radioautographic studies on the fibroblasts of the rat food pad. Am J Anat 168:167–197. https://doi.org/10.1002/aja.1001680206

Marchi F, Leblond CP (1984) Radioautographic characterization of successive compartments along the rough endoplasmic reticulum-Golgi pathway of collagen precursors in foot pad fibroblasts of [3H]proline-injected rats. J Cell Biol 98:1705–1709. https://doi.org/10.1083/jcb.98.5.1705

Markova DZ, Pan T-C, Zhang R-Z, Zhang G, Sasaki T, Arita M et al (2016) Forelimb contractures and abnormal tendon collagen fibrillogenesis in fibulin-4 null mice. Cell Tissue Res 364:637 646. https://doi.org/10.1007/s00441-015-2346-x

Maroudas A, Bayliss MT, Uchitel-Kaushansky N, Schneiderman R, Gilav E (1998) Aggrecan turnover in human articular cartilage: use of aspartic acid racemization as a marker of molecular age. Arch Biochem Biophys 350:61–71. https://doi.org/10.1006/abbi.1997.0492

Marqueti R, Durigan JL, Oliveira A, Mekaro M, Guzzoni V, Aro A et al (2017) Effects of aging and resistance training in rat tendon remodeling. FASEB J 32:fj.201700543R. https://doi.org/10.1096/fj.201700543R

Marsolais D, Duchesne É, Côté CH, Frenette J (2007) Inflammatory cells do not decrease the ultimate tensile strength of intact tendons in vivo and in vitro: protective role of mechanical loading. J Appl Physiol 102:11–17. https://doi.org/10.1152/japplphysiol.00162.2006

Martin CL, Bergman MR, Deravi LF, Paten JA (2020) A role for monosaccharides in nucleation inhibition and transport of collagen. Bioelectricity 2(2):186–197

Matsumoto F, Trudel G, Uhthoff HK, Backman DS (2003) Mechanical effects of immobilization on the Achilles' tendon. Arch Phys Med Rehabil 84:662–667. https://doi.org/10.1016/s0003-9993(02)04834-7

Maximow A (1928) Development of argyrophile and collagenous fibers in tissue cultures. Proc Soc Exp Biol Med 25:439–442

May T, Garmel GM (2020) Rotator cuff injury. In: StatPearls. StatPearls Publishing, Treasure Island

McDonald JA, Kelley DG, Broekelmann TJ (1982) Role of fibronectin in collagen deposition: fab' to the gelatin-binding domain of fibronectin inhibits both fibronectin and collagen organization in fibroblast extracellular matrix. J Cell Biol 92:485–492. https://doi.org/10.1083/jcb.92.2.485

McLaughlin JS, Linsenmayer TF, Birk DE (1989) Type V collagen synthesis and deposition by chicken embryo corneal fibroblasts in vitro. J Cell Sci 94(Pt 2):371–379

McLaughlin PJ, Chen Q, Horiguchi M, Starcher BC, Stanton JB, Broekelmann TJ et al (2006) Targeted disruption of fibulin-4 abolishes elastogenesis and causes perinatal lethality in mice. Mol Cell Biol 26:1700–1709. https://doi.org/10.1128/mcb.26.5.1700-1709.2006

McNeill AR (2002) Tendon elasticity and muscle function. Comp Biochem Physiol A Mol Integr Physiol 133:1001–1011. https://doi.org/10.1016/S1095-6433(02)00143-5

Michna H (1984) Morphometric analysis of loading-induced changes in collagen-fibril populations in young tendons. Cell Tissue Res 236:465–470. https://doi.org/10.1007/bf00214251

Mienaltowski MJ, Birk DE (2014) Structure, physiology, and biochemistry of collagens. Adv Exp Med Biol 802:5–29. https://doi.org/10.1007/978-94-007-7893-1_2

Mienaltowski MJ, Dunkman AA, Buckley MR, Beason DP, Adams SM, Birk DE et al (2016) Injury response of geriatric mouse patellar tendons. J Orthop Res 34:1256–1263. https://doi.org/10.1002/jor.23144

Mienaltowski MJ, Cánovas A, Fates VA, Hampton AR, Pechanec MY, Islas-Trejo A et al (2018) Transcriptome profiles of isolated murine achilles tendon proper- and peritenon- derived progenitor cells. J Orthop Res 37:1409–1418. https://doi.org/10.1002/jor.24076

Miles CA, Ghelashvili M (1999) Polymer-in-a-box mechanism for the thermal stabilization of collagen molecules in fibers. Biophys J 76:3243–3252. https://doi.org/10.1016/S0006-3495(99)77476-X

Miller BF, Olesen JL, Hansen M, Døssing S, Crameri RM, Welling RJ et al (2005) Coordinated collagen and muscle protein synthesis in human patella tendon and quadriceps muscle after exercise. J Physiol 567:1021–1033. https://doi.org/10.1113/jphysiol.2005.093690

Miller KS, Connizzo BK, Feeney E, Tucker JJ, Soslowsky LJ (2012a) Examining differences in local collagen fiber crimp frequency throughout mechanical testing in a developmental mouse supraspinatus tendon model. J Biomech Eng 134:041004. https://doi.org/10.1115/1.4006538

Miller KS, Connizzo BK, Soslowsky LJ (2012b) Collagen fiber re-alignment in a neonatal developmental mouse supraspinatus tendon model. Ann Biomed Eng 40:1102–1110. https://doi.org/10.1007/s10439-011-0490-3

Millis AJ, McCue HM, Kumar S, Baglioni C (1992) Metalloproteinase and TIMP-1 gene expression during replicative senescence. Exp Gerontol 27:425–428. https://doi.org/10.1016/0531-5565(92)90076-c

Minagawa H, Yamamoto N, Abe H, Fukuda M, Seki N, Kikuchi K et al (2013) Prevalence of symptomatic and asymptomatic rotator cuff tears in the general population: from mass-screening in one village. J Orthop 10:8–12. https://doi.org/10.1016/j.jor.2013.01.008

Mirigian LS, Makareeva E, Leikin S (2014) Pulse-chase analysis of procollagen biosynthesis by azidohomoalanine labeling. Connect Tissue Res 55:403–410. https://doi.org/10.3109/03008207.2014.959120

Mori N, Majima T, Iwasaki N, Kon S, Miyakawa K, Kimura C et al (2007) The role of osteopontin in tendon tissue remodeling after denervation-induced mechanical stress deprivation. Matrix Biol 26:42–53. https://doi.org/10.1016/j.matbio.2006.09.002

Morikawa D, Itoigawa Y, Nojiri H, Sano H, Itoi E, Saijo Y et al (2014) Contribution of oxidative stress to the degeneration of rotator cuff entheses. J Shoulder Elb Surg 23:628–635. https://doi.org/10.1016/j.jse.2014.01.041

Morris JL, Cross SJ, Lu Y, Kadler KE, Lu Y, Dallas SL et al (2018) Live imaging of collagen deposition during skin development and repair in a collagen I – GFP fusion transgenic zebrafish line. Dev Biol 441:4–11. https://doi.org/10.1016/j.ydbio.2018.06.001

Mubyana K, Corr DT (2018) Cyclic uniaxial tensile strain enhances the mechanical properties of engineered, scaffold-free tendon fibers. Tissue Eng Part A 24:1808–1817. https://doi.org/10.1089/ten.TEA.2018.0028

Murano S, Thweatt R, Shmookler Reis RJ, Jones RA, Moerman EJ, Goldstein S (1991) Diverse gene sequences are overexpressed in werner syndrome fibroblasts undergoing premature replicative senescence. Mol Cell Biol 11:3905–3914. https://doi.org/10.1128/mcb.11.8.3905

Murphy DJ, Nixon AJ (1997) Biochemical and site-specific effects of insulin-like growth factor I on intrinsic tenocyte activity in equine flexor tendons. Am J Vet Res 58:103–109

N'Diaye E-N, Cook R, Wang H, Wu P, LaCanna R, Wu C et al (2020) Extracellular BMP1 is the major proteinase for C-terminal proteolysis of type I procollagen in lung fibroblasts. Am J Phys Cell Physiol. https://doi.org/10.1152/ajpcell.00012.2020

Na GC (1989) Monomer and oligomer of type I collagen: molecular properties and fibril assembly. Biochemistry 28:7161–7167

Nakagawa Y, Majima T, Nagashima K (1994) Effect of ageing on ultrastructure of slow and fast skeletal muscle tendon in rabbit Achilles tendons. Acta Physiol Scand 152:307–313. https://doi.org/10.1111/j.1748-1716.1994.tb09810.x

Nakamura T, Lozano PR, Ikeda Y, Iwanaga Y, Hinek A, Minamisawa S et al (2002) Fibulin-5/DANCE is essential for elastogenesis in vivo. Nature 415:171–175. https://doi.org/10.1038/415171a

Ng TK, Chen C-B, Xu C, Xu Y, Yao X, Huang L et al (2020) Attenuated regenerative properties in human periodontal ligament-derived stem cells of older donor ages with shorter telomere length and lower SSEA4 expression. Cell Tissue Res 381:71–81. https://doi.org/10.1007/s00441-020-03176-y

Nielsen HM, Skalicky M, Viidik A (1998) Influence of physical exercise on aging rats. III. Life-long exercise modifies the aging changes of the mechanical properties of limb muscle tendons. Mech Ageing Dev 100:243–260

Nirmalanandhan VS, Rao M, Sacks MS, Haridas B, Butler DL (2007) Effect of length of the engineered tendon construct on its structure–function relationships in culture. J Biomech 40:2523–2529. https://doi.org/10.1016/j.jbiomech.2006.11.016

Njieha FK, Morikawa T, Tuderman L, Prockop DJ (1982) Partial purification of a procollagen C-proteinase. Inhibition by synthetic peptides and sequential cleavage of type I procollagen. Biochemistry 21:757–764. https://doi.org/10.1021/bi00533a028

Noda H, Wyckoff RW (1951) The electron microscopy of reprecipitated collagen. Biochim Biophys Acta 7:494–506

Nowack H, Gay S, Wick G, Becker U, Timpl R (1976) Preparation and use in immunohistology of antibodies specific for type I and type III collagen and procollagen. J Immunol Methods 12:117–124. https://doi.org/10.1016/0022-1759(76)90101-0

Nugent GE, Aneloski NM, Schmidt TA, Schumacher BL, Voegtline MS, Sah RL (2006) Dynamic shear stimulation of bovine cartilage biosynthesis of proteoglycan 4. Arthritis Rheum 54:1888–1896. https://doi.org/10.1002/art.21831

Nurminskaya MV, Birk DE (1998) Differential expression of genes associated with collagen fibril growth in the chicken tendon: identification of structural and regulatory genes by subtractive hybridization. Arch Biochem Biophys 350:1–9. https://doi.org/10.1006/abbi.1997.0498

O'Brien C, Marr N, Thorpe C (2020) Microdamage in the equine superficial digital flexor tendon. Equine Vet J. https://doi.org/10.1111/evj.13331

Olesen JL, Heinemeier KM, Haddad F, Langberg H, Flyvbjerg A, Kjaer M et al (2006) Expression of insulin-like growth factor I, insulin-like growth factor binding proteins, and collagen mRNA in mechanically loaded plantaris tendon. J Appl Physiol 1985 101:183–188. https://doi.org/10.1152/japplphysiol.00636.2005

Oryan A, Shoushtari AH (2008) Histology and ultrastructure of the developing superficial digital flexor tendon in rabbits. Anat Histol Embryol 37:134–140. https://doi.org/10.1111/j.1439-0264.2007.00811.x

Pan XS, Li J, Brown EB, Kuo CK (2018) Embryo movements regulate tendon mechanical property development. Philos Trans R Soc B Biol Sci 373:20170325. https://doi.org/10.1098/rstb.2017.0325

Paolillo C, Londin E, Fortina P (2019) Single-cell genomics. Clin Chem 65:972–985. https://doi.org/10.1373/clinchem.2017.283895

Pardes AM, Beach ZM, Raja H, Rodriguez AB, Freedman BR, Soslowsky LJ (2017) Aging leads to inferior Achilles tendon mechanics and altered ankle function in rodents. J Biomech 60:30–38. https://doi.org/10.1016/j.jbiomech.2017.06.008

Parry DA, Craig AS (1977) Quantitative electron microscope observations of the collagen fibrils in rat-tail tendon. Biopolymers 16:1015–1031. https://doi.org/10.1002/bip.1977.360160506

Parry DA, Craig AS (1978) Collagen fibrils and elastic fibers in rat-tail tendon: an electron microscopic investigation. Biopolymers 17:843–845. https://doi.org/10.1002/bip.1978.360170404

Parry DA, Craig AS (1984) Growth and development of collagen fibrils in connective tissue. In: Ultrastructure of the connective tissue matrix. Springer, The Hague, pp 34–64

Parry DA, Craig AS, Barnes GR (1978a) Tendon and ligament from the horse: an ultrastructural study of collagen fibrils and elastic fibres as a function of age. Proc R Soc Lond B Biol Sci 203:293–303. https://doi.org/10.1098/rspb.1978.0106

Parry DA, Barnes GR, Craig AS (1978b) A comparison of the size distribution of collagen fibrils in connective tissues as a function of age and a possible relation between fibril size distribution and mechanical properties. Proc R Soc Lond B Biol Sci 203:305–321

Patel SH, Yue F, Saw SK, Foguth R, Cannon JR, Shannahan JH et al (2019) Advanced glycation end-products suppress mitochondrial function and proliferative capacity of Achilles tendon-derived fibroblasts. Sci Rep 9:12614. https://doi.org/10.1038/s41598-019-49062-8

Paten JA, Siadat SM, Susilo ME, Ismail EN, Stoner JL, Rothstein JP et al (2016) Flow-induced crystallization of collagen: a potentially critical mechanism in early tissue formation. ACS Nano 10:5027–5040. https://doi.org/10.1021/acsnano.5b07756
Paten JA, Martin CL, Wanis JT, Siadat SM, Figueroa-Navedo AM, Ruberti JW et al (2019) Molecular interactions between collagen and fibronectin: a reciprocal relationship that regulates De Novo Fibrillogenesis. Chemistry 5:2126–2145. https://doi.org/10.1016/j.chempr.2019.05.011
Patterson-Kane JC, Firth EC, Goodship AE, Parry DA (1997) Age-related differences in collagen crimp patterns in the superficial digital flexor tendon core region of untrained horses. Aust Vet J 75:39–44. https://doi.org/10.1111/j.1751-0813.1997.tb13829.x
Peach R, Williams G, Chapman JA (1961) Alight and electron optical study of regenerating tendon. Am J Pathol 38:495–513
Peffers MJ, Thorpe CT, Collins JA, Eong R, Wei TKJ, Screen HRC et al (2014) Proteomic analysis reveals age-related changes in tendon matrix composition, with age- and injury-specific matrix fragmentation. J Biol Chem 289:25867–25878. https://doi.org/10.1074/jbc.M114.566554
Peffers M, Fang Y, Cheung K, Wei T, Clegg P, Birch H (2015) Transcriptome analysis of ageing in uninjured human Achilles tendon. Arthritis Res Ther 17:33. https://doi.org/10.1186/s13075-015-0544-2
Petroll WM, Ma L (2003) Direct, dynamic assessment of cell-matrix interactions inside fibrillar collagen lattices. Cell Motil Cytoskeleton 55:254–264. https://doi.org/10.1002/cm.10126
Pickard A, Chang J, Alachkar N, Calverley B, Garva R, Arvan P et al (2019) Preservation of circadian rhythms by the protein folding chaperone, BiP. FASEB J 33:7479–7489. https://doi.org/10.1096/fj.201802366RR
Ploetz C, Zycband EI, Birk DE (1991) Collagen fibril assembly and deposition in the developing dermis: segmental deposition in extracellular compartments. J Struct Biol 106:73–81. https://doi.org/10.1016/1047-8477(91)90064-4
Popov C, Kohler J, Docheva D (2017) Activation of EphA4 and EphB2 reverse signaling restores the age-associated reduction of self-renewal, migration, and actin turnover in human tendon stem/progenitor cells. Front Aging Neurosci 7. https://doi.org/10.3389/fnagi.2015.00246
Porter KR, Pappas GD (1959) Collagen formation by fibroblasts of the chick embryo dermis. J Biophys Biochem Cytol 5:153–166
Privalov PL, Tiktopulo EI, Tischenko VM (1979) Stability and mobility of the collagen structure. J Mol Biol 127:203–216
Prockop DJ, Kivirikko KI, Tuderman L, Guzman NA (1979a) The biosynthesis of collagen and its disorders (first of two parts). N Engl J Med 301:13–23. https://doi.org/10.1056/NEJM197907053010104
Prockop DJ, Kivirikko KI, Tuderman L, Guzman NA (1979b) The biosynthesis of collagen and its disorders (second of two parts). N Engl J Med 301:77–85. https://doi.org/10.1056/NEJM197907123010204
Pryce BA, Brent AE, Murchison ND, Tabin CJ, Schweitzer R (2007) Generation of transgenic tendon reporters, ScxGFP and ScxAP, using regulatory elements of the scleraxis gene. Dev Dyn 236:1677–1682. https://doi.org/10.1002/dvdy.21179
Quigley AS, Bancelin S, Deska-Gauthier D, Légaré F, Kreplak L, Veres SP (2018) In tendons, differing physiological requirements lead to functionally distinct nanostructures. Sci Rep 8. https://doi.org/10.1038/s41598-018-22741-8
Rajpar I, Barrett JG (2019) Optimizing growth factor induction of tenogenesis in three-dimensional culture of mesenchymal stem cells. J Tissue Eng 10. https://doi.org/10.1177/2041731419848776
Ramachandran GN, Chandrasekharan R (1968) Interchain hydrogen bonds via bound water molecules in the collagen triple helix. Biopolymers 6:1649–1658. https://doi.org/10.1002/bip.1968.360061109
Ramachandran GN, Bansal M, Bhatnagar RS (1973) A hypothesis on the role of hydroxyproline in stabilizing collagen structure. Biochim Biophys Acta 322:166–171
Rebelo-Marques A, De Sousa Lages A, Andrade R, Ribeiro CF, Mota-Pinto A, Carrilho F et al (2018) Aging hallmarks: the benefits of physical exercise. Front Endocrinol 9. https://doi.org/10.3389/fendo.2018.00258
Rees SG, Davies JR, Tudor D, Flannery CR, Hughes CE, Dent CM et al (2002) Immunolocalisation and expression of proteoglycan 4 (cartilage superficial zone proteoglycan) in tendon. Matrix Biol 21:593–602. https://doi.org/10.1016/s0945-053x(02)00056-2
Revel JP, Hay ED (1963) An autoradiographic and electron microscopic study of collagen synthesis in differentiating cartilage. Z Zellforsch Mikrosk Anat 61:110–144
Ribitsch I, Gueltekin S, Keith MF, Minichmair K, Peham C, Jenner F et al (2020) Age-related changes of tendon fibril micro-morphology and gene expression. J Anat 236:688–700. https://doi.org/10.1111/joa.13125
Rigozzi S, Müller R, Stemmer A, Snedeker JG (2013) Tendon glycosaminoglycan proteoglycan sidechains promote collagen fibril sliding-AFM observations at the nanoscale. J Biomech 46:813–818. https://doi.org/10.1016/j.jbiomech.2012.11.017
Riley GP, Harrall RL, Constant CR, Chard MD, Cawston TE, Hazleman BL (1994a) Tendon degeneration and chronic shoulder pain: changes in the collagen composition of the human rotator cuff tendons in rotator cuff tendinitis. Ann Rheum Dis 53:359–366. https://doi.org/10.1136/ard.53.6.359
Riley GP, Harrall RL, Constant CR, Chard MD, Cawston TE, Hazleman BL (1994b) Glycosaminoglycans of human rotator cuff tendons: changes with age and in chronic rotator cuff tendinitis. Ann Rheum Dis 53:367–376

Ritty TM, Ditsios K, Starcher BC (2002) Distribution of the elastic fiber and associated proteins in flexor tendon reflects function. Anat Rec 268:430–440. https://doi.org/10.1002/ar.10175

Robbins JR, Evanko SP, Vogel KG (1997) Mechanical loading and TGF-beta regulate proteoglycan synthesis in tendon. Arch Biochem Biophys 342:203–211. https://doi.org/10.1006/abbi.1997.0102

Robinson PS, Huang TF, Kazam E, Iozzo RV, Birk DE, Soslowsky LJ (2005) Influence of decorin and biglycan on mechanical properties of multiple tendons in knockout mice. J Biomech Eng 127:181–185. https://doi.org/10.1115/1.1835363

Robinson KA, Sun M, Barnum CE, Weiss SN, Huegel J, Shetye SS et al (2017) Decorin and biglycan are necessary for maintaining collagen fibril structure, fiber realignment, and mechanical properties of mature tendons. Matrix Biol 64:81–93. https://doi.org/10.1016/j.matbio.2017.08.004

Romanic AM, Adachi E, Kadler KE, Hojima Y, Prockop DJ (1991) Copolymerization of pNcollagen III and collagen I. pNcollagen III decreases the rate of incorporation of collagen I into fibrils, the amount of collagen I incorporated, and the diameter of the fibrils formed. J Biol Chem 266:12703–12709

Rooney SI, Loro E, Sarver JJ, Peltz CD, Hast MW, Tseng W-J et al (2014) Exercise protocol induces muscle, tendon, and bone adaptations in the rat shoulder. Muscles Ligaments Tendons J 4:413–419

Rooney SI, Tobias JW, Bhatt PR, Kuntz AF, Soslowsky LJ (2015) Genetic response of rat supraspinatus tendon and muscle to exercise. PLoS One 10:e0139880. https://doi.org/10.1371/journal.pone.0139880

Rooney SI, Torino DJ, Baskin R, Vafa RP, Kuntz AF, Soslowsky LJ (2017) Rat supraspinatus tendon responds acutely and chronically to exercise. J Appl Physiol 1985 123:757–763. https://doi.org/10.1152/japplphysiol.00368.2017

Ross R, Benditt EP (1961) Wound healing and collagen formation. I. Sequential changes in components of guinea pig skin wounds observed in the electron microscope. J Biophys Biochem Cytol 11:677–700. https://doi.org/10.1083/jcb.11.3.677

Ross R, Benditt EP (1962) Wound healing and collagen formation. III. A quantitative radioautographic study of the utilization of proline-H3 in wounds from normal and scorbutic guinea pigs. J Cell Biol 15:99–108. https://doi.org/10.1083/jcb.15.1.99

Rowson D, Knight MM, Screen HR (2016) Zonal variation in primary cilia elongation correlates with localized biomechanical degradation in stress deprived tendon. J Orthop Res 34:2146–2153. https://doi.org/10.1002/jor.23229

Rumian AP, Draper ER, Wallace AL, Goodship AE (2009) The influence of the mechanical environment on remodelling of the patellar tendon. J Bone Joint Surg (Br) 91:557–564. https://doi.org/10.1302/0301-620x.91b4.21580

Russo V, Mauro A, Martelli A, Di Giacinto O, Di Marcantonio L, Nardinocchi D et al (2015) Cellular and molecular maturation in fetal and adult ovine calcaneal tendons. J Anat 226:126–142. https://doi.org/10.1111/joa.12269

Ruzzini L, Abbruzzese F, Rainer A, Longo UG, Trombetta M, Maffulli N et al (2014) Characterization of age-related changes of tendon stem cells from adult human tendons. Knee Surg Sports Traumatol Arthrosc 22:2856–2866. https://doi.org/10.1007/s00167-013-2457-4

Saadat F, Deymier AC, Birman V, Thomopoulos S, Genin GM (2016) The concentration of stress at the rotator cuff tendon-to-bone attachment site is conserved across species. J Mech Behav Biomed Mater 62:24–32. https://doi.org/10.1016/j.jmbbm.2016.04.025

Samiric T, Ilic MZ, Handley CJ (2004) Large aggregating and small leucine-rich proteoglycans are degraded by different pathways and at different rates in tendon. Eur J Biochem 271:3612–3620. https://doi.org/10.1111/j.0014-2956.2004.04307.x

Samiric T, Ilic MZ, Handley CJ (2006) Sulfated polysaccharides inhibit the catabolism and loss of both large and small proteoglycans in explant cultures of tendon. FEBS J 273:3479–3488. https://doi.org/10.1111/j.1742-4658.2006.05348.x

Schiele NR, Koppes RA, Chrisey DB, Corr DT (2013) Engineering cellular fibers for musculoskeletal soft tissues using directed self-assembly. Tissue Eng Part A 19:1223–1232. https://doi.org/10.1089/ten.tea.2012.0321

Schmitt FO, Hall CE, Jakus MA (1942) Electron microscope investigations of the structure of collagen. J Cell Comp Physiol 20:11–33. https://doi.org/10.1002/jcp.1030200103

Schönherr E, Witsch-Prehm P, Harrach B, Robenek H, Rauterberg J, Kresse H (1995) Interaction of biglycan with type I collagen. J Biol Chem 270:2776–2783. https://doi.org/10.1074/jbc.270.6.2776

Schulz J-N, Nüchel J, Niehoff A, Bloch W, Schönborn K, Hayashi S et al (2016) COMP-assisted collagen secretion – a novel intracellular function required for fibrosis. J Cell Sci 129:706–716. https://doi.org/10.1242/jcs.180216

Schwann T (1839) Mikroskopische Untersuchungen. Sander Berl 268

Schwann TH (1847) Microscopical researches into the accordance in the structure and growth of animals and plants. Рипол Классик

Schweitzer R, Chyung JH, Murtaugh LC, Brent AE, Rosen V, Olson EN et al (2001) Analysis of the tendon cell fate using Scleraxis, a specific marker for tendons and ligaments. Dev Camb Engl 128:3855–3866

Scott JE (1980) Collagen – proteoglycan interactions. Localization of proteoglycans in tendon by electron microscopy. Biochem J 187:887–891. https://doi.org/10.1042/bj1870887

Scott JE (1984) The periphery of the developing collagen fibril. Quantitative relationships with dermatan sulphate and other surface-associated species. Biochem J 218:229–233

Scott JE (1990) Proteoglycan:collagen interactions and subfibrillar structure in collagen fibrils. Implications in the development and ageing of connective tissues. J Anat 169:23–35

Scott JE, Hughes EW (1986) Proteoglycan-collagen relationships in developing chick and bovine tendons. Influence of the physiological environment. Connect Tissue Res 14:267–278

Scott JE, Orford CR (1981) Dermatan sulphate-rich proteoglycan associates with rat tail-tendon collagen at the d band in the gap region. Biochem J 197:213–216

Scott JE, Orford CR, Hughes EW (1981) Proteoglycan-collagen arrangements in developing rat tail tendon. An electron microscopical and biochemical investigation. Biochem J 195:573–581

Screen HRC, Shelton JC, Bader DL, Lee DA (2005a) Cyclic tensile strain upregulates collagen synthesis in isolated tendon fascicles. Biochem Biophys Res Commun 336:424–429. https://doi.org/10.1016/j.bbrc.2005.08.102

Screen HRC, Shelton JC, Chhaya VH, Kayser MV, Bader DL, Lee DA (2005b) The influence of noncollagenous matrix components on the micromechanical environment of tendon fascicles. Ann Biomed Eng 33:1090–1099. https://doi.org/10.1007/s10439-005-5777-9

Segev F, Heon E, Cole WG, Wenstrup RJ, Young F, Slomovic AR et al (2006) Structural abnormalities of the cornea and lid resulting from collagen V mutations. Invest Ophthalmol Vis Sci 47:565–573. https://doi.org/10.1167/iovs.05-0771

Shadwick RE (1990) Elastic energy storage in tendons: mechanical differences related to function and age. J Appl Physiol Bethesda Md 1985 68:1033–1040

Shapiro SD, Endicott SK, Province MA, Pierce JA, Campbell EJ (1991) Marked longevity of human lung parenchymal elastic fibers deduced from prevalence of D-aspartate and nuclear weapons-related radiocarbon. J Clin Invest 87:1828–1834. https://doi.org/10.1172/JCI115204

Sheldon H, Kimball FB (1962) Studies on cartilage. III. The occurrence of collagen within vacuoles of the golgi apparatus. J Cell Biol 12:599–613. https://doi.org/10.1083/jcb.12.3.599

Shepherd JH, Legerlotz K, Demirci T, Klemt C, Riley GP, Screen HRC (2014) Functionally distinct tendon fascicles exhibit different creep and stress relaxation behaviour. Proc Inst Mech Eng H 228:49–59. https://doi.org/10.1177/0954411913509977

Sherratt MJ (2009) Tissue elasticity and the ageing elastic fibre. Age (Dordr) 31:305–325. https://doi.org/10.1007/s11357-009-9103-6

Shoshan S, Segal N, Traub W, Salem G, Kuhn K, Lapiere CM (1974) Normal characteristics of dermatosparactic calf skin collagen fibers following their subcutaneous implantation within a diffusion chamber into a normal calf. FEBS Lett 41:269–274. https://doi.org/10.1016/0014-5793(74)81227-5

Siadat SM et al. (2021a) Measuring collagen fibril diameter with differential interference contrast microscopy. J Struc Biol 213(1):107697. https://doi.org/10.1016/j.jsb.2021.107697

Siadat SM et al. (2021b) Development and validation of fluorescently labelled, functional type I collagen molecules. bioRxiv 2021.03.26.437209. https://doi.org/10.1101/2021.03.26.437209

Sivakumar P (2006) New insights into extracellular matrix assembly and reorganization from dynamic imaging of extracellular matrix proteins in living osteoblasts. J Cell Sci 119:1350–1360. https://doi.org/10.1242/jcs.02830

Sivan SS, Tsitron E, Wachtel E, Roughley PJ, Sakkee N, van der Ham F et al (2006) Aggrecan turnover in human intervertebral disc as determined by the racemization of aspartic acid. J Biol Chem 281:13009–13014. https://doi.org/10.1074/jbc.M600296200

Sivan S-S, Wachtel E, Tsitron E, Sakkee N, van der Ham F, DeGroot J et al (2008) Collagen turnover in normal and degenerate human intervertebral discs as determined by the racemization of aspartic acid. J Biol Chem 283:8796–8801. https://doi.org/10.1074/jbc.M709885200

Skovgaard D, Kjaer A, Heinemeier KM, Brandt-Larsen M, Madsen J, Kjaer M (2011) Use of cis-[18F] fluoro-proline for assessment of exercise-related collagen synthesis in musculoskeletal connective tissue. PLoS One 6:e16678. https://doi.org/10.1371/journal.pone.0016678

Smeets JSJ, Horstman AMH, Vles GF, Emans PJ, Goessens JPB, Gijsen AP et al (2019) Protein synthesis rates of muscle, tendon, ligament, cartilage, and bone tissue in vivo in humans. PLoS One 14:e0224745. https://doi.org/10.1371/journal.pone.0224745

Smith RKW, Zunino L, Webbon PM, Heinegård D (1997) The distribution of Cartilage Oligomeric Matrix Protein (COMP) in tendon and its variation with tendon site, age and load. Matrix Biol 16:255–271. https://doi.org/10.1016/s0945-053x(97)90014-7

Smith RKW, Gerard M, Dowling B, Dart AJ, Birch HL, Goodship AE (2002) Correlation of cartilage oligomeric matrix protein (COMP) levels in equine tendon with mechanical properties: a proposed role for COMP in determining function-specific mechanical characteristics of locomotor tendons. Equine Vet J 34:241–244. https://doi.org/10.1111/j.2042-3306.2002.tb05426.x

Smith SM, Thomas CE, Birk DE (2012) Pericellular proteins of the developing mouse tendon: a proteomic analysis. Connect Tissue Res 53:2–13. https://doi.org/10.3109/03008207.2011.602766

Smith SM, Zhang G, Birk DE (2014) Collagen V localizes to pericellular sites during tendon collagen fibrillogenesis. Matrix Biol 33:47–53. https://doi.org/10.1016/j.matbio.2013.08.003

Södersten F, Ekman S, Schmitz M, Paulsson M, Zaucke F (2006) Thrombospondin-4 and cartilage oligomeric matrix protein form heterooligomers in equine tendon. Connect Tissue Res 47:85–91. https://doi.org/10.1080/03008200600584124

Soeda T, Deng JM, de Crombrugghe B, Behringer RR, Nakamura T, Akiyama H (2010) Sox9-expressing precursors are the cellular origin of the cruciate ligament of the knee joint and the limb tendons. Genes N Y N 2000 48:635–644. https://doi.org/10.1002/dvg.20667

Soslowsky LJ, Fryhofer GW (2016) Tendon homeostasis in hypercholesterolemia. Adv Exp Med Biol 920:151–165. https://doi.org/10.1007/978-3-319-33943-6_14

Sottile J, Hocking DC (2002) Fibronectin polymerization regulates the composition and stability of extracellular matrix fibrils and cell-matrix adhesions. Mol Biol Cell 13:3546–3559. https://doi.org/10.1091/mbc.e02-01-0048

Southall NT, Dill KA, Haymet ADJ (2002) A view of the hydrophobic effect. J Phys Chem B 106:521–533. https://doi.org/10.1021/jp015514e

Spiesz EM, Thorpe CT, Chaudhry S, Riley GP, Birch HL, Clegg PD et al (2015) Tendon extracellular matrix damage, degradation and inflammation in response to in vitro overload exercise. J Orthop Res 33:889–897. https://doi.org/10.1002/jor.22879

Stammers M, Ivanova IM, Niewczas IS, Segonds-Pichon A, Streeter M, Spiegel DA et al (2020) Age-related changes in the physical properties, cross-linking, and glycation of collagen from mouse tail tendon. J Biol Chem 295:10562–10571. https://doi.org/10.1074/jbc.RA119.011031

Starborg T, Kalson NS, Lu Y, Mironov A, Cootes TF, Holmes DF et al (2013) Using transmission electron microscopy and 3View to determine collagen fibril size and three-dimensional organization. Nat Protoc 8:1433–1448. https://doi.org/10.1038/nprot.2013.086

Stearns ML (1940a) Studies on the development of connective tissue in transparent chambers in the rabbit's ear II. Am J Anat 67:55–97

Stearns ML (1940b) Studies on the development of connective tissue in transparent chambers in the rabbit's ear I. Am J Anat 66:133–176

Stopak D, Wessells NK, Harris AK (1985) Morphogenetic rearrangement of injected collagen in developing chicken limb buds. Proc Natl Acad Sci 82:2804–2808. https://doi.org/10.1073/pnas.82.9.2804

Streeter I, de Leeuw NH (2011) A molecular dynamics study of the interprotein interactions in collagen fibrils. Soft Matter 7:3373–3382. https://doi.org/10.1039/C0SM01192D

Strocchi R, De Pasquale V, Guizzardi S, Govoni P, Facchini A, Raspanti M et al (1991) Human Achilles tendon: morphological and morphometric variations as a function of age. Foot Ankle 12:100–104

Sugiyama Y, Naito K, Goto K, Kojima Y, Furuhata A, Igarashi M et al (2019) Effect of aging on the tendon structure and tendon-associated gene expression in mouse foot flexor tendon. Biomed Rep 10:238–244. https://doi.org/10.3892/br.2019.1200

Sullivan BE, Carroll CC, Jemiolo B, Trappe SW, Magnusson SP, Døssing S et al (2009) Effect of acute resistance exercise and sex on human patellar tendon structural and regulatory mRNA expression. J Appl Physiol 1985 106:468–475. https://doi.org/10.1152/japplphysiol.91341.2008

Sun Y, Berger EJ, Zhao C, Jay GD, An K-N, Amadio PC (2006) Expression and mapping of lubricin in canine flexor tendon. J Orthop Res 24:1861–1868. https://doi.org/10.1002/jor.20239

Sun Y-L, Wei Z, Zhao C, Jay GD, Schmid TM, Amadio PC et al (2015a) Lubricin in human achilles tendon: the evidence of intratendinous sliding motion and shear force in achilles tendon. J Orthop Res 33:932–937. https://doi.org/10.1002/jor.22897

Sun M, Connizzo BK, Adams SM, Freedman BR, Wenstrup RJ, Soslowsky LJ et al (2015b) Targeted deletion of collagen V in tendons and ligaments results in a classic Ehlers-Danlos syndrome joint phenotype. Am J Pathol 185:1436–1447. https://doi.org/10.1016/j.ajpath.2015.01.031

Sun M, Luo EY, Adams SM, Adams T, Ye Y, Shetye SS et al (2020) Collagen XI regulates the acquisition of collagen fibril structure, organization and functional properties in tendon. Matrix Biol. https://doi.org/10.1016/j.matbio.2020.09.001

Svensson L, Närlid I, Oldberg Å (2000) Fibromodulin and lumican bind to the same region on collagen type I fibrils. FEBS Lett 470:178–182. https://doi.org/10.1016/s0014-5793(00)01314-4

Swovick K, Welle KA, Hryhorenko JR, Seluanov A, Gorbunova V, Ghaemmaghami S (2018) Cross-species comparison of proteome turnover kinetics. Mol Cell Proteomics 17:580–591. https://doi.org/10.1074/mcp.RA117.000574

Symoens S, Syx D, Malfait F, Callewaert B, De Backer J, Vanakker O et al (2012) Comprehensive molecular analysis demonstrates type V collagen mutations in over 90% of patients with classic EDS and allows to refine diagnostic criteria. Hum Mutat 33:1485–1493. https://doi.org/10.1002/humu.22137

Szczesny SE, Elliott DM (2014) Interfibrillar shear stress is the loading mechanism of collagen fibrils in tendon. Acta Biomater 10:2582–2590. https://doi.org/10.1016/j.actbio.2014.01.032

Taguchi M, Sun Y-L, Zhao C, Zobitz ME, Cha C-J, Jay GD et al (2009) Lubricin surface modification improves tendon gliding after tendon repair in a canine model in vitro. J Orthop Res 27:257–263. https://doi.org/10.1002/jor.20731

Taye N, Karoulias SZ, Hubmacher D (2020) The 'other' 15-40%: the role of non-collagenous extracellular matrix proteins and minor collagens in tendon. J Orthop Res 38:23–35. https://doi.org/10.1002/jor.24440

Taylor SH, Yeung C-YC, Kalson NS, Lu Y, Zigrino P, Starborg T et al (2015) Matrix metalloproteinase 14 is required for fibrous tissue expansion. elife 4:e09345. https://doi.org/10.7554/eLife.09345

Ten Cate AR (1972) Morphological studies of fibrocytes in connective tissue undergoing rapid remodelling. J Anat 112:401–414

Ten Cate AR, Deporter DA (1974) The role of the fibroblast in collagen turnover in the functioning periodontal ligament of the mouse. Arch Oral Biol 19:339–340. https://doi.org/10.1016/0003-9969(74)90199-x

Ten Cate AR, Deporter DA (1975) The degradative role of the fibroblast in the remodelling and turnover of collagen in soft connective tissue. Anat Rec 182:1–13. https://doi.org/10.1002/ar.1091820102

Ten Cate AR, Freeman E (1974) Collagen remodelling by fibroblasts in wound repair. Preliminary observations. Anat Rec 179:543–546. https://doi.org/10.1002/ar.1091790414

Thankam FG, Chandra IS, Kovilam AN, Diaz CG, Volberding BT, Dilisio MF et al (2018) Amplification of mitochondrial activity in the healing response following rotator cuff tendon injury. Sci Rep 8:17027. https://doi.org/10.1038/s41598-018-35391-7

Theodossiou SK, Schiele NR (2019) Models of tendon development and injury. BMC Biomed Eng 1. https://doi.org/10.1186/s42490-019-0029-5

Theodossiou TA, Thrasivoulou C, Ekwobi C, Becker DL (2006) Second harmonic generation confocal microscopy of collagen type I from rat tendon cryosections. Biophys J 91:4665–4677. https://doi.org/10.1529/biophysj.106.093740

Theodossiou SK, Bozeman AL, Burgett N, Brumley MR, Swann HE, Raveling AR et al (2019) Onset of neonatal locomotor behavior and the mechanical development of Achilles and tail tendons. J Biomech 96. https://doi.org/10.1016/j.jbiomech.2019.109354

Thomopoulos S, Williams GR, Gimbel JA, Favata M, Soslowsky LJ (2003) Variation of biomechanical, structural, and compositional properties along the tendon to bone insertion site. J Orthop Res 21:413–419. https://doi.org/10.1016/S0736-0266(03)00057-3

Thomopoulos S, Genin GM, Galatz LM (2010) The development and morphogenesis of the tendon-to-bone insertion what development can teach us about healing.pdf. J Musculoskelet Neuronal Interact 10:35–45

Thomopoulos S, Parks WC, Rifkin DB, Derwin KA (2015) Mechanisms of tendon injury and repair. J Orthop Res 33:832–839. https://doi.org/10.1002/jor.22806

Thompson MS, Bajuri MN, Khayyeri H, Isaksson H (2017) Mechanobiological modelling of tendons: review and future opportunities. Proc Inst Mech Eng H 231:369–377. https://doi.org/10.1177/0954411917692010

Thornton GM, Hart DA (2011) The interface of mechanical loading and biological variables as they pertain to the development of tendinosis. J Musculoskelet Neuronal Interact 11:94–105

Thornton GM, Lemmex DB, Ono Y, Beach CJ, Reno CR, Hart DA et al (2015) Aging affects mechanical properties and lubricin/PRG4 gene expression in normal ligaments. J Biomech 48:3306–3311. https://doi.org/10.1016/j.jbiomech.2015.06.005

Thorpe CT (2010) Extracellular matrix synthesis and degradation in functionally distinct tendons. University College London, London

Thorpe CT, Screen HRC (2016) Tendon structure and composition. Adv Exp Med Biol 920:3–10. https://doi.org/10.1007/978-3-319-33943-6_1

Thorpe CT, Streeter I, Pinchbeck GL, Goodship AE, Clegg PD, Birch HL (2010) Aspartic acid racemization and collagen degradation markers reveal an accumulation of damage in tendon collagen that is enhanced with aging. J Biol Chem 285:15674–15681. https://doi.org/10.1074/jbc.M109.077503

Thorpe CT, Birch HL, Clegg PD, Screen HRC (2013a) The role of the non-collagenous matrix in tendon function. Int J Exp Pathol 94:248–259. https://doi.org/10.1111/iep.12027

Thorpe CT, Udeze CP, Birch HL, Clegg PD, Screen HR (2013b) Capacity for sliding between tendon fascicles decreases with ageing in injury prone equine tendons: a possible mechanism for age-related tendinopathy? Eur Cell Mater 25:48–60

Thorpe CT, Godinho MSC, Riley GP, Birch HL, Clegg PD, Screen HRC (2015a) The interfascicular matrix enables fascicle sliding and recovery in tendon, and behaves more elastically in energy storing tendons. J Mech Behav Biomed Mater 52:85–94. https://doi.org/10.1016/j.jmbbm.2015.04.009

Thorpe CT, Chaudhry S, Lei, Varone A, Riley GP, Birch HL et al (2015b) Tendon overload results in alterations in cell shape and increased markers of inflammation and matrix degradation. Scand J Med Sci Sports 25:e381–e391. https://doi.org/10.1111/sms.12333

Thorpe CT, Peffers MJ, Simpson D, Halliwell E, Screen HRC, Clegg PD (2016a) Anatomical heterogeneity of tendon: fascicular and interfascicular tendon compartments have distinct proteomic composition. Sci Rep 6:20455. https://doi.org/10.1038/srep20455

Thorpe CT, McDermott BT, Goodship AE, Clegg PD, Birch HL (2016b) Ageing does not result in a decline in cell synthetic activity in an injury prone tendon. Scand J Med Sci Sports 26:684–693. https://doi.org/10.1111/sms.12500

Thorpe CT, Riley GP, Birch HL, Clegg PD, Screen HRC (2017) Fascicles and the interfascicular matrix show decreased fatigue life with ageing in energy storing tendons. Acta Biomater 56:58–64. https://doi.org/10.1016/j.actbio.2017.03.024

Tiktopulo EI, Kajava AV (1998) Denaturation of type I collagen fibrils is an endothermic process accompanied by a noticeable change in the partial heat capacity. Biochemistry 37:8147–8152. https://doi.org/10.1021/bi980360n

Timpl R, Glanville RW, Nowack H, Wiedemann H, Fietzek PP, Kuhn K (1975) Isolation, chemical and electron microscopical characterization of neutral-salt-soluble type III collagen and procollagen from fetal bovine skin. Hoppe Seylers Z Physiol Chem 356:1783–1792. https://doi.org/10.1515/bchm2.1975.356.2.1783

Tohidnezhad M, Zander J, Slowik A, Kubo Y, Dursun G, Willenberg W et al (2020) Impact of uniaxial stretching on both gliding and traction areas of tendon explants in a novel bioreactor. Int J Mol Sci 21. https://doi.org/10.3390/ijms21082925

Tourell MC, Momot KI (2016) Molecular dynamics of a hydrated collagen peptide: insights into rotational motion and residence times of single-water bridges in collagen. J Phys Chem B 120:12432–12443. https://doi.org/10.1021/acs.jpcb.6b08499

Tran PHT, Skrba T, Wondimu E, Galatioto G, Svensson RB, Olesen AT et al (2019) The influence of fibrillin-1 and physical activity upon tendon tissue morphology and mechanical properties in mice. Phys Rep 7. https://doi.org/10.14814/phy2.14267

Trelstad RL (1971) Vacuoles in the embryonic chick corneal epithelium, an epithelium which produces collagen. J Cell Biol 48:689–694

Trelstad RL, Hayashi K (1979) Tendon collagen fibrillogenesis: intracellular subassemblies and cell surface changes associated with fibril growth. Dev Biol 71:228–242. https://doi.org/10.1016/0012-1606(79)90166-0

Trelstad RL, Birk DE, Silver FH (1982) Collagen fibrillogenesis in tissues, in a solution and from modeling: a synthesis. J Invest Dermatol 79(Suppl 1):109s–112s. https://doi.org/10.1111/1523-1747.ep12545945

Trivanović D, Jauković A, Popović B, Krstić J, Mojsilović S, Okić-Djordjević I et al (2015) Mesenchymal stem cells of different origin: comparative evaluation of proliferative capacity, telomere length and pluripotency marker expression. Life Sci 141:61–73. https://doi.org/10.1016/j.lfs.2015.09.019

Trotter JA, Chapman JA, Kadler KE, Holmes DF (1998) Growth of sea cucumber collagen fibrils occurs at the tips and centers in a coordinated manner. J Mol Biol 284:1417–1424. https://doi.org/10.1006/jmbi.1998.2230

Trotter JA, Kadler KE, Holmes DF (2000) Echinoderm collagen fibrils grow by surface-nucleation-and-propagation from both centers and ends. J Mol Biol 300:531–540. https://doi.org/10.1006/jmbi.2000.3879

Tsai WC, Hsu CC, Chang HN, Lin YC, Lin MS, Pang JH (2010) Ibuprofen upregulates expressions of matrix metalloproteinase-1, −8, −9, and −13 without affecting expressions of types I and III collagen in tendon cells. J Orthop Res 28:487–491. https://doi.org/10.1002/jor.21009

Tsuzaki M, Guyton G, Garrett W, Archambault JM, Herzog W, Almekinders L et al (2003) IL-1 beta induces COX2, MMP-1, −3 and −13, ADAMTS-4, IL-1 beta and IL-6 in human tendon cells. J Orthop Res 21:256–264. https://doi.org/10.1016/S0736-0266(02)00141-9

Ueda Y, Inui A, Mifune Y, Takase F, Kataoka T, Kurosawa T et al (2019) Molecular changes to tendons after collagenase-induced acute tendon injury in a senescence-accelerated mouse model. BMC Musculoskelet Disord 20:120. https://doi.org/10.1186/s12891-019-2488-1

Van Gulick L, Saby C, Morjani H, Beljebbar A (2019) Age-related changes in molecular organization of type I collagen in tendon as probed by polarized SHG and Raman microspectroscopy. Sci Rep 9:7280. https://doi.org/10.1038/s41598-019-43636-2

Van Remmen H, Ikeno Y, Hamilton M, Pahlavani M, Wolf N, Thorpe SR et al (2003) Life-long reduction in MnSOD activity results in increased DNA damage and higher incidence of cancer but does not accelerate aging. Physiol Genomics 16:29–37. https://doi.org/10.1152/physiolgenomics.00122.2003

van Vijven M, Wunderli SL, Ito K, Snedeker JG, Foolen J (2020) Serum deprivation limits loss and promotes recovery of tenogenic phenotype in tendon cell culture systems. J Orthop Res. https://doi.org/10.1002/jor.24761

Vanamee P, Porter KR (1951) Observations with the electron microscope on the solvation and reconstitution of collagen. J Exp Med 94:255–266

Veis A, Anesey J, Yuan L, Levy SJ (1973) Evidence for an amino-terminal extension in high-molecular-weight collagens from mature bovine skin. Proc Natl Acad Sci USA 70:1464–1467. https://doi.org/10.1073/pnas.70.5.1464

Verzijl N, DeGroot J, Thorpe SR, Bank RA, Shaw JN, Lyons TJ et al (2000) Effect of collagen turnover on the accumulation of advanced glycation end products. J Biol Chem 275:39027–39031. https://doi.org/10.1074/jbc.M006700200

Voelz H (1964) The 'spindle-shaped body' in fibroblasts. J Cell Biol 20:333–337. https://doi.org/10.1083/jcb.20.2.333

Vogel HG (1980) Influence of maturation and aging on mechanical and biochemical properties of connective tissue in rats. Mech Ageing Dev 14:283–292

Vogel KG, Hernandez DJ (1992) The effects of transforming growth factor-beta and serum on proteoglycan synthesis by tendon fibrocartilage. Eur J Cell Biol 59:304–313

Wagenseil JE, Ciliberto CH, Knutsen RH, Levy MA, Kovacs A, Mecham RP (2010) The importance of elastin to aortic development in mice. Am J Physiol Heart Circ Physiol 299:H257–H264. https://doi.org/10.1152/ajpheart.00194.2010

Wall ME, Banes AJ (2005) Early responses to mechanical load in tendon: role for calcium signaling, gap junctions and intercellular communication. J Musculoskelet Neuronal Interact 5:70–84

Wall ME, Dyment NA, Bodle J, Volmer J, Loboa E, Cederlund A et al (2016) Cell signaling in tenocytes: response to load and ligands in health and disease. Adv Exp Med Biol 920:79–95. https://doi.org/10.1007/978-3-319-33943-6_7

Wall M, Butler D, El Haj A, Bodle JC, Loboa EG, Banes AJ (2018) Key developments that impacted the field of mechanobiology and mechanotransduction. J Orthop Res 36:605–619. https://doi.org/10.1002/jor.23707

Wang MX, Wei A, Yuan J, Clippe A, Bernard A, Knoops B et al (2001) Antioxidant enzyme peroxiredoxin 5 is upregulated in degenerative human tendon. Biochem Biophys Res Commun 284:667–673. https://doi.org/10.1006/bbrc.2001.4991

Wang T, Lin Z, Day RE, Gardiner B, Landao-Bassonga E, Rubenson J et al (2013a) Programmable mechanical stimulation influences tendon homeostasis in a bioreactor system. Biotechnol Bioeng 110:1495–1507. https://doi.org/10.1002/bit.24809

Wang T, Gardiner BS, Lin Z, Rubenson J, Kirk TB, Wang A et al (2013b) Bioreactor design for tendon/ligament engineering. Tissue Eng Part B Rev 19:133–146. https://doi.org/10.1089/ten.teb.2012.0295

Wassermann F (1954) Fibrillogenesis in the regenerating rat tendon with special reference to growth and composition of the collagenous fibril. Am J Anat 94:399–437. https://doi.org/10.1002/aja.1000940304

Waters DW, Blokland KEC, Pathinayake PS, Burgess JK, Mutsaers SE, Prele CM et al (2018) Fibroblast senescence in the pathology of idiopathic pulmonary fibrosis. Am J Phys Lung Cell Mol Phys 315:L162–L172. https://doi.org/10.1152/ajplung.00037.2018

Watson RB, Wallis GA, Holmes DF, Viljoen D, Byers PH, Kadler KE (1992) Ehlers Danlos syndrome type VIIB. Incomplete cleavage of abnormal type I procollagen by N-proteinase in vitro results in the formation of copolymers of collagen and partially cleaved pNcollagen that are near circular in cross-section. J Biol Chem 267:9093–9100

Watson RB, Holmes DF, Graham HK, Nusgens BV, Kadler KE (1998) Surface located procollagen N-propeptides on dermatosparactic collagen fibrils are not cleaved by procollagen N-proteinase and do not inhibit binding of decorin to the fibril surface. J Mol Biol 278:195–204. https://doi.org/10.1006/jmbi.1998.1680

Welsh RA (1966) Intracytoplasmic collagen formations in desmoid fibromatosis. Am J Pathol 49:515–535

Wenstrup RJ, Florer JB, Brunskill EW, Bell SM, Chervoneva I, Birk DE (2004) Type V collagen controls the initiation of collagen fibril assembly. J Biol Chem 279:53331–53337. https://doi.org/10.1074/jbc.M409622200

Wenstrup RJ, Florer JB, Davidson JM, Phillips CL, Pfeiffer BJ, Menezes DW et al (2006) Murine model of the Ehlers-Danlos syndrome. col5a1 haploinsufficiency disrupts collagen fibril assembly at multiple stages. J Biol Chem 281:12888–12895. https://doi.org/10.1074/jbc.M511528200

Wenstrup RJ, Smith SM, Florer JB, Zhang G, Beason DP, Seegmiller RE et al (2011) Regulation of collagen fibril nucleation and initial fibril assembly involves coordinate interactions with collagens V and XI in developing tendon. J Biol Chem 286:20455–20465. https://doi.org/10.1074/jbc.m111.223693

Wertz J, Galli M, Borchers JR (2013) Achilles tendon rupture. Sports Health 5:407–409. https://doi.org/10.1177/1941738112472165

Wilkinson DJ (2018) Historical and contemporary stable isotope tracer approaches to studying mammalian protein metabolism. Mass Spectrom Rev 37:57–80. https://doi.org/10.1002/mas.21507

Willett TL, Labow RS, Avery NC, Lee JM (2007) Increased proteolysis of collagen in an in vitro tensile overload tendon model. Ann Biomed Eng 35:1961–1972. https://doi.org/10.1007/s10439-007-9375-x

Willett TL, Labow RS, Lee JM (2008) Mechanical overload decreases the thermal stability of collagen in an in vitro tensile overload tendon model. J Orthop Res 26:1605–1610. https://doi.org/10.1002/jor.20672

Williams BR, Gelman RA, Poppke DC, Piez KA (1978) Collagen fibril formation. Optimal in vitro conditions and preliminary kinetic results. J Biol Chem 253:6578–6585

Wolbach SB (1933) Controlled formation of collagen and reticulum. A study of the source of intercellular substance in recovery from experimental scorbutus. Am J Pathol 9:689–700

Wolbach SB, Howe PR (1926) Intercellular Substances in Experimental Scorbutus. Arch Pathol Lab Med 1:1

Wong MWN, Lui WT, Fu SC, Lee KM (2009) The effect of glucocorticoids on tendon cell viability in human tendon explants. Acta Orthop 80:363–367. https://doi.org/10.3109/17453670902988386

Wood GC (1964) The precipitation of collagen fibers from solution. Int Rev Connect Tissue Res 2:1–31. https://doi.org/10.1016/b978-1-4831-6751-0.50007-0

Wood GC, Keech MK (1960) The formation of fibrils from collagen solutions. 1. The effect of experimental conditions: kinetic and electron-microscope studies. Biochem J 75:588–598. https://doi.org/10.1042/bj0750588

Wren TA, Beaupré GS, Carter DR (2000) Mechanobiology of tendon adaptation to compressive loading through fibrocartilaginous metaplasia. J Rehabil Res Dev 37:135–143

Wu YF, Wang HK, Chang HW, Sun J, Sun JS, Chao YH (2017) High glucose alters tendon homeostasis through downregulation of the AMPK/Egr1 pathway. Sci Rep 7:44199. https://doi.org/10.1038/srep44199

Wunderli SL, Widmer J, Amrein N, Foolen J, Silvan U, Leupin O et al (2017) Minimal mechanical load and tissue culture conditions preserve native cell phenotype and morphology in tendon – a novel ex vivo mouse explant model. J Orthop Res. https://doi.org/10.1002/jor.23769

Wunderli SL, Widmer J, Amrein N, Foolen J, Silvan U, Leupin O et al (2018) Minimal mechanical load and tissue culture conditions preserve native cell phenotype and morphology in tendon – a novel ex vivo mouse explant model. J Orthop Res 36:1383–1390. https://doi.org/10.1002/jor.23769

Wunderli SL, Blache U, Snedeker JG (2020) Tendon explant models for physiologically relevant invitro study of tissue biology – a perspective. Connect Tissue Res 61:262–277. https://doi.org/10.1080/03008207.2019.1700962

Xia Z, Xing Y, Jeon J, Kim Y-P, Gall J, Dragulescu-Andrasi A et al (2011) Immobilizing reporters for molecular imaging of the extracellular microenvironment in living animals. ACS Chem Biol 6:1117–1126. https://doi.org/10.1021/cb200135e

Yan Z, Yin H, Brochhausen C, Pfeifer CG, Alt V, Docheva D (2020) Aged tendon stem/progenitor cells are less competent to form 3D tendon organoids due to cell autonomous and matrix production deficits. Front Bioeng Biotechnol 8:406. https://doi.org/10.3389/fbioe.2020.00406

Yanagisawa H, Davis EC, Starcher BC, Ouchi T, Yanagisawa M, Richardson JA et al (2002) Fibulin-5 is an elastin-binding protein essential for elastic fibre development in vivo. Nature 415:168–171. https://doi.org/10.1038/415168a

Yang GC, Birk DE (1986) Topographies of extracytoplasmic compartments in developing chick tendon fibroblasts. J Ultrastruct Mol Struct Res 97:238–248. https://doi.org/10.1016/s0889-1605(86)80023-4
Yang G, Im HJ, Wang JH (2005) Repetitive mechanical stretching modulates IL-1beta induced COX-2, MMP-1 expression, and PGE2 production in human patellar tendon fibroblasts. Gene 363:166–172. https://doi.org/10.1016/j.gene.2005.08.006
Yee RY, Englander SW, Von Hippel PH (1974) Native collagen has a two-bonded structure. J Mol Biol 83:1–16
Yeung C-YC, Kadler KE (2019) Importance of the circadian clock in tendon development. Curr Top Dev Biol 133:309–342. https://doi.org/10.1016/bs.ctdb.2018.11.004
Yeung C-YC, Gossan N, Lu Y, Hughes A, Hensman JJ, Bayer ML et al (2014) Gremlin-2 is a BMP antagonist that is regulated by the circadian clock. Sci Rep 4:5183. https://doi.org/10.1038/srep05183
Yin Z, Hu J-J, Yang L, Zheng Z-F, An C-R, Wu B-B et al (2016) Single-cell analysis reveals a nestin+ tendon stem/progenitor cell population with strong tenogenic potentiality. Sci Adv 2:e1600874. https://doi.org/10.1126/sciadv.1600874
Yoshida R, Alaee F, Dyrna F, Kronenberg MS, Maye P, Kalajzic I et al (2016) Murine supraspinatus tendon injury model to identify the cellular origins of rotator cuff healing. Connect Tissue Res 57:507–515. https://doi.org/10.1080/03008207.2016.1189910
Young BB, Gordon MK, Birk DE (2000) Expression of type XIV collagen in developing chicken tendons: association with assembly and growth of collagen fibrils. Dev Dyn 217:430–439
Young RD, Knupp C, Pinali C, Png KM, Ralphs JR, Bushby AJ et al (2014) Three-dimensional aspects of matrix assembly by cells in the developing cornea. Proc Natl Acad Sci USA 111:687–692. https://doi.org/10.1073/pnas.1313561110
Yu T-Y, Pang J-HS, Wu KP-H, Chen MJ-L, Chen C-H, Tsai W-C (2013) Aging is associated with increased activities of matrix metalloproteinase-2 and -9 in tenocytes. BMC Musculoskelet Disord 14:2. https://doi.org/10.1186/1471-2474-14-2
Yudoh K, van Trieu N, Nakamura H, Hongo-Masuko K, Kato T, Nishioka K (2005) Potential involvement of oxidative stress in cartilage senescence and development of osteoarthritis: oxidative stress induces chondrocyte telomere instability and downregulation of chondrocyte function. Arthritis Res Ther 7:R380. https://doi.org/10.1186/ar1499
Yue J, Jin S, Li Y, Zhang L, Jiang W, Yang C et al (2016) Magnesium inhibits the calcification of the extracellular matrix in tendon-derived stem cells via the ATP-P2R and mitochondrial pathways. Biochem Biophys Res Commun 478:314–322. https://doi.org/10.1016/j.bbrc.2016.06.108
Zamboulis DE, Thorpe CT, Ashraf Kharaz Y, Birch HL, Screen HRC, Clegg PD (2020) Postnatal mechanical loading drives adaptation of tissues primarily through modulation of the non-collagenous matrix. elife 9. https://doi.org/10.7554/elife.58075
Zapp C, Obarska-Kosinska A, Rennekamp B, Kurth M, Hudson DM, Mercadante D et al (2020) Mechanoradicals in tensed tendon collagen as a source of oxidative stress. Nat Commun 11:2315. https://doi.org/10.1038/s41467-020-15567-4
Zhang J, Wang JH-C (2010) Characterization of differential properties of rabbit tendon stem cells and tenocytes. BMC Musculoskelet Disord 11:10. https://doi.org/10.1186/1471-2474-11-10
Zhang J, Wang JHC (2013) The effects of mechanical loading on tendons – an in vivo and in vitro model study. PLoS One 8:e71740. https://doi.org/10.1371/journal.pone.0071740
Zhang J, Wang JH-C (2015) Moderate exercise mitigates the detrimental effects of aging on tendon stem cells. PLoS One 10:e0130454. https://doi.org/10.1371/journal.pone.0130454
Zhang G, Young BB, Birk DE (2003) Differential expression of type XII collagen in developing chicken metatarsal tendons. J Anat 202:411–420
Zhang G, Young BB, Ezura Y, Favata M, Soslowsky LJ, Chakravarti S et al (2005) Development of tendon structure and function: regulation of coll fibrillogenesis. J Musculoskeletal Neuronal Interact 5:5–11
Zhang G, Ezura Y, Chervoneva I, Robinson PS, Beason DP, Carine ET et al (2006) Decorin regulates assembly of collagen fibrils and acquisition of biomechanical properties during tendon development. J Cell Biochem 98:1436–1449. https://doi.org/10.1002/jcb.20776
Zhang K, Asai S, Yu B, Enomoto-Iwamoto M (2015) IL-1β irreversibly inhibits tenogenic differentiation and alters metabolism in injured tendon-derived progenitor cells in vitro. Biochem Biophys Res Commun 463:667–672. https://doi.org/10.1016/j.bbrc.2015.05.122
Zhou Z, Akinbiyi T, Xu L, Ramcharan M, Leong DJ, Ros SJ et al (2010) Tendon-derived stem/progenitor cell aging: defective self-renewal and altered fate. Aging Cell 9:911–915. https://doi.org/10.1111/j.1474-9726.2010.00598.x
Zitnay JL, Li Y, Qin Z, San BH, Depalle B, Reese SP et al (2017) Molecular level detection and localization of mechanical damage in collagen enabled by collagen hybridizing peptides. Nat Commun 8:14913. https://doi.org/10.1038/ncomms14913
Zitnay JL, Jung GS, Lin AH, Qin Z, Li Y, Yu SM et al (2020) Accumulation of collagen molecular unfolding is the mechanism of cyclic fatigue damage and failure in collagenous tissues. Sci Adv 6:eaba2795. https://doi.org/10.1126/sciadv.aba2795
Zuskov A, Freedman BR, Gordon JA, Sarver JJ, Buckley MR, Soslowsky LJ (2020) Tendon biomechanics and crimp properties following fatigue loading are influenced by tendon type and age in mice. J Orthop Res 38:36–42. https://doi.org/10.1002/jor.24407

Basic Components of Connective Tissues and Extracellular Matrix: Fibronectin, Fibrinogen, Laminin, Elastin, Fibrillins, Fibulins, Matrilins, Tenascins and Thrombospondins

4

Jaroslava Halper

Abstract

Collagens are the most abundant components of the extracellular matrix (ECM) and many types of soft tissues. Elastin is another major component of certain soft tissues, such as arterial walls and ligaments. It is an insoluble polymer of the monomeric soluble precursor tropoelastin, and the main component of elastic fibers in matrix tissue where it provides elastic recoil and resilience to a variety of connective tissues, e.g., aorta and ligaments. Elastic fibers regulate activity of transforming growth factors β (TGFβ) through their association with fibrillin microfibrils. Elastin also plays a role in cell adhesion, cell migration, and has the ability to participate in cell signaling. Mutations in the elastin gene lead to cutis laxa. Many other molecules, though lower in quantity, function as essential, structural and/or functional components of the extracellular matrix in soft tissues. Some of these are reviewed in this chapter. Besides their basic structure, biochemistry and physiology, their roles in disorders of soft tissues are discussed only briefly as most chapters in this volume deal with relevant individual compounds. Fibronectin with its multidomain structure plays a role of "master organizer" in matrix assembly as it forms a bridge between cell surface receptors, e.g., integrins, and compounds such collagen, proteoglycans and other focal adhesion molecules. It also plays an essential role in the assembly of fibrillin-1 into a structured network. Though the primary role of fibrinogen is in clot formation, after conversion to fibrin by thrombin it also binds to a variety of compounds, particularly to various growth factors, and as such, fibrinogen is a player in cardiovascular and extracellular matrix physiology. Laminins contribute to the structure of the ECM and modulate cellular functions such as adhesion, differentiation, migration, stability of phenotype, and resistance towards apoptosis. Fibrillins represent the predominant core of microfibrils in elastic as well as non-elastic extracellular matrixes, and interact closely with tropoelastin and integrins. Not only do microfibrils provide structural integrity of specific organ systems, but they also provide basis for elastogenesis in elastic tissues. Fibrillin is important for the assembly of elastin into elastic fibers. Mutations in the fibrillin-1 gene are closely

J. Halper (✉)
Department of Pathology, College of Veterinary Medicine, and Department of Basic Sciences, AU/UGA Medical Partnership, The University of Georgia, Athens, GA, USA
e-mail: jhalper@uga.edu

J. Halper (ed.), *Progress in Heritable Soft Connective Tissue Diseases*, Advances in Experimental Medicine and Biology 1348, https://doi.org/10.1007/978-3-030-80614-9_4

associated with Marfan syndrome. Latent TGFβ binding proteins (LTBPs) are included here as their structure is similar to fibrillins. Several categories of ECM components described after fibrillins are sub-classified as matricellular proteins, i.e., they are secreted into ECM, but do not provide structure. Rather they interact with cell membrane receptors, collagens, proteases, hormones and growth factors, communicating and directing cell-ECM traffic. Fibulins are tightly connected with basement membranes, elastic fibers and other components of extracellular matrix and participate in formation of elastic fibers. Matrilins have been emerging as a new group of supporting actors, and their role in connective tissue physiology and pathophysiology has not been fully characterized. Tenascins are ECM polymorphic glycoproteins found in many connective tissues in the body. Their expression is regulated by mechanical stress both during development and in adulthood. Tenascins mediate both inflammatory and fibrotic processes to enable effective tissue repair and play roles in pathogenesis of Ehlers-Danlos, heart disease, and regeneration and recovery of musculo-tendinous tissue. One of the roles of thrombospondin 1 is activation of TGFβ. Increased expression of thrombospondin and TGFβ activity was observed in fibrotic skin disorders such as keloids and scleroderma. Cartilage oligomeric matrix protein (COMP) or thrombospondin-5 is primarily present in the cartilage. High levels of COMP are present in fibrotic scars and systemic sclerosis of the skin, and in tendon, especially with physical activity, loading and post-injury. It plays a role in vascular wall remodeling and has been found in atherosclerotic plaques as well.

Keywords

Fibronectin · Fibrinogen · Laminin · Elastin · Fibrillins · LTBPs · Matricellular proteins · Fibulins · Matrilins · Tenascins · Thrombospondins

Abbreviations

ADAMTS	a-Disintegrin-and-etalloproteinase-with-thrombospondin motifs
cbEGF	Calcium-binding epidermal growth factor
CNS	Central nervous system
COMP	Cartilage oligomeric matrix protein
ECM	Extracellular matrix
EFEMP1	EGF-containing fibulin-like ECM protein 1
FGF	Fibroblast growth factor
FN	Fibronectin
LDL	Low-density lipoprotein
LE	Laminin-type epidermal growth factor-like
LOX	Lysyl oxidase
LTBPs	Latent TGFβ binding proteins
MMP	Matrix metalloproteinase
PDGF	Platelet-derived growth factor
SLRP	Small leucine rich proteoglycan
TGFβs	Transforming growth factors β
TIMP	Tissue inhibitor of metalloproteinase
TSP	Thrombospondin
VEGF	Vascular endothelial growth factor

The connective tissue in general is comprised of 3 groups of proteins: collagens, proteoglycans and a variety of different glycoproteins. In addition to the main weight-bearing structural proteins of connective tissue – the fibril forming collagens (discussed in Chap. 2) – as well as the often hydrophilic role of proteoglycan proteins (discussed in Chap. 6), growth factors (see Chap. 7), other proteins are also important for structure and signaling within the matrix tissue of the body. Many of these proteins are currently being identified as having important functions in the developmental phase of the tissue, where these molecules can act as mediators of signaling and/or structural changes in the matrix tissue. Further, many of the glycoproteins have been demonstrated to play important roles in normal tissue physiology, including maintaining tissue homeostasis, and responding and adapting to perturbations such as mechanical loading/unloading, or tissue damage and subsequent regeneration. Furthermore, numerous, if not all, of these glycoproteins are important for pathological tissue response like in e.g. cancer, fibrosis or connective tissue anomalies. Of interest as far as the adaptation of these glycoproteins is that several of them – including collagens and proteoglycans – can be modulated in their level of expression and synthesis by the degree of mechanical loading that the specific tissue exposed to mechanical loading senses (Kjaer 2004). In the following pages some basic information about these glycoproteins is provided. However, as already mentioned above, because many of these glycoproteins are active participants in the pathogenesis of a variety of soft tissue diseases they will be discussed rather briefly in this chapter as they are also described in several chapters dealing with specific disorders of soft tissues.

4.1 Fibronectin

Fibronectin (FN) is a widely distributed multidomain glycoprotein present in most extracellular matrices. It has a molecular weight of 230–270 kD, and can, in addition to its presence in the ECM, be detected also at substantial concentrations in plasma. Fibronectin is composed of types I, II, and III repeating units or modules (FNI, FNII and FNIII) (Pérez-García et al. 2020). Two intramolecular disulfide bonds are formed within type I and type II modules to stabilize the folded structure. Type III modules are formed by seven-stranded β-barrel structures that lack disulfides (Leahy et al. 1996; Potts and Campbell 1994). The FN units or domains mediate self-assembly and ligand binding for collagen/gelatin, integrins, heparin, fibronectin, and other extracellular molecules (Sabatier et al. 2009). The 500-kDa FN dimer is formed through a pair of anti-parallel disulfide bonds at the C terminus. FN exists in multiple isoforms generated by alternative splicing. The single FN gene transcript encodes 12 isoforms in rodents and cows and 20 isoforms in humans. Alternative splicing occurs by exon skipping at EIIIA/EDA and EIIIB/EDB and by exon subdivision at the V region/IIICS. This gives fibronectin considerable diversity in module arrangement resulting in many isoforms (White and Muro 2011). Fibronectin is secreted in the form of soluble inactive dimers with disulfide bonds that must be activated by interaction with $\alpha5\beta1$ and other integrins (Mao and Schwarzbauer 2005; Takahashi et al. 2007).

Fibronectin is widely expressed in embryos and adults, especially in regions of active morphogenesis, cell migration and inflammation. Tumor cells contain reduced levels of fibronectin, whereas fibronectin levels are high in tissues undergoing repair (i.e., wound healing) and/or fibrosis. In the process of matrix assembly, multivalent ECM proteins are induced to self-associate and to interact with other ECM proteins to form fibrillar networks. Matrix assembly is initiated usually by ECM glycoproteins binding to cell surface receptors, such as fibronectin dimers binding to $\alpha5\beta1$ integrin. Receptor binding stimulates fibronectin self-association mediated by the N-terminal assembly domain and organizes the actin cytoskeleton to promote cell contractility. Fibronectin conformational changes expose additional binding sites that participate in fibril formation and in conversion of fibrils into a stabilized, insoluble form. Once assembled, the FN matrix impacts tissue organization by contributing to the assembly of other ECM proteins. Fibronectin plays an important role in fibrillogenesis in regard to initiation, progression and maturation of matrix assembly. The prominent role of fibronectin in matrix assembly lies in fibronectin ability, enabled by its multidomain structure, to bind simultaneously to cell surface receptors, e.g., integrins, and to collagen, proteoglycans and other focal adhesion molecules (Singh and Schwarzbauer 2012). This property also makes it possible to mediate the assembly of several ECM proteins, including type I and III collagen, thrombospondin-1 and microfibrils

(Sabatier et al. 2009). Fibronectin is also called a "master organizer" by some investigators (Sabatier et al. 2009; Dallas et al. 2006). Degradation of fibronectin by proteases activated during a variety of inflammatory processes, including infections leads to unmasking of binding sites within the fibronectin molecule. This triggers binding of fibronectin to different integrin receptors and toll like receptors, ultimately leading to activation of MAPK signaling pathway and transcription factors such as NF-κB, thus further stimulating progression of inflammation (Pérez-García et al. 2020). Perhaps more important in the context of this volume is to emphasize the role fibronectin plays in the assembly of fibrillin-1 into a structured network (see below).

4.2 Fibrinogen

Fibrinogen is a large, complex, fibrous glycoprotein with three pairs of polypeptide chains: Aα, Bβ and γ (Fish and Neerman-Arbez 2012). The chains are linked together by 29 disulfide bonds. Fibrinogen is 45 nm in length, with globular domains at each end and in the middle connected by α-helical coiled-coil rods and has M_r 340 kDa. The E-region consisting of N-terminal ends of the six chains and the D-regions consisting of the C-terminal ends of the Bβ and γ chains and a portion of the Aα chain are separated by a 3-stranded α-helical coiled-coil regions (Doolittle et al. 1978). Both strongly and weakly bound calcium ions are important for maintenance of fibrinogen structure and functions. Fibrinopeptides located in the central region of the molecule are cleaved by thrombin to convert soluble fibrinogen to insoluble fibrin polymer, via intermolecular interactions of the "knobs" exposed by fibrinopeptide removal with "holes" always exposed at the ends of the molecules. Fibrin monomers polymerize via these specific and tightly controlled binding interactions to make half-staggered oligomers that lengthen into protofibrils. The protofibrils aggregate laterally to make fibers, which then branch to yield a three-dimensional network-the fibrin clot-essential for hemostasis. X-ray crystallographic structures of portions of fibrinogen have provided some details on how these interactions occur. Finally, a transglutaminase, Factor XIIIa, covalently binds specific glutamine residues in one fibrin molecule to lysine residues in another fibrin molecule via isopeptide bonds, stabilizing the clot against mechanical, chemical, and proteolytic insults (Ariens et al. 2002). The gene regulation of fibrinogen synthesis and its assembly into multichain complexes proceed via a series of well-defined steps. Alternate splicing of two of the chains yields common variant molecular isoforms. The mechanical properties of clots, which can be quite variable, are essential to fibrin functions in hemostasis and wound healing (Cilia La Corte et al. 2011). The fibrinolytic system, with the zymogen plasminogen binding to fibrin together with tissue-type plasminogen activator to promote activation to the active enzyme plasmin, results in digestion of fibrin at specific lysine residues. Fibrin(ogen) also specifically binds a variety of other proteins, including fibronectin, albumin, thrombospondin, von Willebrand factor, fibulin, fibroblast growth factor-2 (FGF2), vascular endothelial growth factor (VEGF), and interleukin-1. Though its ability to bind to a variety of compounds, particularly to various growth factors makes fibrinogen a player in cardiovascular and extracellular matrix physiology (Fish and Neerman-Arbez 2012; Sahni and Francis 2000; Sahni et al. 1998; Clark et al. 1982; Donaldson et al. 1989), fibrinogen does not appear to play a specific role in pathogenesis of disorders discussed in this volume.

Studies of naturally occurring dysfibrinogenemias and variant molecules have increased our understanding of fibrinogen functions. Fibrinogen binds to activated αIIbβ3 integrin on the platelet surface, forming bridges responsible for platelet aggregation in hemostasis, and also has important adhesive and inflammatory functions through specific interactions with other cells (Armstrong and Peter 2012). Fibrinogen-like domains originated early in evolution, and it is likely that their specific and tightly controlled intermolecular interactions are involved in other aspects of cellular function and developmental biology.

4.3 Laminins

Laminins are a family of large multidomain, heterotrimeric glycoproteins with molecular weights of 500–900 kDa, located in the basement membrane where they function as a bridge between cells and variety of ECM molecules (Chang and Chaudhuri 2019), more specifically, they interact with cellular receptors of cells of the basement membrane (Aumailley 2018). Sixteen trimeric isoforms have been described in mouse and human tissues and these isoforms vary in their cell and tissue specificity (Aumailley 2018). In general, each laminin isoform consists of three chains, α, β, and γ which each exist in five, four, and three genetically distinct forms, respectively (Aumailley et al. 2005; Miner and Yurchenco 2004; Domogatskaya et al. 2012). Most vertebrates have five α, three γ and three to six β genes (Domogatskaya et al. 2012). The large range in size is due to variability in the chain size: the α chains are the largest (M_r ~ 200–400 kDa), both the β and γ chains range in size from 120 to 200 kDa. In addition, all forms of these three chains are highly glycosylated, some have glycosaminoglycan chains attached (Aumailley et al. 2005; Domogatskaya et al. 2012). Homologous tandem repeats of structural motifs are incorporated in all laminins, with more similarities between β and γ chains. Laminins are cross or T-shaped molecules with 2 or 3 short arms and one long arm. The short arms consist of N-terminal parts of one of the three chains and they contain multiple laminin-type epidermal growth factor-like (LE) repeats (Domogatskaya et al. 2012; Hohenester 2019) The long arm contains portions of all 3 chains (Aumailley et al. 2005). Common to all laminins is a coiled-coil domain with about 80 heptad sequence repeats at or close to the C-terminal end. This coiled-coil domain bears homology to segments of β and γ chains and is responsible for proper assembly of the trimer (Domogatskaya et al. 2012; MacDonald et al. 2010). Assembly of the laminin molecule is also controlled to some extent by proteolytic processing prior to laminin binding to its receptors (Domogatskaya et al. 2012).

Laminins adhere to cells primarily via binding of the G domain of the α chains to integrins, dystroglycan, or sulfated glycolipids. The N-terminal globular domains of the α1 and α2 chains as well as the globular domains VI (LN) of the α5 chains can bind to several integrin isoforms (α1β1, α2β1, α3β1, and αVβ3). This process enables cell binding on both ends of laminins containing the three α chains. The laminin γ2 chain has been reported to bind α2β1 integrin. The N-terminal globular domains of some α-chains can also bind sulfatides. This type of binding may link the laminin molecules to the cell surface. Laminins contribute to the structure of ECM and influence associated cells in regards to adhesion, differentiation, migration, stability of phenotype, and resistance towards apoptosis. Laminin molecules interact not only with collagen type IV, integrins and dystroglycans but also with other components of the basal membrane matrix, and thus contribute to the overall structure. They also can interact with components in the underlying interstitial stroma. The cellular effects of laminins are mediated largely via ligand binding to cell membrane receptors, and this signaling can alter transcription levels of genes and even influence chromatin remodeling of gene promoters. The insoluble network formed by laminin and type IV collagen plays a structural and functional role in the basement membrane and cells associated with it. Though at this point we do not know to what extent, if any, laminins play a role in the pathogenesis of connective and soft tissue diseases it is clear that they contribute to normal function of tendons, blood vessels and other connective soft tissues. For example, this network participates in transmission of the contractile force from the skeletal muscle to the tendons (Grounds et al. 2005). A decrease in laminin content in the basement membrane covering the outermost aspect of the tendon was identified in type IV collagen deficient mice. This was accompanied by formation of spontaneous tendon adhesions (Taylor et al. 2011). That laminins are, indeed, required for proper healing of tendons and other connective tissues, such as cornea, has been shown by Molloy et al. (Molloy et al. 2006) and Sato et al. (Sato et al. 1999), respectively. There is some evidence indicating increased expression of β2 chain of laminin in ascending aorta in patients with Marfan syndrome (Della Corte et al. 2006) .

Taken together laminins are not passive adhesion proteins, but rather, they actively modulate cell behavior; influence differentiation, migration, and phenotype stability. They also inhibit apoptosis by signaling via cell membrane receptors such as integrins and dystroglycan. However, the details of laminin signaling are still largely unexplored. Laminins constitute the first ECM component appearing in the developing early embryo, and embryonic laminins have found an important use as culture matrices for stem cells. Other laminins are crucial for normal function of numerous tissues and organs, e.g., nerve, epithelium, blood vessels, and kidney. The commercial unavailability of most laminin isoforms has hampered *in vitro* studies. However, many isoforms have been offered recently by several companies as recombinant proteins, which may enable deeper insight into functional properties. Laminins may find numerous new applications in cell biology and cell therapy research. The vast complexity of laminin effects cannot be explained solely by simple integrin binding and signaling (Domogatskaya et al. 2012).

4.4 Elastin

Elastin is an insoluble polymer of the monomeric soluble precursor tropoelastin. Elastin is the main component of elastic fibers in matrix tissue, and as such it is the main contributor to the elasticity of these fibers (Muiznieks et al. 2010; Mithieux et al. 2012). Tropoelastin is encoded by a single human gene and is secreted as an ~60 kDa unglycosylated protein by a variety of cells, including fibroblasts, endothelial and smooth muscle cells, chondrocytes and keratinocytes (Mithieux et al. 2012). The splicing of the primary tropoelastin transcript is tissue-specific and thus allows for conformational and functional adjustment for each location (Kielty 2006). The primary tropoelastin sequence is an arrangement of hydrophobic domains rich in valine, proline and glycine, providing elasticity to the final product, elastin. These hydrophobic domains alternate with hydrophilic domains which contain lysine residues whose role it is to stabilize elastin microfibrils by cross-linking (Csiszar 2001; Lee and Kim 2006; Kim et al. 2011). However, before this can occur tropoelastin units are initially assembled within the cells or at least on the cell surface (Kozel and Mecham 2019) before they are chaperoned to the extracellular surface (Hinek and Rabinovitch 1994) where they coacervate (Yeo et al. 2011) into protein-dense spherules (Kozel et al. 2006) which then undergo cross-linking and fibril assembly. Ninety per cent of the final product, i.e., of an elastic fiber, consists of a central amorphous core of elastin surrounded by a layer of microfibrils composed mostly of glycoprotein fibrillin, but also of many other proteins, among them fibulins, collagen VIII, and emilins with microfibrils as well (Kielty 2006; Berk et al. 2012; Nakamura 2018). Proteoglycans, including biglycan (Baccarani-Contri et al. 1990) and glycosaminoglycan heparan sulfate (Gheduzzi et al. 2005) have been detected within the elastic core. Moreover, it has been shown that the presence of sulfated proteoglycans within the ECM regulates elastin assembly (Kozel et al. 2004). In addition, water plays an important role not just in the three dimensional organization of elastin molecules but also in the final degree of hydration and elasticity (Gheduzzi et al. 2005). Elastic fibers form an interconnecting fenestrated network of lamellae in the arterial media. The lamellae are layers of elastic fibers surrounded by circumferentially oriented smooth muscle cells and collagen fibers (Wagenseil and Mecham 2012).

The high content of hydrophobic amino acids makes elastin one of the most chemically resistant and durable proteins in the entire body (Mithieux and Weiss 2005). It is distributed throughout the body in the form of tissue-specific elastic networks (Mithieux et al. 2012). Elastin containing fibers provide elastic recoil in tissues where repetitive distention and relaxation is a requirement for their function, and is found typically in skin, lungs, ligaments, tendons and vascular tissues (Chung et al. 2006). The relative content of elastin can vary from around a few percentages in skin, to more than 70% in some ligament structures in the animal kingdom. Elastic fibers are essential for proper function of at least three areas. As a major structural component elastic fibers provide elastic recoil and resilience to a variety of connective tissues, e.g., aorta and ligaments. Elastic fibers regulate activity of TGFβs through their association with fibrillin microfibrils. In addition, elastin also plays a role in cell adhesion, cell migration, survival and differentiation, and can, to some extent, act as a chemotactic agent (Muiznieks et al. 2010; Kielty 2006). Elastin, and for that matter tropoelastin as well, is also a signaling molecule. Tropoelastin inhibits proliferation of arterial smooth muscle cells, induces the formation and organization of actin stress fibers and acts as a chemotactic agent (Karnik et al. 2003).

Elastin and collagen are the dominant components of the ECM in large elastic arteries, such as aorta (Wagenseil and Mecham 2012). The two compounds play different, but complementary roles in arterial physiology: reversible extensibility during cycling loading is provided by elastin (Wagenseil and Mecham 2012; Mecham 1998), whereas strength and the ability to withstand high pressure is the responsibility of collagen (Wagenseil and Mecham 2012; Fung 1993). The assembly of elastic fibers proceeds only during tissue development, and cedes with maturation so older tendons (and other tissues) contain less elastin than young tendons (Wagenseil and Mecham 2009; Kostrominova and Brooks 2013). In effect that means that with aging the stiffness of arterial wall increases due to degradation and fragmentation of elastic fibers (Wagenseil and Mecham 2012; Greenwald 2008). Matrix metalloproteinases (MMPs) are just some of the proteases participating in this destructive process (Wagenseil and Mecham 2012; Li et al. 1999). Increased levels of MMP-1 and MMP-9 have been detected in aortic aneurysms (Tamarina et al. 1999). Local inhibition of MMP activities in animal models either by tissue inhibitor of metalloproteinase 1 (TIMP-1) (Allaire et al. 1998), inhibition of MMP-2 by calpain-1 inhibition (Jiang et al. 2008), or by doxycycline, an inhibitor of MMPs (Castro et al. 2008) shows potential treatment venues. Whether they can be utilized for treatment or even prevention of complications of Marfan syndrome or related disorders remains to be seen. It is thought that production of collagen increases to compensate for the elastin deficit, however, this pushes the arterial wall towards increased stiffness (Wagenseil and Mecham 2012). Increased elastin production has been documented in some animal models of hypertension, but it is either not high enough (Wolinsky 1970) or the new elastin fibers are not assembled properly (Todorovich-Hunter et al. 1988).

Elastin gene mutations can be divided into two groups (Wagenseil and Mecham 2012). Autosomal dominant supravalvular aortic stenosis is a representative of the first group. Besides aortic valve stenosis, patients develop hypertension, increased arterial stiffness leading to congestive heart failure (Wagenseil and Mecham 2012). Hypertrophy and hyperplasia of smooth muscle cells in the media of the affected arteries is due to fragmentation of elastic lamellae and changes in ECM composition (O'Connor et al. 1985). This pathology is due to loss of function mutations in the elastin (*ELN*) gene (Urban et al. 2000). Consequently, the mutant elastin protein is nonfunctional and does not interfere with the production and assembly of normal, functional elastin in heterozygous individuals who are then less affected than homozygous people (Wagenseil and Mecham 2012).

An autosomal dominant form of cutis laxa belongs to the second group which encompasses disorders resulting from missense mutation, usually near the 3′ end of the transcript (Wagenseil and Mecham 2012; Rodriguez-Revenga et al. 2004; Tassabehji et al. 1998). Cutis laxa and related disorders are described in more detail in Chap. 13. The mutant elastin interferes with normal assembly, metabolism and function of elastic fibers (Tassabehji et al. 1998).

Lack of elastin in the body is fatal. Elastin knockout mice (*Eln−/−*) die shortly after birth with subendothelial cell accumulation blocking blood flow and with markedly increased arterial stiffness (Wagenseil and Mecham 2012; Li et al. 1998). The presence of additional lamellar units in heterozygous *Eln+/−* mice indicates an attempt to compensate and to remodel in a response to increased hemodynamic stress during development (Faury et al. 2003). Fibrillin-1 hypomorphic mice (*mgR/mgR*) serve as a model of Marfan syndrome because of aneurysm formation in the ascending aorta and elastolysis in all segments of aorta (Schwill et al. 2013).

4.5 Fibrillins

Because of close association of mutated fibrillin-1 with Marfan syndrome which is being discussed in detail in Chap. 8, only a brief description of fibrillins is provided in this chapter. Fibrillins are a group of large extracellular glycoproteins (~350 kDa) (Kielty 2006) that consists of 3 isoforms, fibrillin-1, -2, and -3. Fibrillin molecules contain 40–80 amino acid residues, several calcium-binding epidermal growth factor (cbEGF)-domains interspersed with several eight-cysteine-containing motifs binding TGFβ (TB) (Sabatier et al. 2009; Kielty 2006; Kielty et al. 2005). No other extracellular proteins contain that much cysteine as fibrillins (Chung et al. 2006). Whereas fibrillin-2 and fibrillin-3 are mostly expressed in embryonic tissues with the exception of peripheral nerves and, to lesser degree, skin and tendon (Zhang et al. 1994; Charbonneau et al. 2003) fibrillin-1 is a protein appearing in both embryonic and adult tissues (Charbonneau et al. 2003; Cain et al. 2006; Robinson et al. 2006).

Fibrillins represent the predominant core of the microfibrils in elastic as well as non-elastic extracellular matrixes, and interact closely with tropoelastin and integrins, e.g., through direct binding. Not only do microfibrils provide structural integrity of specific organ systems, but they also provide a scaffold for elastogenesis in elastic tissues such as skin, lung, and vessels (Wagenseil and Mercham 2009). Thus, fibrillin is important for the assembly of elastin into elastic fibers. The precise arrangement of fibrillin within microfibrils is a matter of speculation; several working models have been suggested to explain the architecture of microfibrils (Kozel and Mecham 2019; Robinson et al. 2006). It is known that different mutations in different regions, including the propeptide sequence encoded by the C-terminal domain, of the fibrillin-1 gene lead to impaired assembly of microfibrils in individuals with Marfan syndrome (Robinson et al. 2006; Milewicz et al. 1995; Raghunath et al. 1999). Robinson et al. provide an excellent, more comprehensive review of these issues, including review of self-assembly of fibrillins and cross-link formation in fibrillin assembly (Robinson et al. 2006). Besides fibrillin and elastin, the two major components, many other proteins participate in the makeup of microfibrils. As noted above fibronectin in particular plays as an essential role in this process, more specifically, through binding of a C-terminal fibrillin-1 region with the fibronectin gelatin-binding region (Dallas et al. 2006). It is interesting to note that homocysteinylation of fibronectin in homocystinuria reduces fibronectin dimers to monomers, and, as a consequence, impairs assembly of fibrillin and microfibrils. Similar impairment is the result of homocysteinylation of fibrillin-1 (Hubmacher et al. 2011).

As already mentioned above, fibrillins contain several TGFβ-binding motifs, this feature makes their structure, and their function. Similar to that of latent-TGFβ-binding proteins (or LTBPs) (see below) (Robinson et al. 2006).

Mutations in genes for fibrillin-1 and -2 lead to several disorders in people. Mutation in fibrillin-1, the most abundant fibrillin, and also the best characterized isoform can result in autosomal dominant Marfan and Weill-Marchesani syndromes (Thomson et al. 2019). It is expressed in embryonic and mature tissues (Ramirez and Sakai 2010). Its involvement in pathogenesis of Marfan syndrome is described in detail in Chap. 9. Above mentioned Weill-Marchesani syndrome leads to pathology of the musculoskeletal, cardiovascular and ocular system. Two forms of Weill-Marchesani syndrome have been identified: autosomal dominant type caused by a fibrillin-1 mutation, and somewhat heterogenous autosomal recessive form caused by mutations in genes for.

ADAMTS-10, ADAMTS-17 or LTBP-2 (Karoulias et al. 2019). Beals syndrome is characterized by congenital contractual arachnodactyly caused by mutation in fibrillin-2 (Sabatier et al. 2009; Robinson et al. 2006; Beals and Hecht 1971; Jaman and Al-Sayegh 2016). Not much is known about fibrillin-3 beyond its expression limited to embryonic extracellular microfibrils (Halper 2021).

4.6 Latent-TGFβ-Binding Proteins (LTBPs)

As described above, LTBPs 1–4 are structural proteins related to fibrillins. They bind to the latency associated peptide (LAP) non-covalently bound to TGFβ (Robertson et al. 2011). The entire complex is embedded in ECM, where it limits bioavailability of TGFβ (Thomson et al. 2019). Phenotypic changes resulting from mutations in individual *LTBP* genes point to the LTBP and TGFβ contribution to proper function of connective tissues. As mentioned above mutation in *LTBP2* leads to an autosomal recessive form of Weill-Marchesani syndrome, and can cause also ectopia lentis. It is deleted completely in a type of congenital glaucoma (Thomson et al. 2019). LTBP-4 together with fibulins -4 and -5 participates in elastogenesis. Persons with mutations and deletions in *LTBP4* and *LTBP3* may present with aortic dilatation and aneurysm (Thomson et al. 2019; Zilberberg et al. 2015). Other mutations in *LTBP4* are behind a form of autosomal recessive cutis laxa (Urban et al. 2009) as described in more detail in Chap. 13.

4.7 Fibulins

Fibulins are a group of eight glycoproteins that are expressed and secreted by many cell types and tissues, and that are tightly connected with basement membranes, elastic fibers and other components of ECM. Interactions with TGFβ and participation in elastic fiber assembly and stability are some of their important functions (Tsuda 2018). Fibulins serve not only as structural ECM components, but also as mediators of several cellular processes, such as cell growth, differentiation, angiogenesis and tumor growth. Thus they serve as modulators of cellular behavior and function (DeVega et al. 2009), and are classified by some investigators as matricellular proteins (Nakamura 2018). All fibulins share a C-terminal module which is preceded by variable number of cbEGF-like domains (Halper 2021). The members of the fibulin family are divided into class I and II, based on their length and domain structures (Yanagisawa and Davis 2010).

Class II consists of fibulins 3, 4, 5 and 7. They are called short because of their small size (M.W. ~60–70 kDa), and are discussed first because most of them (fibulins 3–5) contribute directly to assembly of elastic fibers, and thus mutations in genes encoding them lead to forms of cutis laxa and other disorders of connective tissues. Fibulins 3–5 bind to tropoelastin and are expressed during embryonic development, especially in skeletal and cardiovascular tissues (Yanagisawa and Davis 2010). This is facilitated by Ca^{2+} (Wachi et al. 2008). Fibulin-3 is predominantly found in mesenchyme that develops into heart, placenta, cartilage and bone among other organs (Giltay et al. 1999), fibulin-4 is markedly expressed in heart muscle, and fibulin-5 highly in vasculature. The molecules of short fibulins contain tandem repeats of six cbEGF domains that are connected by one amino acid in a pattern similar to the one found in fibrillin-1 (Hambleton et al. 2004). Human fibulin 3 is encoded by gene called *EFEMP1* (which stands for EGF-containing fibulin-like ECM protein 1). Mutations in this gene are numerous, some lead to autosomal dominant retinal disease (Nakamura 2018). Overexpression of *EFEMP1* is one of the genetic abnormalities identified in Werner syndrome, a form of progeria (Halper 2021; Sarbacher and Halper 2019).

Fibulin 5 contains an arginine-glycine-aspartic acid (RGD) motif which mediates binding to integrin receptors on endothelial cells and vascular smooth muscle cells (Yanagisawa et al. 2009). This step is necessary for elastic fiber assembly (Yanagisawa and Davis 2010). Fibulin-5 also inhibits α5β1 and α4β1 integrin-mediated downstream signaling (Yanagisawa and Davis 2010). The C-terminal fibulin module contains an elastic-binding domain in fibulin-5 (Zheng et al. 2007). The same module in fibulin-5 also interacts with lysyl oxidase-like 1, 2 and 4 (Loxl 1, Loxl 2 and Lox 4), enzymes playing crucial role in cross-linking (Hirai et al. 2007; Liu et al. 2004) whereas it is the N-terminal domain responsible for binding to Lox in fibulin-4 (Horiguchi et al. 2009). Lysyl oxidases, including those binding to fibulin-5 and -4 mediate crosslinking of tropoelastin monomers into insoluble elastin polymer (Sato et al. 2007). The binding between the C-terminal module of fibulin-3 and tissue inhibitor of matrix metalloproteinase 3 is another example of close relationship between a short fibulin and connective tissue metabolism (Klenotic et al. 2004). The level of fibulin-5 is particularly high in the cardiovascular system and lung, though fibulin-4 is expressed in the outer layer of media of large blood vessels, and fibulin-3 appears in capillaries, skin and the basement membrane (Yanagisawa and Davis 2010). The participation of fibulin-5 in elastogenesis is solely due to its exclusive binding to tropoelastin but not to polymerized elastin *in vitro* (Zheng et al. 2007). Its role is inhibition of excessive tropoelastin coacervation into large aggregates, and consequently this allows for integration of microassembles of tropoelastin into the microfibril scaffolding (Yanagisawa and Davis 2010). Together with fibrillins-1 and -2 (see above under Elastin) fibulins are present in microfibrils of scaffolding for elastic fibers as well (Ramirez and Dietz 2007).

Mutations in genes for fibulin 4 and 5, *EFEMP2* and *EVEC*, respectively, are responsible for forms of cutis laxa (see Chap. 13). Fibulin-5 functions also as an inhibitor of angiogenesis (De Vega et al. 2016). Other mutations in gene for fibulin-5 lead to age-related macular degeneration (Tsuda 2018) and even vascular remodeling associated with arterial hypertension (Kartashova and Sarvilina 2019).

The last member of this group, fibulin-7 is not involved in elastic fiber formation. It is highly expressed in teeth, placenta, hair follicles, and cartilage where it functions as a cell adhesion molecule (De Vega et al. 2007). It regulates also calcium and phosphate metabolism in the kidney, and its dysfunction can lead to renal tubule calcification (Tsunezumi et al. 2018).

Group I of so called long fibulins consists of fibulins -1 and -2, and hemicentins -1 and -2 (also known as fibulins -6 and -8) (Fujishima et al. 2017). Though their structure is well described, their functions are less characterized than that those of short fibulins. Fibulin-1 (molecular weight 90 kDa) was originally identified as an intracellular molecule linking cytoskeletal components to β integrins, but later it was shown that fibulin-1 was also present in fibril matrix secreted by fibroblasts in culture (Zhang et al. 1996), and in association with basement membranes and elastic fibers. It appears early in embryonic development, at sites of epithelial-mesenchymal transition (Tsuda 2018). It is expressed in adult blood vessels, lung and skin, i.e., tissues with high content of elastic fibers (Roark et al. 1995). Fibulin-1 participates in ADAMTS-1-induced processing of proteoglycans (Tsuda 2018).

Fibulin-2 is a homodimer of two 195 kDA monomers joined by two disulfide bridges (Sasaki et al. 1997) demonstrates some overlap with fibulin-1, but its expression is more prominent in the developing heart, both aortic and coronary vessels where it binds to tropoelastin and other ECM molecules, and thus contributes to formation of elastic fibers (Tsuda 2018; Tsuda et al. 2001; Timpl et al. 2003). Fibulin-2 null mice have only skin abnormalities, most likely due to compensation of fibulin-1 overexpression (Tsuda 2018). The last two fibulins (6 and 8) AKA hemiceptins -1 and -2 are the largest members of this family (M.W. 600 kDa), and the least characterized. They play roles in mesenchymal cell migration and skin development in zebra fish (Tsuda 2018).

4.8 Matrilins

This group of four matrilins has been included for completeness and future reference as the structural and functional roles of this family of four in musculoskeletal system and in connective tissues have been understood only incompletely. Whereas matrilins -1 and -3 are limited mostly to cartilage, matrilins -2 and -4 were identified in other types of ECM, including loose soft connective tissue (Paulsson and Matrilins 2018). In general, matrilins mediate interactions between collagens and other molecules, such as proteoglycans (aggrecan, small leucine rich proteoglycans (SLRPs)), and other ECM components. Matrilins are trimers or tetramers of units composed of von Willebrand module, EGF-like domain(s) and a C-terminal oligomerization domain. The von Willebrand module is required for protein-protein interactions (Whittaker and Hynes 2002).

Though skeletal disorders due to mutations in genes for matrilins -1 and -3 have been well documented (Paulsson and Matrilins 2018; Jackson et al. 2012; Anthony et al. 2015; Montanaro et al. 2006), so far no physiologically relevant mutations have been identified for matrilins -2 and -4.

4.9 Tenascins

Tenascins are matricellular ECM polymorphic glycoproteins with molecular weight between 150 and 380 kDa. They are a family of multimeric proteins labeled as tenascin-C, -R, -W, -X and -Y (Tucker et al. 2006; Tucker and Chiquet-Ehrismann 2009; Okamoto and Imanaka-Yoshida 2012). Tenascins are composed of identical subunits built from variable numbers of repeated domains, including heptad repeats, EGF-like repeats, fibronectin type III domains and a C-terminal globular domain similar to that seen in fibrinogens (Okamoto and Imanaka-Yoshida 2012). Polymerization of tenascins is facilitated by the heptad repeats. The pattern of arrangement of the domains renders tenascins highly interactive rather than structural proteins in the ECM because of and as such are considered matricellular proteins (Midwood et al. 2016). Whereas the presence of tenascin-R is predominantly limited to the central nervous system (CNS), and then mostly during CNS development, the other members of the tenascin family are found more widespread in connective and soft tissues in the body.

4.9.1 Tenascin-X

Tenascin-X is emerging as a significant player in physiological processes in many systems and in pathogenesis of classic-like type Ehlers-Danlos syndrome (cl-EDS) (Matsumoto and Aoki 2020). Its level rises gradually from undetectable in early embryos into measurable amount in post-natal life. Its presence is ubiquitous, but particularly prominent in skeletal muscle, heart, skin, and gastrointestinal and nervous tissue. It has close association with blood vessels as well (Matsumoto and Aoki 2020). Tenascin-X is less glycosylated than tenascin-C. Post-natal physical activity stimulates the expression of tenascin-X in skeletal muscle as a consequence of acute mechanical loading and is known to be present in tissues that are subjected to high stress (Flück et al. 2000; Chiquet et al. 2009). Tenascin-Y is an avian equivalent of tenascin-X, and it follows tenascin-X expression pattern (Tucker et al. 2006; Hagios et al. 1996).

Tenascin-X is localized in the perineurium and endoneurium of peripheral nerves (Geffrotin et al. 1995), and in leptomeninges of the central nervous system (Matsumoto et al. 2002). Though tenascins are classified as matricellular proteins, tenascin-X has structural function as well. It regulates fibrillogenesis of fibrillar (types I, III and V) and fibril-associated types XII and XIV collagens (Lethias et al. 2006; Veit et al. 2006). This is assisted by its binding to decorin (Elefteriou et al. 2001) and tropoelastin (Egging et al. 2007). These associations would explain how complete absence of tenascin-X due to homozygous or compound heterogenous mutations in both *TNXB* alleles leads to cl-EDS (Matsumoto and Aoki 2020). Patients with cl-EDS present with velvety, hyperextensible skin, joint hypermobility and easy bruising. The diagnosis of this form of EDS can be confirm by the absence of serum form of tenascin-X, a protein of M.W. ~140 kDa which is a cleavage product of the nature 450 kDa form of tenascin-X (Schalkwijk et al. 2001). The deficiency of tenascin-X would extend to the nervous system and would explain chronic pain experienced by many sufferers of cl-EDS in their musculoskeletal and/or gastrointestinal systems (Matsumoto and Aoki 2020). See also Chap. 9 in this volume for more discussion on EDS.

The composition of tenascin-X, more specifically the presence of EGF–like and FNIII-like repeats in its molecule, makes it an angiogenic factor (Demidova-Rice et al. 2011). Interactions with VEGF-B contribute to its angiogenicity (Ikuta et al. 2000). Tenascin-X ability to activate latent TGFβ and TGFβ/Smad signaling pathway promotes epithelial-mesenchymal transition, and may contribute to its function as a matricellular protein (Valcourt et al. 2015).

4.9.2 Tenascin-C

Tenascin-C was the first described tenascin. It is a large molecule of M_r ~300 kDa, assembled into a hexamer. As other tenascins, the molecule consists of an N-teminal domain, EGF-like repeats, several fibronectin type II domains and a C-teminal fibrinogen-like globular domain. The structure of several repeats of the same domain or module enables binding of numerous ligands (Okamoto and Imanaka-Yoshida 2012). Tenascin-C is expressed transiently in the mesenchyme around developing organs such as kidney, teeth and mammary glands. Its expression is associated with epithelial-mesenchymal transition, branching morphogenesis and vascular development (Imanaka-Yoshida et al. 2014; Akbareian et al. 2013). It is present in the periostium, ligaments, tendons, myo-tendinous junctions, smooth muscle and perichondrium both during embryonic development and in adult tissues. However, expression of tenascin-C in the adult tissue is generally low, only to be transiently elevated upon tissue injury (and likely associated with stem cell proliferation) and often down-regulated again after tissue repair is complete (Midwood et al. 2016). Although tenascin-C shares structural relationship to fibronectin, it differs in its adhesive function. Where fibronectin is adhesive in nature, tenascin-C is only weakly adhesive – if at all – for most cells, and it does in fact limit the fibronectin-mediated cell spreading when the two proteins interact in cell cultures (Chiquet-Ehrismann et al. 1988). Tenascin-C interferes with cell spreading by inhibiting binding of fibronectin to its co-receptor syndecan-4, and integrin α5β1 signaling to FAK and RhoA is also impaired whereby focal adhesions are diminished (Huang et al. 2001; Midwood and Schwarzbauer 2002; Chiquet-Ehrismann and Chiquet 2003; Jones and Jones 2000).

As mentioned above, the expression of tenascin-C is regulated by mechanical stress both during development and in adulthood, and its expression is predominantly present in tissues experiencing high tensile stress, such as ligaments, tendons and smooth muscle (Kreja et al. 2012). Mechanical loading of muscle induces tenascin-C mRNA and protein in endomysial fibroblasts of the affected holding muscle (Järvinen et al. 2003). Tenascin-C was found to be over-expressed in hypertensive rat arterial smooth muscle (Mackie et al. 1992) and in the periosteum of rat ulnae loaded in vivo, but tenascin-C expression was low in the osteotendinous interphase of immobilized rat legs (Järvinen et al. 2003). Interestingly, elevated levels of tenascin-C were found in the blood of patients with rheumatoid arthritis (Page et al. 2012), and in synovial fluid after injury to the human and canine knee (Chockalingam et al. 2013).

In relation to ECM tissue damage, tenascin-C has been demonstrated to play different roles that can mediate both inflammatory and fibrotic processes to enable effective tissue repair. For example, tenascin-C makes a prominent appearance in pathological heart conditions. Though barely expressed in the normal adult heart its level increases in the heart after myocardial infarction, during myocarditis, hypertensive heart disease, to name just a few examples (Okamoto and

Imanaka-Yoshida 2012). According to the current hypothesis tenascin-C is directly involved in ventricular remodeling through releasing cardiomyocytes from the adherence to the extracellular matrix and through upregulation of matrix metalloproteinases (Okamoto and Imanaka-Yoshida 2012; Imanaka-Yoshida 2012). A high level of expression of tenascin-C in cardiac tissues correlates with poor patient prognosis (Midwood et al. 2011) Interestingly, tenascin-C was found in calcified aortic valves together with matrix metalloproteinase-12 where they likely contribute to the fragmentation of elastic fibers (Perrotta et al. 2011). Tenascin-C is involved in development of atherosclerosis, and possibly of aortic dissection, though whether its effect stimulating or inhibiting is unclear (Matsumoto and Aoki 2020).

TGFβ and platelet-derived growth factor (PDGF) induce expression of tenascin-C (Midwood et al. 2016). Tenascin-C binding to PDGF receptor or endothelin receptors modulates its inhibition of fibronectin adhesive effect (Midwood et al. 2016).

Similar to tenascin-C, tenascin–W has been identified in a variety of developing tissues, and a large interest has been invested in these tenascins in relation to tumor development and growth, where they play important roles.

In summary tenascin proteins are found to be dysregulated in many pathological conditions like cancer, heart- and vessel disease, as well as in connective tissue diseases with manifestations in skin, tendon and muscle like e.g. special forms of Ehlers-Danlos syndrome (more discussed in Chap. 9) and Dupuytren disease (Berndt et al. 1994). Further, tenascins have been shown to be important in regeneration and recovery of musculo-tendinous tissue, in that they possess a de-adhesive effect whereby they potentially can contribute to a coordinated tissue reorganization and build-up (Mackey et al. 2011). It has been suggested that they "orchestrate" muscle build up after injury (Flück et al. 2008). Thus, it is likely that tenascins are important for ensuring mechanical properties of weight bearing ECM as well as ensuring an optimal recovery of ECM after mechanical injury.

4.10 Thrombospondins

Thrombospondins (TSPs) form the last matricellular group. There are five modular glycoproteins, each one of them encoded by a separate gene (Murphy-Ullrich and Iozzo 2012; Adams and Lawler 2004; Adams and Lawler 2011). Group A consists of TSP-1 and TSP-2, and TSPs 3–5 are in group B. Their binding to various components of the ECM, such as heparan sulfate proteoglycans, and to numerous cell membrane receptors enables TSPs to modulate cell functions in a variety of tissues (Murphy-Ullrich and Iozzo 2012). They are considered to be "adhesion-modulating" components of the ECM (Mosher and Adams 2012).

In particular, we will discuss TSP-1 and TSP-5 in more detail as their involvement in metabolism of the ECM is pertinent to issues discussed in this volume. The activation of latent TGFβ by TSP-1 plays an important role in wound healing, and also in pathogenesis of fibrotic processes in kidney and heart in diabetes (Lu et al. 2011; Belmadani et al. 2007). Increased expression of TSP-1 (accompanied by increased TGFβ activity) was observed in fibrotic skin disorders such as keloids (Chipev et al. 2000) and scleroderma (Mimura et al. 2005).

TSP-1 is released from platelet α-granules where it is stored so it can participate in tissue repair (Sweetwyne and Murphy-Ullrich 2012). It is a homotrimer of three 150 kDa subunits. Each unit is composed of N-terminal laminin G-like domain, and in the last 650 amino acids, of several EGF-like domains, 13 calcium-binding repeats and a globular L-type lectin-like domain. These regions in the last 650 amino acids are usually referred to as the C-terminal or "signature" region (Mosher and Adams 2012). With glycosylation the size of TSP-1 balloons to staggering $M_r \sim 450$ kDa (Rogers et al. 2012). Its expression in adult organism is minimal (except for storage pool in platelets) and is upregulated only as a result of injury (Agah et al. 2002) and/or chronic disease (Rogers et al. 2012; Hohenstein et al. 2008). TSP-1 binds to many cell membrane receptors, including CD47 (Rogers et al. 2012), integrins (Chandrasekaran et al. 1999), also to heparan sulfate and low-density lipoprotein (LDL) (Chen et al. 1996). TSP-1 binds not only to latent TGFβ through thrombospondin repeats, but it also activates this growth factor (Murphy-Ullrich and Poczatek 2000). It is thought that TSP-1 facilitates presentation of TGFβ to the TGFβ receptor (Sweetwyne and Murphy-Ullrich 2012). TSP-1 was shown to upregulate type I collagen expression through its N- and C-terminal domains which may explain the sometimes opposing cellular responses stimulated by TSP-1 (Sweetwyne and Murphy-Ullrich 2012; Elzie and Murphy-Ullrich 2004). TGFβ activity induced by TSP-1 is a normal process during early tissue repair, however, if TSP-1 expression persists in later stages of wound healing fibrosis may prevail (Sweetwyne and Murphy-Ullrich 2012). In addition, TSP-1 regulates activity of several other growth factors, most notably, VEGF, EGF and PDGF. In particular, TSP-1 plays an important role in transactivation of EGF receptors in epithelial and endothelial cells, and thus can disrupt endothelial barrier (Goldblum et al. 1999). TSP-1 is an endogenous inhibitor of angiogenesis conferred by type I repeat domain found only in TSP-1 and TSP-2 (which is also anti-angiogenic – see below) (Bornstein 2009). Though TSP-1 has hypertensive effect on cardiovascular system and is known to play a role in pathogenesis of atherosclerosis and peripheral vascular disease (Robert et al. 2012), the activity is mediated through control of nitric oxide synthesis (and thus increasing arterial resistance), rather than through an impact on or binding to a structural component of the blood vessel wall (Robert et al. 2012). TSP-2 is involved in collagen fibril assembly and is capable of inhibition of angiogenesis and protease activity, but unlike TSP-1 it does not activate TGFβ (Okamoto and Imanaka-Yoshida 2012).

There is at least one syndrome where a mutation in a gene encoding an enzyme responsible for proper TSP-1 function leads to structural changes which form the basis of the so called Peters Plus syndrome. This syndrome is an autosomal recessive disorder phenotypically characterized by eye defects, short stature, developmental delay and cleft lip due to a mutation of a gene encoding a β1,3-glucosyltransferase which adds a glucose to *O*-linked fucose (and producing a rare glucose-β 1,3-fucose disaccharide) and which is responsible for glycosylation of thrombospondin type 1 repeats (Hess et al. 2008; Heinonen and Maki 2009). Beside TSP-1, properdin, F-spondin, some members of a-disintegrin-and-metalloproteinase-with--thrombospondin-like-motifs family (ADAMTS-13 and ADAMTSL-1) carry the same disaccharide (Hess et al. 2008; Heinonen and Maki 2009). Heart defects, such as hypoplastic left heart syndrome (Shimizu et al. 2010), patent ductus arteriosus, and atrial septal defect are present is some variants (Hanna et al. 2010). Though the eye involvement is usually characterized by anterior eye chamber defects leading to glaucoma (Hess et al. 2008; Hanna et al. 2010), corneal pathology has been recognized in some cases as well, and then it consists of intracorneal fibrosis (Eberwein et al. 2010) and keratolenticular adhesions (Hess et al. 2008; Hanna et al. 2010).

4.10.1 Cartilage Oligomeric Matrix Protein (COMP) or Thrombospondin-5

COMP or thrombospondin-5 belongs to the family of five extracellular calcium- and glycosaminoglycan-binding proteins that play a role predominantly during development, angiogenesis and wound healing. It consists of 5 identical subunits that are linked together at their N-terminal pentamerization end to result in an almost “star-like” structure and has $M_r \sim 524$ kDa (Oldberg et al. 1992). COMP shares a conserved multidomain architecture in its C-terminal region with TSP-1 (Mosher and Adams 2012). It also contains eight calmodulin units, four EGF-like repeats, and a globular C-terminal domain (Oldberg et al. 1992; Rock et al. 2010), and the 5 “arms” have on their C-terminal end high affinity binding sites for type I, II and IX collagen (Holden et al. 2001; Rosenberg et al. 1998), and for fibronectin (Di Cesare et al. 1994). Thrombospondin-5/COMP is present primarily in cartilage, and has been suggested to be important in relation to cartilage turnover and pathogenesis of osteoarthritis (Heinegård 2009). It is also expressed in other connective tissues like tendon, especially if the tissue has undergone strenuous mechanical loading, but also in cardiac cells and activated platelets (Smith et al. 1997; Södersten et al. 2013; Posey et al. 2018). The exact role of COMP in the fibril formation and assembly in the ECM is becoming better understood, and it is thought that COMP facilitates the joining of collagen molecules during formation of fibril structures (Södersten et al. 2013; Halasz et al. 2007). It has been shown that high levels of COMP are present in fibrotic scars and systemic sclerosis of the skin (Smith et al. 1997; Hesselstrand et al. 2008). It has been suggested that a very high concentration of COMP can in fact inhibit collagen fibril formation (115).

COMP is expressed in normal tendon where its mRNA is confined to tenocytes and the protein itself is located in the normally aligned fiber structures together with type I collagen. Virtually no COMP (and no type I collagen), but only type III collagen was found in the normal endotenon (Södersten et al. 2013). Physical activity leads to increased expression of COMP, at least in the equine tendon (Smith et al. 1997), as do pathological processes. High levels of COMP were identified in the synovial fluid obtained from the sheaths of the equine superficial digital flexor tendons diagnosed with synovitis (Smith et al. 2011). Likewise, injury to superficial digital flexor tendons leads to increased expression of COMP, and type I and III collagens in the endotenon and high levels of all three molecules can be visualized in the injured and granulation tissue (Södersten et al. 2013). Rock et al. have shown that COMP promotes attachment of ligament cells and chondrocytes to components of the ECM using two mechanisms which involves CD47 and integrins. Such data indicate an important role for COMP in formation of structural scaffolding, an essential step in cell attachment to the ECM and in matrix-cell signaling (Rock et al. 2010).

In addition, new data indicate that COMP, and its degradation by ADAMTS-7, plays an important role in vascular remodeling (Wang et al. 2010). COMP has been found in atherosclerotic plaques and lesions forming in arteries undergoing re-stenosis (Riessen et al. 2001), together with SLRPs, such as decorin (Riessen et al. 1994). It has been suggested that COMP promotes differentiation of vascular smooth muscle cells and that binding and degradation of COMP by ADAMTS-7 in injured arteries enables migration of vascular smooth muscle cells and neointima formation. The hope is that ADAMTS-7 may be a suitable therapeutic agent in combating restenosis of atherosclerotic blood vessels after angioplasties and related procedures (Wang et al. 2010). More recent study from the same laboratory shows that COMP inhibits vascular smooth muscle calcification by interacting with bone morphogenetic protein 2 (BMP2) and that the COMP in atherosclerotic arteries story is a little bit more complicated than initially thought (Du et al. 2011).

Though COMP has been involved in metabolism of multiple tissues, including cartilage, tendons and blood vessels the only mutations in the COMP gene known to be responsible for pathological conditions identified so far, are those affecting the skeleton, such as pseudoachondroplasia and multiple epiphyseal dysplasia (Rock et al. 2010; Posey and Hecht 2008). COMP is a good biomarker of cartilage turnover and was found to be elevated in osteoarthritis and rheumatoid arthritis (Posey et al. 2018).

References

Adams JC, Lawler J (2004) The thrombospondins. Int J Biochem Cell Biol 36:961–968

Adams JC, Lawler J (2011) The thrombospondins. Cold Spring Harb Perspect Biol 3:a00971

Agah A, Kyriakides TR, Lawler J, Bornstein P (2002) The lack of thrombospondin-1 (TSP1) dictates the course of wound healing in double-TSP1/TSP/2-null mice. Am J Pathol 161:831–839

Akbareian SE et al (2013) Enteric neueal crest-derived cells promote their migration by modifying their microenvironment through tenascin-C production. Dev Biol 382:446–456

Allaire E, Forough R, Clowes M, Starcher B, Clowes AW (1998) Local overexpression of TIMP-1 prevents aortic aneurysm degeneration and rupture in a rat model. J Clin Invest 102:1413–1420

Anthony S et al (2015) Multiple epiphyseal dysplasia. J Am Acad Orthop Surg 23:164–172
Ariens RA, Lai TS, Weisel JW, Greenberg CS, Grant PJ (2002) Role of factor XIII in fibrin clot formation and effects of genetic polymorphisms. Blood 100:743–754
Armstrong PC, Peter K (2012) GPIIb/IIIa inhibitors: from bench to bedside and back to bench again. Thromb Haemost 107:808–814
Aumailley M (2018) Isolation and purification of laminins. Methods Cell Biol 143:187–205
Aumailley M, Bruckner-Tuderman L, Carter WG, Deutzmann R, Edgar D, Ekblom P, Engel J, Engvall E, Hohenester E, Jones JCR, Kleinman HK, Marinkovich MP, Martin GR, Mayer U, Meneguzzi G, Miner JH, Miyazaki M, Patarroyo M, Paulsson M, Quaranta V, Sanes JR, Sasaki T, Sekiguchi K, Sorokin LM, Talts JF, Tryggvason K, Uitto J, Virtanen I, von der Mark K, Wewer UM, Yamada Y, Yurchenco PD (2005) A simplified laminin nomenclature. Matrix Biol 24:326–332
Baccarani-Contri M, Vincenzi D, Cicchetti F, Mori G, Pasquali-Ronchetti I (1990) Immunocytochemical localization of proteoglycans within normal elastin fibers. Eur J Cell Biol 53:305–312
Beals RK, Hecht F (1971) Congenital contractural arachnodactyly: a heritable disorder of connective tissue. Bone Joint Surg Am 53:987–993
Belmadani S, Bernal J, Wei CC, Pallero MA, Dell'italia L, Murphy-Ullrich JE, Brecek KH (2007) A thrombospondin-1 antagonist of transforming growth factor-beta activation blocks cardiomyopathy in rats with diabetes and elevated angiotensin II. Am J Pathol 171:777–789
Berk DR, Bentley DD, Bayliss SJ, Lind A, Urban Z (2012) Cutis laxa: a review. J Am Acad Dermatol 66:842.e1-17
Berndt A, Kosmehl H, Katenkamp D, Tauchmann V (1994) Appearance of the myofibroblastic phenotype in Dupuytren's disease is associated with a fibronectin, laminin, collagen type IV and tenascin extracellular matrix. Pathobiology 62:55–58
Bornstein P (2009) Matricellular proteins: an overview. J Cell Commun Signal 3:163–165
Cain SA, Morgan A, Sherratt MJ, Ball SG, Shuttleworth CA, Kielty CM (2006) Proteomic analysis of fibrillin-rich microfibrils. Proteomics 6:111–122
Castro MM, Rizzi E, Figueiredo-Lopes L, Fernandes K, Bendhack LM, Pitol DL, Gerlach RF, Tanus-Santos JE (2008) Metalloproteinase inhibition ameliorates hypertension and prevents vascular dysfunction and remodeling in renovascular hypertensive rats. Atherosclerosis 198:320–331
Chandrasekaran S, Guo NH, Rodrigues RG, Kaiser J, Roberts DD (1999) Pro-adhesive abd chemotactic activities of thrombospondin-1 for breast carcinoma cells are mediated by alpha3beta1 integrin and regulated by insulin-like growth factor-1 and CD98. J Biol Chem 274:11408–11416
Chang J, Chaudhuri O (2019) Beyond proteases: basement membrane mechanics and cancer invasion. J Cell Biol 218:2456–2469
Charbonneau NL, Dzamba BJ, Ono RN, Keene DR, Corson GM, Reinhardt DP, Sakai LY (2003) Fibrillins can co-assemble in fibrils, but fibrillin fibril composition displays cell-specific differences. J Biol Chem 278:2740–2749
Chen H, Sottile J, Strickland DK, Mosher DF (1996) Binding and degradation of thrombospondin-1 mediated through heparan sulfate proteoglucans and low-density-lipoprotein receptor-related protein: localization of the functional activity to the trimeric N-terminal heparin-binding region of thrombospondin-1. Biochem J 318:959–963
Chipev CC, Simman R, Hatch G, Katz AE, Siegel DM, Simon M (2000) Myofibroblast phenotype and apoptosis in keloid and palmar fibroblasts in vitro. Cell Death Differ 7:166–176
Chiquet M, Gelman L, Lutz R, Maier S (2009) From mechanostransduction to extracellular matrix gene expression in fibroblasts. Biochim Biophys Acta 1793:911–920
Chiquet-Ehrismann R, Chiquet M (2003) Regulation and putative functions during pathological stress. J Pathol 200:488–499
Chiquet-Ehrismann R, Kalla P, Pearson CA, Beck K, Chiquet M (1988) Tenascin interferes with fibronectin action. Cell 53:383–390
Chockalingam PS, Glasson SS, Lohmander LS (2013) Tenascin-C levels in synovial fluid are elevated after injury to the human and canine joint and correlate with markers of inflammation and matrix degradation. Osteoarthr Cartil 21:339–345
Chung MI, Miao M, Stahl RJ, Chan E, Parkinson J, Keeley FW (2006) Sequences and domain structures of mammalian, avian, amphibian, and teleost tropoelastins: clues to the evolutionary history of elastin. Matrix Biol 25:495–504
Cilia La Corte AL, Philippou H, Ariëns RA (2011) Role of fibrin structure in thrombosis and vascular disease. Adv Protein Chem Struct Biol 83:75–127
Clark RA, Lanigan JM, DellaPelle P, Manseau E, Dvorak HF, Colvin RB (1982) Fibronectin and fibrin provide a provisional matrix for epidermal cell migration during wound reepithelialization. J Invest Dermatol 79:264–269
Csiszar K (2001) Lysyl oxidases: a novel multifunctional amine oxidase family. Prog Nucleic Acid Res Mol Biol 70:1–32
Dallas SL, Chen Q, Sivakumar P (2006) Dynamics of assembly and reorganization of extracellular matrix proteins. Curr Top Dev Biol 75:1–24
De Vega S et al (2007) TM14 is a new member of the fibulin family (fibulin-7) that interacts with extracellular matrix molecules and is active for cell binding. J Biol Chem 282:30878–30888
De Vega S et al (2016) Identification of peptides derived from the C-terminal domain of fibulin-7 active for endothelial cell adhesion and tube formation disruption. Biopolymers 106:184–195
Della Corte A, De Santo LS, Montagnani S, Quarto C, Romano G, Amarelli C, Scardone M, De Feo M,

Cotrufo M, Caianiello G (2006) Spatial patterns of matrix protein expression in dilated ascending aorta with aortic regurgitation: congenital bicuspid valve versus Marfan's syndrome. J Heart Valve Dis 15:20–27
Demidova-Rice TN, Geevarghese A, Herman IM (2011) Bioactive peptides derived from vascular endothelial cell extracellular matrices promote microvascular morphogenesis and wound healing in vitro. Wound Repair Regen 19(1):59–70
DeVega A, Iwamoto T, Yamada Y (2009) Fibulins: multiple roles in matrix structures and tissue functions. Cell Mol Life Sci 66:1890–1902
Di Cesare P, Hauser N, Lehman D, Pasumarti S, Paulsson M (1994) Cartilage oligomeric matrix protein (COMP) is an abundant component of tendon. FEBS Lett 354:237–240
Domogatskaya A, Rodin S, Tryggvason K (2012) Functional diversity of laminins. Annu Rev Cell Dev Biol 28:523–553
Donaldson DJ, Mahan JT, Amrani D, Hawiger J (1989) Fibrinogen-mediated epidermal cell migration: structural correlates for fibrinogen function. J Cell Sci 94:101–108
Doolittle RF, Goldbaum DM, Doolittle LR (1978) Designation of sequences involved in the "coiled-coil" interdominal connections in fibrinogen: constructions of an atomic scale model. J Mol Biol 120:311–325
Du Y, Wang Y, Wang L, Liu B, Tian Q, Liu CJ, Zhang T, Xu Q, Zhu Y, Ake O, Qi Y, Tang C, Kong W, Wang X (2011) Cartilage oligomeric matrix protein inhibits vascular smooth muscle calcification by interacting with bone morphogenetic protein-2. Circ Res 108:917–928
Eberwein P, Reinhard T, Agostini H, Poloschek CM, Guthoff R, Auw-Haedrich C (2010) Intensive intracorneal keloid formation in a case of Peters plus syndrome and in Peters anomaly with maximum manifestation. Ophthalmologe 107:178–181
Egging D et al (2007) Interactions of human tenascin-X domains with dermal extracellular matrix molecules. Arch Dermatol Res 298(8):389–396
Elefteriou F et al (2001) Binding of tenascin-X to decorin. FEBS Lett 495(1–2):44–47
Elzie CA, Murphy-Ullrich JE (2004) The N-terminus of thrombospondin: the domain stands apart. Int J Biochem Cell Biol 36:1090–1101
Faury G, Pezet M, Knutsen RH, Boyle WA, Heximer SP, MacLean SE, Minkes RK, Blumer KJ, Kovacs A, Kelly DP, Li DY, Starcher B, Mecham RP (2003) Developmental adaptation of the mouse cardiovascular system to elastin haploinsufficiency. J Clin Invest 112:1419–1428
Fish RJ, Neerman-Arbez M (2012) Fibrinogen gene regulation. Thromb Haemost 108:419–426
Flück M, Tunc-Civelek V, Chiquet M (2000) Rapid and reciprocal regulation of tenascin-C and tenascin-Y expression by loading of skeletal muscle. J Cell Sci 113:3583–3591
Flück M, Mund SI, Schittny JC, Klossner S, Durieux AC, Giraud MN (2008) Mechano-regulated tenascin-C orchestrates muscle repair. Proc Natl Acad Sci U S A 105:13662–13667
Fujishima Y et al (2017) Adiponectin association with T-cadherin protects against neointima proliferation and atherosclerosis. FASEB J 31(4):1571–1583
Fung YC (1993) Biomechanics: mechanical properties of living tissues, 2nd edn. Springer, New York
Geffrotin C et al (1995) Distinct tissue distribution in pigs of tenascin-X and tenascin-C transcripts. Eur J Biochem 231(1):83–92
Gheduzzi D, Guerra D, Bochicchio B, Pepe A, Tamburro AM, Quaglino D, Mithieux S, Weiss AS, Pasquali Ronchetti I (2005) Heparan sulphate interacts with tropoelastin, with some tropoelastin peptides and is present in human dermis elastic fibers. Matrix Biol 24:15–25
Giltay R, Timpi R, Kostka G (1999) Sequence, recombinant expression and tissue localization of two novel extracellular matrix proteins, fibulin-3 and fibulin-4. Matrix Biol 18:469–480
Goldblum SE, Young BA, Wang P, Murphy-Ullrich JE (1999) Thrombospondin-1 induces tyrosine phosphorylation of adherens junction proteins and regulates an endothelial paracellular pathway. Mol Biol Cell 10:1537–1551
Greenwald SJ (2008) Ageing of the conduit arteries. J Pathol 211:157–172
Grounds MD, Sorokin L, White J (2005) Strength at the extracellular matrix-muscle interface. Scand J Med Sci Sports 15:381–391
Hagios C, Koch M, Spring J, Chiquet M, Chiquet-Ehrismann R (1996) Tenascin-Y: a protein of novel domain structure is secreted by differentiated fibroblasts of muscle connective tissue. J Cell Biol 134:1499–1512
Halasz K, Kassner A, Morgelin M, Heinegård D (2007) COMP as a catalyst in collagen fibrillogenesis. J Biol Chem 282:31166–31173
Halper J (2021) Major proteins of the extracellular proteins. In: Allewell NM (ed) Encyclopedia of biological chemistry. Elsevier
Hambleton S, Valeyev NV, Muranyi A, Knott V, Werner JM, McMichael AJ, Handford PA, Downing AK (2004) Structural and functional properties of the human notch-1 ligand binding region. Structure 12:2173–2183
Hanna NN, Eickholt K, Agamanolis D, Burnstine R, Edward DP (2010) Atypical Peters plus syndrome with new associations. J AAPS 14:181–183
Heinegård D (2009) Proteoglycans and more – from molecules to biology. Int J Exp Path 70:575–586
Heinonen TY, Maki M (2009) Peters'-plus syndrome is a congenital disorder of glycosylation caused by a defect in the beta1,3-glucosyltransferase that modifies thrombospondin type 1 repeats. Ann Med 41:2–10
Hess D, Keusch JJ, Lesnik Oberstein SA, Hennekam RC, Hofsteenge J (2008) Peter Plus syndrome is a new congenital disorder of glycosylation and involves defective O-glycosylation of thrombospondin type 1 repeats. J Biol Chem 283:7354–7360

Hesselstrand R, Kassner A, Heinegård D, Saxne T (2008) COMP: a candidate molecule in the pathogenesis of systemic sclerosis with a potential as a disease marker. Ann Rheum Dis 67:1242–1248
Hinek A, Rabinovitch M (1994) 67-kD elastin-binding protein is a protective "companion" of extracellular insoluble elastin and intracellular tropoelastin. J Cell Biol 126:563–574
Hirai M, Ohbayashi T, Horiguchi M, Okawa K, Hagiwara A, Chien KR, Kita T, Nakamura T (2007) Fibulin-5/DANCE has an elastogenic organizer activity that is abrogated by proteolytic cleavage in vivo. J Cell Biol 176:1061–1071
Hohenester E (2019) Structural biology of laminins. Essays Biochem 63:285–295
Hohenstein B, Daniel C, Hausknecht B, Boehmer K, Riess R, Amann KU, Hugo CP (2008) Correlation of enhanced thrombospondin-1 expression, TGF-beta signalling, and proteinuria in human type-2 diabetic nephropathy. Nephrol Dial Transplant 23:3880–3887
Holden P, Meadows RS, Chapman KL, Grant ME, Kadler KE, Briggs MD (2001) Cartilage oligomeric matrix protein interacts with type IX collagen, and disruptions to these interactions identify a pathogenetic mechanism in a bone dysplasia family. J Biol Chem 276:6046–6055
Horiguchi M, Inoue T, Ohbayashi T, Hirai M, Noda K, Marmorstein LY, Yabe D, Takagi K, Akama TO, Kita T, Kimura T, Nakamura T (2009) Fibulin-4 conducts proper elastogenesis via interaction with cross-linking enzyme lysyl oxidase. Proc Natl Acad Sci U S A 106:19029–19034
Huang W, Chiquet-Ehrismann R, Moyano JV, Garcia-Pardo V, Orend G (2001) Interference of tenascin-C with syndecan-4 binding to fibronectin blocks cell adhesion and stimulates tumor cell proliferation. Cancer Res 61:8586–8594
Hubmacher D, Sabatier L, Annis DS, Mosher DF, Reinhardt DP (2011) Homocysteine modifies structural and functional properties of fibronectin and interferes with the fibronectin-fibrillin-1 interaction. Biochemistry 50:5322–5332
Ikuta T, Ariga H, Matsumoto K (2000) Extracellular matrix tenascin-X in combination with vascular endothelial growth factor B enhances endothelial cell proliferation. Genes Cells 5(11):913–927
Imanaka-Yoshida K (2012) Tenascin-C in cardiovascular tissue remodeling: from development to inflammation and repair. Circ J 76:2513–2520
Imanaka-Yoshida K, Yoshida T, Miyagawa-Tomita S (2014) Tenascin-C in development and disease of blood vessels. Anat Rec (Hoboken) 297:1747–1757
Jackson GC et al (2012) Pseudoachondroplasia and multiple epiphyseal dysplasia: a 7-year comprehensive analysis of the known disease genes identify novel and recurrent mutations and procides an accurate assessment of their relative contribution. Hum Mutat 33:144–157
Jaman NB, Al-Sayegh A (2016) Seizures as an atypical feature of Beal's syndrome. Sultan Qaboos Univ Med J 16:e375–e378
Järvinen TA, Józsa L, Kannus P, Järvinen TL, Hurme T, Kvist M, Pelto-Huikko M, Kalimo H, Järvinen M (2003) Mechanical loading regulates the expression of tenascin-C in the myotendinous junction and tendon but does not induce de novo synthesis in the skeletal muscle. J Cell Sci 116:857–866
Jiang L, Wang M, Zhang J, Monticone RE, Telljohann R, Spinnetti G, Pintus G, Lakatta EG (2008) Increased calpain-1 activity mediates age-associated angiotensin II signaling of vascular smooth muscle cells. PLoS One 3:e2231
Jones FS, Jones PL (2000) The tenascin family of ECM glycoproteins: structure, function, and regulation during embryonic development and tissue remodeling. Dev Dyn 218
Karnik SK, Brooke BS, Bayes-Genis A, Sorensen L, Wythe JD, Schwartz RS, Keating MT, Li DY (2003) A critical role for elastin signaling in vascular morphogenesis and disease. Development 130:411–423
Karoulias SZ et al (2019) A novel ADAMTS17 variant that causes Weill-Marchesani syndrome 4 alters fibrillin-1 and collagen type I deposition in the extracellular matrix. Matrix Biol. (in press)
Kartashova EA, Sarvilina IV (2019) About the prognostic role of fibulin-5 protein in the progression of pathological vascular remodeling in patients with isolated sistolic arterial hypertension. Adv Gerontol 32(6):1003–1010
Kielty CM (2006) Elastic fibres in health and disease. Expert Rev Mol Med 8:1–23
Kielty CM, Sherratt MJ, Marson A, Baldock C (2005) Fibrillin microfibrils. Adv Protein Chem 70:405–436
Kim YM, Kim EC, Kim Y (2011) The human lysyl oxidase-like 2 protein functions as an amine oxidase toward collagen and elastin. Mol Biol Rep 38:145–149
Kjaer M (2004) Role of extracellular matrix in adaptation of tendon and skeletal muscle to mechanical loading. Physiol Rev 84:649–698
Klenotic PA, Munier FL, Marmorstein LY, Anand-Apte B (2004) Tissue inhibitor of metalloproteinase-3 (TIMP-3) is a binding partner of epithelial growth factor-containing fibulin-like extracellular matrix protein 1 (EFEMP1): implications for macular degenerations. J Biol Chem 279:30469–30473
Kostrominova TY, Brooks SV (2013) Age-related changes in structure and extracellular matrix protein expression levels in rat tendons. Age
Kozel BA, Mecham RP (2019) Elastic fiber ultrastructure and assembly. Matrix Biol 84:31–40
Kozel BA, Ciliberto CH, Mecham RP (2004) Deposition of tropoelastin into the extracellular matrix requires a competent elastic fiber scaffold but not live cells. Matrix Biol 23:23–34
Kozel BA, Rongish BJ, Czirok A, Zach J, Little CD, Davis EC, Knutsen RH, Wagenseil JE, Levy MA, Mecham RP (2006) Elastic fiber formation: a dynamic view of

extracellular matrix assembly using timer reporters. J Cell Physiol 207:87–96

Kreja L, Liedert A, Schlenker H, Brenner RE, Fiedler J, Friemert B, Dürselen L, Ignatius A (2012) Effects of mechanical strain on human mesenchymal stem cells and ligament fibroblasts in a textured poly(L-lactide) scaffold for ligament tissue engineering. J Mater Sci Mater Med 23:2575–2582

Leahy DJ, Aukhil I, Erickson HP (1996) 2.0 A crystal structure of a four-domain segment of human fibronectin encompassing the RGD loop and synergy region. Cell 84:155–164

Lee JE, Kim Y (2006) A tissue-specific variant of the human lysyl oxidase-like protein 3 (LOXL3) functions as an amine oxidase with substrate specificity. J Biol Chem 281:37282–37290

Lethias C et al (2006) A model of tenascin-X integration within the collagenous network. FEBS Lett 580(26):6281–6285

Li DY, Brooke D, Davis EC, Mecham RP, Sorensen LK, Boak KK, Eichwald E, Keating MT (1998) Elastin is an essential determinant of arterial morphogenesis. Nature 393:276–289

Li Z, Froehlich J, Galis ZS, Lakatta EG (1999) Increased expression of matrix metalloproteinase-2 in the thickened intima of aged rats. Hypertension 33:116–123

Liu X, Zhao Y, Gao J, Pawlyk B, Starcher B, Spencer JA, Yanagisawa H, Zuo J, Li T (2004) Elastic fiber homeostasis requires lysyl oxidase-like 1 protein. Nat Genet 36:178–182

Lu A, Miao M, Schoeb TR, Agarwal A, Murphy-Ullrich JE (2011) Blockade of TSP-1 dependent TGF-beta activity reduces renal injury and proteinuria in a murine model of diabetic nephropathy. Am J Pathol 178:2573–2586

MacDonald PR, Lustig A, Steinmetz MO, Kammerer RA (2010) Laminin chain assembly is regulated by specific coiled-coil interactions. J Struct Biol 170:398–405

Mackey AL, Brandstetter S, Schjerling P, Bojsen-Moller J, Qvortrup K, Pedersen MM, Doessing S, Kjaer M, Magnusson SP, Langberg H (2011) Sequences response of extracellular matrix de-adhesion and fibrotic regulators after muscle damage is involved in protection against future injury in human skeletal muscle. FASEB J 25:1943–1959

Mackie EJ, Scott-Burden T, Hahn AW, Kern F, Bernhardt J, Regenass S, Weller A, Bühler FR (1992) Expression of tenascin by vascular smooth muscle cells. Alterations in hypertensive rats and stimulation by angiotensin II. Am J Pathol 141:377–388

Mao Y, Schwarzbauer J (2005) Fibronectin fibrillogenesis, a cell-mediated matrix assembly process. Matrix Biol 24:389–399

Matsumoto KI, Aoki H (2020) The roles of tenascins in cardiovascular, inflammatory, and heritable connective tissue diseases. Front Immunol 11:609752

Matsumoto K et al (2002) Distribution of extracellular matrix tenascin-X in sciatic nerves. Acta Neuropathol 104(5):448–454

Mecham RP (1998) Overview of extracellular matrix, Current protocols in cell biology. Wiley, New York

Midwood KS, Schwarzbauer JS (2002) Tenascin-C modulates matrix contraction via focal adhesion kinase- and Rho-mediated signaling pathways. Mol Biol Cell 13:3601–3613

Midwood KS, Hussenet T, Langlois B, Orend G (2011) Advances in tenascin-C biology. Cell Mol Life Sci 68:3175–3199

Midwood KS et al (2016) Tenascin-C at a glance. J Cell Sci 129:4321–4327

Milewicz DM, Grossfield J, Cao SN, Kielty C, Covitz W, Jewett T (1995) A mutation in FBN1 disrupts profibrillin processing and results in isolated skeletal features of the Marfan syndrome. J Clin Invest 95:2373–2378

Mimura Y, Ihn H, Jinnin M, Assano Y, Yamane K, Tamaki K (2005) Constitutive thrombospondin-1 overexpression contributes to autocrine transforming growth factor-beta signalin in cultured scleroderma fibroblasts. Am J Pathol 166:1451–1463

Miner JH, Yurchenco PD (2004) Laminin functions in tissue morphogenesis. Annu Rev Cell Dev Biol 20:255–284

Mithieux SM, Weiss AS (2005) Elastin. Adv Protein Chem 70:437–461

Mithieux SM, Wise SG, Weiss AS (2012) Tropoelastin — a multifaceted naturally smart material. Adv Drug Del Rev. (in press)

Molloy TJ, de Bock CE, Wang Y, Murrell GA (2006) Gene expression changes in SNAP-stimulated and iNOS-transfected tenocytes--expression of extracellular matrix genes and its implications for tendon-healing. J Orthop Res 24:1869–1882

Montanaro L et al (2006) Evidence of a linkage between matrilin-1 gene (MATN1) and idiopathic scoliosis. Scoliosis 1:21

Mosher DF, Adams JC (2012) Adhesion-modulating/matricellular ECM protein families: a structural, functional and evolutionary appraisal. Matrix Biol 31:155 161

Muiznieks LD, Weiss AS, Keeley FW (2010) Structural disorder and dynamics of elastin. Biochem Cell Biol 88:239–250

Murphy-Ullrich JE, Iozzo RV (2012) Thrombospondins in physiology and disease: new tricks for old dogs. Matrix Biol 31:152–154

Murphy-Ullrich JE, Poczatek M (2000) Activation of latent TGF-beta by thrombospondin-1: mechanism and physiology. Cytokine Growth Factor Rev 11

Nakamura T (2018) Role of short fibulins, a family of matricellular proteins, in lung matrx assembly and disease. Matrix Biol 73:21–33

O'Connor WN, Davis JB Jr, Geissler R, Cottrill CM, Noonan JA, Todd EP (1985) Supravalvular aortic stenosis. Clinical and pathological observations in six patients. Arch Pathol Lab Med 109:179–185

Okamoto H, Imanaka-Yoshida K (2012) Matricellular proteins: new molecular targets to prevent heart failure. Cardiovasc Ther 30:e198–e209

Oldberg Å, Antonssen P, Lindholm K, Heinegård D (1992) COMP (cartilage oligomeric matrix protein) is structurally related to thrombospondins. J Biol Chem 267:22346–22350

Page TH, Charles PJ, Piccinini AM, Nicolaidou V, Taylor PC, Midwood KS (2012) Raised circulating tenascin-C in rheumatoid arthritis. Arthritis Res Ther 14:R260

Paulsson M, Matrilins WR (2018) Methods Cell Biol 143:429–446

Pérez-García S et al (2020) Profile of matrix-remodeling proteinases in osteoarthritis: impact of fibronectin. Cell 9:40

Perrotta I, Russo E, Camastra C, Filice G, Di Mizio G, Colosimo F, Ricci P, Tripepi S, Amorosi A, Triumbari F, Donato G (2011) New evidence for a critical role of elastin in calcification of native heart valves: immunohistochemical and ultrastructural study with literature review. Histopathology 59:504–513

Posey KL, Hecht JT (2008) The role of cartilage oligomeric matrix protein (COMP) in skeletal disease. Curr Drug Targets 9:869–877

Posey KL, Coustry F, Hecht JT (2018) Cartilage oligomeric matrix protein: COMPopathies and beyond. Matrix Biol 71-72:161–173

Potts JR, Campbell ID (1994) Fibronectin structure and assembly. Curr Opin Cell Biol 6:648–655

Raghunath M, Putnam EA, Ritty T, Hamstra D, Park ES, Tschodrich-Rotter M, Peters P, Rehemtulla A, Milewicz DM (1999) Carboxy-terminal conversion of profibrillin to fibrillin at a basic site by PACE/furin-like activity required for incorporation in the matrix. J Cell Sci 112:1093–1100

Ramirez F, Dietz HC (2007) Fibrillin-rich microfibrils: structural determinants of morphogenetic and homeostatic events. J Cell Physiol 213:326–330

Ramirez F, Sakai LV (2010) Biogenesis and function of fibrillin assemblies. Cell Tissue Res 339:71–82

Riessen R, Isner JM, Blessing E, Loushin C, Nikol S, Wight TN (1994) Regional differences in the distribution of the proteoglycans biglycan and decorin in the extracellular matrix of atherosclerotic and restenotic human coronary arteries. Am J Pathol 144:962–974

Riessen R, Fenchel M, Chen H, Axel DL, Karsch KR, Lawler J (2001) Cartilage oligomeric matrix protein (thrombospondin-5) is expressed by human vascular smooth muscle cells. Arteriosc Thromb Vasc Biol 21:47–54

Roark EF et al (1995) The association of human fibulin-1 with elastic fibers: an immunohistological, ultrastructural, and RNA study. J Histochem Cytochem 43:401–411

Robert DD, Miller TW, Rogers NM, Yao M, Isenberg JS (2012) The matricellular protein thrombospondin-1 globally regulates cardiovascular function and responses to stress via CD47. Matrix Biol 31:162–169

Robertson I, Jensen S, Handford P (2011) TB domain proteins: evolutionary insights into the multifaceted roles of fibrillins and LTBPs. Biochem J 433:263–276

Robinson PN, Arteaga-Solis E, Baldock C, Collod-Béroud G, Booms P, De Paepe A, Dietz HC, Guo G, Handford PA, Judge DP, Kielty CM, Loeys B, Milewicz DM, Ney A, Ramirez F, Reinhardt DP, Tiedemann K, Whiteman P, Godfrey M (2006) The molecular genetics of Marfan syndrome and related disorders. J Med Genet 43:769–787

Rock MJ, Holden P, Horton WA, Cohn DH (2010) Cartilage oligometric matrix protein promotes cell attachment via two independent mechanisms involving CD47 and αVβ3 integrin. Mol Cell Biochem 338:215–224

Rodriguez-Revenga L, Iranzo P, Badenas C, Puig S, Carrio A, Mila M (2004) A novel elastin gene mutation resulting in an autosomal dominant from of cutis laxa. Arch Dermatol 149:1135–1139

Rogers NM, Yao M, Novelli EM, Thomson AW, Roberts DD, Isenberg JS (2012) Activated CD47 regulates multiple vascular and stress responses: implications for acute kidney injury and its management. Am J Physiol Renal Physiol 303:F1117–F1125

Rosenberg K, Olsson H, Mörgelin M, Heinegård D (1998) Cartilage oligomeric matrix protein shows high affinity zinc-dependent interaction with triple helical collagen. J Biol Chem 273:20397–20403

Sabatier L, Chen D, Fagotto-Kaufmann C, Hubmacher D, McKee MD, Annis DS, Mosher DF, Reinhardt DP (2009) Fibrillin assembly requires fibronectin. Mol Biol Cell 20:846–858

Sahni A, Francis CW (2000) Vascular endothelial growth factor binds to fibrinogen and fibrin and stimulates endothelial cell proliferation. Blood 96:3772–3778

Sahni A, Odrljin T, Francis CW (1998) Binding of basic fibroblast growth factor to fibrinogen and fibrin. J Biol Chem 273:7554–7559

Sarbacher CA, Halper JT (2019) Connective tissue and age-related diseases. Subcell Biochem 91:281–310

Sasaki T et al (1997) Dimer model for the microfibrillar protein fibulin-2 and identification of the connecting disulfide bridge. EMBO J 16:3035–3043

Sato N, Nakamura M, Chikama T, Nishida T (1999) Abnormal deposition of laminin and type IV collagen at corneal epithelial basement membrane during wound healing in diabetic rats. Jpn J Ophthalmol 43:343–347

Sato F, Wachi H, Ishida M, Nonaka R, Onoue S, Urban Z, Starcher BC, Seyama Y (2007) Distinct steps of cross-linking, self-association, and maturation of tropoelastin are necessary for elastic fiber formation. J Mol Biol 369:841–851

Schalkwijk J et al (2001) A recessive form of the Ehlers-Danlos syndrome caused by tenascin-X deficiency. N Engl J Med 345(16):1167–1175

Schwill S, Seppelt P, Grünhagen J, Ott CE, Jugold M, Ruhparwar A, Robinson PN, Karck M, Kallenbach K (2013) The fibrillin-1 hypomorphic mgR/mgR murine model of Marfan syndrome shows severe elastolysis in all segments of the aorta. J Vasc Surg. in press

Shimizu R, Saito R, Hoshino K, Ogawa K, Negishi T, Nishimura J, Mitsui N, Osawa M, Ohashi H (2010) Severe Peters Plus syndrome-like phenotype with anterior eye staphyloma and hypoplastic left heart syn-

drome: proposal of a new syndrome. Congenit Anom (Kyoto) 50:197–199
Singh P, Schwarzbauer JE (2012) Fibronectin and stem cell differentiation – lessons from chondrogenesis. J Cell Sci 125:3703–3712
Smith RKW, Zunino L, Webbon PM, Heinegård D (1997) The distribution of cartilage oligomeric matrix protein (COMP) in tendon and its variation with tendon site, age and load. Matrix Biol 16:255–271
Smith MR, Wright IM, Minshall GJ, Dudhia J, Verheyen K, Heinegård D, Smith RK (2011) Increased cartilage oligomeric matrix protein concentrations in equine digital flexor tendon sheath synovial fluid predicts interthecal tendon damage. Vet Surg 40:54–58
Södersten F, Hultenby K, Heinegård D, Johnston C, Ekman S (2013) Immunolocalization of collagens (I and III) and cartilage oligomeric matrix protein in the normal and injured equine superficial digital flexor tendon. Connect Tissue Res 54:62–69
Sweetwyne MT, Murphy-Ullrich JE (2012) Thrombospondin1 in tissue repair and fibrosis: TGF-β-dependent and independent mechanisms. Matrix Biol 31:178–186
Takahashi S, Leiss M, Moser M, Ohashi T, Kitao T, Heckmann D, Pfeifer A, Kessler H, Takagi J, Erickson HP, Fässler R (2007) The RGD motif in fibronectin is essential for development but dispensable for fibril assembly. J Cell Biol 178:167–178
Tamarina NA, McMillan WD, Shively VP, Pearce WH (1999) Expression of matrix metalloproteinases and their inhibitors in anuerysm and normal aorta. Surgery 122:264–271
Tassabehji M, Metcalfe K, Hurst J, Ashcroft GS, Kielty C, Wilmot C, Donnai D, Read AP, Jones CJP (1998) An elastin gene mutation producing abnormal tropoelastin and abnormal elastic fibres in a patient with autodomal dominant cutis laxa. Hum Mol Genet 7:1021–1028
Taylor SH, Al-Youha S, Van Agtmael T, Lu Y, Wong J, McGrouther DA, Kadler KE (2011) Tendon is covered by a basement membrane epithelium that is required for cell retention and the prevention of adhesion formation. PLoS One 6:e16337
Thomson J et al (2019) Fibrillin microfibrils and elastic fibre proteins: functional interactions and extracellular regulation of growth factors. Semin Cell Dev Biol 89:109–117
Timpl R et al (2003) Fibulins: a versatile family of extracellular matrix proteins. Nat Rev Mol Cell Biol 4:479–489
Todorovich-Hunter L, Johnson D, Ranger P, Keeley F, Rabinovitch M (1988) Altered elastin and collagen synthesis associated with progressive pulmonary hypertension induced by monocrotaline. A biochemical and ultrastructural study. Lab Investig 58:184–195
Tsuda T (2018) Extracellular interactions between fibulins and transforming growth factor (TGF)-β in physiological and pathological conditions. Int J Mol Sci 19
Tsuda T, Wang H, Timpl R, Chu ML (2001) Fibulin-2 expression marks transformed mesenchymal cells in developing cardiac valves, aortic arch vessels and coronary vessels. Dev Dyn 222:89–100
Tsunezumi J et al (2018) Fibulin-7, a heparin binding matricellular protein, promotes renal tubular calcification in mice. Matrix Biol 74:5–20
Tucker RP, Chiquet-Ehrismann R (2009) The regulation of tenascin expression by tissue microenvironments. Biochim Biophys Acta 1793:888–892
Tucker RP, Drabikowski K, Hess JF, Ferralli J, Chiquet-Ehrismann R, Adams JC (2006) Phylogenetic analysis of the tenascin gene family: evidence of origin early in the chordate lineage. BMC Evol Biol 6:60
Urban Z, Michels VV, Thibodeau SN, Davis EC, Bonnefont J-P, Munnich A, Eyskens B, Gewillig M, Devriendt K, Boyd CD (2000) Isolated supravalvular aortic stenosis: functional haploinsufficiency of the elastin gene as a result of nonsense-mediated decay. Hum Genet 106:577–588
Urban Z et al (2009) Mutations in LTBP4 cause a syndrome of impaired pulmonary, gastrointestinal, genitourinary, musculoskeletal, and dermal development. Am J Hum Genet 85:593–605
Valcourt U et al (2015) Tenascin-X: beyond the architectural function. Cell Adhes Migr 9(1–2):154–165
Veit G et al (2006) Collagen XII interacts with avian tenascin-X through its NC3 domain. J Biol Chem 281(37):27461–27470
Wachi H, Nonaka R, Sato F, Shibata-Sato K, Ishida M, Iketani S, Maeda I, Okamoto K, Urban Z, Onoue S, Seyama Y (2008) Characterization of the molecular interaction between tropoelastin and DANCE/fibulin-5. J Biochem 143:633–639
Wagenseil JE, Mecham RP (2009) Vascular extracellular matrix and arterial mechanics. Physiol Rev 89:957–989
Wagenseil JE, Mecham RP (2012) Elastin in large artery stiffness and hypertension. J Cardiovasc Transl Res 5:264–273
Wagenseil JE, Mercham RP (2009) Vascular extracellular matrix and arterial mechanisms. Physiol Rev 89:957–989
Wang L, Wang X, Kong W (2010) ADAMTS-7, a novel proteolytic culprit in vascular remodeling. Sheng Li Xue Bao 62:285–294
White ES, Muro AF (2011) Fibronectin splice variants. Understanding their multiple roles in health and diases using engineered mouse models. IUBMB Life 63:538–546
Whittaker CA, Hynes RO (2002) Distribution and evolution of of von Willebrand/integrin A domains: widely dispersed domains with roles in cell adhesion and and elsewhere. Mol Biol Cell 13:3369–3387
Wolinsky H (1970) Response of the rat aortic media to hypertension. Morphological and chemical studies. Circ Res 26:507–522
Yanagisawa H, Davis EC (2010) Unraveling the mechanism of elastic fiber assembly: the roles of short fibulins. Int J Biochem Cell Biol 42:1084–1093

Yanagisawa H, Schluterman MK, Brekkcn RA (2009) Fibulin-5, an integrin-binding matricellular protein: its function in development and disease. J Cell Commun Signal 3:337–347

Yeo GC, Keeley FW, Weiss AS (2011) *Coacervation of tropoelastin*. Adv Colloid Interface Science 167:94–103

Zhang H, Apfelroth SD, Hu W, Davis EC, Sanguineti C, Bonadio J, Mecham RP, Ramirez F (1994) Structure and expression of fibrillin-2, a novel microfibrillar component preferentially located in elastic matrices. J Cell Biol 124:855–863

Zhang HY, Timpl R, Sasaki T, Chu ML, Ekblom P (1996) Fibulin-1 and fibulin-2 expression during organogenesis in the developing mouse embryo. Dev Dyn 205:348–364

Zheng Q, Davis EC, Richardson JA, Starcher BC, Li T, Gerard RD, Yanagisawa H (2007) Molecular analysis of fibulin-5 function during de novo synthesis of elastic fibers. Mol Cell Biol 27:1083–1095

Zilberberg L et al (2015) Genetic analysis of the contribution of LTBP-3 to thoracic aneurysm in Marfan syndrome. Proc Natl Acad Sci U S A 112:14012–14017

Proteoglycans and Diseases of Soft Tissues

5

Chloe Taejoo Hwang and Jaroslava Halper

Abstract

Proteoglycans consist of protein cores to which at least one glycosaminoglycan chain is attached. They play important roles in the physiology and biomechanical function of tendons, ligaments, cardiovascular system, and other systems through their involvement in regulation of assembly and maintenance of extracellular matrix, and through their participation in cell proliferation together with growth factors. They can be divided into two main groups, small and large proteoglycans. The small proteoglycans are also known as small leucine-rich proteoglycans (SLRPs) which are encoded by 18 genes and are further subclassified into Classes I-V. Several members of Class I and II, such as decorin and biglycan from Class I, and Class II fibromodulin and lumican, are known to regulate collagen fibrillogenesis. Decorin limits the diameter of collagen fibrils during fibrillogenesis. The function of biglycan in fibrillogenesis is similar to that of decorin. Though biomechanical function of tendon is compromised in decorin-deficient mice, decorin can substitute for lack of biglycan in biglycan-deficient mice. New data also indicate an important role for biglycan in disorders of the cardiovascular system, including aortic valve stenosis and aortic dissection. Two members of the Class II of SLRPs, fibromodulin and lumican bind to the same site within the collagen molecule and can substitute for each other in fibromodulin- or lumican-deficient mice.

Aggrecan and versican are the major representatives of the large proteoglycans. Though they are mainly found in the cartilage where they provide resilience and toughness, they are present also in tensile portions of tendons and, in slightly different biochemical form in fibrocartilage. Degradation by aggrecanase is responsible for the appearance of different forms of aggrecan and versican in different parts of the tendon where these cleaved forms play different roles. In addition, they are important components of the ventricularis of cardiac valves. Mutations in the gene for versican or in the gene for elastin (which binds to versican) lead to severe disruptions of normal developmental of the heart at least in mice.

C. T. Hwang
Department of Pathology, College of Veterinary Medicine, The University of Georgia, Athens, GA, USA

J. Halper (✉)
Department of Pathology, College of Veterinary Medicine, and Department of Basic Sciences, AU/UGA Medical Partnership, The University of Georgia, Athens, GA, USA
e-mail: jhalper@uga.edu

J. Halper (ed.), *Progress in Heritable Soft Connective Tissue Diseases*, Advances in Experimental Medicine and Biology 1348, https://doi.org/10.1007/978-3-030-80614-9_5

Keywords

Small Leucine-Rich Proteoglycans (SLRPs) · Decorin · Biglycan · Fibromodulin · Lumican · Collagen fibrillogenesis · Aggrecan · Versican · Knockout mice · Aoric valve · Aorta · Marfan syndrome · Ehlers-Danlos syndrome · Heart development

Abbreviations

ADAMTS	a Disintegrin and metalloproteinase with thrombospondin motifs
BMP	Bone morphogenetic protein
CS	Chondroitin sulfate
DS	Dermatan sulfate
ECM	Extracellular matrix
EGF	Epidermal growth factor
FGF	Fibroblast growth factor
KS	Keratan sulfate
LDL	Low density lipoprotein
LRRs	Leucine-rich repeats
MMP	Matrix metalloproteinase
PRELP	Proline and arginine-rich end leucine-rich repeat protein
SLRP	Small leucine-rich proteoglycan
TGFβ	Transforming growth factor β
TNF	Tumor necrosis factor

Proteoglycans are ubiquitous molecules consisting of protein cores to which one or more glycosaminoglycan chains are attached. Their functions are numerous: regulation of assembly and maintenance of extracellular matrix, and participation, together with growth factors, in cell proliferation. They also regulate collagen fibrillogenesis and tensile strength of skin and tendons, affect tumor cell growth and invasion, influence corneal transparency, and neurite outgrowth among many other functions (Yoon and Halper 2005; Iozzo 1998). In the context of this volume two functions, the heavy involvement of proteoglycans in regulation of assembly and maintenance of extracellular matrix (ECM), and their participation in cell proliferation, particularly in certain tissues such as tendons, heart valves and arteries are perhaps the most pertinent to the topic and of most interest to us. We do include involvement of proteoglycans in seemingly isolated cardiovascular pathologies as some of them are affected also in many soft-tissue diseases, described in several chapters of this volume.

Proteoglycans play a more important role in the physiology and biomechanical function of tendons and ligaments than would be obvious from their low content in the tendon: they comprise less than 1% of the dry weight of tendons and ligaments (mostly decorin and biglycan, but also fibromodulin and aggrecan) but their presence is instrumental for proper assembly and organization of the structural backbone of tendons and ligaments (Carter and Beaupré 2007). Because of very low aggrecan content tendons have a lower water content (55% of wet weight) and higher collagen content (38%) than cartilage (Carter and Beaupré 2007; Vogel et al. 1994). With tendon injury the water content of the tendon increases as the composition of proteoglycans in the tendons changes, and the content of sulfated proteoglycans, including aggrecan increases (Samiric et al. 2009). Changes in proteoglycan makeup of tendons, including cartilaginous metaplasia also occur with age and loading. Calcification occurs in tendons of adult turkeys (Engel and Zerlotti 1967; Spiesz et al. 2012) and also in tendons of other species that undergo high tensile forces; for example, ossified tendons were found adjacent to posterior spinal elements of dinosaurs (Manning et al. 2009). Formation of fibrocartilage with local chondrometaplasia was found also in regions of mammalian, including human, tendons wrapped around bony prominences as an effect of transverse compressive forces (Berenson et al. 1996; Parkinson et al. 2011).

Proteoglycans are usually divided into two large groups of small and large proteoglycans (Yoon and Halper 2005). The small proteoglycans are also known as small leucine-rich proteoglycans (or SLRPs) (Yoon and Halper 2005; Iozzo 1999, 2007; Iozzo and Schaefer 2010).

These compounds consist of relatively small core proteins (~ 40 kDa) to which, or at least to most of them, glycosaminoglycan chains are attached. The core proteins contains numerous leucine-rich repeats (LRRs), which are 20–30 amino acid long with leucine situated in conserved positions (Yoon and Halper 2005).

SLRPs are a group of 18 compounds, and can be further divided into 5 classes. The first three classes, I-III, are canonical as they bind glycosaminoglycan chains (Halper 2021). These chains could be one to two chondroitin or dermatan sulfate (CS or DS), or several keratan sulfate (KS) (Yoon and Halper 2005; Iozzo and Murdoch 1996). The members of the first two classes are involved in collagen fibrillogenesis and related processes (and so we will focus on only those SLRPs). Decorin and biglycan, members of Class I, bind to the same site on collagen I. Class I SLRPs contain at least 10 leucine rich regions, and have chondroitin and/or dermatan sulfate attached to the protein core (Iozzo 2007; Pietraszek-Gremplewicz et al. 2019). Fibromodulin and lumican, members of Class II, bind to a different site on type I collagen molecule (Rada et al. 1993; Schönherr et al. 1995a, b; Svensson et al. 2000). The members of class II have 10 leucine rich regions in their central domain, and carry keratan sulfate chains attached to the leucine rich regions. All of these SLRPs are present in tendons where they participate in fibrillogenesis and thus contribute to the development of proper biomechanical function (Zhang et al. 2006). The effect of SLRPs on fibrillogenesis is not limited to soft tissues, but more specifically to tendon and its fibrillar collagens. Tanaka et al. have shown that decreased expression of decorin, biglycan and fibromodulin in osteoarthritic cartilage lead to reduced expression of type IX and XI procollagens, and thus to impaired fibrillogenesis and to thickening of the fibrils (Tanaka et al. 2019). The composition of amino acid sequences intervening between four cysteine residues comprising N-terminal cysteine clusters determines which class the particular SLRP belongs to (Schaefer and Iozzo 2008). In addition, proteoglycans in Classes I-III and extracellular matrix protein 2 (ECM2) contain a so called ear repeat, an LRR elongated by several residues (McEwan et al. 2006). McEwan et al. proposed that the ear-repeat C-terminal motif is the hallmark for members of the true SLRP family (McEwan et al. 2006).

Based on their protein homology Class II members can be further subdivided into three smaller groups. Class II SLRPs contain N-terminus clusters of tyrosine sulfate residues, and are, therefore, polyanionic. They are fibromodulin, lumican, proline and arginine-rich end leucine-rich repeat protein or prolargin (PRELP), keratocan and osteocadherin and they bind keratan sulfate and polyalactosamine (a nonsulfated form of keratan sulfate). Their exonic organization is similar among them with three exons (Iozzo and Schaefer 2010). All of them except for PRELP contain an N-terminal extension with at least one sulfate tyrosine residue (Heinegård 2009). The N-terminal region of PRELP binds to heparin instead (Bengtsson et al. 2000).

Three compounds are included in Class III. Epiphycan, and opticin have 6 LRRs each, whereas osteoglycin has seven LRRs (Pietraszek-Gremplewicz et al. 2019; Schaefer and Iozzo 2008). They also possess a consensus sequence for glycosylation. They usually appear in tissues as glycoproteins rather than as proteoglycans (Schaefer and Iozzo 2008).

Chondroadherin, nyctalopin and tsukushi form a new non-canonical Class IV, which should not be considered true SLRPs because of a different C-terminal capping (McEwan et al. 2006). They have 11 homologous LRRs flanked by cysteine rich N-terminus. Like members of Class I, tsukushi is an inhibitor of several bone morphogenetic proteins (BMP-2, −4 and −7) (Schaefer and Iozzo 2008; Ohta et al. 2006).

Another non-canonical group, Class V, includes podocan and podocan-like protein. Both are characterized by 20 LRRSs with homology to members of Class I and II. The ability of podocan to bind type I collagen makes this molecule even more similar to other SLRPs (Schaefer and Iozzo 2008; Shimizu-Hirota et al. 2004).

Schaefer and Iozzo suggested that many SLRPs, particularly those in Classes I – III, arose by duplication of chromosomal segments. This would explain the similarity and redundancy in their function (Schaefer and Iozzo 2008). A complete list of all proteoglycans and summary of their functions and activity can be found in Iozzo and Schaefer summary (Iozzo and Schaefer 2015).

The development of knockout mice had brought increased understanding of the role SLRPs play in normal physiology and of the role of mutated or deficient SLRPs in disease. Those SLRPs secreted into ECM are indispensable participants in proper assembly and function of ECM. This is the case especially for those belonging to Classes I and III which bind to several type of collagen, and thus play important roles in the assembly of collagen fibrils, extracellular matrix function, and by extension in physiology of most, if not all, tissues and organs (Iozzo and Schaefer 2015).

As pointed out by Schaefer and Iozzo, SLRP-linked genetic diseases lead to eye abnormalities. Interestingly, though point mutations in the human decorin gene result in congenital stromal corneal dystrophy, an autosomal dominant disorder limited to the cornea (Bredrup et al. 2005, 2010), decorin-deficient mice exhibit abnormal collagen fibril structure in the tendon and skin, but not in the cornea (Danielson et al. 1997). The fact that the decorin in congenital stromal corneal dystrophy was found to be truncated and prone to form high molecular weight complexes, at least *in vitro*, and opacities *in vivo* (Bredrup et al. 2010), and thus present, albeit defective whereas the absence of decorin mRNA or a protein itself in decorin-deficient mice (Danielson et al. 1997) likely accounts for the difference in phenotype due to complete decorin deficiency.

An N-terminus with a typical cluster of cysteine residues forming two sulfide bonds is considered a marker of Class I with decorin, biglycan and asporin as representative members. The Class I members have a similar gene organization with eight exons and highly conserved intron/exon junctions. ECM2, though much larger protein, has 33% of its LRR identical to the corresponding domains of decorin, and its gene is actually linked to the gene for asporin on chromosome 9 (Schaefer and Iozzo 2008).

Decorin is the main proteoglycan in tendon. It consists of three domains (Reed and Iozzo 2003). A single dermatan/chondroitin sulfate chain binds to an N-terminal region. A central region contains 10 leucine rich repeats forming repeating parallel beta sheets which contain a binding site for fibrils of several collagen types (type I, II, III, and VI) at D-periods which maintains fibril diameters and biomechanical stability (Kamma-Loger et al. 2016). It contains also binding sites for several other proteins such as for transforming growth factor β (TGFβ) (Schönherr et al. 1998) and epidermal growth factor (EGF) and its receptor (Santra et al. 2002). In addition, the attached glycosaminoglycan chain also has binding sites for TGFβ and tumor necrosis factor α (TNFα) (Hildebrand et al. 1994; Tiedemann et al. 2005). A C-terminal region contains, just like the N-terminus, several cysteine residues (Reed and Iozzo 2003). Decorin is detected in tendons during the first stage of collagen fibrillogenesis (Kuo et al. 2008), and remains expressed in tendons (and other connective tissues) throughout the life of an individual. Decorin regulates the size of collagen fibrils during fibrillogenesis with the purpose to limit the diameter of fibrils and modify the rate of the process (Heinegård 2009). Decorin knockout mice have collagen fibrils varying in dimension with thicker diameters and widely uneven shapes, especially in skin and tendon. As a consequence their skin (and tendon) was fragile with reduced tensile strength (Danielson et al. 1997). As decorin expression increases with age the tendon worsens biomechanically because of changes in fibril structure and diameter. Paradoxically, tendons from decorin-null mice show less decline in their biomechanical parameters with aging than tendons from intact mice (Dunkman et al. 2013).

Decorin has many other functions besides its regulation of fibrillogenesis. It inhibits TGFβ activity, presumably by sequestering the growth factor (Hildebrand et al. 1994), and this mechanism might account for inhibition of fibrosis (and counteracting disease progression) in recessive dystrophic epidermolysis bullosa in mice (Cianfarani et al. 2019). Decorin participates in regulation of cell proliferation and immune response as well (Yoon and Halper 2005). Decorin can substitute for absent biglycan, at least in animal models (Ameye and Young 2002).

Two chondroitin or dermatan sulfate chains are attached to the protein core of biglycan (Krusius and Ruoslahti 1986). As a proper member of Class I SLRP its protein core consists of 10 leucine-rich repeats (Iozzo 2007; Heegaard et al. 2012). Its central domain contains binding sites for type I and VI collagens (Schönherr et al. 1995a; Wiberg et al. 2002). The affinity of biglycan binding to collagen is similar to decorin and is independent of *N*-linked oligosaccharides present on the biglycan molecule (Schönherr et al. 1995a). However, the two glycosaminoglycan chains attached to the biglycan core protein potentiate the binding of the core protein alone to collagens (Kram et al. 2020). Similarly to decorin, it interacts with TGFβ1-3 and TNFα through binding sites on the core protein and the dermatan sulfate chain attached to the core protein (Hildebrand et al. 1994; Tiedemann et al. 2005; Kram et al. 2020). Biglycan is present at particularly high levels in developing tendon. At E16-18 of mouse embryonic development, tendons are already quite prominent and express high levels of biglycan where it is coexpressed with collagen VI, and though the level of biglycan then declines it remains detectable postnatally (Lechner et al. 2006). Together with aggrecan (see below) biglycan is present in higher concentrations in the fibrocartilaginous region of tendon (Rees et al. 2009). Several studies have shown that abnormal collagen fibrils and their assembly in bone, dermis and tendon lead to impaired function in these systems in biglycan- deficient mice (Ameye and Young 2002; Corsi et al. 2002; Ameye et al. 2002; Young et al. 2002). The stimulation of BMP-4 expression by biglycan leads to proper tendon development in the embryo (Iozzo and Schaefer 2010). Biglycan also modulates BMP-4-induced differentiation of osteoblasts (Iozzo and Schaefer 2015). The core protein of biglycan stimulates activity of BMP-2 (Kram et al. 2020). In biglycan (and fibromodulin) deficient mice BMP-2 which is released from biglycan-controlled inhibition (at least in tendon) drives inhibition of tendon development (Schaefer and Iozzo 2008; Bi et al. 2007). Both decorin and biglycan are also expressed in healthy arteries (Bianco et al. 1990; Yeo et al. 1995). The adventitia of aorta is a major site for biglycan deposition (Heegaard et al. 2012). However, biglycan plays a somewhat sinister role in the development of atherosclerotic plaques and aortic valve stenosis. Biglycan promotes lipid deposition through binding to Toll-like receptors, with induction of cytokine production and inflammation. It binds to Toll-like receptors 2 and 4 located on macrophages, and thus functions as one of the mediators of innate immunity (Schaefer and Iozzo 2008). It also induces the expression of BMP-2, TGFß and alkaline phosphatase in interstitial cells of human aortic valve primarily through Toll-like receptor 2 (Song et al. 2012, 2015). In turn, both BMP-2 and TGFß enhance the osteogenic process and calcifications leading to aortic valve stenosis (Song et al. 2015). In addition, hyperglycemia of diabetes mellitus contributes to upregulating of biglycan in aortic valves as well (Barth et al. 2019).

A diminished low density lipoprotein (LDL) binding ability of biglycan (and versican, another LDL-binding proteoglycan) and the subsequent release of LDL from human aortic lesions can be induced by a disintegrin and metalloproteinase with thrombospondin motifs 5 (ADAMTS-5) activity (Didangelos et al. 2012). The risk for aortic dissection is X-linked (Coady et al. 1999), and aortic dissection and rupture are, indeed, more frequent in men with loss-of-function mutation in the X-linked biglycan gene (Meester et al. 2017) and in women with Turner syndrome (Lin et al. 1986). Similarly, it is the biglycan-deficient male mice rather than the females which suffer aortic rupture due to abnormal collagen fibrils with great variation in size and shape (Heegaard et al. 2012). Sporadic aortic dissection can also be induced by ADAMTS-4 activity. ADAMTS-4 can lead to the inflammation and degradation of versican, inducing smooth muscle cell apoptosis (Ren et al. 2017).

Asporin, another member of Class I, competes with decorin for collagen binding but because of its ability to bind calcium (Kalamajski et al. 2009) it plays a role in pathogenesis of osteoarthritis, and other joint and bone diseases rather than in tendon disorders (Ikegawa 2008). Interestingly, it has no glycosaminoglycan chain attached to its protein core. Asporin-mediated decreased calcification of aortic valve cell cultures is probably the result of negative regulation of activities of BMP-2, and likely of TGFβ (Ikegawa 2008). This suggests that asporin may function as a counterpoint to the osteogenic effect of biglycan on aortic valve (Polley et al. 2020). Though the skin in asporin deficient chicken (*Aspn*$^{-/-}$) appears normal on histology, including immunohistochemistry, the skin feels tougher, likely as a consequence of upregulation of genes for several collagens and matrix metalloproteinases (MMPs) and for decorin and biglycan, accompanied by abundance of chondroitin/dermatan sulfate,. Those two SLRPs were likely compensatory and responsible for the normal histology pattern (Maccarana et al. 2017).

All known members of Class II bind to fibrillar collagens through the leucine-rich region, and both fibromodulin and lumican bind to the same site within the collagen molecule (Heinegård 2009). Fibromodulin is associated with types I and II collagen fibrils, as it binds to the gap regions of both collagens, at sites different from collagen binding sites for decorin and biglycan (Hedlund et al. 1994; Viola 2007). It has multiple binding sites for type I collagen, but only one binding site for type II collagen (Svensson et al. 2000). It impedes fibril formation of these two collagens (Hedbom and Heinegård 1989). The keratan sulfate chains bound to fibromodulin protein core confer negative charge onto this proteoglycan. In addition, the presence of sulfated tyrosine residues together with numerous acidic amino acids at the N-terminus endows this region with properties similar to negatively charged heparin (Önnerfjord et al. 2004). As a consequence, this domain of fibromodulin binds growth factors and cytokines, such as fibroblast growth factor 2 (FGF2), interleukin 10 (IL-10) and oncostatin M and likely plays a role in interactions with collagen (Heinegård 2009).

Lumican, an SLRP closely related to fibromodulin, also present in tendons, plays a major role in connective tissues of cornea, skin, liver, articular cartilage and muscle, including cardiac muscle, where by virtue of multiple keratan sulfate chains attached to its core protein it retains water, and thus provides hydration and resilience of these tissues to stressful stimuli. It is involved in cell proliferation, migration, angiogenesis and Toll-like receptor 4 signaling. Lumican increases gene expression of many ECM molecules, including fibrillar collagens, MMP9 and TGFß (Xiao et al. 2020). Lumican-deficient mice have corneal clouding and skin laxity, but only minimal tendon impairment (Chakravarti et al. 1998). Several recent studies have shown the importance of lumican in cardiac health and dysfunction. Like fibromodulin, lumican is involved in collagen fibrillogenesis and cross-linking (Svensson et al. 2000; Mohammadzadeh et al. 2019) and substitutes for fibromodulin in fibromodulin-deficient mice (Svensson et al. 1999). Similar to biglycan, lumican binds to Toll-like receptor as part of the antigen recognition process (Schaefer and Iozzo 2008). Its presence is increased in hearts of patients and mice with heart failure where it contributes to pressure overload induced cardiac remodeling, i.e., to cardiac fibrosis (Mohammadzadeh et al. 2019). The suspected contribution of lumican to cardiac fibrosis is supported by attenuation of cardiac fibrosis in response to pressure overload, and development of cardiac dilatation in lumican knockout mice (Mohammadzadeh et al. 2020).

Studies on mice deficient in one or two the genes for the four SLRPs (i.e., decorin, biglycan, fibromodulin and lumican) showed somewhat overlapping role for these proteoglycans in fibrillogenesis. Deficiency in at least one of them has an effect on morphology and diameter of collagen fibrils. For example, the most dramatic effect of decorin deficiency is the emergence of irregular fibrils with large diameter and a decrease in skin and tendon tensile strength (Danielson et al. 1997). This is due to uncontrolled lateral fusion of thin and thick fibrils (Reed and Iozzo 2003). Abnormal fibril formation is also seen in biglycan-deficient mice (Young et al. 2002), and such abnormalities are accentuated in double knockout mice for both decorin and biglycan (Corsi et al. 2002). Such findings confirm that these proteoglycans play a major role in at least the third stage of fibrillogenesis where they limit lateral fibril growth. Biglycan deficient mice have normal skin but show reduced bone density, possibly due to modulation of BMP-4 effect on osteoblast differentiation described above (Iozzo and Schaefer 2015; Corsi et al. 2002). Lumican is able to substitute functionally for fibromodulin in fibromodulin-deficient mice. In contrast, large diameter collagen fibrils forming disorganized matrix in cornea and skin were found in lumican-deficient mice. It is interesting to note that this did not lead to loss in tendon biomechanical function and that fibromodulin did not substitute for loss of lumican function (Jepsen et al. 2002).

Keratocan, a less described member of Class II, has been identified in the tendon in the fibrocartilage, in the tensional region and in the endotenon (Rees et al. 2009). It is interesting that tendon keratocan is poorly sulfated in the tendon and highly glycosylated in the cornea, perhaps reflecting differences in its function in those two tissues (Rees et al. 2009). Mutations in keratocan can lead to cornea plana, a rare eye disease characterized by small flat corneas, reduced visual activity, and corneal clouding among other eye problems (Huang et al. 2019).

Aggrecan and versican, two representatives of the group of large proteoglycans binding hyaluronan, are the main proteoglycans synthesized in cartilage where they provide resilience and toughness. Brevican and neurocan are two other members of this group of so called hyalectans, designation derived from their ability to bind hyaluronan and lectin (Halper 2021). Aggrecan binding to hyaluronan is mediated by a link protein binding to the globular G1 domain of aggrecan. The second domain or G2 though quite homologous to G1 does not have the ability bind hyaluronan, and its function is not understood. G3 is located at the C-terminal end of the molecule and has homology to C-type lectin (Heinegård 2009). This domain binds and interacts with extracellular matrix proteins with EGF-repeats in their molecules, such as fibrillins, fibulins and tenascins (Day et al. 2004). A point mutation in this lectin region leads to chondritis dissecans due to loss of interactions with fibulin-1, fibulin-2, and tenascin-R, and it is characterized by fragmentation of articular cartilage and subsequent dislocation of subchondral bone from the joint surface early in life (Stattin et al. 2010). Aggrecan is a highly glycosylated molecule with numerous CS and KS chains attached to its large core protein ($M_r \sim 220$ kDa). In addition, aggrecan contains a variable number of *O*- and *N*-linked oligosaccharides (Iozzo and Murdoch 1996). Up to 100, somewhat heterogenous CS chains are attached to the CS1 and CS2 domains and even undergo further increase in diversity in diseases such as rheumatoid arthritis (Heinegård 2009). Because the numerous negatively charged keratan sulfate chains attached to aggrecan protein core constitute the collagen binding region of aggrecan, the ECM acquires high osmotic pressure. The tissue then attracts water. The resulting high tissue hydration renders the tissue resistant to compressive loading (Valhmu et al. 1998). This resilience is important not only in cartilage but also in regions of weight bearing tendons experiencing compression (Vogel et al. 1994), and in the walls of atherosclerotic blood vessels experiencing high shear (Ström et al. 2004).

Several aggrecanases cleaving specific sites on the protein core participate in degradation of aggrecan. They are members of the ADAMTS family of proteinases and are primarily active in the cartilage (Munteanu et al. 2002). The activity of ADAMTS was also demonstrated in tendons where some of them, particularly ADAMTS-4 and -5 cleave aggrecan at specific sites, especially in tendons with abnormally increased content of aggrecan (Rees et al. 2000; Tsuzaki et al. 2003; Plaas et al. 2011).

New acquisition of aggrecan happens also with age (Rees et al. 2009) and it is a sign of tendons or their regions undergoing remodeling into fibrocartilage (Vogel et al. 1994). In addition, a different form of aggrecan has been found in tensile regions of tendon: it lacks the G1 domain due to proteolytic degradation by aggrecanases (Vogel et al. 1994; Rees et al. 2009; Samiric et al. 2004). Interestingly, comparative explant cultures of tendons and of articular cartilage have shown much higher turnover of aggrecan in tendons than in cartilage. This is associated with release of aggrecan G1-containing metabolites and high levels of hyaluronan from tendon matrix (Rees et al. 2009). This is likely due to differences in composition between tendon and cartilage aggrecans. It has been suggested that tendon aggrecan is more susceptible to aggrecanase degradation due to the higher content of non-sulfated chondroitin sulfate disaccharide isomers and less keratan sulfate (Rees et al. 2009).

Versican is another member of the hyaluronan-binding PGs (Iozzo and Murdoch 1996; Wight 2002). It is expressed in many tissues, especially in fast growing cells of soft tissues. Versican is found in the dermis of skin, in the media of the aorta, and in lower portions in tendon. It appears transiently during mesenchymal condensation in developing chicken limb buds (Choocheep et al. 2010). An increase in versican content leads to expansion of ECM and to increased viscoelasticity of pericellular matrix that supports cell-shape changes necessary for cell proliferation and migration. ADAMTS-1 and -4 were shown to cleave also versican, in addition to aggrecan (Corps et al. 2008). Similar to aggrecan, versican is present in tensile regions of tendons in a catabolized form, presumably the result of aggrecanase activity. Like aggrecan, versican in this location lacks the G1 domain (Samiric et al. 2004). The presence of G3 domain which contains a C-type lectin sequence and binds multiple components of the extracellular matrix including fibulins -1 and -2 (Aspberg et al. 1999; Olin et al. 2001), and tenascin C (Day et al. 2004) indicates that versican is involved in structural organization of the tendon (Rees et al. 2009). Versican interacts also with CD44, integrin and EGF receptors through another binding site (Wu et al. 2005). It stabilizes also the presence of TGFβ in the extracellular matrix and regulates TGFβ signaling (Choocheep et al. 2010).

Though neither aggrecan nor versican have been implicated as direct culprits in the pathogenesis of systemic diseases affecting soft connective tissues, including those involving cardiac valves and aorta, such as Marfan or Ehlers-Danlos syndrome these two proteoglycans do play important roles in the structure and physiology of the cardiovascular system. So far we know that mutations in fibrillin -1 lead to Marfan syndrome and mitral valve prolapse (for more details see Chap. 8). Mutations in type III collagen and tenascin X may contribute to mitral valve prolapse and pulmonary valve stenosis (Hinton et al. 2006). Because these molecules are tightly associated with aggrecan and versican, it is likely that the products of the mutated genes affect the function of these proteoglycans as well. The extracellular matrix in cardiac valves is organized into three overlapping layers (fibrosa, spongiosa and ventricularis). Collagen fibers form the fibrosa, or arterial layer of the valvular cusp. The central layer, called spongiosa, is composed of loosely deposited proteoglycans, including both aggrecan and versican. Finally, the ventricularis, facing the ventricle, has a high content of elastin (Hinton et al. 2006; Lincoln et al. 2006). The precise composition of these layers changes during embryonic remodeling. The expression of aggrecan increases during cushion elongation and this proteoglycan is also present in the fibrosa and spongiosa during the cusp remodeling. This is associated with differential expression of other components of cardiac valves such as type III collagen, tenascin and elastin as already referred to above (Hinton et al. 2006). The expression of aggrecan and versican is controlled by at least two transcription factors, Twist1 (Shelton and Yutzey 2008) and TBx20 (Shelton and Yutzey 2007). Versican plays an important role in cardiac development. Several enzymes, such as MMP-2, MMP-9, ADAMTS-5 and ADAMTS-9 process and cleave versican which is particularly relevant in the development of certain experimental valve defects (see below) (Krishnamurthy et al. 2012; Visse and Nagase 2003). Interestingly, though ADAMTS-5 deficient mice on high fat diet (and obese) the lack of versican cleavage leads to increase in diastolic posterior wall thickness and left ventricle volume, cardiac function remain intact (Hemmeryckx et al. 2019). The significance of these findings for people suffering from obesity has not been determined. A lethal mutation in the versican gene was described in mice which die *in utero* due to hypoplastic endocardial cushions and malformations of the right cardiac chambers (Mjaatvedt et al. 1998). However, some mutant mouse embryos with a mixed background and lacking only the A subdomain of the G1 domain survived to the neonatal stage with a ventricular septal defect (Hatano et al. 2012). Versican in these embryos had a decreased ability to bind hyaluronan. The versican splicing forms were decreased in the atrioventricular canal cushion and the ventricular septa. The normal endocardial mesenchymal transition from highly proliferative endocardial cells to less proliferative remodeling mesenchymal cells occurring as part of normal cardiac development (Hinton et al. 2006) was disrupted as endocardial cells persisted in their proliferative state due to the lack of the versican A subdomain (Hatano et al. 2012).

The defect or mutation does not have to necessarily reside in versican or its gene for core protein of versican to disrupt normal heart development. Versican interacts closely with elastin, the major component of the ventricularis layer of cardiac valves. Versican limits elastogenesis as it interferes with elastic fiber assembly (Wu et al. 2005; Huang et al. 2006). Excessive proteoglycan accumulation was found in the annulus region of the developing aortic valve of elastin-insufficient mice (heterogenous for elastin deficiency, designated as *Eln+/−*) due to increase in intact and cleaved versican accompanied by an increase in MMP-2 and MMP-9. This was in sharp contrast to wild type mice where the proteoglycan in the annulus was identified as aggrecan, and no versican, intact or cleaved was detected (Krishnamurthy et al. 2012). As Krishnamurthy et al. point out, versican and aggrecan complement each other and so may substitute for each other in function if necessary (Krishnamurthy et al. 2012).

Whether this would be the case in other cardiac valves and whether disorders of other components of the extracellular matrix expressed in Marfan, Ehlers-Danlos and other related syndromes lead to imbalance between aggrecan and versican expression and function needs to be demonstrated in the future.

References

Ameye L, Young MF (2002) Mice deficient in small leucine-rich proteoglycans: novel in vivo models for osteoporosis, osteoarthritis, Ehlers-Danlos syndrome, muscular dystrophy, and corneal diseases. Glycobiology 12(9):107R–116R

Ameye L et al (2002) Abnormal collagen fibrils in tendons of biglycan/fibromodulin-deficient mice lead to gait impairment, ectopic ossification, and osteoarthritis. FASEB J 16(7):673–680

Aspberg A, Adam S, Kostka G, Timpl R, Heinegård D (1999) Fibulin-1 is a ligand for the C-type lectin domains of aggrecan and versican. J Biol Chem 274:20444–20449

Barth M et al (2019) Degenerative aortic valve disease and diabetes: implications for a link between proteoglycans and diabetic disorders in the aortic valve. Diab Vasc Dis Res 16(3):254–269

Bengtsson E, Apberg A, Heinegård D, Sommarin Y, Spillmann D (2000) The amino-terminal part of PRELP binds to heparin and heparan sulfate. J Biol Chem 275:40695–40702

Berenson MC, Blevins FT, Plaas AH, Vogel KG (1996) Proteoglycans of human rotator cuff tendons. J Orthop Res 14:518–525

Bi Y, Ehirchiou D, Kilts TM, Inkson CA, Embree MC, Sonoyama W, Li L, Leet AI, Seo BM, Zhang L, Shi S, Young MF (2007) Identification of tendon stem/progenitor cells and the role of the extracellular matrix in their niche. Nat Med 13:1219–1227

Bianco P, Fiaher LR, Young MF, Termine JD, Robey PG (1990) Expression and localization of of the two small proteoglycans biglycan and decorin in developing human skeletal and non-skeletal tissues. J Histochem Cytochem 38:1549–1563

Bredrup C, Knappskog PM, Majewski J, Rødahl E, Boman H (2005) Congenital stromal dystrophy of the cornea caused by a mutation in the decorin gene. Invest Ophthalmol Vis Sci 46:420–426

Bredrup C, Stang E, Bruland O, Palka BP, Young RD, Haavik J, Knappskog PM, Rødahl E (2010) Decorin accumulation contributes to the stromal opacities found in congenital stromal corneal dystrophy. Invest Ophthalmol Vis Sci 51:5578–5582

Carter DR, Beaupré GS (eds) (2007) Skeletal function and form. Cambridge University Press, Cambridge

Chakravarti S, Magnuson T, Lass JH, Jepsen KJ, LaMantia C, Carroll H (1998) Lumican regulates collagen fibril assembly: skin fragility and corneal opacity in the absence of lumican. J Cell Biol 141:1277–1286

Choocheep K, Hatano S, Takagi H, Watanabe H, Kimata K, Kongtawelert P, Watanabe H (2010) Versican facilitates chondrocyte differentiation and regulates joint morphogenesis. J Biol Chem 285:21114–21125

Cianfarani F et al (2019) Decorin counteracts disease progression in mice with recessive dystrophic epidermolysis bullosa. Matrix Biol 81:3–16

Coady MA, Davies RR, Roberts M, Goldstein LJ, Rogalski MJ, Rizzo JA, Hammond GL, Kopf GS, Elefteriades JA (1999) Familial patterns of thoracic aortic aneurysms. Arch Surg 134:361–367

Corps AN, Jones GC, Harrall RL, Curry VA, Hazleman BL, Riley GP (2008) The regulation of aggrecanase ADAMTS-4 expression in human Achilles tendon and tendon-derived cells. Matrix Biol 27:393–401

Corsi A, Xu T, Chen XD, Boyde A, Liang J, Mankani M, Summer B, Iozzo RV, Eichstetter I, Robey PG, Bianco P, Young MF (2002) Phenotypic effects of bioblycan deficiency are linked to collagen fibril abnormalities, are synergized by decorin deficiency, and mimic Ehlers-Danlos-like changes in bone and other connective tissues. J Bone Miner Res 17:1180–1189

Danielson KG, Baribault H, Holmes DF, Graham H, Kadler KE, Iozzo RV (1997) Targeted disruption of decorin leads to abnormal collagen fibril morphology and skin fragility. J Cell Biol 136:729–743

Day JM, Olin AL, Murdoch AD, Canfield A, Sasaki T, Timpl R, Hardingham TE, Aspberg A (2004) Alternative splicing in the aggrecan G3 domain influences binding interactions with tenascin-C and other extracellular matrix proteins. J Biol Chem 279:12511–12518

Didangelos A, Mayr U, Monaco C, Mayr M (2012) Novel role of ADAMTS-5 protein in proteoglycan turnover and lipoprotein retention in atherosclerosis. J Biol Chem 287:19341–19345

Dunkman AA, Buckley MR, Mienaltowski MJ, Adams SM, Thomas SJ, Satchell L, Kumar A, Pathmanathan L, Beason DP, Iozzo RV, Birk DE, Soslowsky LJ (2013) Decorin expression is important for age-related changes in tendon structure and mechanical properties. Matrix Biol 32:3–13

Engel MB, Zerlotti E (1967) Changes in cells, matrix and water of calcifying turkey leg tendons. Am J Anat 120:489–525

Halper J (2021) Major proteins of the extracellular proteins. In: Allewell NM (ed) Encyclopedia of biological chemistry. Elsevier, Amsterdam

Hatano S, Kimata K, Hiraiwa N, Kusakabe M, Isogai Z, Adachi E, Shinomura T, Watanabe H (2012) Versican/PG-M is essential for ventricular septal formation subsequent to cardiac atrioventricular cushion development. Glycobiology 22:1268–1277

Hedbom E, Heinegård D (1989) Interaction of a 59-dDa connective tissue matrix protein with collagen I and collagen II. J Biol Chem 264:6898–6905

Hedlund H, Mengarelli-Widholm S, Heinegård D, Reinholt FP, Svensson O (1994) Fibromodulin distribution and association with collagen. Matrix Biol 14:227–232

Heegaard AM, Corsi A, Danielsen CC, Nielsen KL, Jorgensen HL, Riminucci M, Young MF, Bianco P (2012) Biglycan deficiency causes spontaneous dissection and rupture in mice. Circulation 115:2731–2738

Heinegård D (2009) Proteoglycans and more – from molecules to biology. Int J Exp Pathol 70:575–586

Hemmeryckx B, Carai P, Roger Lijnen H (2019) ADAMTS5 deficiency in mice does not affect cardiac function. Cell Biol Int 43(6):593–604

Hildebrand A, Romaris M, Rasmussen IM, Heinegard D, Twardzik DR, Border WA, Ruoslahti E (1994) Interaction of the small interstitial proteoglycans biglycan, decorin and fibromodulin with transforming growth beta. Biochem J 302:527–534

Hinton RBJ, Lincoln J, Deutsch GH, Osinska H, Manning PB, Benson DW, Yutzey KE (2006) Extracellular matrix remodeling and organization in developing and diseased aortic valves. Circ Res 98:1431–1438

Huang R, Merrilees MJ, Braun K, Beaumont B, Lemire J, Clowes AW, Hinek A, Wight TN (2006) Inhibition of versican synthesis by antisense alters smooth muscle cell phenotype and induces elastic fiber formation in vitro and in neointima after vessel injury. Circ Res 98:370–377

Huang C et al (2019) Novel variants in the KERA gene cause autosomal recessive cornea plana in a Chinese family: a case report. Mol Med Rep 19(6):4711–4718

Ikegawa S (2008) Expression, regulation and function of asporin, a susceptibility gene in common bone and joint diseases. Curr Med Chem 15:724–728

Iozzo RV (1998) Matrix proteoglycans: from molecular design to cellular function. Annu Rev Biochem 67:609–652

Iozzo RV (1999) The biology of the small leucine-rich proteoglycans. Functional network of interactive proteins. J Biol Chem 274:18843–18846

Iozzo RV (2007) The family of the small leucine-rich proteoglycans: key regulators of matrix assembly and cellular growth. Mol Biol Cell 32:141–174

Iozzo RV, Murdoch AD (1996) Proteoglycans of the extracellular environment: clues from the gene and protein side offer novel perspectives in molecular diversity of function. FASEB J 10:598–614

Iozzo RV, Schaefer L (2010) Proteoglycans in health and disease: novel regulatory signaling mechanisms evoked by the small leucine-rich proteoglycans. FEBS J 277:3864–3875

Iozzo RV, Schaefer L (2015) Proteoglycan form and function: a comprehensive nomenclature of proteoglycans. Matrix Biol 42:11–55

Jepsen KJ, Wu F, Peragallo JH, Paul J, Roberts L, Ezura Y, Oldberg Å, Birk DE, Chakravarti S (2002) A syndrome of joint laxity and impaired joint integrity in lumican- and fibromodulin-deficient mice. J Biol Chem 277:35532–35540

Kalamajski S, Aspberg A, Lindblom K, Heinegård D, Oldberg Å (2009) Asporin competes with decorin for collagen bidnign, binds calcium and promotes osteoblast collagen mineralization. Biochem J 423:534–539

Kamma-Loger CS et al (2016) Role of decorin core protein in collagen organisation in congenital stromal corneal dystrophy (CSCD). PLoS One 11:e0147948

Kram V et al (2020) Biglycan in the skeleton. J Histochem Cytochem 68(11):747–762

Krishnamurthy VK, Opoka AM, Kern CB, Guilak F, Narmoneva DA, Hinton RB (2012) Maladaptive matrix remodeling and regional biomechanical dysfunction in a mouse model of aortic valve disease. Matrix Biol 31:197–205

Krusius T, Ruoslahti E (1986) Primary structure of an extracellular matrix proteoglycan core protein deduced from cloned cDNA. Proc Natl Acad Sci U S A 83:7683–7687

Kuo CK, Petersen BC, Tuan RS (2008) Spatiotemporal protein distribution of TGFβs, their receptors, and extracellular matrix molecules during embryonic tendon development. Dev Dyn 237:1477–1489

Lechner BE, Lim JH, Mercado ML, Fallon JR (2006) Developmental regulation of biglycan expression in muscle and tendon. Muscle Nerve 34:347–355

Lin AE, Lippe BM, Geffner ME, Gomes A, Lois JF, Barton CW, Rosenthal A, Friedman WF (1986) Aortic dilation, dissection, and rupture in patients with Turner syndrome. J Pediatr 109:820–826

Lincoln J, Lange AW, Yutzey KE (2006) Hearts and bones: shared regulatory mechanisms in heart valve, cartilage, tendon, and bone development. Dev Biol 294:292–302

Maccarana M et al (2017) Asporin-deficient mice have tougher skin and altered skin glycosaminoglycan content and structure. PLoS One 12(8):e0184028

Manning PL, Morris PM, McMahon A, Jones E, Gize A, Macquaker JH, Wolff G, Thompson A, Marshall J, Taylor KG, Lyson T, Gaskell S, Reamtong O, Sellers WI, van Dongen BE, Buckley M, Wogelius RA (2009) Mineralized soft-tissue structure and chemistry in a mummified hadrosaur from the Hell Creek Formation, North Dakota (USA). Proc Biol Sci 276:3429–3437

McEwan PA, Scott PG, Bishop PN, Bella J (2006) Structural correlations in the family of small leucine-rich repeat proteins and proteoglycans. J Struct Biol 155:294–305

Meester JA et al (2017) Loss-of-function mutations in the X-linked biglycan gene cause a severe syndromic form of thoracic aortic aneurysms and dissections. Genet Med 19(4):386–395

Mjaatvedt CH, Yamamura H, Capehart AA, Turner D, Markwald RR (1998) The Cspg2 gene, disrupted in the hdf mutant, is required for right cardiac chamber and endocardial cushion formation. Dev Biol 202:56–66

Mohammadzadeh N et al (2019) The extracellular matrix proteoglycan lumican improves survival and counteracts cardiac dilatation and failure in mice subjected to pressure overload. Sci Rep 9(1):9206

Mohammadzadeh N et al (2020) Moderate loss of the extracellular matrix proteoglycan Lumican attenuates cardiac fibrosis in mice subjected to pressure overload. Cardiology 145(3):187–198

Munteanu SE, Ilic MZ, Handley CJ (2002) Highly sulfated glycosminoglycans inhibit aggrecanase degradation of aggrecan by bovine articular cartilage explant culture. Matrix Biol 21:429–440

Ohta K, Kuriyama S, Okafuji T, Gejima R, Ohnuma S, Tanaka H (2006) Tsukushi cooperates with VG1 to induce primitive streak and Hensen's node formation in the chick embryo. Development 133:3777–3786

Olin AI, Mörgelin M, Sasaki T, Timpl R, Heinegård D, Aspberg A (2001) The proteoglycans aggrecan and versican form networks with fibulin-2 through their lectin domain binding. J Biol Chem 276:1253–1261

Önnerfjord PH, Heathfield TF, Heinegård D (2004) Identification of tyrosine sulfation in extracellular leucine-rich repeat proteins using mass spectrometry. J Biol Chem 279:26–33

Parkinson J, Samiric T, Ilic MZ, Cook J, Handley CJ (2011) Involvement of proteoglycans in tendinopathy. J Musculoskelet Neuronal Interact 11:86–93

Pietraszek-Gremplewicz K et al (2019) Small leucine-rich proteoglycans and matrix metalloproteinase-14: key partners? Matrix Biol 75–76:271–285

Plaas A, Sandy JD, Liu H, Diaz MA, Schenkman D, Magnus RP, Bolam-Bretl C, Kopesky PW, Wang VM, Galante JO (2011) Biochemical identification and immunolocalization of aggrecan, ADAMTS5 and inter-alpha-trypsin-inhibitor in equine degenerative suspensory ligament desmitis. J Orthop Res 29:900–906

Polley A et al (2020) Asporin reduces adult aortic valve interstitial cell mineralization induced by osteogenic media and Wnt signaling manipulation in vitro. Int J Cell Biol 2020:2045969

Rada JA, Schrecengost PK, Hassell JR (1993) Regulation of corneal collagen fibrillogenesis in vitro by corneal proteoglycan (lumican and decorin) core protein. Exp Eye Res 56:635–648

Reed CC, Iozzo RV (2003) The role of decorin in collagen fibrillogenesis and skin homeostasis. Glycoconj J 19:249–255

Rees SG, Flannery CR, Little CB, Hughes CE, Caterson B, Dent CM (2000) Catabolism of aggrecan, decorin and biglycan in tendon. Biochem J 350:181–188

Rees SG, Dent CM, Caterson B (2009) Metabolism of proteoglycans in tendon. Scand J Med Sci Sports 19:470–478

Ren P et al (2017) Critical role of ADAMTS-4 in the development of sporadic aortic aneurysm and dissection in mice. Sci Rep 7(1):12351

Samiric T, Ilic MZ, Handley CJ (2004) Characterisation of proteoglycans and their catabolic products in tendon and explant cultures of tendon. Matrix Biol 23:127–140

Samiric T, Parkinson J, Ilic MZ, Cook J, Feller JA, Handley CJ (2009) Changes in the composition of the extracellular matrix in patellar tendinopathy. Matrix Biol 28:230–236

Santra M, Reed CC, Iozzo RV (2002) Decorin binds to a narrow region of the EGF receptor, partially overlapping with, but distinct from, the EGF-binding epitope. J Biol Chem 277:35671–35681

Schaefer L, Iozzo RV (2008) Biological functions of the small leucine-rich proteoglycans: from genetics to signal transduction. J Biol Chem 283:21305–21309

Schönherr E, Witsch-Prehm P, Harrach B, Robenek H, Rauterberg J, Kresse H (1995a) Interaction of biglycan with type I collagen. J Biol Chem 270:2776–2783

Schönherr E, Hausser E, Beavan L, Kresse H (1995b) Decorin-type I collagen interaction. Presence of separate core protein-binding domains. J Biol Chem 270:8877–8883
Schönherr E, Broszat M, Brandan E, Bruckner P, Kresse H (1998) Decorin core protein fragment Leu155-Val260 interacts with TGF-beta but does not compete for decorin binding to type I collagen. Arch Biochem Biophys 355:241–248
Shelton EL, Yutzey KE (2007) Tbx20 regulation of endocardial cushion cell proliferation and extracellular matrix gene expression. Dev Biol 302:376–388
Shelton EL, Yutzey KE (2008) Twist1 function in endocardial cushion cell proliferation, migration, and differentiation during heart valve development. Dev Biol 317:282–295
Shimizu-Hirota R, Sasamura H, Kuroda M, Kobayashi E, Saruta T (2004) Functional characterization of podocan, a member of a new class in the small leucine-rich repeat protein family. FEBS Lett 563:69–74
Song R, Zeng Q, Ao L, Yu JA, Cleveland JC, Zhao KS, Fullerton DA, Meng X (2012) Biglycan induces the expression of osteogenic factors in human aortic valve interstitial cells via Toll-like receptor-2. Arterioscler Thromb Vasc Biol 32:2711–2720
Song R et al (2015) BMP-2 and TGF-beta1 mediate biglycan-induced pro-osteogenic reprogramming in aortic valve interstitial cells. J Mol Med (Berl) 93(4):403–412
Spiesz EM, Roschger P, Zysset PK (2012) Influence of mineralization and microporosity on tissue elasticity: experimental and numerical investigation on mineralized turkey leg tendons. Calcif Tissue Int 90:319–329
Stattin EL, Wiklund F, Lindblom K, Onnerfjord P, Jonsson BA, Tegner Y, Sasaki T, Struglics A, Lohmander S, Dahl N, Heinegård D, Aspberg A (2010) A missense mutation in the aggrecan C-type lectin domain disrupts extracellular matrix interactions and causes dominant familial osteochondritis dissecans. Am J Hum Genet 86:126–137
Ström A, Ahlqvist E, Franzen A, Heinegård D, Hultgårdh-Nilsson A (2004) Extracellular matrix components in atherosclerotic arteries of Apo E/LDL receptor deficient mice: an immunohistochemical study. Histol Histopathol 19:337–347
Svensson L, Aszódi A, Reinholt FP, Fässler R, Heinegård D, Oldberg A (1999) Fibromodulin-null mice have abnormal collagen fibrils, tissue organization, and altered lumican deposition in tendon. J Biol Chem 274:9636–9647
Svensson L, Närlid I, Oldberg Å (2000) Fibromodulin and lumican bind to the same region on collagen type I fibrils. FEBS Lett 470:178–182
Tanaka N et al (2019) Expression of minor cartilage collagens and small leucine rich proteoglycans may be relatively reduced in osteoarthritic cartilage. BMC Musculoskelet Disord 20(1):232
Tiedemann K, Olander B, Eklund E, Todorova L, Bengtsson M, Maccarana M, Westergren-Thorsson G, Malmström A (2005) Regulation of the chondroitin/dermatan fine structure by transforming growth factor-β1 through effects on polymer-modifying enzymes. Glycobiology 15:1277–1285
Tsuzaki M, Guyton G, Garrett W, Archambault JM, Herzog W, Almekinders L, Bynum D, Yang X, Banes AJ (2003) IL-1 beta induces COX2, MMP-1,-3 and −13, ADAMTS-4, IL-1 beta and IL-6 in human tendon cells. J Orthop Res 21:256–264
Valhmu WB, Stazzone EJ, Bachrach NM, Saed-Nejad F, Fischer SG, Mow VC, Ratcliffe A (1998) Load-controlled compression of articular cartilage induces a transient stimulation of aggrecan gene expression. Arch Biochem Biophys 353:29–36
Viola M (2007) Fibromodulin interactions with type I and II collagens. Connect Tissue Res 48:141–148
Visse MR, Nagase H (2003) Matrix metalloproteinases and tissue inhibitors of metalloproteinases: structure, function, and biochemistry. Circ Res 92:827–839
Vogel KS, Sandy JD, Pogány G, Robbins JR (1994) Aggrecan in bovine tendon. Matrix Biol 14:171–179
Wiberg C, Heinegard D, Wenglen C, Timpl R, Morgelin M (2002) Biglycan organizes collagen VI into hexagonal-like networks resembling tissue structures. J Biol Chem 277:49120–49126
Wight T (2002) Versican: a versatile extracellular matrix proteoglycan in cell biology. Curr Opin Cell Biol 14:617–623
Wu YJ, Lapierre D, Wu J, Yee AJ, Yang BB (2005) The interaction of versican with its binding partners. Cell Res 15:483–494
Xiao D et al (2020) Lumican promotes joint fibrosis through TGF-beta signaling. FEBS Open Biol 10(11):2478–2488
Yeo TK, Torok MA, Kraus HL, Evans SA, Zhou Y, Marcum JA (1995) Distribution of biglycan and its propeptide form in rat and bovine aortic tissue. J Vasc Res 32:175–182
Yoon JH, Halper J (2005) Tendon proteoglycans: biochemistry and function. J Musculoskelet Neuronal Interact 5:22–34
Young MF, Bi Y, Ameye L, Chen XD (2002) Biglycan knockout mice: new models for musculoskeletal diseases. Glycoconj J 19:257–262
Zhang G, Ezura Y, Chervoneva I, Robinson PS, Beason DP, Carine ET, Soslowsky LJ, Iozzo RV, Birk DE (2006) Decorin regulates assembly of collagen fibrils and acquisition of biomechanical properties during tendon development. J Cell Biochem 98:1436–1449

Growth Factor Roles in Soft Tissue Physiology and Pathophysiology

6

Jennifer H. Roberts and Jaroslava Halper

Abstract

Repair and healing of injured and diseased tendons has been traditionally fraught with apprehension and difficulties, and often led to rather unsatisfactory results. The burgeoning research field of growth factors has opened new venues for treatment of tendon disorders and injuries, and possibly for treatment of disorders of the aorta and major arteries as well. Several chapters in this volume elucidate the role of transforming growth factor β (TGFß) in pathogenesis of several heritable disorders affecting soft tissues, such as aorta, cardiac valves, and tendons and ligaments. Several members of the bone morphogenetic group either have been approved by the FDA for treatment of non-healing fractures or have been undergoing intensive clinical and experimental testing for use of healing bone fractures and tendon injuries. Because fibroblast growth factors (FGFs) are involved in embryonic development of tendons and muscles among other tissues and organs, the hope is that applied research on FGF biological effects will lead to the development of some new treatment strategies providing that we can control angiogenicity of these growth factors. The problem, or rather question, regarding practical use of imsulin-like growth factor I (IGF-I) in tendon repair is whether IGF-I acts independently or under the guidance of growth hormone. FGF2 or platelet-derived growth factor (PDGF) alone or in combination with IGF-I stimulates regeneration of periodontal ligament: a matter of importance in Marfan patients with periodontitis. In contrast, vascular endothelial growth factor (VEGF) appears to have rather deleterious effects on experimental tendon healing, perhaps because of its angiogenic activity and stimulation of matrix metalloproteinases-proteases whose increased expression has been documented in a variety of ruptured tendons. Other modalities, such as local administration of platelet-rich plasma (PRP) and/or of mesenchymal stem cells have been explored extensively in tendon healing. Though treatment with PRP and mesenchymal stem cells has met with some success in horses (who experience a lot of tendon injuries and other tendon problems), the use of PRP and mesenchymal stem cells in people has been more problematic and requires more

J. H. Roberts
Department of Pathology, College of Veterinary Medicine, The University of Georgia, Athens, GA, USA
e-mail: Jennifer.Roberts@uga.edu

J. Halper (✉)
Department of Pathology, College of Veterinary Medicine, and Department of Basic Sciences, AU/UGA Medical Partnership, The University of Georgia, Athens, GA, USA
e-mail: jhalper@uga.edu

J. Halper (ed.), *Progress in Heritable Soft Connective Tissue Diseases*, Advances in Experimental Medicine and Biology 1348, https://doi.org/10.1007/978-3-030-80614-9_6

studies before PRP and mesenchymal stem cells can become reliable tools in management of soft tissue injuries and disorders.

Keywords

Tendon repair · Transforming growth factor β (TGFβ) family · Bone morphogenetic proteins (BMPs) · Fibroblast growth factors (FGFs) · Insulin-like growth factor I (IGF-I) · Platelet-derived growth factor (PDGF) · Vascular endothelial growth factor (VEGF) · Platelet-rich plasma (PRP) · Bioengineering · Bone marrow- or peripheral blood-derived stem cells · Tendon-derived progenitor cells

Abbreviations

ALK	Activin receptor-like kinase
BMP	Bone morphogenetic protein
BMPR	Bone morphogenetic protein receptor
DPP	Decapentaplegic
FGF	Fibroblast growth factors
FSH	Follicle stimulating hormone
GDF	Growth differentiation factor
GH	Growth hormone
IGF-I	Insulin-like growth factor I
PDGF	Platelet-derived growth factor
PRP	Platelet-rich plasma
TGFβ	Transforming growth factor β
VEGF	Vascular endothelial growth factor

6.1 Basics of Tendon Repair

One of the challenges facing patients with tendon injuries either due to trauma or joint laxity in Ehlers-Danlos syndrome is the lack of good options for management and treatment of such disorders. Repair in the tendon generally follows along the classical pathways for wound healing marked by a sequence of inflammation, proliferation and remodeling. As a result of remodeling, type I and type III collagen are laid down first in the form of small diameter fibrils (Södersten et al. 2013). Tensile strength increases when crosslinks are re-established at 3 weeks after the injury (Jones et al. 2006a), though the healed tendon does not regain the original organization or biomechanical function even years later in most cases (Woo et al. 1999; Graham et al. 2019). Surgical repair is usually incomplete as well, resulting in decreased stiffness and adhesion formation (Greenough 1996; Noguchi et al. 1993). Various suture techniques have been tested on the flexor digitorum profundus tendon in the dog with mixed results depending on the technique that was utilized (Zhao et al. 2001). Fibrin glue has been used in rabbits but no significant difference in range of motion was detected after 8 weeks post-treatment (He et al. 2013). Suture techniques in tendon repair have also been studied in rats, with similar insignificant results (Sardenberg et al. 2015). Patients with Marfan syndrome also suffer from joint problems and from periondontitis, or inflammation of the gingiva, periodontal ligament and alveolar bone loss which does not accompany dental caries (Suda et al. 2009).

Though less obvious than in other tissues, healing of tendons and ligaments occurs through two pathways. In the extrinsic pathway cells participating in tendon repair are recruited from outside the injury site. In this case the process ends with a scar, and, in many cases, with adhesions hindering the normal gliding motion of tendon in the sheath (Ingraham et al. 2003; Manske et al. 1985; Singh et al. 2015). The intrinsic pathway leads, at least in theory, to regeneration of the tendon without scarring and adhesion formation. It consists mostly of proper organization of collagen deposited into the wound bed during its fibroblastic/proliferative and remodeling stages (Ingraham et al. 2003; Hazard et al. 2011). During the initial or inflammatory phase, lasting about 2 weeks, neutrophils and fibroblasts migrate into the wounds. Accumulation of hyaluronan makes tendon repair distinct from repair in other tissues (Mast et al. 1992). It is the first macromolecule appearing in the wound bed. It promotes cell differentiation and growth (Chen et al. 2008a; Mast et al. 1995), and contributes to scarless healing (Ingraham et al. 2003). The proliferative or fibroblastic stage lasts for several weeks and it is marked by proliferation and differentiation of epitenon cells until they form a multicellular layer encasing collagen bundles at severed or injured ends of the tendon. Those fibroblasts then synthesize and secrete growth factors, such as TGFβ and IGF-I which in turn stimulate the production of collagen fibrils which then undergo proper assembly (Ingraham et al. 2003; Singh et al. 2015; Mass et al. 1993; Mass and Tuel 1990; Ohashi et al. 2018). At this time type III collagen appears before type I collagen and participates in fibrillogenesis (Ingraham et al. 2003). Inherently, the emphasis on preservation of biomechanical function of intact and healing tendon results in its poor vascularization (Fenwick et al. 2002; Pufe et al. 2005a).

In addition, a distinction has to be made between acute and chronic injuries of both tendons and ligaments as the pathogenesis and pathology differ (Paoloni et al. 2011). Whereas tearing of collagen fibers, hemorrhage, and inflammation with progression through standard stages of tissue repair are typical for the appearance and subsequent healing of acute (or first-time) injuries of tendons and ligaments (Paoloni et al. 2011; Murray et al. 2007; Kokubu et al. 2020), chronic (or repeated) injuries of tendons and ligaments are less well defined (Paoloni et al. 2011; Kokubu et al. 2020; Loiacono et al. 2019) Their incidence increases with age and they are associated with overuse (Wilson and Best 2005). Their hallmarks are degenerative change such as collagen disruption, mucoid or proteoglycan accumulation, and neovascularization. Inflammation is conspicuously absent (Khan et al. 1999). Chronic ligament injuries are thought to arise from chronic instability secondary to failed healing of acute injuries, and often involve fiber stretching or tearing and joint laxity, though inflammatory changes are sometimes present (Hayler and Guffre 2009; Yastrebov and Lobenhoffer 2009).

As a consequence, even a seemingly well-healed tendon (or ligament) does not return to its original structural integrity and full function (Loiacono et al. 2019; Yoon and Halper 2005; Frank et al. 1997; Achari et al. 2011; Nayak et al. 2020). Besides traditional measures such as surgery, exercise or immobility, or administration of painkillers, advances in basic research on growth factors markedly expanded our knowledge of tissue repair, and thus started bringing solutions to practical problems. Growth factors are numerous, and it seems that each year we add more of them to our already extensive catalog of small, mostly glycosylated proteins involved not only in cellular proliferation, but also in embryonic development and in maintaining or modulating differentiated functions of tissues and organs throughout the entire lifespan (Halper 2010). Most growth factors are ubiquitous and pleiotropic, and a comprehensive review of their activities would go beyond the scope of this volume. Growth factors, such as transforming growth β1 (TGFβ1), bone morphogenetic proteins (BMP), epidermal growth factor (EGF), platelet-derived growth factor (PDGF), vascular endothelial growth factor (VEGF), fibroblast growth factors (FGFs) and insulin-like growth factor I (IGF-I) are produced by tenocytes, inflammatory white blood cells, and are released from platelets during wound-induced degranulation (Halper 2010). These growth factors (and some others as well) actively participate in proliferation and formation of extracellular matrices and connective tissues, including tendons and ligaments (Molloy et al. 2003).

Because of their sheer number and complex activity and effects, only a few whose role in tendon healing and whose function in soft tissue disorders has been understood the most will be described here.

6.2 Transforming Growth Factor β (TGFβ) Family

This numerous and complex group of growth factors is indispensable during embryonic development and beyond, and is usually divided into four groups (Halper 2010):

1. Mullerian inhibitory substance group: regulates Mullerian duct regression in male embryos.
2. Inhibin/activin group: includes inhibin (which blocks follicle stimulating hormone (FSH) release by pituitary cells) and activin (which stimulates FSH release by pituitary cells).
3. Vg-related proteins: regulate primarily embryonic development and cell differentiation. This group includes bone morphogenetic proteins or BMPs (which regulate bone and tendon development) and closely related (and in some instances identical) growth differentiation factors or GDFs; dorsalin (which regulates neural tube differentiation); DPP or decapentaplegic transcript (regulates dorsal-ventral patterning in Drosophila); and VgI gene (inducer of mesoderm development from ectoderm in Xenopus).
4. TGFβ family proper: contains three isoforms, TGFβ1–3.

All members of the TGFβ superfamily transduce their signals through serine/threonine kinase type I and type II receptors, designated as TβR-I and TβR-II or their equivalent cell membrane compounds (Connolly et al. 2012). Initial binding of TGFβ to TβR-II leads to recruitment of TβR-I, and formation of a heterotetramer. This configuration triggers a phosphorylation cascade starting with the TβR-II mediated activation of the TβR-I kinase, and of several Smad proteins and other molecules (Hinck 2012). As an alternative to the division into the 4 groups outlined above the members of the TGFβ family can be divided into two large clusters depending on the type of TβR-I they bind to and the Smad proteins they activate (Hinck 2012). According to this classification, TGFβs 1–3, activins, GDFs 8, 9 and 11, BMP3 and *nodal* form the first class which, through TβR-II binding, turns on one type of TβR-I receptor which then activates Smad 2 and 3. The second class includes presumably more distantly related proteins, which include BMPs 2, 4–10, and GDFs 1, 3, 5, 6 and 7 (Hinck 2012). These growth factors bind to two types of serine/threonine kinase receptors which have diversified somewhat from the original TβR-I and TβR-II scheme. There are as many as three type I receptors (BMPR-1a, BMPR-1b and ActR-1), and three type II receptors (ActR-II, ActR-III and BMPR-II). The type I receptors to which the BMP/GDF ligands bind activate Smads 1, 5, and 8 (Hinck 2012; Massague and Wotton 2000). In addition, these ligands are known to bind to multiple type I and type II receptors (Hinck 2012). According to another classification BMPR-1a receptor is also known as activin receptor-like kinase 3 (ALK3) and BMPR-1b is identical to ALK6 (Cai et al. 2012; Dyer et al. 2014). More complete review of TGFβ signaling is provided in Chap. 7.

Only BMPs and TGFβ isoforms are pertinent to the topic of disorders of soft tissues as discussed in this volume. TGFβ1–3 isoforms share 75% amino acid homology and are secreted as homodimers in their latent precursor form. The latency-associated peptide prevents TGFβ from binding to its receptor, and proteolytic cleavage is necessary for TGFβ activation. Active TGFβ isoforms are homodimers of M_r ~ 25 kDa with several disulfide bonds forming a cystic knot (Kingsley 1994; O'kane and Ferguson 1997; Lichtman et al. 2016).

TGFβ1 is expressed in most cells, but primarily in endothelial, hematopoietic and connective tissue cells. TGFβ1 stimulates the proliferation of mesenchymal cells (though, on occasion it inhibits their proliferation), and is a potent inhibitor of epithelial and endothelial cell proliferation, and a potent immunosuppressant. TGFβ1 is also a strong fibrogenic agent: it induces the production of collagen and other components of extracellular matrix, downregulates the expression of matrix metalloproteinases (which degrade extracellular matrix and connective tissues), and upregulates the synthesis of protease inhibitors (Halper 2010).

TGFβ1 has been shown to play crucial roles in pathogenesis and pathophysiology of several syndromes and diseases of soft tissues. Though its involvement in Marfan syndrome is secondary due to its dysregulation by mutated fibrillin, its role as a mediator of substantial pathology in Marfan syndrome makes this disorder an excellent model for studies on the role of TGFβ1 (and to lesser extent of BMPs) in proper function of the cardiovascular system, and other soft tissue systems as well. We will not go into more detail regarding TGFβ1 as Chapters 7 and 8 in this volume provide comprehensive overview of TGFβ activity and participation in pathogenesis of Marfan syndrome and related disorders. Though the authors concentrate on the effect of TGFβ1 on cardiovascular system, TGFβ1 has similar impact on bones and ligament as the two systems (cardiovascular and musculoskeletal) share certain developmental and regulatory mechanisms, and similar extracellular matrix structure and composition (Lincoln et al. 2006a; Zhao et al. 2007). However, it has been shown that TGFβ1 null mice suffer primarily from overwhelming inflammatory processes and from vasculogenesis defects rather than from musculoskeletal problems, and that they die early in life (Dickson et al. 1995; Kruithof et al. 2012). BMP2 null mice fare even worse as they suffer from severe heart defects or even death *in utero* because of these defects (Kruithof et al. 2012; Zhang and Bradley 1996; Yoon et al. 2004; Dronkers et al. 2020). Smad 1, 2, 4 and 5 deficient mice meet the same fate *in utero* because of cardiac defects (Kruithof et al. 2012; Goumans and Mummery 2000; Bosman et al. 2006). As those reports were all focused on cardiovascular systems, no description of musculoskeletal system involvement in these mice was provided.

Because TGFβ is known to be a very powerful promoter not only of collagen synthesis, but of other components of ECM as well, it is not surprising that it plays an important role in tissue repair. It appears in early stages of tendon repair during the first week post-injury (Dahlgren et al. 2005; Chen et al. 2008b). Exogenous TGFβ1 promotes tendon and ligament development either alone or in combination with other growth factors, such as VEGF or BMP12 (Hou et al. 2009; Heisterbach et al. 2011), and though so far it has not been used in clinical practice, it does have a promise as a treatment agent in tendon and ligament injuries, at least in some studies described here. The addition of TGFβ-neutralizing antibody to cultures of acellular tendon matrix alone, and of acellular tendon matrix co-cultured with tendon-derived fibroblast media had negative effect on biomechanical parameters of the tissue (Azuma et al. 2007). Implantation of rat muscle tissue transduced first with recombinant adenovirus encoding for TGFβ1 into rat transected tendons of Achilles led to accelerated healing (Majewski et al. 2012). However, other studies present a more nuanced picture (Nixon et al. 2012). The overexpression of TGFβ1 and TGFβ2 in post-injury rabbit flexor tendons correlated with fibrosis and scarring which could be controlled with anti-TGFβ antibodies (Chang et al. 2000; Chang et al. 1997; Ngo et al. 2001).

Our understanding of TGFβ dysfunctional activity in pathogenesis of Marfan and Marfan-related syndromes might lead to development of treatment modalities which would focus on blocking TGFβ1 action rather than promoting it. This could be achieved using an approach employed in chemotherapy of cancers with biologics, e.g., with monoclonal antibodies to the growth factor itself, or with an agent blocking a step in the TGFβ1/Smad signaling pathway.

BMPs were first discovered as inducers of bone and cartilage formation (Urist 1965). Today, we know that this rather large group with more than 20 members is involved in organ patterning and in formation of tissue architecture and structure during embryonic development (Guo and Wu 2012; Miyazono et al. 2005). The group of growth and differentiation factors (GDFs) is closely related to BMPs and regulates cartilage and skeletal development (King et al. 1996; Guenther et al. 2008; Pregizer and Mortlock 2015). Some of the GDFs turned out to be identical to certain BMPs (see below). The so called heterodimeric BMPs are composites of monomers derived from different BMP isoforms and they are primarily involved in osteogenesis (Guo and Wu 2012; Wang et al. 1988). Both natural and recombinant BMPs have found use in orthopedics and dental surgery as they are highly osteogenic (Guo and Wu 2012). A recent study added that inhibition of EZH2 (an epigenetic suppressor) can serve as a primer for osteogenic differentiation mediated by BMP2 (Dudakovic et al. 2020). Several laboratories have shown that three BMPs (BMPs12–14), first described as growth and differentiation factors in mice and rats (GDFs) 5–7, were able to induce tendon- or ligament-like tissue *in vivo* (Dudakovic et al. 2020; Aspenberg and Forslund 1999; D'Souza and Patel 1999). GDF5 is also known as BMP14, GDF6 as BMP13, and GDF7 as BMP12 (Lee et al. 2011). In another study, ectopic administration of the same BMPs led to formation of tendon-like structures and expression of a tenocytic marker gene (Wolfman et al. 1997). GDF5 (BMP14) stimulated healing of Achilles tendon in rats (Forslund et al. 2003; Dines et al. 2007; Bolt et al. 2007). GDF5 also induces tenogenic differentiation in human mesenchymal stem cells derived from bone marrow as judged from upregulation of genes encoding for tendon markers such as scleraxis, tenascin and type I collagen (Tan et al. 2012). The application of BMP14 also supported periodontal regeneration through stimulation of both bone and ligament formation (Moore et al. 2010). BMP13 in particular appears to be a potent promoter of ectopic tendon tissue formation (though with some focal ossification foci occurring in the newly formed tendon tissue) upon intramuscular injection of BMP13 (Helm et al. 2001). BMP12 is another tendon-differentiating agent which has been used for *in vivo* (Lou and Tu 2001) and *in vitro* (Wang et al. 2005; Violini et al. 2009) tendon healing and development either by gene transfer (Lou and Tu 2001; Wang et al. 2005) or by *in vitro* exposure to BMP12 (Violini et al. 2009). The healing action of BMP12 was enhanced when combined with bFGF and TGFβ1 (Majewski et al. 2018). Furthermore, whether the final product of BMP activity is tendon or bone (or cartilage) depends not only on the BMP isoform used but also on the type of mechanical stimulation. Tension or shear applied to undifferentiated mesenchymal tissues leads to the development of fibrous (tendinous) tissues, whereas cartilage appears with the application of hydrostatic compression (Forslund and Aspenberg 2002).

BMP2 is a powerful osteogenic agent which has been approved by the FDA for treatment of acute open tibial shaft fractures, and has been used in the off-label treatment of non-union and acute fractures with considerable success (Starman et al. 2012; Ronga et al. 2013), though it has some tenogenic qualities as well. Mesenchymal stem cells overexpressing BMP2 and Smad8 (a mediator of signaling of TGFβ superfamily) have been found to differentiate into tendon cells upon expressing scleraxis, a factor stimulating tendon formation in the embryo (Hoffmann et al. 2006). In the same study, those cells had also the capacity to induce Achilles tendon repair in rats (Hoffmann et al. 2006). However, *in vivo* experiments have shown that BMP2 stimulated healing at the tendon-bone interface due to accelerated new bone formation rather than due to tendon formation (Chen et al. 2008a; Rodeo et al. 1999). And, indeed, in yet another study BMP2 had rather detrimental effect on tendon-to-bone healing in a canine flexor tendon repair model likely due to lack of effect on tendon itself (Thomopoulos et al. 2012). Such data suggest that there may be a fine line between the osteogenic and tendon promoting activity of BMP2. BMP7 (identical to OP1 growth factor in literature) likely plays an important role during embryonic development as it is strongly expressed in developing chicken tendons (Macias et al. 1997). BMP7 is also a more potent promoter for cell proliferation and type I collagen synthesis in tendon cell cultures established from both young and aged male tendons than BMP2 which performs these functions well in cell cultures from young but not older tendons (Klatte-Schulz et al. 2012). BMP2 might be more suitable for clinical use as a tendon healing agent in non-union bone fractions because of its osteogenic effect (Haubruck et al. 2018)

As pointed out by Lincoln et al. heart valves and parts of the musculoskeletal system, including tendons, share regulatory mechanisms (Lincoln et al. 2006a). The BMP2-Sox9-aggrecan axis is an example of a pathway connecting regulation of cardiac and skeletal development. This pathway is instrumental in endocardial cushion formation, and in the differentiation of chondrogenic cell lineages (Lincoln et al. 2006a; Nishimura et al. 2012) (Jo et al. 2014). Transcription factor Sox9 regulates expression of chondrogenic genes and their protein products such as type II and type XI collagens, aggrecan, cartilage link protein and some ECM proteins produced in chicken heart valves (Nishimura et al. 2012; Bell et al. 1997; Chimal-Monroy et al. 2003; Kou and Ikegawa 2004; Lefebvre et al. 1997). It is interesting that aggrecan-expressing cartilage like cells have been found in the spongiosa layer of heart valves. The spongiosa layer is sandwiched between the fibrosa and atrialis layers where it exhibits compressibility, similar to a layer of cartilage (Lincoln et al. 2006b).

On the other hand, it is the FGF-scleraxis-tenascin which regulates the development of semilunar valve precursor cells into valve cups and chordae tendineae (the "tendons" of heart valves), and its expression is equally important during tendon cell lineage differentiation (Lincoln et al. 2006a; Zhao et al. 2007).

6.3 Fibroblast Growth Factors

Like most other families of growth factors, fibroblast growth factors (FGFs) participate in embryonic development; more specifically, they regulate gastrulation, neurulation, anteroposterior specification of body segments and organ morphogenesis (Böttcher and Niehrs 2005; Coutu and Galipeau 2011; Takebayashi-Suzuki and Suzuki 2020). In postnatal life they function as classical growth factors, including their function as powerful promoters of angiogenesis. This group consists of at least 23 growth factors binding to four high-affinity transmembrane receptors with tyrosine kinase activity (FGFRs). The receptors are subject to alternative splicing which modulates their activity (Kosaka et al. 2009; Beenken and Mohammadi 2009; Hui et al. 2018).

Several FGFs are involved in myogenesis and tendon development during the embryonic stage (Volk 1999; Kieny and Chevallier 1979; Chen et al. 2020; Theodossiou and Schiele 2019). FGF2, known previously as basic FGF, is synthesized and secreted without a signal peptide. It is a potent mitogen, and a potent angiogenic agent as well (Halper 2010; Beenken and Mohammadi 2009). It also induces regeneration of the chicken limb bud after application of FGF2-loaded heparin gel beads to the amputation site (Taylor et al. 1994), or of gelatin beads loaded with a mixture of FGF2 and FGF8 (Makanae and Satoh 2018). FGF2 directs development of tenocytes from multipotent mouse cells (Ker et al. 2011), and stimulates proliferation of mouse tendon cell lines together with induction of scleraxis, a transcription factor and an early tendon marker (Salingcarnboriboon et al. 2003). Ide et al. have shown that FGF2 accelerated initial tendon-to-bone healing due to stimulation of formation of tendon and fibrocartilage tissues but not bone, something what BMP2 failed to provide (Ide et al. 2009).

FGF4, unlike FGF2, is synthesized with a signal peptide which is cleaved off after extracellular secretion (Taira et al. 1987; Yoshida et al. 1987). Though originally discovered as an oncogene, FGF4 is expressed in preimplantation mouse blastocysts. Its multiple roles during embryonic development include facilitation of epithelial-mesenchymal communication (vital for morphogenesis) and maintenance of self-renewal and pluripotency of stem cells. More pertinent to the topic under discussion here is that FGF4 also promotes early cardiac development (Kosaka et al. 2009), limb development and, as already described above, expression of scleraxis (Lincoln et al. 2006a; Kosaka et al. 2009). Scleraxis in turn induces expression of tenascin, another tendon marker. Both scleraxis and tenascin are also expressed in valve progenitor cells derived from chicken embryos (Lincoln et al. 2006b).

Their avid binding to heparan sulfate and other glycans present in the extracellular matrix enables FGFs to be stored in the extracellular matrix and to be readily available for use in tissue repair (Li et al. 2016). Whether FGFs are suitable for accelerating or improving healing of tendon injuries in a clinical setting remains to be evaluated because their potent angiogenic capacity may interfere with proper assembly of collagen fibrils, especially in the rather avascular tendon and ligament tissues (and in the avascular cardiac valves as well). The angiogenic ability of these growth factors spurred experimental application of recombinant FGF1, FGF2 and FGF4 gene therapy aimed at treatment of cardiovascular problems though clinical use is still far away in the future (Beenken and Mohammadi 2009).

6.4 Role of IGF-I in Tendon Healing

Unlike the huge TGFβ superfamily, the IGF family has only two members: IGF-I and IGF-II. Both of them retain their close relationship to insulin as they evolved from an ancestral insulin-like gene by duplication (Kelley et al. 2002). IGF-I has been much better characterized and thus will be the target of discussion in this review. It is mainly produced by the liver under the control of growth hormone (GH) (Messier and Teutenberg 2005). It has several important roles including mediation of growth hormone activity and stimulation of proliferation of various cell types of animal and human origin, including tenocytes (Halper 2010; Jones and Clemmons 1995; Tsuzaki et al. 2000; Ramos et al. 2019). It can bind to any of the six IGF binding proteins which are also synthesized in the liver and their main function is to regulate IGF-I activity (Halper 2010; Brown and Halper 1990; Juul 2003). IGF-I receptor bears partial homology to insulin receptor and c-ros oncogene. The mature form of this receptor is a heterotetramer $\alpha_2\beta_2$ of 350 kDa upon which ligand binding leads to activation of tyrosine kinase (Cheatham and Kahn 1995; LeRoith et al. 1995; Janssen 2020). Tsuzaki et al. have found IGF-I mRNA, IGF-I and receptors for IGF-I in both tendon surface cells (residing in the epitenon) and tendon internal fibroblasts (embedded in collagen bundles and fascicles), and they noted an increase in the levels of IGF-I and its mRNA in actively proliferating cells (Tsuzaki et al. 2000). Hansson et al. have observed the induction of IGF-I synthesis in tendon cells after injury and with mechanical loading (Hansson et al. 1988a, b). IGF-I is known to stimulate collagen synthesis in fibroblasts, including tenoblasts (Halper 2010; Hansen et al. 2013) which is closely associated with and dependent on mechanical loading of the tendon (Hansen et al. 2013; Banes et al. 1995; Heinemeier et al. 2012). It is not clear whether IGF-I stimulates collagen directly through the PI-3 kinase pathway alone (Chetty et al. 2006) or through the PI-3 kinase and ERK complex (Svegliati-Baroni et al. 1999). Some studies even indicate that IGF-I stimulates collagen production indirectly through TGFβ1 (Ghahary et al. 2000; Hernandez et al. 2020). In any case, increased collagen synthesis in tendons appears to be a part of a systemic response to exercise, especially to repeated and sustained training in people. Regular exercise leads to sustained elevation of blood levels of GH and IGF-I due to an increase of GH release from the pituitary, and this is accompanied by increased expression of IGF-I mRNA in human skeletal muscle and tendon (Heinemeier et al. 2012). However, it remains speculative whether there is a causal relationship between elevated circulating GH/IGF-I and stimulation of collagen synthesis in the musculoskeletal system (Heinemeier et al. 2012) as other investigators have not seen stimulatory effects of GH on tendon or ligament healing at least in some animal models (Andersson et al. 2012; Provenzano et al. 2007). The stimulatory effect of IGF-I alone (i.e., without involvement of GH) in these models suggests only a local rather than systemic effect of IGF-I. Similarly, Yamaguchi et al. showed that loading-induced IGF-I expression in skeletal muscle occurs independently of release of GH from the pituitary (Yamaguchi et al. 2006). Banes et al. have shown that mechanical load alone did not stimulate DNA synthesis, but only when both IGF-I and PDGF-BB were added to cultures of avian tendon cells (Banes et al. 1995). Ramos et al. have shown that insulin added to mesenchymal stem cells seeded in electron-spun poly (lactic acid-co-glycolic acid) scaffolds stimulated not only cell proliferation but expression of tendon markers as well. If this is the case, then insulin can serve as a cheap and effective substitute for IGF-I (Ramos et al. 2019). In a similar experiment, a combination of growth factors, such as IGF-I, bFGF and PDGF stimulated repopulation by host cells in injured rat tendons implanted with tendon-specific hydrogel seeded with adipose-derived stem cells (Farnebo et al. 2017). These studies are good examples of bioengineering efforts currently paving the way for future approach to tendon healing and reconstruction.

The presence of several isoforms of IGF-I, the result of alternative slicing of IGF-I pre-mRNA, compensates for the smallness of the IGF family and helps explain the role of IGF-I in stimulation of collagen synthesis in the tendon as a consequence of mechanical loading. So far, three isoforms of IGF were identified: IGF-IEa, IGF-IEb and IGF-IEc. Rat isoform IGF-IEb is identical to human IGF-IEc, and to mechano-growth factor (MGF) which got its designation because of its association with mechanical loading in the muscle and tendon (Dai et al. 2010). MGF, unlike standard IGF-I, is a strictly autocrine/paracrine factor produced locally upon proper stimulation (McCoy et al. 1999). Upregulation of MGF and IGF-Ea7 has been observed in the Achilles tendon after eccentric and concentric training (Heinemeier et al. 2007). A more recent study showed that resistance training caused no differences in MGF and IGF-Ea protein expression in both young and old men (Ahtiainen et al. 2016). However, prolonged hindlimb suspension led to an increase in MGF mRNA in rat Achilles tendon (Heinemeier et al. 2009). One can only speculate that the application of MGF could alleviate tendon problems and joint laxity in patients with Ehlers-Danlos syndrome, cutis laxa and related disorders.

IGF-I binding proteins (IGF-BPs) are involved in tendon healing as well, though their specific roles have not been yet established. Whereas the levels of mRNA for IGF-I and IGF-BPs 2–4 increased following injury, especially later, at 4 weeks after injury (Dahlgren et al. 2005), those for IGF-BP 5 and 6 decreased (Dahlgren et al. 2006). Both IGF-I and IGF- BP4, and their respective mRNAs were co-localized to tendon fibroblasts and vascular endothelium in the human Achilles tendon (Olesen et al. 2006). The increase in IGF-BP 4 level seen in the Achilles tendons during long distance running can be attributed to an increase in mechanical loading (Olesen et al. 2006).

The outcome of rather common tendinopathies and injuries of superficial digital flexor and other weight-bearing tendons in racehorses is not satisfactory when only more traditional treatment and management are applied (Durgam et al. 2012a; Genovese 1993; Peloso et al. 1994). Though initial studies suggested that local administration of IGF-I enhances collagen synthesis in equine tendon explants (Murphy and Nixon 1997) and improves equine tendon healing *in vivo* (Durgam et al. 2012a, b; Dahlgren et al. 2002), more recent reports were less enthusiastic about the long-term prognosis of IGF-I-treated tendon injuries in racehorses (Witte et al. 2011), while a more recent review of literature concluded that hormonal treatment of tendon injuries shows improved tendon healing, but not biomechanical function (Abate et al. 2016). This includes the role of the growth hormone (GH)/IGF-I system in tendon healing as extrapolated from findings in patients with acromegaly who have increased expression of mRNAs for collagen and IGF-I. IFG-I stimulates stem cells and its effects are overall anabolic in nature and it is considered to be a survival factor. Effects of GH/IGF-I are mediated through endocrine and paracrine/endocrine pathways (Abate et al. 2016).

6.5 PDGF

Growth factors, such as PDGF and VEGF, its relative, have been shown to be synthesized and likely to play similar roles in tissue repair (Aspenberg 2007), though perhaps not in tendon healing. These two growth factors, or rather these two families of growth factors share 20% homology of their respective amino acid sequences (Halper 2010). Like many other growth factors PDGF/VEGF growth factors contain cysteine knots in their molecules (McDonald and Hendrickson 1993; Hearn and Gomme 2000).

PDGF is instrumental in repair mechanism of vasculature (Andrae et al. 2008). It is stored in platelet α granules so it gets released and ready for action during wounding or bleeding when restoration of vasculature becomes essential. PDGF is synthesized by endothelial cells, placental trophoblasts, smooth muscle cells, macrophages and sarcoma cells. PDGF is a 29–33 kDa glycoprotein, usually composed of 2 chains (at least in humans), either A or B which are joined by disulfide bridges. There are several other isoforms, but heterodimer PDGF-AB and homodimers PDGF-AA and PDGF-BB are the best described. PDGF and its isoforms bind to two types of receptors with five extracellular immunoglobulin loops and with an intracellular tyrosine kinase domain. PDGF is required by fibroblasts, smooth muscle and glial cells for their optimal growth and cell proliferation. Its role in wound healing and as a chemotactic agent for fibroblasts and smooth muscle cells, neutrophils and macrophages has been well documented (Kim et al. 2015). Besides its involvement in blood vessel repair it plays a role in the pathogenesis of atherosclerosis, pulmonary fibrosis and glomerulonephritis (Halper 2010). PDGF-AA and PDGF-BB are important mitogens for normal and neoplastic mesenchymal tissues. For example, exogenous PDGF-BB had a modest effect on healing of equine superficial digital flexor tendons (Haupt et al. 2006) and of rat Achilles tendons (Aspenberg and Virchenko 2004). A more recent study has shown that PDGF-AA (which binds to PDGF receptor α) plays an important role both in proper healing of the tendon and also in fibrotic scarring. As Harvey et al. have documented the outcome depends on the target cell of PDGF-AA: stimulation of quiescent stem cells leads to proper healing and repair during the first several weeks post tendon injury, whereas PDFF-AA stimulation of fibro-adipogenic progenitors coexistent with the stem cells is responsible for scaring and fibrosis (Harvey et al. 2019).

PDGF cooperates with other growth factors. The classic paradigm is that while PDGF renders cells "competent" to replicate their DNA, and thus prepares cells for IGF-I activity as the actual growth factors moving cell through the cell cycle into the S phase when cells start synthesizing DNA (Stiles et al. 1979). This explains why, as described above, avian tendon cell cultures required both growth factors in addition to mechanical loading for DNA synthesis (Banes et al. 1995). The excellent review by Andrae et al. provides more information on all aspects of PDGF biology and role in physiology (Andrae et al. 2008).

PDGF delivered via *in vivo* gene transfection is capable to enhance healing of rat Achilles tendons as determined by histological and biomechanical parameters. The gene transfer was superior to direct administration to the tendons (Suwalski et al. 2010). PDGF alone, in particular the homodimer PDGF- BB, or in combination with IGF-I stimulated regeneration of periodontal ligament and its attachment to the alveolar bone (Raja et al. 2009; Ammar et al. 2018), a matter of importance in Marfan patients with periodontitis. By the way, FGF2, discussed in more detail above, had similar effect on cells of the periodontal ligament (Suda et al. 2009). PDGF was shown to improve healing and biomechanical parameters of acute injury of the medial collateral ligament in rabbits and rats, especially, if administered shortly after injury (Hildebrand et al. 1998; Batten et al. 1996; Kim et al. 2019). In a recent study, growth factors, such as PDGF, IGF-I and basic FGF present in processed amniotic/chorionic membranes, contributed to production of ECM in cultured tenocytes (McQuilling et al. 2019).

6.6 VEGF

VEGF and its isoforms bind to three receptors, VEGF-R1–3, which belong to a subfamily of tyrosine-kinase receptors within the PDGF-receptor class. Just like PDGF, VEGFs are composed of two subunits linked by disulfide bonds. These growth factors are usually considered exclusive angiogenic factors, particularly VEGF-A (Halper 2010). VEGF-A (designated just as VEGF for the sake of simplicity from now on) is the main VEGF representative (Halper 2010). More recent studies have shown that VEGF has the ability to function as an autocrine mitogen for cancer cells (Roy and Peretz 2009; Saman et al. 2020).

Whereas splice variants of VEGF, $VEGF_{121}$ and $VEGF_{165}$, have been identified in fetal tendons, healthy adult tendons possess little VEGF (Pufe et al. 2005a). Instead, adult tendons express endostatin, an antiangiogenic peptide which downregulates VEGF-induced Erk1/2 kinase phosphorylation (Pufe et al. 2003, 2004; Zhang et al. 2019). VEGF presence and neovascularization appearance in the tendon are signs of tendon overuse injuries (Pufe et al. 2005a), and they impede rather than stimulate proper healing of tendons (Savitskaya et al. 2011; Sahin et al. 2012a). This is due to a sequence of several events. Injury to rat patellar tendon results in hypoxia, and in increased expression of hypoxia-inducible factor-1 (HIF-1) which in turn induces transcription of VEGF gene (Hofer et al. 2001). In addition to hypoxia, mechanical overload, an important factor in overuse tendon injuries, is another stimulator of VEGF synthesis (Pufe et al. 2005a) as is the production of epidermal growth factor and inflammatory cytokines (IL-1β, IL-6, IL-8) (Pufe et al. 2005a; Savitskaya et al. 2011). To facilitate angiogenesis VEGF stimulates the production of matrix metalloproteinases (MMPs), enzymes actively degrading connective tissues so to assist sprouting of new blood vessels. VEGF also inhibits the production of tissue inhibitor metalloproteinases (TIMPs), enzymes counteracting MMP activity (Pufe et al. 2005a, b; Sahin et al. 2012a). It is likely that VEGF regulates the production of MMPs and TIMPs by endothelial cells in blood vessels of injured tendons as well as by tenocytes as those latter cells express VEGF-R2 (Petersen et al. 2004). Increase and decrease in MMP-3 synthesis correlates well with timing of increase in VEGF synthesis at 7 days post-injury, and with decrease in VEGF synthesis 14 and 28 days post-injury, respectively (Sahin et al. 2012b). MMP-3 is not the only metalloproteinase associated with VEGF production. For example, expression of MMP-1 and MMP-13 is regulated also by VEGF, at least in chondrocytes (Pufe et al. 2005a). MMP-9 secreted by inflammatory cells follows the same cycle. It is elevated early in tendon healing, its decrease 7 days after wounding is associated with rapid increase first in type III collagen, which is followed by an increase in type I collagen. MMP-2 and MMP-14 were found to be elevated 21 days after wounding when granulation tissues starts undergoing extensive remodeling into organized collagen tissue, and scarring (Loiselle et al. 2009). Increased or decreased levels of many MMPs were found in patients with ruptured rotator cuff, supraspinatus, Achilles and other tendons (Riley et al. 2002; Riley 2008; Attia et al. 2013; Fleischhacker et al. 2020). The induction of MMP activity and angiogenesis by VEGF then leads to impaired biomechanical parameters of tendon due to matrix degradation (Sahin et al. 2012b). These tissue changes also lead to pain, especially, in the case of neovascularization, in mid-portion of chronic Achilles tendinosis (Pufe et al. 2005a; Ohberg et al. 2001).

Another group of proteinases, members of a-disintegrin-and-metalloproteinase-with--thrombospondin-like-motifs (ADAMTS) activity family or aggrecanases are expressed in tendons as well (Riley 2008; Jones et al. 2006b) . For example, upregulation of ADAMTS-4 was observed in injured tendons and in tendon cell cultures exposed to TGFβ, whereas no differences in ADAMTS-1 levels were noted between normal and diseased tendons (Corps et al. 2008) . Because expression of ADAMTS forms is tied with degradation of aggrecan they are discussed in more detail in the preceeding Chapter of this volume.

6.7 Other Modalities

The fact that combinations of several growth factors have been found beneficial to tendon healing should not surprise anybody (Hou et al. 2009). Not only are such preparations able to stimulate cell proliferation, they lead also to improvement of mechanical properties of tendons (Hou et al. 2009; Lyras et al. 2010; Raghavan et al. 2012; Bachl et al. 2009; Hagerty et al. 2012). Such findings have been essential in tissue engineering where the current immediate goal is to improve cellular repopulation of acellularized scaffolds (Raghavan et al. 2012).

Rather than using purified recombinant or natural IGF-I or mixture of other growth factors, a somewhat expensive undertaking, to treat tendon injuries, investigators have been using platelet-rich plasma (PRP) in experiments, animal trials and treatment of chronic leg ulcers, and in tissue engineering among other applications

(Lyras et al. 2011). The use of PRP provides several advantages – it is rich in numerous growth factors, it provides other, some of them still uncharacterized compounds, promoting tissue repair. It is much easier and cheaper to manufacture than individual growth factors. PRP stimulates cell proliferation and collagen synthesis or arrangement in human tendon cell cultures (de Mos MvdW et al. 2008; Nishio et al. 2020; Kobayashi et al. 2020). A study in rabbits showed accelerated healing of a ruptured Achilles tendon accompanied by an upsurge in IGF-I expression (Lyras et al. 2011). The effect of autologous conditioned serum on tendon healing was comparable perhaps because the conditioned serum increased expression of several growth factors (bFGF, BMP12 and TGFβ1) in an *in vivo* model of rat Achilles tendon healing (Heisterbach et al. 2011). PRP, in combination with BMP2, even accelerated healing of bone-tendon injuries, with PRP improving tendon healing and BMP enhancing bone healing in rabbits with their Achilles tendon surgically separated from the bone insertion (Kim et al. 2011). Similarly, PRP had beneficial effect on healing of porcine ligaments, resulting in improvement of several biomechanical parameters, such as load to failure, maximum load and stiffness. However, this effect was rather limited as it did not improve laxity, maximum tensile strength or linear stiffness (Murray et al. 2009). In a study of PRP effect on healing of surgically transected tendon of Achilles in rats, local application of PRP led to increased expression of type I and III collagens, and to improved ultimate tensile strength (Kaux et al. 2012).

PRP products have been used in human medicine, mostly for healing of athletic injuries (Paoloni et al. 2011; Bava and Barber 2011). Unfortunately, the use of PRP has not met with much success in human patients. Why this is so is not quite obvious. For one, pathogenesis of human tendinopathies may be quite different from artificial tendinopathies induced under highly simulated conditions, or from Achilles tendon rupture in rat or rabbits; two, there is no standardization for PRP preparations; three, the content of growth factors, cytokines and other compounds may differ between human and animal PRP (Abate et al. 2012; Del Buono et al. 2011). It is also possible that the presence of TGFβ in PRP might vary from batch to batch, and, if present in higher concentrations, would lead to fibrosis, and thus to worse outcome (Nixon et al. 2012). Perhaps most importantly, PRP is used on injuries affecting a wide variety of tendons, ligaments and even muscle strains without much, if any prior clinical evidence of effectiveness (Paoloni et al. 2011; Bava and Barber 2011). However, more recent studies indicate that certain preparations of PRP have positive effect on healing of injured skeletal muscles, especially those with enhanced ability to promote anti-fibrotic healing (Chellini et al. 2019).

As noted above, not only the presence of specific growth factors, but also the milieu and mechanical parameters determine the final fate of tissue differentiation. This is particularly the case with stem cells which can be induced to commit toward bone or tendon (or many other different) lineages with cell culture on properly engineered substrates (Sharma and Snedeker 2010). These substrates functionalized with type I collagen and fibronectin were able to direct bone marrow-derived stem cells toward tenogenic differentiation and to production of BMP2/Smad8. The authors speculate that BMP2 facilitates healing at the bone-tendon interface by its ability to regulate both osteogenic and tenogenic differentiation depending on the degree of stiffness environment and through stimulating signaling of other growth factors (Sharma and Snedeker 2012). Tenogenic differentiation required the presence of type I collagen with fibronectin being optional. The substrate itself had to be within a narrow range of elastic moduli. An intermediate elastic modulus then stimulated production of BMP2/Smad8, and ultimately of scleraxis. Osteogenic development demanded stiffer substrate and fibronectin, type I collagen was inhibitory, especially with less stiff substrates (Sharma and Snedeker 2012). It has also been shown that mechanical stimulation of bioengineered constructs *in vitro* modulates stiffness, and such constructs upon application *in vivo* enhance tendon repair, at least in rabbits (Juncosa-Melvin et al. 2007; Lim et al. 2019).

It is likely that bioengineering of soft tissues will play an important role in treatment of disorders such as Marfan or Ehlers-Danlos. Replacement of diseased portion of aorta by bioengineered constructs would prevent the development and/or rupture of aortic aneurysm. Similarly, bioengineered tissues have a great potential to serve as patches in other organs with potential to rupture (e.g., uterus in women with Ehlers-Danlos), or as replacement for affected tendons in Ehlers-Danlos.

As noted above injuries of the superficial digital flexor and other weight bearing tendons are common in horses of all breeds, but particularly in horses of racing breeds. Because the full recovery and return to races is rather rare after a tendon injury, and because of financial considerations, such as investments into racehorses, the field of stem cell therapy for equine tendons has been quite active and has met with some success. Some investigators have identified progenitor cells residing in the superficial digital flexor tendon and found these cells comparable to bone marrow derived stem cells. The tendon-derived progenitor cells expand quickly in cell culture and express the same markers as bone marrow-derived cells. The advantage is that tendon-derived progenitor cells have high clonogenic properties and limited chondrogenic differentiation potential (Lovati et al. 2011; Lin et al. 2019). There is one caveat: though tendons have multipotent progenitor cells, likely residing in the endotenon between the collagen fascicles and vasculature, it is very difficult to activate such cells *in vivo*, especially, in more mature tendons (Godwin et al. 2011). However, autologous bone marrow- or peripheral blood-derived stem cells are easier to obtain, and they have been used for treatment of equine tendon injuries for several years. Almost flawless repair of these tendons with fibers perfectly aligned parallel to the long axis of the tendon itself was achieved in three injured horses with a single injection of autologous "deprogrammed" peripheral blood-derived stem cells (Marfe et al. 2012). Three other horses with similar tendon injuries who served as controls and received traditional therapy ended up with tendon scarring (documented by ultrasound) and re-injury only a few months later (Marfe et al. 2012). Godwin et al. harvested mesenchymal stem cells from a bone marrow aspirate which was suspended in culture medium. The bone marrow-derived mesenchymal stem cells attached themselves to the plastic of the tissue culture vessels and proliferated until confluent. These cells were then detached from the plastic and implanted directly into the injury site of 141 racehorses (Godwin et al. 2011). As in the study of Marfe et al. (Marfe et al. 2012), many horses were able to re-enter racing with low re-injury rate. The ultrasonic appearance indicated complete healing without any evidence of ossification in the healed tendon (Godwin et al. 2011). Similar results were achieved by other groups as well (Borjesson and Peroni 2011; Kol et al. 2013), though treatment with bone marrow-derived stem cells had no effect on the diameter of collagen fibrils in the healing tendon (Caniglia et al. 2012). Equine adipose-derived stromal vascular fraction was particularly effective in stimulation cell proliferation, and application of equine cord blood-derived mononuclear cells led to increased production of interleukin 6, TGFβ1 and prostaglandin E_2 at least *in vitro* (Kol et al. 2013), though Uysal et al. have shown the effectiveness of adipose-derived stem cells on healing of rabbit primary tendon healing *in vivo* (Uysal et al. 2012).

6.8 Epilogue

Though we do have a better understanding of the mechanism of tendon injury and subsequent repair, we still have a long way to go before we will feel confident that we have master the art of tendon healing. Progress in growth factor and stem cell research is bringing us closer not only to the development of new strategies for treatment of tendon injuries, but also to better understanding of pathogenesis of heritable diseases of soft tissues. What we have to do now is to figure out how to use our knowledge to improve the lives of patients afflicted with these disorders.

References

Abate M, Di Gregorio P, Schiavone C, Salini V, Tosi U, Muttini A (2012) Platelet rich plasma in tendinopathies: how to explain the failure. Int J Immunopathol Pharmacol 25:325–334

Abate M, Guelfi M, Pantalone A, Vanni D, Schiavone C, Andia I et al (2016) Therapeutic use of hormones on tendinopathies: a narrative review. Muscles Ligaments Tendons J 6:445–452

Achari Y, Chin JW, Heard BJ, Rattner JB, Shrive NG, Frank CB et al (2011) Molecular events surrounding collagen fibril assembly in the early healing rabbit medial collateral ligament--failure to recapitulate normal ligament development. Conn Tissue Res. 52:301–312

Ahtiainen JP, Hulmi JJ, Lehti M, Kraemer WJ, Nyman K, Selänne H et al (2016) Effects of resistance training on expression of IGF-I splice variants in younger and older men. Eur J Sport Sci 16:1055–1063

Ammar MM, Waly GH, Saniour SH, Moussa TA (2018) Growth factor release and enhanced encapsulated periodontal stem cells viability by freeze-dried platelet concentrate loaded thermo-sensitive hydrogel for periodontal regeneration. Saudi Dent J 30:355–364

Andersson T, Eliasson P, Apenberg P (2012) Growth hormone does not stimulate early healing in rat tendons. Int J Sports Med 33:240–243

Andrae J, Gallini R, Betsholtz C (2008) Role of platelet-derived growth factors in physiology and medicine. Genes Dev 22:1276–1312

Aspenberg P (2007) Stimulation of tendon repair: mechanical loading. GDFs and platelets Int Orthop 31:783–789

Aspenberg P, Forslund C (1999) Enhanced tendon healing with GDF 5 and 6. Acta Orthop Scand 70:51–54

Aspenberg P, Virchenko O (2004) Platelet concentrate injection improves Achilles tendon repair in rats. Acta Orthop Scand 75:93–99

Attia M, Huet E, Gossard C, Menashi S, Tassoni MC, Martelly I (2013) Early events of overused supraspinatus tendons involve matrix metalloproteinases and EMMPRIN/CD147 in the absence of inflammation. Am J Sports Med 41(4):908–917

Azuma C, Tohyama H, Nakamura H, Kanaya F, Yasuda K (2007) Antibody neutralization of TGF-β enhances the deterioration of collagen fascicles in a tissue-cultured tendon matrix with ex vivo fibroblast infiltration. J Biomech 40:2184–2190

Bachl N, Derman W, Engebretsen L, Goldspink G, Kinzlbauer M, Tschan H, Volpi P, Venter D, Wessner B (2009) Therapeutic use of growth factors in the musculoskeletal system in sports-related injuries. J Sports Med Phys Fitness 49:346–357

Banes AJ, Tsuzaki M, Hu P, Brigman B, Brown T, Almekinders L, Lawrence WT, Fischer T (1995) PDGF-BB, IGF-I and mechanical load stimulate DNA synthesis in avian tendon fibroblasts in vitro. J Biomech 28:1505–1513

Batten ML, Hansen JC, Dahners LE (1996) Influence of dosage and timing of application of platelet-derived growth factor on early healing of the rat medial collateral ligament. J Orthop Res 14:736–741

Bava ED, Barber FA (2011) Platelet-rich plasma products in sports medicine. Phys Sportsmed 39:94–95

Beenken A, Mohammadi M (2009) The FGF family: biology, pathophysiology and therapy. Nat Rev Drug Discov 8:235–253

Bell DM, Leung KK, Wheatley SC, Ng LJ, Zhou S, Ling KW, Sham MH, Koopman P, Tam PP, Cheah KS (1997) SOX9 directly regulates the type-II collagen gene. Nat Genet 16:174–178

Bolt P, Clerk AN, Luu HH, Kang Q, Kummer JL, Deng ZL, Olson K, Primus F, Montag AG, He TV, Haydon RC, Toolan BC (2007) BMP-14 gene therapy increases tendon tensile strength in a rat model of Achilles tendon injury. J Bone Joint Surg Am 89:1315–1320

Borjesson DL, Peroni JF (2011) The regenerative medicine laboratory: facilitating stem cell therapy for equine disease. Clin Lab Med 31:109–123

Bosman EA, Lawson KA, Debruyn J, Beek L, Francis A, Schoonjans L et al (2006) Smad5 determines murine amnion fate through the control of bone morphogenetic protein expression and signalling levels. Development 133:3399–3409

Böttcher RT, Niehrs C (2005) Fibroblast growth factor signaling during early vertebrate development. Endocr Rev 26:63–77

Brown CA, Halper J (1990) Mitogenic effects of transforming growth factor type e on epithelial and fibroblastic cells--comparison with other growth factors. Exp Cell Res 190:233–242

Cai J, Pardali E, Sánchez-Duffhues G, ten Dijke P (2012) BMP signaling in vascular diseases. FEBS Lett 586:1993–2002

Caniglia CJ, Schramme MC, Smith RK (2012) The effect of intralesional injection of bone marrow derived mesenchymal stem cells and bone marrow supernatant on collagen fibril size in a surgical model of equine superficial digital flexor tendonitis. Equine Vet J 44:587–593

Chang J, Most D, Stelnicki E, Siebert JW, Longaker MT, Hui K, Lineaweaver WC (1997) Gene expression of transforming growth factor beta-1 in rabbit zone II flexor tendon wound healing: evidence for dual mechanisms of repair. Plast Reconstr Surg 100:937–944

Chang J, Thunder R, Most D, Longaker MT, Lineaweaver WC (2000) Studies in flexor tendon wound healing: neutralizing antibody to TGF-beta1 increases postoperative range of motion. Plast Reconstr Surg 105:148–155

Cheatham B, Kahn CR (1995) Insulin action and the insulin signalling network. Endocr Rev 16:117–142

Chellini F, Tani A, Zecchi-Orlandini S, Sassoli C (2019) Influence of platelet-rich and platelet-poor plasma on endogenous mechanisms of skeletal muscle repair/regeneration. Int J Mol Sci 20(3):683

Chen CH, Liu HW, Tsai CL, Yu CM, Lin IH, Hsiue GH (2008a) Photoencapsulation of bone morphogenetic protein-2 and periosteal progenitor cells improve tendon graft healing in a bone tunnel. Am J Sports Med 36:461–472

Chen CH, Cao Y, Wu YF, Bais AJ, Gao JS, Tang JB (2008b) Tendon healing in vivo: gene expression and production of multiple growth factors in early tendon healing period. J Hand Surg Am 33:1834–1842

Chen JW, Niu X, King MJ, Noedl MT, Tabin CJ, Galloway JL (2020) The mevalonate pathway is a crucial regulator of tendon cell specification. Development 147:dev185389

Chetty A, Cao GJ, Nielsen HC (2006) Insulin-like growth factor signaling mechanisms, type I collagen and alpha smooth muscle actin in human fetal lung fibroblasts. Pediatr Res 60:389–394

Chimal-Monroy J, Rodriguez-Leon J, Montero JA, Ganan Y, Macias D, Merino R, Hurle JM (2003) Analysis of the molecular cascade responsible for mesodermal limb chondrogenesis: sox genes and BMP signaling. Dev Biol 257:292–301

Connolly EC, Freimuth J, Akhurst RJ (2012) Complexities of TGF-β targeted cancer therapy. Int J Bio Sci 8:964–978

Corps AN, Jones GC, Harrall RL, Curry VA, Hazleman BL, Riley GP (2008) The regulation of aggrecanase ADAMTS-4 expression in human Achilles tendon and tendon-derived cells. Matrix Biol 27:393–401

Coutu DL, Galipeau J (2011) Roles of FGF signaling in stem cell self-renewal, senescence and aging. Aging 3:920–933

Dahlgren LA, va der Meulen MC, Bertram JE, Starrak GS, Nixon AJ (2002) Insulin-like growth factor-I improves cellular and molecular aspects of healing in a collagenase-induced model of flexor tendinitis. J Orthop Res 20:909–910

Dahlgren LA, Mohammed HO, Nixon AJ (2005) Temporal expression of growth factors and matrix molecules in healing tendon lesions. J Orthop Res 23:84–92

Dahlgren LA, Mohammed HO, Nixon AJ (2006) Expression of insulin-like growth factor binding proteins in healing tendon lesions. J Orthop Res 24:183–192

Dai Z, Wu F, Yeung EW, Li Y (2010) IGF-IEc expression, regulation and biological function in different tissues. Growth Hormon IGF Res 20:275–281

de Mos MvdW AE, Jahr H, van Schie HT, Weinans H, Verhaar JA, van Osch GJ (2008) Can platelet-rich plasma enhance tendon repair? A cell culture study. Am J Sports Med 36:1171–1178

Del Buono A, Papalia R, Denaro V, Maccauro G, Maffulli N (2011) Platelet rich plasma and tendinopathy: state of the art. Int J Immunopathol Pharmacol 24:79–83

Dickson MC, Martin JS, Cousins FM, Kulkarni AB, Karlsson S, Akhurst RJ (1995) Defective haematopoiesis and vasculogenesis in transforming growth factor-beta 1 knock out mice. Development 121:1845–1854

Dines JC, Weber L, Razzano P, Prajapati R, Timmer M, Bowman S, Bonasser L, Dines DM, Grande DP (2007) The effect of growth differentiation factor-5 coated sutures on tendon repair in a rat model. J Shoulder Elb Surg 16:S215–8S25

Dronkers E, Wauters MMM, Goumans MJ, Smits AM (2020) Epicardial TGFβ and BMP signaling in cardiac regeneration: what lesson can we learn from the developing heart? Biomol Ther 10:404

D'Souza D, Patel K (1999) Involvement of long- and short-range signaling during early tendon development. Anat Embryol (Berl) 200:367–375

Dudakovic A, Samsonraj RM, Paradise CR, Galeano-Garces C, Mol MO, Galeano-Garces D et al (2020) Inhibition of the epigenetic suppressor EZH2 primes osteogenic differentiation mediated by BMP2. J Biol Chem 295:7877–7893

Durgam SS, Stewart AA, Pondenis HC, Yates AC, Evans RB, Stewart MC (2012a) Responses of equine tendon- and bone marrow–derived cells to monolayer expansion with fibroblast growth factor-2 and sequential culture with pulverized tendon and insulinlike growth factor-I. Am J Vet Res 73:162–170

Durgam SS, Stewart AA, Pondenis HC, Gutierrez-Nibeyro SM, Evans RB, Stewart MC (2012b) Comparison of equine tendon- and bone marrow–derived cells cultured on tendon matrix with or without insulin-like growth factor-I supplementation. Am J Vet Res 73:153–161

Dyer LA, Pi X, Patterson C (2014) The role of BMPs in endothelial cell function and dysfunction. Trends Endocrinol Metab 25:472–480

Farnebo S, Farnebo L, Kim M, Woon C, Pham H, Chang J (2017) Optimized repopulation of tendon hydrogel: synergistic effects of growth factor combinations and adipose-derived stem cells. Hand (NY) 12:68–77

Fenwick SA, Hazleman BL, Riley GP (2002) The vasculature and its role in the damaged and healing tendon. Arthritis Res 4:252–260

Fleischhacker V, Klatte-Schulz F, Minkwitz S, Schmock A, Rummler M, Seliger A et al (2020) In vivo and in vitro mechanical loading of mouse Achilles tendons and tenocytes-a pilot study. Int J Mol Sci 21:1313

Forslund C, Aspenberg P (2002) CDMP-2 induces bone or tendon-like tissue depending on mechanical stimulation. J Orthop Res 20:1170–1174

Forslund C, Rueger D, Aspenberg P (2003) A comparative dose-response study of cartilage-derived morphogenetic protein (CDMP)-1, −2 and −3 for tendon healing in rats. J Orthop Res 21:617–621

Frank C, McDonla D, Shrive N (1997) Collagen fibril diameters in the rabbit medial collateral ligament scar. Conn Tissue Res 36:261–269

Genovese RL (1993) Prognosis of superficial digital flexor tendon and suspensory ligament injuries. Proc Am Assoc Equine Pract:17–19

Ghahary A, Tredget EE, Shen Q, Kilani RT, Scott PG, Houle Y (2000) Mannose-6-phosphate/IGF-II receptors mediate the effects of IGF-I induced latent transforming growth factor beta 1 on expression of type I collagen and collagenase in dermal fibroblasts. Growth Factors 17:167–176

Godwin EE, Young NJ, Dudhia J, Beamish IC, Smith RKW (2011) Implantation of bone marrow-derived mesenchymal stem cells demonstrates improved outcome in horses with overstrain injury of the superficial digital flexor tendon. Equine Vet J 44:25–32
Goumans MJ, Mummery C (2000) Functional analysis of the TGFbeta receptor/Smad pathway through gene ablation in mice. Int J Dev Biol 44:253–265
Graham JG, Wang ML, Rivlin M, Beredkjiklian PK (2019) Biologic and mechanical aspects of tendon fibrosis after injury and repair. Conn Tissue Res 60:10–20
Greenough CG (1996) The effect of pulsed electromagnetic fields on flexor tendon healing in the rabbit. J Hand Surg 21:808–812
Guenther C, Pantalena-Filho L, Kingsley DM (2008) Shaping skeletal growth by modular regulatory elements in the Bmp5 gene. PLoS Genet 4:e1000308
Guo J, Wu G (2012) The signaling and functions of heterodimeric bone morphogenetic proteins. Cytokine Growth Factor Rev 23:61–67
Hagerty P, Lee A, Calve S, Lee CA, Vidal M, Baar K (2012) The effect of growth factors on both collagen synthesis and tensile strength of engineered human ligaments. Biomaterials 33:6355–6361
Halper J (2010) Growth factors as active participants in carcinogenesis: a perspective. Vet Pathol 47:77–97
Hansen M, Boesen A, Holm L, Flyvbjerg A, Landberg H, Kjaer M (2013) Local administration of insulin-like growth factor-I (IGF-I) stimulates tendon collagen synthesis in humans. Scand J Sports Med 23:614–618
Hansson HA, Dahlin LB, Lundborg G, Löwenadler B, Paleus S, Skottner A (1988a) Transiently increased insulin-like growth factor I immunoreactivity in tendons after vibration trauma: an immunohistochemical study on rats. Scand J Plast Reconstr Surg Hand Surg 22:1–6
Hansson HA, Engström AM, Holm S, Rosenqvist AL (1988b) Somatomedin C immunoreactivity in the Achilles tendon varies in a dynamic manner with the mechanical load. Acta Physiol Scand 134:199–208
Harvey T, Flamenco S, Fan CM (2019) A Tppp3+Pdgfra+ tendon stem cell population contributes to regeneration and reveals a shared role for PDGF signalling in regeneration and fibrosis. Nat Cell Biol 21:1490–1503
Haubruck P, Tanner MC, Vlachopoulos W, Hagelskamp S, Miska M, Ober J et al (2018) Comparison of the clinical effectiveness of bone Morphogenic protein (BMP) -2 and −7 in the adjunct treatment of lower limb nonunions. Orthop Traumatol Surg Res 104:1241–1248
Haupt JL, Donnelly BP, Nixon AJ (2006) Effects of platelet-derived growth factor-BB on the metabolic function and morphologic features of equine tendon in explant culture. Am J Vet Res 67:1595–1600
Hayler CI, Guffre BM (2009) Overuse of isolated and multiple ligament injuries of the elbow. Magn Reson Imaging Clinic N Am 17:617–638
Hazard SW, Myers RL, Ehrlich HP (2011) Demonstrating collagen tendon fibril segments involvement in intrinsic tendon repair. Exp Mol Pathol 91:660–663
He M, Gan AW, Lim AY, Goh JC, Hui JH, Lee EH et al (2013) The effect of fibrin glue on tendon healing and adhesion formation in a rabbit model of flexor tendon injury and repair. J Plast Surg Hand Surg 47:509–512
Hearn MT, Gomme PT (2000) Molecular architecture and biorecognition processes of the cystine knot protein superfamily: part I. the glycoprotein hormones. J Mol Recognit 13:223–278
Heinemeier KM, Olesen JL, Schjerling P, Haddad F, Langberg H, Baldwin KM, Kjaer M (2007) Short-term strength training and the expression of myostatin and IGF-I isoforms in rat muscle and tendon: differential effects of specific contraction types. J Appl Physiol 102:573–581
Heinemeier KM, Olesen JL, Haddad F, Schjerling P, Baldwin KM, Kjaer M (2009) Effect of unloading followed by reloading on expression of collagen and related growth factors in rat tendon and muscle. J Appl Physiol 106:178–186
Heinemeier KM, Mackey AL, Doessing S, Hansen M, Bayer ML, Nielsen RH, Herchenhan A, Malmgaard-Clausen NM, Kjaer M (2012) GH/IGF-I axis and matrix adaptation of the musculotendinous tissue to exercise in humans. Scand J Med Sci Sports 22:1–7
Heisterbach PE, Todorov A, Flückiger R, Evans CH, Majewski M (2011) Effect of BMP-12, TGF-β1 and autologous conditioned serum on growth factor expression in Achilles tendon healing. Knee Surg Sports Traumatol Arthrosc 20:1907–1914
Helm GA, Li JZ, Alden TD, Hudson SB, Beres EJ, Cunningham M, Mikkelsen MM, Pittman DD, Kerns KM, Kallmes DF (2001) A light and electron microscopic study of ectopic tendon and ligament formation induced by bone morphogenetic protein–13 adenoviral gene therapy. J Neurosurg 95:298–307
Hernandez DM, Kang JH, Choudhury M, Andrianifahanana M, Yin X, Limper AH et al (2020) IPF pathogenesis is dependent upon TGFβ induction of IGF-1. FASEB J 34:5363–5388
Hildebrand KA, Woo SL, Smith DW, Allen CR, Deie M, Taylor BJ, Schmidt CC (1998) The effects of platelet-derived growth factor-BB on healing of the rabbit medial collateral ligament. An in vivo study. Am J Sports Med 26:549–554
Hinck AP (2012) Structural studies of the TGF-bs and their receptors – insights into evolution of the TGF-b superfamily. FEBS Lett 586:1860–1870
Hofer T, Desbaillets I, Höpfl G, Gassmann M, Wenger RH (2001) Dissecting hypoxia-dependent and hypoxia-independent steps in the HIF-1alpha activation cascade: implications for HIF-1alpha gene therapy. FASEB J 15:2715–2717
Hoffmann A, Pelled G, Turgeman G, Eberle P, Zilberman Y, Shinar H, Keinan-Adamsky K, Winkel A, Shahab S, Navon G, Gross G, Gazit D (2006) Neotendon formation induced by manipulation of the Smad8 signalling pathway in mesenchymal stem cells. J Clin Invest 116:940–952

Hou Y, Mao ZB, Wei XL, Lin L, Chen LX, Wang HJ, Fu X, Zhang JY, Yu C (2009) Effects of transforming growth factor-β1 and vascular endothelial growth factor 165 gene transfer on Achilles tendon healing. Matrix Biol 28:324–335

Hui Q, Jin Z, Li X, Liu C, Wang X (2018) FGF family: from drug development to clinical application. Int J Mol Sci 19:1875

Ide J, Kikukawa K, Hirose J, Iyama KI, Sakamoto H, Eng DR, Fujimoto T, Mizuta H (2009) The effect of a local application of fibroblast growth factor-2 on tendon-to-bone remodeling in rats with acute injury and repair of the supraspinatus tendon. J Shoulder Elb Surg 18:391–398

Ingraham JM, Hauck RM, Ehrlich HP (2003) Is the tendon embryogenesis process resurrected during tendon healing? Plast Reconstr Surg 112:844–854

Janssen JAMJL (2020) New insights from IGF-IR stimulating activity analyses: pathological ConsiderationsNew insights from IGF-IR stimulating activity analyses: pathological considerations. Cell 9:864

Jo A, Denduluri S, Zhang B, Wang Z, Yin L, Yan Z et al (2014) The versatile functions of Sox9 in development, stem cells, and human diseases. Genes Dis 1:149–161

Jones JI, Clemmons DR (1995) Insulin-like growth factors and their binding proteins: biological actions. Endocr Rev 16:3–34

Jones GC, Corps AN, Pennington CJ, Clark IM, Edwards DR, Bradley MM, Hazleman BL, Riley GP (2006a) Expression profiling of metalloproteinases and tissue inhibitors of metalloproteinases in normal and degenerate human Achilles tendon. Arthritis Rheumatol 54:832–842

Jones GC, Corps AN, Pennington CJ, Clark IM, Edwards DR, Bradley MM, Hazleman BL, Riley GP (2006b) Expression profiling of metalloproteinases and tissue inhibitors of metalloproteinases in normal and degenerate human Achilles tendon. Arthritis Rheumatol 54:832–842

Juncosa-Melvin N, Matlin KS, Holdcraft RW, Nirmalanandhan VS, Butler DL (2007) Mechanical stimulation increases collagen type I and collagen type III gene expression of stem cell-collagen sponge constructs for patellar tendon repair. Tissue Eng 13:1219–1226

Juul A (2003) Serum levels of insulin-like growth factor I and its binding protein in health and disease. Growth Hormon IGF Res 13:113–120

Kaux JF, Drion PV, Colige A, Pascon F, Libertiaux V, Hoffmann A, Janssen L, Heyers A, Nusgens BV, Le Goff C, Gothot A, Cescotto S, Defraigne JO, Rickert M, Crielaard JM (2012) Effects of platelet-rich plasma (PRP) on the healing of Achilles tendons of rats. Wound Repair Regen 20:748–756

Kelley KM, Schmidt KE, Berg L, Sak K, Galima MM, Gillespie C, Balogh L, Hawayek A, Reyes JA, Jamison M (2002) Comparative endocrinology of the insulin-like growth factor-binding protein. J Endocrinol 175:3–18

Ker ED, Chu B, Phillippi JA, Gharaibeh B, Huard J, Weiss LE, Campbell PG (2011) Engineering spatial control of multiple differentiation fates within a stem cell population. Biomaterials 32:3413–3422

Khan KM, Cook JL, Bonar F, Harcourt P, Astrom M (1999) Histopathology of common tendinopathies. Update and implications for clinical management. Sports Med 27:393–408

Kieny M, Chevallier A (1979) Autonomy of tendon development in the embryonic chick wing. J Embryol Exp Morphol 49:153–165

Kim HJ, Nam HW, Hur CY, Park M, Yang HS, Kim BS, Park JH (2011) The effect of platelet rich plasma from bone marrow aspirate with added bone morphogenetic protein-2 on the Achilles tendon-bone junction in rabbits. Clin Orthop Surg 3:325–331

Kim WS, Park HS, Sung JH (2015) The pivotal role of PDGF and its receptor isoforms in adipose-derived stem cells. Histol Histopathol 30:793–799

Kim SE, Kim JG, Park K (2019) Biomaterials for the treatment of tendon injury. Tissue Eng Regen Med 16:467–477

King JA, Storm EE, Marker PC, Dileone RJ, Kingsley DM (1996) The role of BMPs and GDFs in development of region-specific skeletal structures. Ann N Y Acad Sci 785:70–79

Kingsley DM (1994) The TGFβ superfamily: new members, new receptors, and new genetic tests of function on different organisms. Genes Dev 8:133–146

Klatte-Schulz F, Pauly S, Scheibel M, Greiner S, Gerhardt C, Schmidmaier G, Wildemann B (2012) Influence of age on the cell biological characteristics and the stimulation potential of male human tenocyte-like cells. Eur Cell Mater 24:74–89

Kobayashi Y, Saita Y, Takaku T, Yokomizo T, Nishio H, Ikeda H et al (2020) Platelet-rich plasma (PRP) accelerates murine patellar tendon healing through enhancement of angiogenesis and collagen synthesis. J Exp Orthop 7(1):49

Kokubu S, Inaki R, Koshi K, Hikita A (2020) Adipose-derived stem cells improve tendon repair and prevent ectopic ossification in tendinopathy by inhibiting inflammation and inducing neovascularization in the early stage of tendon healing. Regen Ther 14:103–110

Kol AW, Walker NJ, Galuppo LD, Clark KC, Buerchler S, Bernanke A, Borjesson DL (2013) Autologous point-of-care cellular therapies variably induce equine mesenchymal stem cell migration, proliferation and cytokine expression. Equine Vet J 45:193–198

Kosaka N, Sakamoto H, Terada M, Ochiya T (2009) Pleiotropic function of FGF-4: its role in development and stem cells. Dev Dyn 238:265–276

Kou I, Ikegawa S (2004) SOX9-dependent and -independent transcriptional regulation of human cartilage link protein. J Biol Chem 279:50942–50948

Kruithof BPT, Duim SN, Moerkamp AT, Goumans MJ (2012) TGFβ and BMP signaling in cardiac cushion formation: lessons from mice and chicken. Differentiation 84:89–102

Lee JY, Zhou Z, Taub PJ, Ramcharan M, Li Y, Akinbiyi T, Maharam ER, Leong DJ, Laudier DM, Ruike T, Torina PJ, Zaidi M, Majeska RJ, Schaffler MB, Flatow EL, Sun HB (2011) BMP-12 treatment of adult mesenchymal stem cells in vitro augments tendon-like tissue formation and defect repair in vivo. PLoS One 6:e17531
Lefebvre V, Huang W, Harley VR, Goodfellow PN, de Crombrugghe B (1997) SOX9 is a potent activator of the chondrocyte-specific enhancer of the pro alpha1(II) collagen gene. Mol Cell Biol 17:2336–2346
LeRoith D, Werner H, Beitner JD, Roberts CJ (1995) Molecular and cellular aspects of the insulin-like factor I receptor. Endocr Rev 16:143–163
Li X, Wang C, Xiao J, McKeehan WL, Wang F (2016) Fibroblast growth factors, old kids on the new block. Semin Cell Dev Biol 53:155–167
Lichtman MK, Otero-Vinas M, Falanga Y (2016) Transforming growth factor beta (TGF-β) isoforms in wound healing and fibrosis. Wound Repair Regen 24:215–222
Lim WL, Liau LL, Ng MH, Chowdhury SR, Law JX (2019) Current Progress in tendon and ligament tissue engineering. Tissue Eng Regen Med. 16(6):549–571
Lin Y, Zhang L, Liu NQ, Yao Q, Van Handel B, Xu Y et al (2019) In vitro behavior of tendon stem/progenitor cells on bioactive electrospun nanofiber membranes for tendon-bone tissue engineering applications. Int J Nanomedicine 14:5831–5848
Lincoln J, Lange AW, Yutzey KE (2006a) Hearts and bones: shared regulatory mechanisms in heart valve, cartilage, tendon, and bone development. Dev Biol 294:292–302
Lincoln J, Alfieri CM, Yutzey KE (2006b) BMP and FGF regulatory pathways control cell lineage diversification of heart valve precursor cells. Dev Biol 292:290–302
Loiacono C, Palermi S, Massa B, Belviso I, Romano V, Di Gregorio A et al (2019) Tendinopathy: pathophysiology, therapeutic options, and role of nutraceutics. A narrative literature review. Medicina (Kaunas) 55:447
Loiselle AE, Bragdon GA, Jacobson JA, Hasslund S, Cortes ZE, Schwarz EM, Mitten DJ, Awad HA, O'Keefe RJ (2009) Remodeling of murine intrasynovial tendon adhesions following injury: MMP and neotendon gene expression. J Orthop Res 27:833–840
Lou J, Tu Y (2001) BMP-12 gene transfer augmentation of lacerated tendon repair. J Orthop Res 19:1199–1202
Lovati AB, Corradetti B, Lange Consiglio A, Recordati C, Bonacina E, Bizzaro D, Cremonesi E (2011) Characterization and differentiation of equine tendon-derived progenitor cells. J Biol Regul Homeost Agents 25:S75–S84
Lyras DN, Kazakos K, Veretas D, Chronopoulos E, Folaranmi S, Agrogiannis G (2010) Effect of combined administration of transforming growth factor-b1 and insulin-like growth factor I on the mechanical properties of a patellar tendon defect model in rabbits. Acta Orthop Belg 76:380–386
Lyras DN, Kazakos K, Georgiadis G, Mazis G, Middleton R, Richards S, O'Connor D (2011) Does a single application of PRP alter the expression of IGF-I in the early phase of tendon healing? J Foot Ankle Surg 50:276–282
Macias D, Ganan Y, Sampath TK, Piedra ME, Ros MA, Hurle JM (1997) Role of BMP-2 and OP-1 (BMP-7) in programmed cell death and skeletogenesis during chick limb development. Development 124:1109–1117
Majewski M, Porter RM, Betz OB, Betz VM, Clahsen H, Flückiger R, Evans CH (2012) Improvement of tendon repair using muscle grafts transduced with TGF-β1 cDNA. Eur Cell Mater 23:94–101
Majewski M, Heisterbach P, Jaquiéry C, Dürselen L, Todorov A, Martin IE et al (2018) Improved tendon healing using bFGF, BMP-12 and TGFβ1, in a rat model. Eur Cell Mater 13:318–334
Makanae A, Satoh A (2018) Ectopic Fgf signaling induces the intercalary response in developing chicken limb buds. Zoological Lett 4:8
Manske PR, Gelberman RH, Lesker PA (1985) Flexor tendon healing. Hand Clin 1:25–34
Marfe G, Rotta G, De Martino L, Tafani M, Fiorito F, Di Stefano C, Polettini M, Ranalli M, Russo MA, Gambacurta A (2012) A new clinical approach: use of blood-derived stem cells (BDSCs) for superficial digital flexor tendon injuries in horses. Life Sci 90:825–830
Mass DP, Tuel RJ (1990) Participation of human superficialis flexor tendon segments in repair in vitro. J Orthop Res 8:21–34
Mass DP, Tuel RJ, Labarbera M, Greenwald DP (1993) Effects of constant mechanical tension on the healing of rabbit flexor tendons. Clin Orth Rel Res 296:301–306
Massague J, Wotton D (2000) Transcriptional control by the TGF-beta/Smad signaling system. EMBO J 19:1745–1754
Mast BA, Haynes JH, Krummel TM, Diegelmann RF, Cohen IK (1992) In vivo degradation of fetal wound hyaluronic acid results in increased fibroplasia, collagen deposition, and neovascularization. Plast Reconstr Surg 89:503–509
Mast BA, Frantz FW, Diegelmann RF, Krummel TM, Cohen IK (1995) Hyaluronic acid degradation products induce neovascularization and fibroplasia in fetal rabbit wounds. Wound Repair Regen 3:66–72
McCoy G, Ashley W, Mander J, Yang SY, Williams N, Russell B, Goldspink G (1999) Expression of insulin growth factor-1 splice variants and structural genes in rabbit skeletal muscle induced by stretch and stimulation. J Physiol 516:583–592
McDonald NQ, Hendrickson WA (1993) A structural superfamily of growth factors containing a cystine knot motif. Cell 73:421–424
McQuilling JP, Kimmerling KA, Staples MC, Mowry KC (2019) Evaluation of two distinct placental-derived membranes and their effect on tenocyte responses in vitro. J Tissue Eng Regen Med 13:1316–1330
Messier C, Teutenberg K (2005) The role of insulin, insulin-like growth factor, and insulin-degrading

enzyme in brain aging and Alzheimer's disease. Neural Plast 12:311–328
Miyazono K, Maeda S, Imamura T (2005) BMP receptor signaling: transcriptional targets, regulation of signals, and signaling cross-talk. Cytokine Growth Factor Rev 16:251–263
Molloy T, Wang Y, Murrell G (2003) The roles of growth factors in tendon and ligament healing. Sports Med 33:381–394
Moore YR, Dickinson DP, Wikesjö UME (2010) Growth/differentiation factor-5: a candidate therapeutic agent for periodontal regeneration? A review of pre-clinical data. J Clin Periodontol 37:288–298
Murphy DJ, Nixon AJ (1997) Biochemical and site-specific effects of insulin-like growth factor I on intrinsic tenocyte activity in equine flexor tendons. Am J Vet Res 58:103–109
Murray MM, Spindler KP, Ballard P, Welch TP, Zurakowski D, Nanney LB (2007) Enhanced histologic repair in a central wound in the anterior cruciate ligament with a collagen-platelet-rich plasma scaffold. J Orthop Res 25:1007–1017
Murray MM, Palmer M, Abreu E, Spindler KP, Zurakowski D, Fleming BC (2009) Platelet-rich plasma alone is not sufficient to enhance suture repair of the ACL in skeletally immature animals: an in vivo study. J Orthop Res 27:639–645
Nayak M, Nag HL, Nag TC, Digge V, Yadav R (2020) Ultrastructural and histological changes in tibial remnant of ruptured anterior cruciate ligament stumps: a transmission electron microscopy and immunochemistry-based observational study. Musculoskelet Surg 104:67–74
Ngo M, Pham H, Longaker MT, Chang J (2001) Differential expression of transforming growth factor-beta receptors in a rabbit zone II flexor tendon wound healing model. Plast Reconstr Surg 108:1260–1267
Nishimura R, Hata K, Matsubara T, Wakabayashi M, Yoneda T (2012) Regulation of bone and cartilage development by network between BMP signalling and transcription factors. J Biochem 151:247–254
Nishio H, Saita Y, Kobayashi Y, Takaku T, Fukusato S, Uchino S et al (2020) Platelet-rich plasma promotes recruitment of macrophages in the process of tendon healing. Regen Ther 14:262–270
Nixon AJ, Watts AE, Schnabel LV (2012) Cell- and gene-based approaches to tendon regeneration. J Shoulder Elb Surg 21:278–294
Noguchi M, Seiler JG, Gelberman RH, Sofranko RA, Woo SL-Y (1993) In vitro biomechanical analysis of suture methods for flexor tendon repair. J Orthop Res 11:603–611
O'kane S, Ferguson MWJ (1997) Transforming growth factor βs and wound healing. Int J Biochem Cell Biol 29:63–78
Ohashi Y, Nakase J, Shimozaki KT, Tsuchiya H (2018) Evaluation of dynamic change in regenerated tendons in a mouse model. J Exp Orthop. 5:37
Ohberg L, Lorentzon R, Alfredson H (2001) Neovascularisation in Achilles tendons with painful tendinosis but not in normal tendons: an ultrasonographic investigation. Knee Surg Sports Traumatol Arthrosc 9:235–238
Olesen JL, Heinemeier KM, Langberg H, Magnusson SP, Kjaer M (2006) Expression, content, and localization of insulin-like growth factor I in human Achilles tendon. Conn Tissue Res 37:200–206
Paoloni J, De Vos RJ, Hamilton B, Murrell GA, Orchard J (2011) Platelet-rich plasma treatment for ligament and tendon injuries. Clin J Sport Med 21:37–45
Peloso JG, Mundy GD, Cohen ND (1994) Prevalence of, and factors associated with musculoskeletal racing injuries of thoroughbreds. J Am Vet Med Assoc 204:620–626
Petersen W, Pufe T, Zantop T, Tillmann B, Tsokos M, Mentlein R (2004) Expression of VEGFR-1 and VEGFR-2 in degenerative Achilles tendon. Clin Orthop 420:286–291
Pregizer SK, Mortlock DP (2015) Dynamics and cellular localization of Bmp2, Bmp4, and noggin transcription in the postnatal mouse skeleton. J Bone Miner Res 30:64–70
Provenzano PP, Alejandro-Osorio AL, Grorud KW, Martinez DA, Vailas AC, Grindenland RE, Vanderby R Jr (2007) Systemic administration of IGF-I enhances healing in collagenous extracellular matrices: evaluation of loaded and unloaded ligaments. BMC Physiol 7:2
Pufe T, Petersen WJ, Kurz B, Tsokos M, Tillmann BN, Mentlein R (2003) Mechanical factors influence the expression of endostatin - an inhibitor of angiogenesis - in tendons. J Orthop Res 21:610–616
Pufe T, Petersen W, Goldring MB, Mentlein R, Varoga D, Tsokos M, Tillmann B (2004) Endostatin - an inhibitor of angiogenesis - is expressed in human cartilage and fibrocartilage. Matrix Biol 23:267–276
Pufe T, Petersen WJ, Mentlein R, Tillmann BN (2005a) The role of vasculature and angiogenesis for the pathogenesis of degenerative tendon disease. Scand J Med Sci Sports 15:211–222
Pufe T, Kurz B, Petersen W, Varoga D, Mentlein R, Kulow S, Lemke A, Tillmann B (2005b) The influence of biomechanical parameters on the expression of VEGF and endostatin in the bone and joint system. Ann Anat 187:461–472
Raghavan SS, Woon, C.Y., Kraus, A., Megerle, K., Pham, H., Chang, J. Optimization of human tendon tissue engineering: synergistic effects of growth factors for use in tendon scaffold repopulatio. Plast Reconstr Surg 2012;129:479–489
Raja S, Byakod G, Pudakalkatti P (2009) Growth factors in periodontal regeneration. Int J Dent Hygiene 7:82–89
Ramos DM, Abdulmalik S, Arul MR, Rudraiah S, Laurencin CT, Mazzocca AD et al (2019) Insulin immobilized PCL-cellulose acetate micro-nanostructured fibrous scaffolds for tendon tissue engineering. Polym Adv Technol 30:1205–1215
Riley G (2008) Tendinopathy – from basic science to treatment. Nat Clin Pract Rheumatol 4:82–89

Riley GP, Curry V, DeGroot J, van El B, Verzijl N, Hazleman BL, Bank RA (2002) Matrix metalloproteinase activities and their relationship with collagen remodeling in tendon pathology. Matrix Biol 21:185–195

Rodeo SA, Suzuki K, Deng XH, Wozney J, Warren RF (1999) Use of recombinant human bone morphogenetic protein-2 to enhance tendon healing in a bone tunnel. Am J Sports Med 27

Ronga M, Fagetti A, Canton G, Paiusco E, Surace MF, Cherubino P (2013) Clinical applications of growth factors in bone injuries: experience with BMPs. Injury 44(Suppl):534–539

Roy V, Peretz EA (2009) Biologic therapy of breast cancer: focus on co-inhibition of endocrine and angiogenesis pathways. Breast Cancer Res Treat 116:31–38

Sahin H, Tholema N, Petersen W, Raschke MJ, Stange R (2012a) Impaired biomechanical properties correlate with neoangiogenesis as well as VEGF and MMP-3 expression during rat patellar tendon healing. J Orthopaed Res 30(12):1952–1957

Sahin H, Tholema N, Raschke MJ, Stange R (2012b) Impaired biomechanical properties correlate with neoangiogenesis as well as VEGF and MMP-3 expression during rat patellar tendon healing. J Orthop Res 30:1952–1957

Salingcarnboriboon RY, Tsuji K, Obinata M, Amagasa T, Nifuji A, Noda M (2003) Establishment of tendon-derived cell lines exhibiting pluripotent mesenchymal stem cell-like property. Exp Cell Res 287:289–300

Saman H, Raza SS, Uddin S, Rasul K (2020) Inducing angiogenesis, a key step in Cancer vascularization, and treatment approaches. Cancers (Basel) 12:1172

Sardenberg T, Muller SS, Garms LZ, Miduati FB (2015) Immediate and late effects of sutures in extrasynovial tendons: biomechanical study in rats. Rev Bras Ortop 46:305–308

Savitskaya YA, Izaguirre A, Sierra L, Perez F, Cruz F, Villalobos E, Almazan A, Ibarra C (2011) Effect of angiogenesis-related cytokines on rotator cuff disease: the search for sensitive biomarkers of early tendon degeneration. Clin Med Insights Arthritis Musculoskelet Disord 4:43–53

Sharma RI, Snedeker JG (2010) Biochemical and biomechanical gradients for directed bone marrow stromal cell differentiation toward tendon and bone. Biomaterials 31:7695–7704

Sharma RI, Snedeker JG (2012) Paracrine interactions between mesenchymal stem cells affect substrate driven differentiation toward tendon and bone phenotypes. PLoS One 7:e31504

Singh R, Rymer B, Theobald P, Thomas PB (2015) A review of current concepts in flexor tendon repair: physiology, biomechanics, surgical technique and rehabilitation. Orthop Rev (Pavia) 7:6125

Södersten F, Hultenby K, Heinegård D, Johnston C, Ekman S (2013) Immunolocalization of collagens (I and III) and cartilage oligomeric matrix protein (COMP) in the normal and injured equine superficial digital flexor tendon. Conn Tissue Res. 54:62–69

Starman JS, Bosse MJ, Cates CA, Norton HJ (2012) Recombinant human bone morphogenetic protein-2 use in the off-label treatment of nonunions and acute fractures: a retrospective review. J Trauma Acute Care Surg 72:676–681

Stiles CD, Capone GT, Scher CD, Antoniades HN, Van Wyk JJ, Pledger WJ (1979) Dual control of cell growth by somatomedins and platelet-derived growth factor. Proc Natl Acad Sci U S A 76:1279–1283

Suda N, Shiga M, Ganburged G, Moriyama K (2009) Marfan synfrome and its disorder in periodontal tissues. J Exp Zool (Mol Dev Evol) 312B:503–509

Suwalski A, Dabboue H, Delalande A, Bensamoun SF, Canon F, Midoux P, Saillant G, Klatzmann D, Salvetat JP, Pichon C (2010) Accelerated Achilles tendon healing by PDGF gene delivery with mesoporous silica nanoparticles. Biomaterials 31:5237–5245

Svegliati-Baroni G, Ridolfi F, Di SA, Casini A, Marucci L, Gaggiotti G, Orlandoni P, Macarri G, Perego L, Benedetti A, Folli F (1999) Insulin and insulin-like growth factor-I stimulate proliferation and type I accumulation by hepatic stellate cells; differential effects on signal transduction pathways. Hepatology 29:1743–1751

Taira M, Yoshida T, Miyagawa K, Sakamoto H, Terada M, Sugimura T (1987) cDNA sequence of human transforming gene hst and identification of the coding sequence required for transforming activity. Proc Natl Acad Sci U S A 84:2980–2984

Takebayashi-Suzuki K, Suzuki A (2020) Intracellular communication among morphogen signaling pathways during vertebrate body plan formation. Genes (Basel) 11:341

Tan SL, Ahmad RE, Ahmad TS, Merican AM, Abbas AA, Ng WM, Kamarul T (2012) Effect of growth differentiation factor 5 on the proliferation and tenogenic differentiation potential of human mesenchymal stem cells in vitro. Cells Tissues Organs 196:325–338

Taylor GP, Anderson R, Reginelli AD, Muneola K (1994) FGF-2 induces regeneration of the chick limb bud. Dev Biol 163:282–284

Theodossiou SK, Schiele NR (2019) Models of tendon development and injury. BMC Biomed Eng 1:10

Thomopoulos S, Kim HM, Silva MJ, Ntouvali E, Manning CN, Potter R, Seeherman H, Gelberman RH (2012) Effect of bone morphogenetic protein 2 on tendon-to-bone healing in a canine flexor tendon model. J Orthop Res 30:1702–1709

Tsuzaki M, Brigman BE, Yamamoto J, Lawrence WT, Simmons JG, Mohapatra NK, Lund PK, Van Wyk J, Hannafin JA, Bhargava MM, Banes AJ (2000) Insulin-like growth factor-I is expressed by avian flexor tendon cells. J Orthop Res 18:546–556

Urist MR. Bone: formation by autoinduction. Science 1965;150:893–0

Uysal CA, Tobita M, Hyakusoku H, Mizuno H (2012) Adipose-derived stem cells enhance primary tendon repair: biomechanical and immunohistochemical evaluation. J Plast Reconstr Aesthetic Surg 65:1712–1719

Violini S, Ramelli P, Pisani LF, Gorni C, Mariani P (2009) Horse bone marrow mesenchymal stem cells express embryo stem cell markers and show the ability for tenogenic differentiation by in vitro exposure to BMP-12. BMC Cell Biol 10:29

Volk T (1999) Singling out Drosophila tendon cells: a dialogie between two distinct cell types. Trend Genet 15:448–453

Wang EA, Rosen V, Cordes P, Hewick RM, Kriz MJ, Luxenberg DP, Sibley BS, Wozney JM (1988) Purification and characterization of other distinct bone-inducing factors. Proc Natl Acad Sci U S A 85:9484–9488

Wang QW, Chen ZL, Piao YJ (2005) Mesenchymal stem cells differentiate into tenocytes by bone morphogenetic protein (BMP) 12 gene transfer. J Biosci Bioeng 100:418–422

Wilson JJ, Best TM (2005) Common overuse tendon problems: a review and recommendations for treatment. Am Fam Physician 72:711–718

Witte TH, Yeager AE, Nixon AJ (2011) Intralesional injection of insulin-like growth factor-I for treatment of superficial digital flexor tendonitis in thoroughbred racehorses: 40 cases (2000–2004). J Am Vet Med Assoc 239:992–997

Wolfman NM, Hattersley G, Cox A, Celeste AJ, Nelson R, Yamaji N, Dube JL, DiBlasio-Smith E, Nove J, Song JJ, Wozney JM, Rosen V (1997) Ectopic induction of tendon and ligament in rats by growth and differentiation factors 5, 6, and 7, members of the TGF-β gene family. J Clin Invest 100:321–330

Woo SL-Y, Hildebrand K, Watanabe N, Fenwick JA, Papageorgiou CD, Wang JH-C (1999) Tissue engineering of ligament and tendon healing. Clin Orth Rel Res 357S:S312–S323

Yamaguchi AF, T., Shimada S, Kanbayashi I, Tateoka M, Soya H, Takeda H, Morita I, Matsubara K, Hirai T (2006) Muscle IGF-I Ea, MGF, and myostatin mRNA expressions after compensatory overload in hypophysectomized rats. Pflugers Arch 453:203–210

Yastrebov O, Lobenhoffer I (2009) Treatment of isolated and multiple ligament injuries of the knee: anatomy, biomechanics, diagnosis, indications for repair, surgery. Othopaed 38:563–580

Yoon JH, Halper J (2005) Tendon proteoglycans: biochemistry and function. J Musculoskelet Neuronal Interact 5(1):22–34

Yoon JH, Brooks RL Jr, Zhao JZ, Isaacs D, Halper J (2004) The effects of enrofloxacin on decorin and glycosaminoglycans in avian tendon cell cultures. Arch Toxicol 78(10):599–608

Yoshida T, Miyagawa K, Odagiri H, Sakamoto H, Little PF, Terada M, Sugimura T (1987) Genomic sequence of hst, a transforming gene encoding a protein homologous to fibroblasts growth factors and the int-2-encoded proteins. Proc Natl Acad Sci U S A 84:7305–7309

Zhang H, Bradley A (1996) Mice deficient for BMP2 are nonviable and have defects in amnion/chorion and cardiac development. Development 122:2977–2986

Zhang J, Nie D, Williamson K, Rocha JL, Hogan MV, Wang JH (2019) Selectively activated PRP exerts differential effects on tendon stem/progenitor cells and tendon healing. J Tissue Eng 10

Zhao C, Amadio PC, Zobitz ME, An KN (2001) Gliding characteristics of tendon repair in canine flexor digitorum profundus tendons. J Orthop Res 19:580–586

Zhao B, Etter L, Hinton RB Jr, Benson DW (2007) BMP and FGF regulatory pathways in semilunar valve precursor cells. Dev Dyn 236:971–980

Connective Tissue Disorders and Cardiovascular Complications: The Indomitable Role of Transforming Growth Factor-β Signaling

7

Jason B. Wheeler, John S. Ikonomidis, and Jeffrey A. Jones

Abstract

Marfan Syndrome (MFS) and Loeys-Dietz Syndrome (LDS) represent heritable connective tissue disorders that segregate with a similar pattern of cardiovascular defects (*thoracic aortic aneurysm, mitral valve prolapse/regurgitation, and aortic dilatation with regurgitation*). This pattern of cardiovascular defects appears to be expressed along a spectrum of severity in many heritable connective tissue disorders and raises suspicion of a relationship between the normal development of connective tissues and the cardiovascular system. With overwhelming evidence of the involvement of aberrant Transforming Growth Factor-beta (TGF-β) signaling in MFS and LDS, this signaling pathway may represent the common link in the relationship between connective tissue disorders and their associated cardiovascular complications. To further explore this hypothetical link, this chapter will review the TGF-β signaling pathway, the heritable connective tissue syndromes related to aberrant TGF-β signaling, and will discuss the pathogenic contribution of TGF-β to these syndromes with a primary focus on the cardiovascular system.

Keywords

Aorta · Aneurysm · Extracellular matrix · Collagen · Metalloproteinase · Shprintzen-Goldberg syndrome · Thoracic aortic aneurysm and dissection syndrome · Hereditary hemorrhagic telangiectasia (HHT) · Marfan syndrome (MFS) · Loeys-Dietz syndrome (LDS) · Ehlers-Danlos syndrome (EDS) · Aortic aneurysm thoracic (AAT) · Aneurysm-osteoarthritis syndrome (AOS) · Arterial tortuosity syndrome (ATS) · Primary pulmonary hypertension · Fibrodysplasia Ossificans progressive (FOP) · Familial thoracic aortic aneurysm and dissection syndrome (FTAAD) · Transforming growth factor-β (TGF-β) · Endoglin · Mitral valve · Arteriovenous

J. B. Wheeler
Division of Cardiology, Cleveland Clinic, Cleveland, OH, USA
e-mail: wheelerjb@cclm.org

J. S. Ikonomidis
Division of Cardiothoracic Surgery, University of North Carolina, Chapel Hill, NC, USA
e-mail: john_ikonomidis@med.unc.edu

J. A. Jones (✉)
Division of Cardiothoracic Surgery,
Medical University of South Carolina and Ralph H. Johnson Veterans Affairs Medical Center,
Charleston, SC, USA
e-mail: jonesja@musc.edu

J. Halper (ed.), *Progress in Heritable Soft Connective Tissue Diseases*, Advances in Experimental Medicine and Biology 1348, https://doi.org/10.1007/978-3-030-80614-9_7

malformation · SMAD · TGF-β receptor · BMP receptor · Activin receptor-like kinase (ALK) · Mitogen-activated protein kinase (MAPK) · Extracellular signal related kinase (ERK) · Fibrillin · Curacao diagnostic criteria · Genetic testing · Beta blockers · Losartan · Doxycycline · Integrins

Abbreviations

AAT	Aortic Aneurysm Thoracic
ACE	Angiotensin Converting Enzyme
AKT	Protein Kinase B
ALK-1	Activin Receptor-like Kinase 1
ALK-3	Activin Receptor-like Kinase 3
ALK-5	Activin Receptor-like Kinase 5
AOS	Aneurysm-Osteoarthritis Syndrome
ATIIR	Angiotensin II Receptor Type II
ATIR	Angiotensin II Receptor Type I
ATS	Arterial Tortuosity Syndrome
AVM	Arteriovenous Malformation
BMP	Bone Morphogenetic Protein
BMPR1A	Bone Morphogenetic Protein Receptor 1A
BMPR2	Bone Morphogenetic Protein Receptor 2
Co-SMAD	Common SMAD; SMAD-4
CTGF	Connective Tissue Growth Factor
ECM	Extracellular Matrix
ERK1/2	Extracellular Signal-Regulated Kinase 1/2
FBN1	Fibrillin-1
FOP	Fibrodysplasia Ossificans Progressiva
FTAAD	Familial Thoracic Aortic Aneurysm and Dissection Syndrome
GI	Gastrointestinal
HHT1	Hereditary Hemorrhagic Telangiectasia, Type 1
HHT2	Hereditary Hemorrhagic Telangiectasia, Type 2
JNK	c-Jun N-terminal Kinase
LAP	Latent Associated Protein
LDS	Loeys-Dietz Syndrome
LLC	TGF-β Large Latent Complex
LTBP	Latent Transforming Growth Factor-β Binding Protein
MAPK	Mitogen-Activated Protein Kinase
MFS	Marfan Syndrome
MMPs	Matrix Metalloproteinases
MVP	Mitral Valve Prolapse
OMIM	Online Mendelian Inheritance in Man
PAH	Pulmonary Artery Hypertension
PI3K	Phosphoinositide-3-Kinase
RA	Rheumatoid Arthritis
R-SMAD	Receptor SMAD
SARA	SMAD Anchor for Receptor Activation
SGS	Shprintzen-Goldberg Syndrome
SLC	TGF-β Small Latent Complex
SLC2A10	Solute Carrier Family 2, Facilitated Glucose Transporter Member 10
SLE	Systemic Lupus Erythematosus
Smurf	SMAD Ubiquitination Regulatory Factor
TAA	Thoracic Aortic Aneurysm
TAK1	Transforming Growth Factor-Beta Associated Kinase 1
TGFBR1	Transforming Growth Factor-Beta Receptor, Type-I gene
TGFBR2	Transforming Growth Factor-Beta Receptor, Type-I gene
TGF-β	Transforming Growth Factor-Beta
TGF-βRI	Transforming Growth Factor-Beta Receptor, Type-I protein
TGF-βRII	Transforming Growth Factor-Beta Receptor, Type-II protein
TIMPs	Tissue Inhibitors of Matrix Metalloproteinases
TRAF6	Tumor Necrosis Factor Receptor Associated Factor 6

7.1 Introduction

The connective tissue comprises the most abundant and widely distributed primary tissues within the body. Its key functions include protecting, supporting, and insulating our major organs, as well as serving as a fuel reserve by sequestering important growth factors and metabolites. Due to its key role in multi-organ support, it is not surprising that there are over 200 disorders, both non-genetic and hereditary (including some autoimmune disorders), that impact the structure, function, and integrity of the connective tissue. The Transforming Growth Factor-Beta (TGF-β) superfamily of signaling molecules, is a large structurally related, and functionally overlapping family of signaling proteins including the TGF-βs, the bone morphogenetic proteins, the growth and differentiation factors, and the activins and inhibins. Overall, this family includes approximately 30 different ligands, 7 different type-I receptors, 5 different type-II receptors, and 8 different signaling intermediates (SMAD proteins) (Jones et al. 2009). Among other functions, these factors play key roles in regulating the cellular responses that maintain the balance between extracellular matrix deposition and degradation. Perturbations in homeostatic signaling have been associated with numerous human diseases involving multiple organ systems, of which disorders of the cardiovascular system often carry grave life-threatening consequences.

Marfan syndrome (MFS) is a well described connective tissue disorder characterized by musculoskeletal, ocular, and cardiovascular defects including: ascending aortic aneurysm with dissection, mitral valve prolapse (MVP)/regurgitation, and aortic root dilatation with aortic valve regurgitation (Judge and Dietz 2005). In the early 90's, Dietz and Pyeritz (Dietz and Pyeritz 1995) identified key mutations in the fibrillin-1 (FBN-1) gene that were associated with MFS and its related disorders. Fibrillin-1 is a 350 kDa fibular glycoprotein comprised of a series of epidermal growth factor-like motifs, many of which contain calcium-binding sequences, termed calcium-binding EGF (cbEGF) repeats (Yuan et al. 1997). These cbEGF modules are arranged in tandem, separated by cysteine-rich motifs (8-cys/TB) that have high homology to the family of latent transforming growth factor-beta (TGF-β) binding proteins (LTBPs) (Yuan et al. 1997; Ramirez and Pereira 1999). Fibrillin monomers self-assemble into macroaggregates forming the basic structures on which mature elastin fibers are synthesized from tropoelastin subunits. Accordingly, it was postulated that the FBN-1 mutations within the aorta result in a weakened and disordered microfibril network connecting the elastic lamellae to the adjacent interstitial cells, and that this weakening pre-disposed patients to the primary cardiovascular manifestation of MFS, ascending aortic aneurysm, the primary cause of mortality in patients with MFS (Jones and Ikonomidis 2010; Pereira et al. 1999). Dietz and Pyeritz (Dietz and Pyeritz 1995) however, went further to suggest that MFS may be caused by more than just a disordered microfibril matrix, suggesting that the inability of fibrillin to sequester latent TGF-β may play a prominent role in its many pathological manifestations. This was followed by a series of seminal studies in which a TGF-β neutralizing antibody was used in animal models to demonstrate that antagonizing the activation of the TGF-β signaling could reverse the lung, skeletal muscle, mitral valve, and aortic dysfunction associated with MFS (Habashi et al. 2006; Neptune et al. 2003; Ishibashi et al. 2004).

Subsequent to this revelation, in 2005, Loeys and Dietz described a cohort of patients with a connective tissue disorder that significantly overlapped with the phenotype of MFS but included a greater cardiovascular risk (Loeys et al. 2005). This disorder was designated Loeys-Dietz syndrome (LDS, OMIM #609192), and was characterized by the enhanced development of aortic aneurysms and dissections that occur at a younger age and smaller aortic diameter. Both disorders exhibit a marfanoid habitus (*pectus deformity*), arachnodactyly (*elongated fingers*), scoliosis, and dolichostenomelia (*elongated limbs*), heart valve prolapse with significant regurgitation, and aortic aneurysm with dissection. Interestingly, this group of patients presented with mutations in the type-I (TGF-βRI) or type-II (TGF-βRII) TGF-β receptors (Loeys et al. 2005). Moreover, despite mutated receptors incapable of propagating signals, patients with Loeys-Dietz syndrome (LDS) paradoxically exhibited indications of increased TGF-β signaling: enhanced phosphorylation of SMAD-2, a key intracellular mediator of TGF-β signaling, as well as elevated expression of collagen and connective tissue growth factor (CTGF), common indicators of TGF-β pathway activation (Loeys et al. 2005).

As examples, MFS and LDS represent connective tissue disorders that are significantly affected by altered TGF-β signaling and display significant cardiovascular defects. This pattern of cardiovascular defects appears to be expressed along a spectrum of severity in many heritable connective tissue disorders and raises suspicion of a relationship between the normal development of connective tissues and the cardiovascular system. Given the evidence of increased TGF-β signaling in MFS and LDS, this signaling pathway may represent the common link in this relationship.

To further explore this hypothetical link, this chapter will review the TGF-β signaling pathway, with respect to heritable connective tissue syndromes that present with significant cardiovascular complications and will discuss the pathogenic contribution of TGF-β to these syndromes.

7.2 TGF-β, Signaling Pathways, and Physiological Effects

Transforming growth factor-β is a soluble cytokine secreted by many cell-types within the body. The different three TGF-β isoforms (TGF-β 1, 2, 3) are produced as 55 kDa precursor proteins that homodimerize and bind to an inhibitory peptide fragment called the Latency Associated Peptide, or LAP, to form the Small Latent Complex (SLC). The SLC then associates with a Latent Transforming Growth Factor-Beta Binding Protein (LTBP), to form the Large Latent Complex (LLC), which is then secreted from the cell and targeted for sequestration in the extracellular matrix (Gentry et al. 1988; Lawrence et al. 1984; Zeyer and Reinhardt 2015). Fibrillin-1, comprised of multiple cbEGF motifs arranged in tandem with 8-cys/TB repeats, has high homology to the family of LTBPs, and provides a landing point to sequester the LLC within the matrix (Yuan et al. 1997; Pereira et al. 1999; Isogai et al. 2003). Thus, the ECM serves to sequester and concentrate TGF-β in locations where it may be rapidly activated when needed (Ramirez and Rifkin 2003). Indeed, the ECM is no longer thought to be a passive structural support but rather a dynamic regulator of growth factor bioavailability and signaling (Brekken and Sage 2001; Hynes 2009).

7.2.1 "Canonical" TGF-β Signaling

Mature TGF-β is activated through several different known mechanisms including: proteolysis of the LTBP or LAP by the action of several matrix metalloproteinases (MMP-2, MMP-9, MMP-13, MMP-14); interaction of the LAP with thrombospondin-1; exposure of the LAP to extracellular reactive oxygen species; a decrease in extracellular matrix pH effectively causing degradation of the LAP; or through interaction with a family of transmembrane protein receptors called the integrins, including $\alpha_v\beta_1$, $\alpha_v\beta_3$, $\alpha_v\beta_6$, and $\alpha_v\beta_8$ (Annes et al. 2003; Costanza et al. 2017; Patsenker et al. 2009; Ramirez and Rifkin 2009; Reed et al. 2015; Roth et al. 2013). Once released from the LAP, the TGF-β homodimer interacts with a family of TGF-β receptors initiating a signaling cascade. The TGF-β receptors have been subdivided into three families. The type-I and type-II receptors belong to a superfamily of tissue- and cell- specific transmembrane serine/threonine kinase receptors. The type-III receptors, also known as accessory receptors, including endoglin and betaglycan, contain transmembrane monomers that lack the kinase domain, but function to bind and present the TGF-β homodimer to the type-I and type-II receptor complex (Wrana et al. 1992).

In the "Canonical TGF-β Signaling Pathway" TGF-β binds to a homodimer of the type-II receptor, which auto-activates its serine/threonine kinase domain, resulting in the autophosphorylation of the type-II receptor. The phosphorylated type-II receptor, recruits a homodimer of the type-I receptor, and activates the type-I receptor through a transphosphorylation event. The activated type-I receptor then recruits and phosphorylates one of a class of cytoplasmic mediators, the receptor regulated SMADs (R-SMADs), these intracellular signaling intermediates were named for their homologues in *Caenorhabditis elegans* (SMA genes) and *Drosophila* (MAD genes; mothers against decapentaplegic). The activated R-SMAD then recruits and binds to a common-mediator SMAD (co-SMAD; SMAD-4) forming a complex revealing a nuclear localization signal, and is then shuttled into the nucleus where, upon interaction with transcriptional modifiers (activators or repressors), it forms a competent transcription complex capable of inducing or repressing numerous TGF-β-dependent genes (Fig. 7.1) (Jones et al. 2009; Shi and Massague 2003).

7.2.2 Alternate "Noncanonical" TGF-β Signaling

In addition to the canonical signaling pathway, a growing body of evidence now supports the hypothesis that TGF-β signaling can proceed by alternative mechanisms that bypass key mediators in the classical pathway (Moustakas and Heldin 2005). For example, these noncanonical signaling responses include: 1) signals propagated directly by type-II receptors without type-I receptor

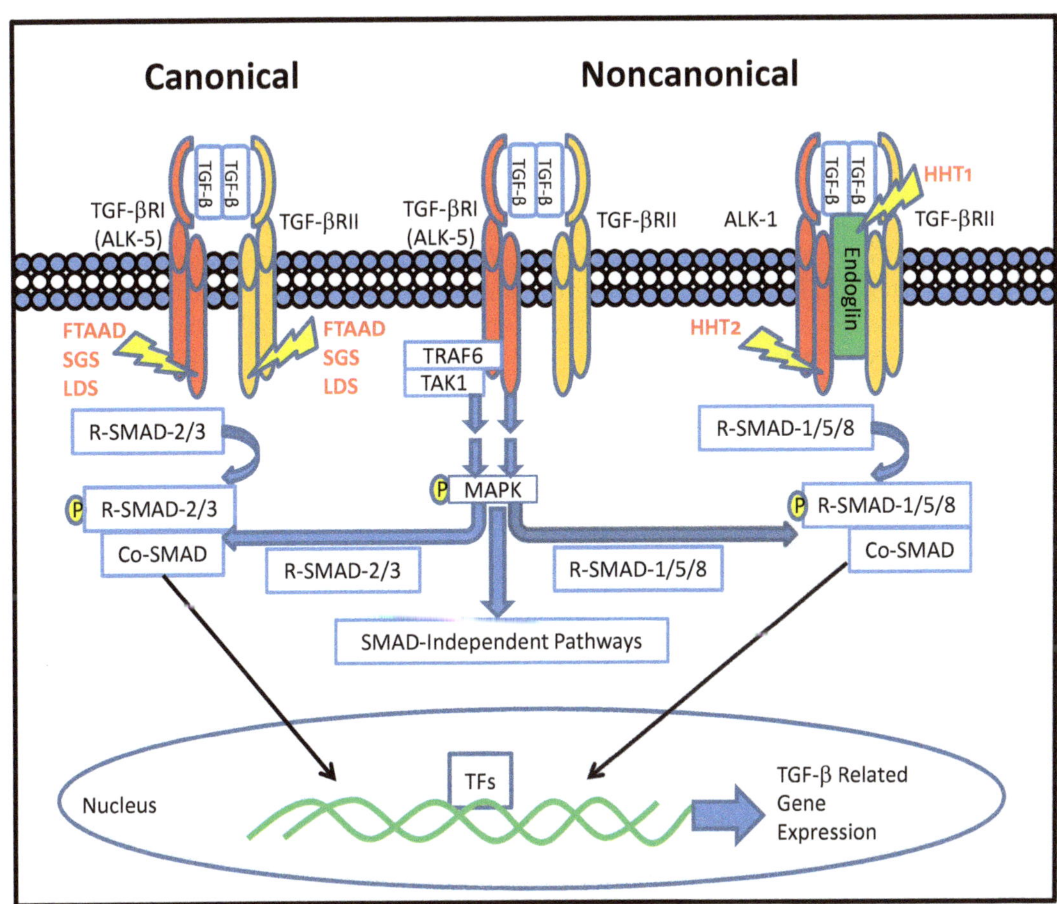

Fig. 7.1 Canonical, Noncanonical, and Endoglin/ALK-1 signaling pathways. Both Canonical and Endoglin/ALK-1 TGF-β signaling is mediated by the phosphorylation of distinct R-SMAD proteins. Nuclear translocation requires Co-SMAD binding in both pathways. The R-SMAD and Co-SMAD complex, once translocated to the nucleus, associates with other transcriptional regulatory factors to either repress or activate TGF-β-dependent gene expression. The noncanonical pathway is mediated by TGF-βRI, TRAF6, and TAK1, resulting in the phosphorylation of other intracellular signaling intermediates such as ERK1/2, JNK, and p38 MAPK. These MAPK-dependent signals can re-enter the SMAD-dependent pathway through direct activation of the SMADs, or they can mediate downstream signaling through SMAD-independent pathways. The lightning bolts represent mutations to the indicated proteins causative of the syndromes listed in red

involvement; 2) direct type-I receptor signals in the absence of R-SMAD activation; 3) R-SMAD signaling to parallel pathways in the absence of co-SMAD involvement; and 4) activation of R-SMADs by other signaling mediators in response to TGF-β, but not as a result of direct interaction with TGF-β receptors (Jones et al. 2009).

The downstream intracellular mediators of these alternative pathways are not as well understood as the role of the SMAD proteins in the classical pathway. Studies of noncanonical signaling in FBN1 deficient mice have proven helpful in this regard. Carta et al. demonstrated *in vivo* that p38 mitogen-activated protein kinase (p38 MAPK) mediated the phosphorylation of SMAD-2/3, which was attenuated by treatment with p38 MAPK inhibitors. This suggested that R-SMAD activation/phosphorylation could occur in a manner independent of TGF-βRI activity (Carta et al. 2009). Subsequent studies examining the role of Tumor Necrosis Factor-Receptor Associated Factor 6 (TRAF6), an E3 ubiquitin ligase, demonstrated association with TGF-βRI in a TGF-β-dependent manner (Yamashita et al. 2008; Sorrentino et al. 2008), however, this complex was found to recruit and activate a secondary intracellular kinase called TGF-β Associated Kinase 1 (TAK1), which was shown to subsequently activate p38 MAPK (Fig. 7.1) (Yamashita et al. 2008; Sorrentino et al. 2008). In addition to p38 MAPK, TGF-β can activate many other signal pathways independent of R-SMAD mediation including: Extracellular-signal Regulated Kinase 1 and 2 (ERK1/2) (Lee et al. 2007), c-Jun N-terminal Kinase (JNK) (Yamashita et al. 2008; Sorrentino et al. 2008), and the Phosphoinositide-3-Kinase (PI3K), in the protein kinase B (AKT)-mediated cellular survival pathway (Wilkes et al. 2005). Importantly, the canonical and noncanonical pathways appear to exert differential effects on the connective tissues within the ECM. Thus, through combinatorial receptor interactions, multiple regulatory mechanisms, and alternative signaling pathways, the complexity of signaling through this pathway is dramatically amplified, contributing to the myriad complex signaling outcomes reported for TGF-β.

7.2.3 Cellular Responses to TGF-β Signaling

Transforming growth factor-β signaling is known to contribute to a number of disparate and opposing physiologic processes including angiogenesis, proliferation, differentiation, apoptosis, and wound healing, and is a well-established modulator of ECM structure and composition (Bertolino et al. 2005; Massague 2000). Stimulation of the canonical TGF-β pathway has been well accepted as a primary driver of increased extracellular matrix protein production and deposition (*collagen and elastin*) (Ignotz and Massague 1986). In addition, while driving matrix production the pathway simultaneously functions to attenuate the expression of matrix proteolytic enzymes (*matrix metalloproteinases (MMPs)*) (Yan and Boyd 2007), and enhances the expression of the tissue inhibitors of MMPs (TIMPs) (Kwak et al. 2006). Alternatively, stimulation of the noncanonical pathways has been associated with enhanced matrix degradation through increased MMP-dependent proteolysis (Kim et al. 2004) and enhanced MMP activation via plasminogen activators (Laiho et al. 1986). This is supported by work from Kim et al. demonstrating that the activation of p38 MAPK-mediated signaling has been associated with increased MMP-2 and -9 production and release in breast cancer cells (Kim et al. 2004). In similar fashion, *in vitro* expression of MMP-13 in rat osteoblasts was found to be dependent on p38 MAPK, involving SMAD-2 activation and ERK1/2 signaling (Selvamurugan et al. 2004).

The complexity of this signaling pathway highlights the importance of TGF-β as a key regulatory factor in maintaining the homeostatic balance within the ECM, and further underscores how dysregulation of this pathway could play a dominant role in connective tissue disorders and disease. Accordingly, regulation of this pathway is highly critical at multiple levels. An example of this regulation is the inhibitory feedback pathway involving I-SMADs (*inhibitory SMAD6 and SMAD7*) (Park 2005). SMAD6 exerts its effects by binding directly to type-I receptors and blunting R-SMAD phosphorylation (Imamura et al. 1997). SMAD6 also inhibits signaling by competing with SMAD-4 for R-SMAD binding sites, reducing nuclear translocation (Hata et al. 1998). In addition, SMAD7 inhibits TGF-β signaling by targeting both type-I and type-II receptors for ubiquitination and subsequent degradation, through the recruitment of SMAD Ubiquitination Regulatory Factors 1 and 2 (Wicks et al. 2006; Ebisawa et al. 2001; Kavsak et al. 2000). Lastly, there are many other regulatory proteins that influence the bioavailability of TGF-β, such as proteoglycans decorin and biglycan, which bind and scavenge its availability for signaling (Droguett et al. 2006; Lopez-Casillas et al. 1994; Stander et al. 1999).

This group of TGF-β receptors and intermediates is a small part of a larger "superfamily" of growth factors and receptors (Feng and Derynck 2005). These superfamily members also occupy roles in normal connective tissue development and repair. Like the TGF-β family, dysregulation of BMP signaling has been implicated in heritable connective tissue disorders. Mutations within Activin receptor-Like Kinase 2 (ALK-2), a type-I BMP receptor, are associated with fibrodysplasia ossificans progressive (FOP), a skeletal dysplasia characterized by progressive heterotopic bone formation (Pignolo et al. 2011). Dysregulation of the activin and inhibin signaling has yet to be linked to heritable connective tissue disorders. However, activin signaling has been implicated in the regulation of wound healing and scar formation, processes dependent upon normal connective tissue remodeling (Werner and Alzheimer 2006).

7.3 TGF-β Signaling and Connective Tissue Disorders

In addition to MFS and LDS, several heritable connective tissue disorders have been associated with mutations in the TGF-β signaling pathway including Ehlers-Danlos Syndrome (EDS), Familial Thoracic Aortic Aneurysms and Dissections (FTAAD), Shprintzen-Goldberg Syndrome (SGS), and Hereditary Hemorrhagic Telangiectasia (HHT), among others (Table 7.1). Interestingly, each of these disorders also display unique cardiovascular manifestations, resulting in a spectrum of disorders ranging from heart valve defects to thoracic aortic aneurysms. These disorders will be explored in more detail below.

7.3.1 Ehlers-Danlos Syndrome

The Ehlers-Danlos syndromes (EDS) are a clinically diverse and genetically heterogeneous group of inherited connective tissue disorders primarily characterized by causal mutations in genes encoding collagen, its modifying enzymes, or other proteins involved in extracellular matrix formation (Cortini et al. 2019). These mutations lead to a loss of structural integrity within different organ systems, and are commonly characterized by joint hypermobility, skin hyperextensibility, and tissue fragility. To date, there have been 13 different subtypes of EDS defined based on diverse clinical presentation and genetic heterogeneity (Malfait et al. 2017). Of these 13 subtypes, two have been associated with altered TGF-β signaling, hypermobile EDS (Type-III) and Vascular EDS (Type-IV). Ehlers-Danlos syndrome type III (EDS, OMIM #130020), also known as Hypermobile Ehlers-Danlos Syndrome (hEDS), is the most prevalent subtype of EDS, and is primarily characterized by marked joint hypermobility in the absence of skeletal deformity, with a milder involvement of the skin as typically observed in the classical and vascular types of EDS. Commonly, patients with hEDS present with softer "velvety" skin, that displays reduced thickness, and increased fragility. In addition, an impaired wound healing response often results in atrophic scarring. While the primary underlying genetic basis of hEDS has yet to be identified, several candidate genes have been identified, most without definitive evidence proving causality (Gensemer et al. 2020). Of these candidate genes, TNXB (Tenascin X) may further highlight the association between aberrant TGF-β signaling and cardiovascular complications in hEDS.

Table 7.1 TGF-β Related Heritable Connective Tissue Disorders

Connective Tissue Syndrome	Associated Mutations	Connective Tissue Manifestations	Cardiovascular Manifestations	References
Marfan syndrome (MFS)	FBN1 OMIM#154700	Marfanoid habitus: dolichostenomelia, reduced upper:lower body ratio, scoliosis, pectus excavatum or carinatum; protrusio acetabuli; ectopia lentis; high arched palate; dural ectasia; lax joints	Ascending aortic aneurysm involving sinuses of Valsalva and dissection; aortic root dilatation with possible valve insufficiency; mitral valve prolapse and regurgitation	Judge and Dietz (2005)) Dietz (1993))
Loeys-Dietz syndrome (LDS)	TGFBRI & II OMIM#609192	Bifid uvula; cleft palate; clubfoot; hypertelorism; thin/velvety skin; blue sclera; cervical anomaly/ instability; craniosynostosis; scoliosis; dural ectasia; protrusion acetabuli; lax joints	Ascending aortic aneurysm and dissection; diffuse arterial tortuosity and aneurysms; easy bruising; mitral valve prolapse and regurgitation	Loeys et al. (2005); Loeys et al. (2006))
Ehlers-Danlos Syndrome (EDS), Type-III and Type-IV	Tenascin X TGF-βRII COL3A1 OMIM# 130020 OMIM# 130050	Hypermobile joints, reduced skin thickness, increased skin fragility, atrophic scarring, aberrant TGF-β signaling	Enhanced levels of circulating TGF-β, aortic and small vessel aneurysm, left ventricular dysfunction, heart failure, mitral valve insufficiency	Gensemer et al. (2020)) Morissette et al. (2014)) Sylvan et al. (2013)) Superti-Furga et al. (1988) Boutouyrie et al. (2004)
Shprintzen-Goldberg syndrome (SGS)	Reported FBN1, TGFBR I & II OMIM#182212 (FBN1)	Marfanoid habitus: dolichostenomelia, reduced upper:lower body ratio, scoliosis, pectus excavatum or carinatum; club foot; flat foot; hernias; scaphocephaly; craniosynostosis; digital contractures; Chiari-I; osteopenia	Mitral valve prolapse and regurgitation; aortic valve insufficiency; occasional aortic root dilatation	Greally et al. (1998) Van Steensel et al. (2008)
Familial thoracic aortic aneurysm and dissection syndrome (FTAAD)	AAT1–11 AAT5: TGFBRI OMIM#610380 AAT3: TGFBRII OMIM#608967	Marfanoid Habitus: dolichostenomelia, reduced upper:lower body ratio, scoliosis, pectus excavatum or carinatum; livedo reticularis; iris flocculi	Ascending and aortic root aneurysm and dissection; mitral valve prolapse and regurgitation	Gleason (2005) Pannu et al. (2005)
Aneurysm-osteoarthritis syndrome (AOS)	SMAD3	Early onset osteoarthritis; osteochondritis dissecans; mild hypertelorism; abnormal uvula	Aortic aneurysms and dissection; tortuosity of large and medium sized vessels, even intracranial	Van De Laar et al. (2011, 2012)

(continued)

Table 7.1 (continued)

Connective Tissue Syndrome	Associated Mutations	Connective Tissue Manifestations	Cardiovascular Manifestations	References
Arterial tortuosity syndrome (ATS)	SLC2A10 OMIM#208050	High palate; skin and joint laxity; hernias; keratoconus; facies; micrognathia; contractures; arachnodactyly	Large and medium vessel tortuosity; diffuse aneurysms; aortic regurgitation; telangiectasias; pulmonary artery stenoses and aneurysms	Callewaert et al. (2008) Coucke et al. (2006)
Hereditary hemorrhagic telangiectasia (HHT)	HHT1: TGFBR Type-III (Endoglin; ENG) OMIM#187300 HHT2: TGFBR Type-I (Activin receptor-like kinase-1/ALK-1) OMIM#600376	Specialized Connective Tissues: Blood-Thrombophilia; Lymphatic tissue-immunodeficiency	Diffuse GI and mucocutaneous telangiectasias; arteriovenous malformations in lungs, brain and liver; nosebleeds; easy bleeding and bruising; iron deficiency anemia; pulmonary artery hypertension	Govani and Shovlin (2009) Fernandez et al. (2006)

Tenascin X is an extracellular matrix protein that plays an essential role in collagen deposition by regulating the spacing between collagen fibrils. Interestingly, Tenascin X-deficiency has been associated with increased expression and abundance of TGF-β ligands, elevated TGF-β signaling, activation of signaling intermediates (SMAD-1/5/8), and increased MMP-13 production (Morissette et al. 2014). While enhanced levels of circulating TGF-β resulting in signaling through the classical TGF-β pathway, are hard to reconcile with the aberrant wound healing and atrophic scarring observed in hEDS, these results may suggest that the signaling proceeds through a non-canonical TGF-β pathway, dominated by the activation of SMAD-1/5/8. Reports by Bertolino and Jones, have suggested that TGF-β ligands (TGF-β 1, 2, 3) can signal through a second Type-I TGF-β receptor, Activin receptor-Like Kinase 1 (ALK-1), when recruited to TGF-βRII, resulting in the activation of a different set of intracellular R-SMADs (SMAD-1/5/8) (Jones et al. 2009; Bertolino et al. 2005; Jones et al. 2008). Interestingly, activation of this alternative pathway may be associated with ECM degradation by leading to enhanced MMP-13 production, resulting in the diminished collagen network.

In another report by Sylvan and coworkers, a patient with a complex medical history including a previous diagnosis of type-III EDS, mitral valve replacement surgery at age 18, left ventricular dysfunction, congestive heart failure, and pulmonary issues was identified (Sylvan et al. 2013). The patient presented with multiple small vessel aneurysms, and upon further genetic testing, a mutation in the TGF-βRII was identified, while confirming the absence of any COL3A1 mutations. While this report represents a single case study, the presence of a TGF-β receptor mutation in the absence of COL3A1 mutations (*commonly observed in Type-IV EDS; vascular EDS*) and in the absence of an LDS phenotype, may also suggest TGF-β receptor mutations can contribute to hEDS.

Ehlers-Danlos Syndrome type IV (EDS, OMIM #130050), also known as vascular type EDS, primarily affects the skin and large arteries, and can lead to medial degenerative disease of the aorta resulting in acute dissection. The original cause was linked to a 3.3-kb DNA deletion in one allele of the type III procollagen gene (COL3A1). This mutation, which results in a truncated procollagen monomer is characterized by decreased thermal stability, resistance to proteolytic processing, and reduced secretion

(Superti-Furga et al. 1988). Ultimately, in the aorta, it results in a diminished collagen network, with a low intimal-medial thickness, increased wall stress, and an increased propensity for acute dissection and rupture (Boutouyrie et al. 2004). As with many connective tissue disorders on this spectrum, diagnosis of EDS type IV can be difficult since there is significant phenotypic overlap with patients presenting with LDS. Loeys et al. (Loeys et al. 2006), while characterizing 52 LDS-affected families for mutations in TGFBR1 and TGFBR2 genes also assessed a cohort of EDS type IV patients that lacked the COL3A1 gene mutations and the craniofacial features of the typical LDS patient. Twelve EDS type IV probands were identified containing TGF-βRI or TGF-βRII mutations, suggesting a possible reclassification of this group as LDS type 2. One could speculate that EDS type IV patients with COL3A1 mutations display enhanced TGF-β signaling to compensate for the loss of type III collagen within the aorta. Nonetheless, these results further bolster the association between cardiovascular complications in connective tissue disorders involving aberrant TGF-β signaling.

7.3.2 Familial Thoracic Aortic Aneurysm and Dissection Syndrome (FTAAD)

Classically, FTAAD was defined as a collection of inherited genetic disorders, leading to thoracic aortic aneurysms or aortic dissections, characterized by medial necrosis of the proximal ascending aorta (Nicod et al. 1989). Diagnosed patients typically had a first degree relative who also had aortopathy, however diagnosis was often clouded by dominate inheritance with variable penetrance (Milewicz et al. 1998). Medial necrosis is described as degeneration and fragmentation of elastic fibers, loss of smooth muscle cells, with the presence of interstitial pools of basophilic ground substance. This typically occurs in the absence of a known genetic syndrome, like MFS. Aortic disease in these families is characterized by aneurysms involving the ascending aorta often leading to type I and II aortic dissections in the absence of hypertension (Guo et al. 2001). Since its original description, FTAAD has been localized to multiple genetic loci, now with at least 11 independently associated genes identified in the Online Mendelian Inheritance in Man (OMIM) database: Aortic Aneurysm Thoracic (AAT) 1–11. Interestingly, four of these mutations, AAT3 (OMIM#608967; TGF-βRII), AAT5 (OMIM#610380; TGF-βRI), AAT9 (OMIM#601103; MFAP5) and AAT10 (OMIM#617168; LOX), were found in genes capable of affecting the connective tissue directly. AAT3 and AAT5 were found to be associated with mutations in TGF-β-receptors (receptors -II and -I respectively) and have subsequently been grouped under the heading of Loeys-Dietz Syndrome as LDS2 and LDS1, to simplify classification overall. AAT9, was associated with a mutation in the Microfibrillar Associated Protein 5 (MFAP5) gene, which encodes Microfibril-Associated Glycoprotein 2, a small proteoglycan involved in the assembly and/or maintenance of elastic fibers (Barbier et al. 2014), while AAT10, was found to encode Lysyl Oxidase (LOX), a protein that plays a critical role in initiating the crosslinking of collagens and elastin. In each case, the initial precipitating event was often incidentally discovered as aortic dilatation, dissection, or sudden death (Attias et al. 2009; Bruno et al. 1984; Von Kodolitsch et al. 2004). Subsequently, an aortopathy syndrome is suspected when a family history of early aortic disease or sudden death is revealed or the constellation of unique connective tissue symptoms (marfanoid habitus like MFS and LDS and/or the FTAAD specific iris flocculi) provokes suspicion. While these seemingly non-syndromic TAAs and dissections may be exacerbated by contributing risk factors such as hypertension, atherosclerosis, it has been reported that almost 20% of these patients have a first degree relative with a similar presentation, suggesting a genetic predisposition (Pannu et al. 2005). Diagnosing this group of patients remains complicated because of the variable penetrance within a family group.

With new and refined genetic technologies, additional genes are being added to this list on a regular basis, underscoring the importance of the connective tissue compartment, and further defining the link between dysregulation of extracellular matrix and cardiovascular complications.

7.3.3 Shprintzen-Goldberg Syndrome (SGS)

In 1982, Shprintzen and Goldberg first described their eponymous heritable connective tissue syndrome in two patients (Shprintzen and Goldberg 1982). Shprintzen-Goldberg syndrome is characterized by anomalies of the head/face, skeleton, brain, and cardiovascular system (Greally et al. 1998). Shprintzen-Goldberg syndrome has since been recognized as part of a group of phenotypically overlapping syndromes associated with TGF-β receptor mutations (i.e. LDS and FTAAD) affecting connective tissues and the cardiovascular system (Akutsu et al. 2007). However, SGS has been linked to mutations in TGF-β RI and RII, as well as fibrillin-1 (OMIM #182212). Thus, unlike LDS and MFS, it is not yet known whether the connective tissue and cardiovascular manifestations of SGS are due to a defect in a TGF-β receptor, or a connective tissue component like fibrillin-1. Independent of the initiating event, the defect lies somewhere in the TGF-β pathway creating a heterogeneous range of symptoms, making a definite genotype-phenotype correlation difficult. Consequently, the clinical presentation of SGS is not well defined and is still developing. Intellectual impairment may be the only regularly occurring symptom, with all documented patients presenting with a range from moderate retardation to learning disabilities (Greally et al. 1998). These impairments are known to occur simultaneously with brain abnormalities: communicating hydrocephalus, dilated lateral ventricles, and Arnold-Chiari formation type-I (Robinson et al. 2005). Ocular defects may also be present in SGS patients. Lens dislocation, while seen in MFS, does not appear to be a typical feature of SGS (Loeys et al. 2006), however, hypertelorism (seen in LDS), myopia, and exophthalmos are commonly observed (Greally et al. 1998). In addition, several skeletal anomalies have been identified, appearing in early childhood (Ades et al. 1995). The major skeletal characteristic is scaphocephaly (*boat shaped skull*) with craniosynostosis (*premature fusion of skull*) (Ades et al. 1995). In fact, SGS has been referred to as marfanoid habitus with craniosynostosis (Ades et al. 2006). Many of the other skeletal findings associated with MFS and LDS are likewise observed in SGS: dolichostenomelia, arachnodactyly, scoliosis, pectus excavatum or carinatum (*hollowed or pigeon chest*), joint hypermobility, and contracture of the proximal joints of the hand (Loeys et al. 2006). Regarding facial dysmorphic features, SGS may produce micrognathia, midface hypoplasia, low-set ears, and palatal soft tissue hyperplasia (*pseudocleft palate*) that may be noted as early as the first year of life and become more pronounced with time (Greally 1993). Additional characteristic findings include: minimal subcutaneous fat, hypotonia, obstructive apnea, defects in the abdominal wall musculature with hernias, hyperelastic skin, and cryptorchidism (Greally 1993).

The cardiovascular defects in SGS are mostly limited to the heart valves. Mitral and/or aortic valve regurgitation is commonly observed (Greally et al. 1998). Mitral Valve Prolapse (MVP), often seen in MFS, occurs commonly, as well (Greally 1993). Given that FBN1 mutations have been associated with an increase in TGF-β release and signaling, Ng et al. examined the association of aberrant TGF-β pathway signaling and the pathogenesis of MVP using a mouse model of MFS (Ng et al. 2004). Changes in mitral architecture were observed to be temporally and spatially linked with increased TGF-β activation resulting in enhanced proliferation/growth of valve cells and decreased apoptosis. Furthermore, normal valve phenotype was restored with the administration of a TGF-β neutralizing antibody. This study provided a potential pathogenic mechanism for MVP in MFS/LDS and perhaps SGS as well. While common in MFS, LDS, and FTAAD, aortic root dilatation and aneurysm has been previously described in SGS but is not present in most affected individuals (Loeys et al. 2005). The presence of aortic dilatation may therefore represent overlap with one of these phenotypically similar syndromes. Aortic valve pathology has also been linked to TGF-βRII mutations. An SGS patient with a bicuspid aortic valve and an ascending aortic aneurysm, which later dissected, was found to have a mutation in TGF-βRII (Girdauskas et al. 2011). Thus, both the mitral and aortic valvular manifestations of SGS may be due to mutations in TGF-β receptors, and alterations in the signaling pathway.

7.3.4 Hereditary Hemorrhagic Telangiectasia (HHT)

Originally described in the nineteenth century by Osler, Weber, Rendu, and Hanes, HHT is an autosomal dominant disorder characterized by vascular malformations and dilated small blood vessels which are fragile due to thin supporting connective tissue (Macri et al. 2020; Sys and Van Den Hoogen 2005). HHT is most commonly caused by mutations within TGF-β receptors, disrupting TGF-β signaling, and inducing the characteristic vascular and connective tissue defects. Epidemiologic reports estimate the prevalence of HHT between 1 in 5000 to 1 in 8000, though some reports believe HHT may be underreported due to many patients being unaware of their diagnosis (Bideau et al. 1989; Kjeldsen et al. 1999; Shovlin et al. 2008). The diagnosis is often difficult due to its variable penetrance and severity, as well as its relatively slow progression. Manifestations of HHT typically are not present at birth and develop with time. Clinical signs and symptoms may be present in childhood though generally are noted after puberty with an estimated 7 in 10 HHT patients developing at least one clinical symptom or sign by age 16 and almost 100 percent by 40 years of age (Bourdeau et al. 1999; Cole et al. 2005; Wallace and Shovlin 2000).

Initially, HHT patients will develop telangiectasias, small blood vessels that dilate near the surface of the skin, mucous membranes and gastrointestinal tract. These telangiectasias increase in number and size with age (Pasculli et al. 2005; Plauchu et al. 1989). Nosebleeds (*also known as epistaxis*), the most common clinical manifestation of HHT, result from ruptured telangiectasias of the nasal mucosa. Epistaxis and telangiectasias within the gastrointestinal tract, commonly in the duodenum, are the two major mechanisms of iron deficiency anemia secondary to hemorrhage in this population. Most HHT patients experience only these three symptoms: nosebleeds, mucocutaneous telangiectasias, and iron deficiency anemia. These symptoms are relatively minor, in terms of their contribution to the morbidity and mortality associated with HHT, while the primary concern results from arteriovenous malformations (AVMs), vascular abnormalities resulting from malformed connections between arteries and veins in the visceral organs (Kjeldsen et al. 1999).

While AVMs may occur sporadically in the general population, AVMs occur in high numbers in multiple organs in HHT patients. The most clinically relevant locations are distributed among the lungs (50%), the liver (30%) and the brain (10%) (Cottin et al. 2004; Fulbright et al. 1998; Piantanida et al. 1996). A further pulmonary manifestation of HHT is severe pulmonary artery hypertension (PAH) arising mainly from 2 sources: (1) high output heart failure secondary to hepatic AVM shunting and (2) primary PAH without signs of heart failure (Govani and Shovlin 2009).

Additionally, HHT patients may also exhibit pathologic defects within specialized connective tissues such as the blood and the immune system/lymphoid tissue. Elevated clotting factor VIII and von Willebrand factor was measured in the blood of HHT patients versus normal controls and associated with venous thromboembolism (Shovlin et al. 2007). Reports of defects in adaptive immunity and a mononuclear cell infiltrate around telangiectasias spawned a suspicion of immune system involvement in HHT. These reports were further supported by an analysis of the oxidative burst activity of HHT monocytes and polymorphonuclear cells, which found single or multiple deficits in both cell groups in 20 of 22 HHT patients (Cirulli et al. 2006).

The Curacao diagnostic criteria are based on international consensus and used to diagnose HHT with a score that gauges the likelihood of its presence (Faughnan et al. 2011; Shovlin et al. 2000). The criteria include a first degree relative with HHT, the presence of several telangiectasias on the skin and mucous membranes, recurrent and spontaneous epistaxis, and visceral AVMs. One point is scored for each of the criteria present. If only 1 of the criteria is noted, HHT is "unlikely." Two criteria indicate "suspected" HHT. The presence of more than two criteria is evidence of "definitive" HHT disease. The diagnosis of HHT is made clinically, without requiring genetic testing to identify a potentially causative mutation. If desired, genetic testing may be employed to confirm the diagnosis.

Mutations in at least five genes have been directly associated with the development of HHT. These are subdivided based on the gene loci involved (Govani and Shovlin 2009). HHT1 and 2 are the major subtypes linked to mutations within endoglin (HHT1, OMIM#187300) (Mcallister et al. 1994) and ALK-1 (HHT2, OMIM#600367) (Berg et al. 1997; Johnson et al. 1996). Additionally, mutations in SMAD4 result in HHT with juvenile polyposis. Interestingly, juvenile polyposis in the general population results from mutation in activin receptor-like kinase 3 (ALK-3; also known as BMP Receptor-1A), which signals through SMAD-4 (Govani and Shovlin 2009). However, a family with a history of juvenile polyposis, aortopathy, and mitral valve dysfunction, was described by Andrabi and coworkers, suggesting co-segregation of these phenotypes with a mutation in SMAD-4 (Andrabi et al. 2011). Aortopathy and mitral valve defects are more typically related to MFS and LDS, not HHT, though case reports of large vessel aneurysms in HHT do exist (Andrabi et al. 2011) The presence of these features associated with SMAD4 mutations further supports the role of dysfunctional TGF-β signaling in the common pathogenesis of disorders. Furthermore, it provides a spectral link between the vascular features of MFS and LDS (aortopathy, aneurysm, and mitral valve defects) and those seen in HHT (AVMs, small vessel dilatation, and juvenile polyposis).

Activin receptor-like kinase 1 and endoglin are expressed on the surface of vascular endothelial cells, suggesting that dysregulated TGF-β signaling in endothelial cell plays a major role in inducing telangiectasia/dilatation and AVM formation (Letarte et al. 2005). Interestingly, homozygous ALK-1 mutations in zebrafish and mice produce embryonic lethality and exhibit severely dilated vessels (*including the aorta*) and abnormal vessel fusion (Oh et al. 2000; Roman et al. 2002). These vascular defects were associated with increased endothelial cell number, enhanced expression of angiogenic factors and proteases, and deficient differentiation and recruitment of smooth muscle cells. Thus, the small vessel dilatation in HHT represents a phenotypic microcosm of the aortic and extra-aortic dilatation seen in MFS and LDS. Furthermore, Seki et al. demonstrated in mice, that ALK-1 is highly expressed in the developing endothelium of arteries (Seki et al. 2003). Taken together, these observations support the role of TGF-β signaling in early vascular development and dilatation.

As mentioned above, TGF-β ligands are capable of signaling through two distinct type-I receptors, TGF-βRI and ALK-1, when complexed with TGF-βRII, resulting in different groups of SMAD proteins being activated intracellularly; TGF-βRI activates SMAD-2/3, whereas ALK-1 activates SMAD-1/5/8 (Fig. 7.1) (Bertolino et al. 2005). Endoglin plays an interactive role with ALK-1 and is required for TGF-β-dependent ALK-1 signaling, through mediating/facilitating the binding of TGF β to the ALK-1 receptor (Bertolino et al. 2005). TGF-β stimulation of the endoglin/ALK-1 pathway activates SMAD-1/5/8 and has been associated with endothelial proliferation and migration, both essential to angiogenesis. Alternatively, signaling through ALK-5 pathway activates SMAD-2/3 and produces an opposite response, inhibiting proliferation and migration (*quiescence*) (Stefansson et al. 2001). This would seem to suggest that mutated ALK-1 or endoglin would result in a quiescent endothelium, opposite to that seen in HHT. However, the ALK-1 pathway can regulate the expression of ALK-5, such that decreased ALK-1 signaling leads a reduction in ALK-5 signaling, as evidenced by an 80% decrease in ALK-5 mRNA transcripts as measured in peripheral blood endothelial cells from

HHT1 and HHT2 patients (Fernandez et al. 2005). This adaptive compensation may produce an imbalance favoring dysregulated angiogenesis and the formation of AVMs. Evidence suggests that the ratio of ALK-1 to ALK-5 determines whether or not endothelial cells will become quiescent or actively proliferate and migrate (Goumans et al. 2003). Alternatively, similar to LDS in which increased TGF-β signaling was described despite TGF-β -RI and -RII mutations, endoglin and ALK-1 mutations may paradoxically increase signaling through angiogenic pathways thus resulting in AVMs.

In comparison to what is known about the mechanisms of AVMs and dilatation, little was known about the mechanism of primary PAH in HHT until recently. The primary PAH phenotype (*without heart failure*) was observed in <2% of HHT patients, and was identical to inherited primary PAH due to a loss of function mutation in Bone Morphogenetic Protein Receptor 2 (BMPR2); another type-II receptor in the TGF-β superfamily (Govani and Shovlin 2009). The mechanism is thought to involve a loss of pulmonary artery endothelial and smooth muscle cell apoptosis mediated by BMPR2 that results in abnormally elevated growth and proliferation (Davies et al. 2012; Kimura et al. 2000; West et al. 2004). When examining families with both HHT and PAH, a suggestive linkage between BMPR2 and ALK-1 was identified (Trembath et al. 2001). However, the HHT patients only exhibited mutations in ALK-1, not BMPR2. This suggests a common signaling pathway downstream of BMPR2 and ALK-1 is involved in the pathogenesis of primary PAH. Mutations in endoglin, facilitating signaling through ALK-1, may also result in PAH. Type-III TGF-β receptors have been implicated in the TGF-β-mediated growth inhibition in myoblasts, with dependence on SMAD-3 activation and p38 MAPK signaling (Roman et al. 2002; You et al. 2007). Thus, interrupted TGF-β signaling via a mutated ALK-1 or endoglin gene, may remove a TGF-β mediated growth inhibitory effect on vascular endothelial or smooth muscle cells and contribute to the development of PAH and other cardinal manifestations of HHT-telangiectasias and AVMs.

7.3.5 Other Connective Tissue Disorders with TGF-β Involvement

TGF-β is also implicated in several other connective tissue disorders which are not commonly defined by gene abnormalities. Most include a hyperactive immune system as a component and are referred to as autoimmune connective tissue diseases. These diseases include systemic sclerosis, or scleroderma, rheumatoid arthritis (RA), and systemic lupus erythematosus (SLE). The severity of scleroderma, a disease of excessive fibrosis of vessels, organs, and particularly the skin, has been associated with increased levels of TGF-β signaling (Hawinkels and Ten Dijke 2011). Patients with rheumatoid arthritis (RA), often present with elevated plasma levels of both thrombospondin-1 and TGF-β; both of which are associated with early onset atherosclerosis commonly observed in RA (Rico et al. 2008; Rico et al. 2010). Other studies have provided evidence that TGF-β can play a role in suppressing immune function directly and stimulating T-cell conversion to a suppression phenotype (Wahl and Chen 2005). Interestingly, TGF-β production is decreased in SLE (Ohtsuka et al. 1998). Thus, a lack of TGF-β may contribute to SLE through diminished ability to suppress the immune system (Mageed and Prud'homme 2003). Due to its relationship with fibrosis and immune modulation, TGF-β may plausibly be involved in many autoimmune connective tissue disorders though its exact roles remain to be clarified.

Further lending support to the pivotal role of TGF-β in connective tissues and the cardiovascular system, heritable mutations in downstream and upstream mediators of TGF-β signaling display symptoms overlapping with the other TGF-β receptor mutation syndromes. A mutation in SMAD-3 was recently linked to a heritable syndrome of vascular aneurysms, arterial tortuosity (*twisted, corkscrew like arteries*), skeletal/craniofacial abnormalities with osteoarthritis, and is referred to as Aneurysm-Osteoarthritis Syndrome (AOS) (Van De Laar et al. 2011). Similarly, arterial tortuosity syndrome (ATS), characterized by tortuosity of medium to large vessels and aneurysms, has associated with loss-of-function mutations in SLC2A10 (Solute Carrier Family 2, Facilitated Glucose Transporter Member 10) (Coucke et al. 2006; Loeys and De Paepe 2008). Increased TGF-β signaling is also believed to be associated with this syndrome, as vascular smooth muscle cells from ATS patients exhibit decreased production of decorin, a small leucine rich proteoglycan known to bind and sequester TGF-β in the ECM (Coucke et al. 2006). The decorin promoter contains a glucose response element which is less active, resulting in fewer functional glucose transporters, creating a decorin deficiency that can result in increased TGF-β ligand abundance and signaling (Coucke et al. 2006).

7.4 TGF-β-Directed Therapy as a Prime Target for Connective Tissue Disorders

The evidence presented above highlights the indominable role of TGF-β in connective tissue disorders with cardiovascular complications. Understanding the mechanistic underpinnings of these disorders may allow for the development of specific therapies designed to interrupt aberrant TGF-β signaling and perhaps diminish the primary complication leading to mortality in many of these disorders. The following examples are provided as potential evidence to support therapeutic options targeted at attenuating the TGF-β signaling.

In a very short period of time, significant progress has been made in understanding the cause, clinical pathogenesis, and clinical management of MFS (Pyeritz 2016). The dilation of the aortic root and its dissection or rupture, are the primary manifestations of MFS that drive mortality in these patients. Thus, much effort has been focused on how to treat these sequelae in hope of attenuating aneurysm disease. As a first line therapy for all aneurysm patients (with MFS or not) β-blockers (e.g. atenolol) are used to regulate systemic blood pressure as well as dP/dt (rate–rise time) of the cardiac pulse wave as it moves through the aorta. This therapy was chosen based on the rationale that regulating systemic blood pressure is likely advantageous to prevent rupture and dissection of an aneurysm. In an effort to build on this principle, Habashi et al. attempted treating MFS mice (C1039G/+), which spontaneously develop ascending/aortic root aneurysms, with another class of antihypertensives, the angiotensin-II receptor blockers (ARBs), specifically losartan (Habashi et al. 2006). Losartan became a drug of interest for several reasons. Previous studies by Daugherty and coworkers, discovered that treating mice that had been induced to develop abdominal aortic aneurysms (AAAs) with Angiotensin-II (AngII) infusion, using an AngII **type-I** specific receptor (AT1-receptor) blocker (losartan), could attenuate AAA formation, while treatment with an Ang-II **type-II** specific receptor (AT2-receptor) inhibitor (PD123319), enhanced aortic pathology and accelerated aneurysm development (Daugherty et al. 2001). Taken together, this suggested that signaling through the AT1-receptor had deleterious consequences, accelerating aortic pathology, while signaling through the AT2-receptor may have been protective. More importantly, many studies had suggested that stimulation of the AT1-receptor could enhance the expression and activation of TGF-β ligands and receptors (Everett et al. 1994; Fukuda et al. 2000; Naito et al. 2004; Wolf et al. 1999), therefore, inhibition of AT1-receptors may be advantageous toward reducing overall TGF-β signaling. Not surprisingly, when Habashi and colleagues treated MFS (C1039G/+) mice with losartan, aortic dilation and the phosphorylation of SMAD-2 were both attenuated (Habashi et al. 2006). Furthermore, losartan was also shown to rescue the TGF-β-dependent skeletal muscle defects and the alveolar septation defects observed in the MFS mice (Judge et al. 2011). Subsequent studies by Carta et al. (Carta et al. 2009) and Rodriguez-Vita et al. (Rodriguez-Vita et al. 2005) went further to demonstrate that AngII could stimulate the phosphorylation of SMAD-2, independent of signaling through TGF-βRI, resulting in enhanced production of CTGF. Importantly, losartan treatment was able to attenuate both the phosphorylation of SMAD-2 and the accumulation of CTGF, implicating an additional role for an unknown downstream mediator in the AT1-receptor pathway. These data not only linked TGF-β to aortic pathology, but also clearly demonstrate that the inhibition of the AT1-receptor is capable of attenuating TGF-β signaling.

Xiong and colleagues discovered that doxycycline, an FDA approved tetracycline antibiotic with non-specific MMP-inhibitory activity, could delay rupture of the ascending aorta in fibrillin-1 under-expressing mice (mgR/mgR mouse model) (Xiong et al. 2008). This was directly associated with doxycycline's ability to inhibit matrix metalloproteinases (MMP) -2 and – 9. Further, the study showed that doxycycline enhanced the preservation of elastic fiber integrity, normalized aortic wall stiffness, and prevented vessel wall weakening when compared to β-blocker therapy alone (Chung et al. 2008). When compared head-to-head with losartan, doxycycline was shown to inhibit both MMP-2 activation and the phosphorylation of ERK1/2 (Xiong et al. 2012). Interestingly, the action of doxycycline wasn't attributed to its effects on MMP activity, but rather to its secondary effect of inhibiting MMP-2-dependent TGF-β release. Accordingly, further studies of doxycycline alone and in combination with β-blockers or ARBs, should be considered in the treatment of other TGF-β-dependent connective tissue disorders.

Finally, progress in understanding the process of TGF-β release and activation as mediated by membrane integrins should be further explored as a potential therapy in the treatment of TGF-β-dependent connective tissue disorders. Recent reports have demonstrated roles for integrin αvβ1 and αvβ3 in the activation and release of TGF-β, in processes leading to the development of fibrosis. Newly developed integrin-specific inhibitors such as C8 (αvβ1 specific) (Reed et al. 2015) or Cilengitide (αvβ3/αvβ5 specific) (Patsenker et al. 2009; Roth et al. 2013; Li et al. 2013) have been used to attenuate TGF-β activation and its downstream signaling response.

Together, as the role of TGF-β signaling in connective tissue disorders continues to be elucidated, it is important to note that therapeutic advances targeted at attenuating aberrant TGF-β signaling events may provide significant benefit by enhancing quality and quantity of life, given that the cardiovascular-associated disorders are often the most life-threatening manifestations of these disorders.

7.5 Genetic Testing

As reported herein, identifying and diagnosing heritable connective tissue disorders is not without significant challenges. As noted, many of the disorders have significantly overlapping symptomology. This is further complicated by their often-dominant inheritance with variable penetrance, leaving the patient and their physician to examine family history and the overt phenotypic symptoms in an effort to make an accurate diagnosis. Accordingly, the use of genetic testing through new technologies such as Genome Wide Association Studies (GWAS) and whole exome Next Generation Sequencing (NGS), provides some hope in identifying mutations that may be associated with a given disorder. The interpretation of the genetic sequencing data alone, however, continues to provide significant challenges due to a large number of sequence variants and private mutations, further complicated by the fact that many of the missense mutations responsible for disease often resemble benign genetic variation. Thus, clinical symptoms and common phenotypes are often used in combination with NGS results to narrow down the possibilities, identifying the specific underlying genetic variants that cause disease. As reported by Pope and coworkers, the presence of cardiovascular disease with a documented family history, is a suitable indication for NGS, based on the presence of the large number of causative sequence variants that are well associated with heritable connective tissue disorders (Pope et al. 2019). Together, the phenotype and genotype can be used diagnostically with a much greater chance of identifying the underlying cause of disease. Once identified, then familial inheritance can be back confirmed, and specific treatment plans can be made. Thus, this section will discuss the role of genetic testing in some of the heritable connective tissue disorders identified above.

7.5.1 Genetic Testing for FTAAD

Due to its phenotypic overlap with other inherited aortopathies, FTAAD should be confirmed with genetic testing. Genetic testing in cases of inherited aortopathy can be beneficial in several ways. Identification of an associated mutation can change follow-up and medical management of the affected patient. Furthermore, identification of the mutation present in the proband will narrow and facilitate testing in potentially affected relatives as well as potential prenatal testing. Beginning with identification of the first family member (proband) with an FTAAD mutation, guidelines recommend all first-degree relatives be genetically counseled and screened (Milewicz et al. 2010). Those relatives found to have the genetic mutation should obtain baseline aortic imaging immediately, and second-degree relatives could reasonably be notified. If aortic disease is found in any first-degree relatives, imaging of second-degree relatives would be warranted (Milewicz et al. 2010). If a patient with aneurysm/dissection does not have any of the major gene mutations associated with heritable aortic disease, first-degree family members are recommended to seek aortic imaging rather than genetic testing (Milewicz et al. 2010). This recommendation is particularly relevant because only ~30% percent of FTAAD patients will have one of the 11 known associated mutations (Biddinger et al. 1997; Coady et al. 1999; Pomianowski and Elefteriades 2013).

7.5.2 Genetic Testing for SGS

SGS is clinically suspected when an individual present with a combination of the major characteristics: marfanoid skeletal features, craniosynostosis, craniofacial dysmorphism, left sided heart valve prolapse or regurgitation, intellectual disability with delayed milestones, and brain abnormalities (Grealy et al. 1998). No specific diagnostic criteria or scoring rubric exists for SGS as for MFS with the Ghent criteria and HHT with the Curacao criteria. Genetic diagnosis of SGS is difficult due to the limited number of SGS patients, the range of mutations associated, and phenotypic overlap of related syndromes with other heritable connective tissue disorders, and the proclivity for variability in presentation. Fibrillin-1 mutations were initially reported in 3 clinically diagnosed SGS patients, two of whom had a mutation atypical of MFS, and exhibited an overlapping phenotype between SGS and MFS (Kosaki et al. 2006; Sood et al. 1996). A later genetic study of multiple SGS patients found no FBN1 mutations (Ades et al. 2006). These observations suggest that a similar signaling pathway is involved in both SGS and MFS, even though more than one gene may be affected. A patient described by van Steensel et al., with a TGFBR2 mutation, displayed a significant phenotypic overlap between SGS and LDS (Van Steensel et al. 2008). In a study describing TGF-β receptor mutation phenotypes, a TGFBR1 mutation was identified in a patient with clinically diagnosed SGS (Stheneur et al. 2008). Current criteria do not require the identification of a specific mutation to diagnose SGS, though identification of an FBN1 or TGFBR mutation may help to confirm the diagnosis.

7.5.3 Genetic Testing for HHT

In those cases where HHT is symptomatic, patient management is based on the standard of care for each of the individual symptoms/conditions as if they occurred alone in a normal patient. Arteriovenous malformations are treated dependent on their location and based upon an expert clinical consultation pertaining to the organ involved, the primary diagnosis of HHT, or both. Embolotherapy is the preferred and most definitive therapy for AVMs (Shovlin et al.

2008), though surgical resection or arterial ligation are alternative options (Faughnan et al. 2011).

Patients with HHT are typically identified in childhood with a progression of subtle clinical signs increasing with age. Therefore, determining whether a newborn has inherited HHT is not possible clinically and if suspected, requires genetic testing for confirmation (Cohen et al. 2005). Genetic testing for HHT involves the examination of the endoglin, ALK-1, and SMAD-4 genes; these mutations account for more than 80% of all HHT cases (Govani and Shovlin 2009). However, this means the diagnosis for a substantial portion, approximately 20% of individuals with HHT, cannot be confirmed or excluded by molecular genetic testing. In most instances, a positive genetic test does not alter the recommended course of treatment or screening. However, it may alert the patient to be more vigilant. For example, if a SMAD-4 mutation (*associated with juvenile polyposis*) was detected, it would be critical to know whether there was a family history of GI polyps and/or malignancy. This would instruct the patient to be more rigorous about GI screening (Abdalla and Letarte 2006). For this reason, genetic testing and counseling are of great potential benefit to HHT suspected and affected families.

7.6 Summary

We have reviewed several heritable connective tissue syndromes associated with mutations in TGF-β receptors I and II, as well as accessory receptors, and related pathway intermediates. Many of these syndromes including MFS, LDS, and EDS exhibit concomitant cardiovascular manifestations that provide a phenotypic readout of the presence of disease and are described in more detail in dedicated chapters in this volume. While significant progress has been made in understanding their underlying mechanisms, the refinement of treatment strategies remains an area of critical need. In particular, the study of TGF-β receptor mutation syndromes holds great promise in this regard for the treatment of both connective tissue and cardiovascular disorders. Mutations in FBN1 or TGF-β receptors appear to result in a number of phenotypically overlapping connective tissue/cardiovascular syndromes involving dysregulation of the TGF-β signaling pathway. These TGF-β dysregulation syndromes (MFS, LDS, FTAAD, EDS, HHT, SGS, AOS and ATS) exhibit a spectrum of cardiovascular defects including aortic or arterial aneurysm, dilatation or dissection, mitral valve disease, arterial tortuosity, and primary PAH (Fig. 7.2). Their pathogeneses emphasize a common theme, that normal TGF-β signaling is integral to the normal development and homeostasis of connective tissues and the cardiovascular system. This signaling superfamily contains potent regulators of many cell-types within the mesoderm-derived tissues. Thus, perturbations within the signaling pathway are uniquely situated to produce defects in these tissue types. In Marfan syndrome and its related disorders, characteristic abnormalities of these syndromes that were once thought to result from purely structural deficiencies (e.g. FBN1) are now attributed to disruptions of normal TGF-β signaling. The undeniable overlap in connective tissue disorders with cardiovascular complications involving fibrillinopathies and mutated TGF-β receptor syndromes, supports this notion of a common signaling pathway (Loeys et al. 2005). Indeed, these mutated receptor phenotypes are even recapitulated by mutations in downstream TGF-β pathway components (e.g. SMAD-4 in HHT, SMAD-3 in LDS and FTAAD) (Van De Laar et al. 2011; Abdalla and Letarte 2006).

Our new understanding of causal signaling disturbances in these disorders significantly improves the treatment prospects for highly morbid cardiovascular manifestations and debilitating connective tissue defects. Through amassing evidence derived from a combination of understanding the symptomology, the familial inheritance, and sequencing results, it may be the indominable role of TGF-β that holds the key to discovering novel therapeutic strategies for these patients.

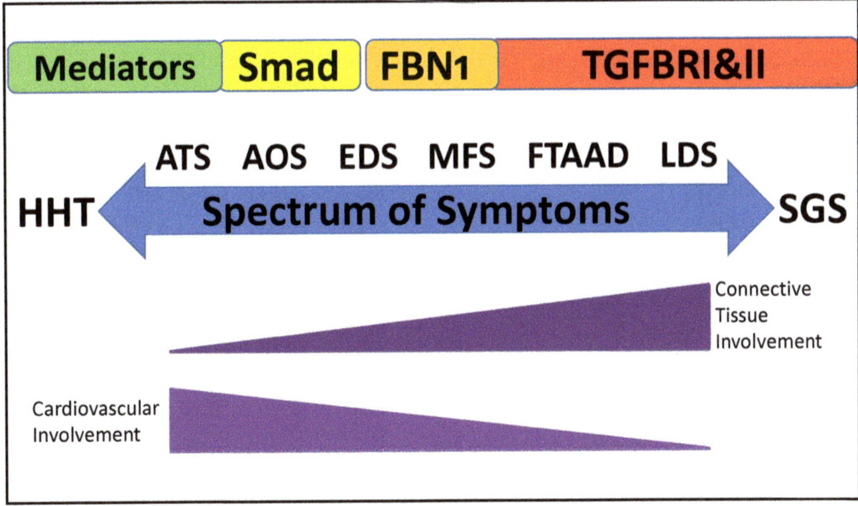

Fig. 7.2 Known heritable connective tissue disorders with cardiovascular involvement that associate with gene mutations related to aberrant TGF-β signaling. These disorders are arranged based on their increasing level of either connective tissue involvement or cardiovascular involvement, and notably share a spectrum of common symptoms, which support their related pathophysiologies

References

Abdalla SA, Letarte M (2006) Hereditary haemorrhagic telangiectasia: current views on genetics and mechanisms of disease. J Med Genet 43:97–110

Ades LC, Morris LL, Power RG et al (1995) Distinct skeletal abnormalities in four girls with Shprintzen-Goldberg syndrome. Am J Med Genet 57:565–572

Ades LC, Sullivan K, Biggin A et al (2006) FBN1, TGFBR1, and the Marfan-craniosynostosis/mental retardation disorders revisited. Am J Med Genet A 140:1047–1058

Akutsu K, Morisaki H, Takeshita S et al (2007) Phenotypic heterogeneity of Marfan-like connective tissue disorders associated with mutations in the transforming growth factor-beta receptor genes. Circulat J Off J Japanese Circulat Soc 71:1305–1309

Andrabi S, Bekheirnia MR, Robbins-Furman P et al (2011) SMAD4 mutation segregating in a family with juvenile polyposis, aortopathy, and mitral valve dysfunction. Am J Med Genet A 155A:1165–1169

Annes JP, Munger JS, Rifkin DB (2003) Making sense of latent TGFbeta activation. J Cell Sci 116:217–224

Attias D, Stheneur C, Roy C et al (2009) Comparison of clinical presentations and outcomes between patients with TGFBR2 and FBN1 mutations in Marfan syndrome and related disorders. Circulation 120:2541–2549

Barbier M, Gross MS, Aubart M et al (2014) MFAP5 loss-of-function mutations underscore the involvement of matrix alteration in the pathogenesis of familial thoracic aortic aneurysms and dissections. Am J Hum Genet 95:736–743

Berg JN, Gallione CJ, Stenzel TT et al (1997) The activin receptor-like kinase 1 gene: genomic structure and mutations in hereditary hemorrhagic telangiectasia type 2. Am J Hum Genet 61:60–67

Bertolino P, Deckers M, Lebrin F et al (2005) Transforming growth factor-beta signal transduction in angiogenesis and vascular disorders. Chest 128:585S–590S

Biddinger A, Rocklin M, Coselli J et al (1997) Familial thoracic aortic dilatations and dissections: a case control study. J Vasc Surg 25:506–511

Bideau A, Plauchu H, Brunet G et al (1989) Epidemiological investigation of Rendu-Osler disease in France: its geographical distribution and prevalence. Population 44:3–22

Bourdeau A, Dumont DJ, Letarte M (1999) A murine model of hereditary hemorrhagic telangiectasia. J Clin Invest 104:1343–1351

Boutouyrie P, Germain DP, Fiessinger JN et al (2004) Increased carotid wall stress in vascular Ehlers-Danlos syndrome. Circulation 109:1530–1535

Brekken RA, Sage EH (2001) SPARC, a matricellular protein: at the crossroads of cell-matrix communication. Matrix Biol 19:816–827

Bruno L, Tredici S, Mangiavacchi M et al (1984) Cardiac, skeletal, and ocular abnormalities in patients with Marfan's syndrome and in their relatives. Comparison with the cardiac abnormalities in patients with kyphoscoliosis. Br Heart J 51:220–230

Callewaert BL, Loeys BL, Casteleyn C et al (2008) Absence of arterial phenotype in mice with homozygous slc2A10 missense substitutions. Genesis 46:385–389

Carta L, Smaldone S, Zilberberg L et al (2009) p38 MAPK is an early determinant of promiscuous Smad2/3 signaling in the aortas of fibrillin-1 (Fbn1)-null mice. J Biol Chem 284:5630–5636

Chung AW, Yang HH, Radomski MW et al (2008) Long-term doxycycline is more effective than atenolol to prevent thoracic aortic aneurysm in marfan syndrome through the inhibition of matrix metalloproteinase-2 and -9. Circ Res 102:e73–e85

Cirulli A, Loria MP, Dambra P et al (2006) Patients with hereditary hemorrhagic Telangectasia (HHT) exhibit a deficit of polymorphonuclear cell and monocyte oxidative burst and phagocytosis: a possible correlation with altered adaptive immune responsiveness in HHT. Curr Pharm Des 12:1209–1215

Coady MA, Davies RR, Roberts M et al (1999) Familial patterns of thoracic aortic aneurysms. Arch Surg 134:361–367

Cohen JH, Faughnan ME, Letarte M et al (2005) Cost comparison of genetic and clinical screening in families with hereditary hemorrhagic telangiectasia. Am J Med Genet A 137:153–160

Cole SG, Begbie ME, Wallace GM et al (2005) A new locus for hereditary haemorrhagic telangiectasia (HHT3) maps to chromosome 5. J Med Genet 42:577–582

Cortini F, Villa C, Marinelli B et al (2019) Understanding the basis of Ehlers-Danlos syndrome in the era of the next-generation sequencing. Arch Dermatol Res 311:265–275

Costanza B, Umelo IA, Bellier J et al (2017) Stromal modulators of TGF-beta in Cancer. J Clin Med 6

Cottin V, Plauchu H, Bayle JY et al (2004) Pulmonary arteriovenous malformations in patients with hereditary hemorrhagic telangiectasia. Am J Respir Crit Care Med 169:994–1000

Coucke PJ, Willaert A, Wessels MW et al (2006) Mutations in the facilitative glucose transporter GLUT10 alter angiogenesis and cause arterial tortuosity syndrome. Nat Genet 38:452–457

Daugherty A, Manning MW, Cassis LA (2001) Antagonism of AT2 receptors augments angiotensin II-induced abdominal aortic aneurysms and atherosclerosis. Br J Pharmacol 134:865–870

Davies RJ, Holmes AM, Deighton J et al (2012) BMP type II receptor deficiency confers resistance to growth inhibition by TGF-beta in pulmonary artery smooth muscle cells: role of proinflammatory cytokines. Am J Physiol Lung Cell Mol Physiol 302:L604–L615

Dietz H (1993) Marfan Syndrome. In: Adam MP, Ardinger HH, Pagon RA, Wallace SE, Bean LJH, Mirzaa G, Amemiya A (eds) GeneReviews((R)). Seattle (WA)

Dietz HC, Pyeritz RE (1995) Mutations in the human gene for fibrillin-1 (FBN1) in the Marfan syndrome and related disorders. Hum Mol Genet 4:1799–1809

Droguett R, Cabello-Verrugio C, Riquelme C et al (2006) Extracellular proteoglycans modify TGF-beta bioavailability attenuating its signaling during skeletal muscle differentiation. Matrix biology : journal of the International Society for Matrix Biology 25:332–341

Ebisawa T, Fukuchi M, Murakami G et al (2001) Smurf1 interacts with transforming growth factor-beta type I receptor through Smad7 and induces receptor degradation. J Biol Chem 276:12477–12480

Everett AD, Tufro-Mcreddie A, Fisher A et al (1994) Angiotensin receptor regulates cardiac hypertrophy and transforming growth factor-beta 1 expression. Hypertension 23:587–592

Faughnan ME, Palda VA, Garcia-Tsao G et al (2011) International guidelines for the diagnosis and management of hereditary haemorrhagic telangiectasia. J Med Genet 48:73–87

Feng XH, Derynck R (2005) Specificity and versatility in TGF-beta signaling through Smads. Annu Rev Cell Dev Biol 21:659–693

Fernandez LA, Sanz-Rodriguez F, Zarrabeitia R et al (2005) Blood outgrowth endothelial cells from hereditary Haemorrhagic telangiectasia patients reveal abnormalities compatible with vascular lesions. Cardiovasc Res 68:235–248

Fernandez LA, Sanz-Rodriguez F, Blanco FJ et al (2006) Hereditary hemorrhagic telangiectasia, a vascular dysplasia affecting the TGF-beta signaling pathway. Clin Med Res 4:66–78

Fukuda N, Hu WY, Kubo A et al (2000) Angiotensin II upregulates transforming growth factor-beta type I receptor on rat vascular smooth muscle cells. Am J Hypertens 13:191–198

Fulbright RK, Chaloupka JC, Putman CM et al (1998) MR of hereditary hemorrhagic telangiectasia: prevalence and spectrum of cerebrovascular malformations. AJNR Am J Neuroradiol 19:477–484

Gensemer C, Burks R, Kautz S et al (2020) Hypermobile Ehlers-Danlos syndromes: complex phenotypes, challenging diagnoses, and poorly understood causes. Dev Dyn

Gentry LE, Lioubin MN, Purchio AF et al (1988) Molecular events in the processing of recombinant type 1 pre-pro-transforming growth factor beta to the mature polypeptide. Mol Cell Biol 8:4162–4168

Girdauskas E, Schulz S, Borger MA et al (2011) Transforming growth factor-beta receptor type II mutation in a patient with bicuspid aortic valve disease and intraoperative aortic dissection. Ann Thorac Surg 91:e70–e71

Gleason TG (2005) Heritable disorders predisposing to aortic dissection. Semin Thorac Cardiovasc Surg 17:274–281

Goumans MJ, Valdimarsdottir G, Itoh S et al (2003) Activin receptor-like kinase (ALK)1 is an antagonistic mediator of lateral TGFbeta/ALK5 signaling. Mol Cell 12:817–828

Govani FS, Shovlin CL (2009) Hereditary haemorrhagic telangiectasia: a clinical and scientific review. Eur J Hum Genet 17:860–871

Greally MT (1993) Shprintzen-Goldberg Syndrome. In: Adam MP, Ardinger HH, Pagon RA, Wallace SE, Bean LJH, Mirzaa G, Amemiya A (eds) GeneReviews((R)). Seattle (WA)
Greally MT, Carey JC, Milewicz DM et al (1998) Shprintzen-Goldberg syndrome: a clinical analysis. Am J Med Genet 76:202–212
Guo D, Hasham S, Kuang SQ et al (2001) Familial thoracic aortic aneurysms and dissections: genetic heterogeneity with a major locus mapping to 5q13-14. Circulation 103:2461–2468
Habashi JP, Judge DP, Holm TM et al (2006) Losartan, an AT1 antagonist, prevents aortic aneurysm in a mouse model of Marfan syndrome. Science 312:117–121
Hata A, Lagna G, Massague J et al (1998) Smad6 inhibits BMP/Smad1 signaling by specifically competing with the Smad4 tumor suppressor. Genes Dev 12:186–197
Hawinkels LJ, Ten Dijke P (2011) Exploring anti-TGF-beta therapies in cancer and fibrosis. Growth Factors 29:140–152
Hynes RO (2009) The extracellular matrix: not just pretty fibrils. Science 326:1216–1219
Ignotz RA, Massague J (1986) Transforming growth factor-beta stimulates the expression of fibronectin and collagen and their incorporation into the extracellular matrix. J Biol Chem 261:4337–4345
Imamura T, Takase M, Nishihara A et al (1997) Smad6 inhibits signalling by the TGF-beta superfamily. Nature 389:622–626
Ishibashi M, Egashira K, Zhao Q et al (2004) Bone marrow-derived monocyte chemoattractant protein-1 receptor CCR2 is critical in angiotensin II-induced acceleration of atherosclerosis and aneurysm formation in hypercholesterolemic mice. Arterioscler Thromb Vasc Biol 24:e174–e178
Isogai Z, Ono RN, Ushiro S et al (2003) Latent transforming growth factor beta-binding protein 1 interacts with fibrillin and is a microfibril-associated protein. J Biol Chem 278:2750–2757
Johnson DW, Berg JN, Baldwin MA et al (1996) Mutations in the activin receptor-like kinase 1 gene in hereditary haemorrhagic telangiectasia type 2. Nat Genet 13:189–195
Jones JA, Ikonomidis JS (2010) The pathogenesis of aortopathy in Marfan syndrome and related diseases. Curr Cardiol Rep 12:99–107
Jones JA, Barbour JR, Stroud RE et al (2008) Altered transforming growth factor-beta signaling in a murine model of thoracic aortic aneurysm. J Vasc Res 45:457–468
Jones JA, Spinale FG, Ikonomidis JS (2009) Transforming growth factor-beta signaling in thoracic aortic aneurysm development: a paradox in pathogenesis. J Vasc Res 46:119–137
Judge DP, Dietz HC (2005) Marfan's syndrome. Lancet 366:1965–1976
Judge DP, Rouf R, Habashi J et al (2011) Mitral valve disease in Marfan syndrome and related disorders. J Cardiovasc Transl Res 4:741–747
Kavsak P, Rasmussen RK, Causing CG et al (2000) Smad7 binds to Smurf2 to form an E3 ubiquitin ligase that targets the TGF beta receptor for degradation. Mol Cell 6:1365–1375
Kim ES, Kim MS, Moon A (2004) TGF-beta-induced upregulation of MMP-2 and MMP-9 depends on p38 MAPK, but not ERK signaling in MCF10A human breast epithelial cells. Int J Oncol 25:1375–1382
Kimura N, Matsuo R, Shibuya H et al (2000) BMP2-induced apoptosis is mediated by activation of the TAK1-p38 kinase pathway that is negatively regulated by Smad6. J Biol Chem 275:17647–17652
Kjeldsen AD, Vase P, Green A (1999) Hereditary haemorrhagic telangiectasia: a population-based study of prevalence and mortality in Danish patients. J Intern Med 245:31–39
Kosaki K, Takahashi D, Udaka T et al (2006) Molecular pathology of Shprintzen-Goldberg syndrome. Am J Med Genet A 140:104–108. author reply 109-110
Kwak HJ, Park MJ, Cho H et al (2006) Transforming growth factor-beta1 induces tissue inhibitor of metalloproteinase-1 expression via activation of extracellular signal-regulated kinase and Sp1 in human fibrosarcoma cells. Mol Cancer Res 4:209–220
Laiho M, Saksela O, Andreasen PA et al (1986) Enhanced production and extracellular deposition of the endothelial-type plasminogen activator inhibitor in cultured human lung fibroblasts by transforming growth factor-beta. J Cell Biol 103:2403–2410
Lawrence DA, Pircher R, Kryceve-Martinerie C et al (1984) Normal embryo fibroblasts release transforming growth factors in a latent form. J Cell Physiol 121:184–188
Lee MK, Pardoux C, Hall MC et al (2007) TGF-beta activates Erk MAP kinase signalling through direct phosphorylation of ShcA. EMBO J 26:3957–3967
Letarte M, Mcdonald ML, Li C et al (2005) Reduced endothelial secretion and plasma levels of transforming growth factor-beta1 in patients with hereditary hemorrhagic telangiectasia type 1. Cardiovasc Res 68:155–164
Li C, Flynn RS, Grider JR et al (2013) Increased activation of latent TGF-beta1 by alphaVbeta3 in human Crohn's disease and fibrosis in TNBS colitis can be prevented by cilengitide. Inflamm Bowel Dis 19:2829–2839
Loeys B, De Paepe A (2008) New insights in the pathogenesis of aortic aneurysms. Verh K Acad Geneeskd Belg 70:69–84
Loeys BL, Chen J, Neptune ER et al (2005) A syndrome of altered cardiovascular, craniofacial, neurocognitive and skeletal development caused by mutations in TGFBR1 or TGFBR2. Nat Genet 37:275–281
Loeys BL, Schwarze U, Holm T et al (2006) Aneurysm syndromes caused by mutations in the TGF-beta receptor. N Engl J Med 355:788–798
Lopez-Casillas F, Payne HM, Andres JL et al (1994) Betaglycan can act as a dual modulator of TGF-beta access to signaling receptors: mapping of ligand binding and GAG attachment sites. J Cell Biol 124:557–568

Macri A, Wilson AM, Shafaat O et al. (2020) Osler-Weber-Rendu Disease. In: StatPearls. Treasure Island (FL)

Mageed RA, Prud'homme GJ (2003) Immunopathology and the gene therapy of lupus. Gene Ther 10:861–874

Malfait F, Francomano C, Byers P et al (2017) The 2017 international classification of the Ehlers-Danlos syndromes. Am J Med Genet C Semin Med Genet 175:8–26

Massague J (2000) How cells read TGF-beta signals. Nat Rev Mol Cell Biol 1:169–178

Mcallister KA, Grogg KM, Johnson DW et al (1994) Endoglin, a TGF-beta binding protein of endothelial cells, is the gene for hereditary haemorrhagic telangiectasia type 1. Nat Genet 8:345–351

Milewicz DM, Chen H, Park ES et al (1998) Reduced penetrance and variable expressivity of familial thoracic aortic aneurysms/dissections. Am J Cardiol 82:474–479

Milewicz DM, Carlson AA, Regalado ES (2010) Genetic testing in aortic aneurysm disease: PRO. Cardiol Clin 28:191–197

Morissette R, Merke DP, Mcdonnell NB (2014) Transforming growth factor-beta (TGF-beta) pathway abnormalities in tenascin-X deficiency associated with CAH-X syndrome. Eur J Med Genet 57:95–102

Moustakas A, Heldin CH (2005) Non-Smad TGF-beta signals. J Cell Sci 118:3573–3584

Naito T, Masaki T, Nikolic-Paterson DJ et al (2004) Angiotensin II induces thrombospondin-1 production in human mesangial cells via p38 MAPK and JNK: a mechanism for activation of latent TGF-beta1. Am J Physiol Renal Physiol 286:F278–F287

Neptune ER, Frischmeyer PA, Arking DE et al (2003) Dysregulation of TGF-beta activation contributes to pathogenesis in Marfan syndrome. Nat Genet 33:407–411

Ng CM, Cheng A, Myers LA et al (2004) TGF-beta-dependent pathogenesis of mitral valve prolapse in a mouse model of Marfan syndrome. J Clin Invest 114:1586–1592

Nicod P, Bloor C, Godfrey M et al (1989) Familial aortic dissecting aneurysm. J Am Coll Cardiol 13:811–819

Oh SP, Seki T, Goss KA et al (2000) Activin receptor-like kinase 1 modulates transforming growth factor-beta 1 signaling in the regulation of angiogenesis. Proc Natl Acad Sci U S A 97:2626–2631

Ohtsuka K, Gray JD, Stimmler MM et al (1998) Decreased production of TGF-beta by lymphocytes from patients with systemic lupus erythematosus. J Immunol 160:2539–2545

Pannu H, Tran-Fadulu V, Milewicz DM (2005) Genetic basis of thoracic aortic aneurysms and aortic dissections. Am J Med Genet C Semin Med Genet 139:10–16

Park SH (2005) Fine tuning and cross-talking of TGF-beta signal by inhibitory Smads. J Biochem Mol Biol 38:9–16

Pasculli G, Quaranta D, Lenato GM et al (2005) Capillaroscopy of the dorsal skin of the hands in hereditary hemorrhagic telangiectasia. QJM 98:757–763

Patsenker E, Popov Y, Stickel F et al (2009) Pharmacological inhibition of integrin alphavbeta3 aggravates experimental liver fibrosis and suppresses hepatic angiogenesis. Hepatology 50:1501–1511

Pereira L, Lee SY, Gayraud B et al (1999) Pathogenetic sequence for aneurysm revealed in mice underexpressing fibrillin-1. Proc Natl Acad Sci U S A 96:3819–3823

Piantanida M, Buscarini E, Dellavecchia C et al (1996) Hereditary haemorrhagic telangiectasia with extensive liver involvement is not caused by either HHT1 or HHT2. J Med Genet 33:441–443

Pignolo RJ, Shore EM, Kaplan FS (2011) Fibrodysplasia ossificans progressiva: clinical and genetic aspects. Orphanet J Rare Dis 6:80

Plauchu H, De Chadarevian JP, Bideau A et al (1989) Age-related clinical profile of hereditary hemorrhagic telangiectasia in an epidemiologically recruited population. Am J Med Genet 32:291–297

Pomianowski P, Elefteriades JA (2013) The genetics and genomics of thoracic aortic disease. Ann Cardiothorac Surg 2:271–279

Pope MK, Ratajska A, Johnsen H et al (2019) Diagnostics of hereditary connective tissue disorders by genetic next-generation sequencing. Genet Test Mol Biomarkers 23:783–790

Pyeritz RE (2016) Recent Progress in understanding the natural and clinical histories of the Marfan syndrome. Trends Cardiovascul Med 26:423-428

Ramirez F, Pereira L (1999) The fibrillins. Int J Biochem Cell Biol 31:255–259

Ramirez F, Rifkin DB (2003) Cell signaling events: a view from the matrix. Matrix Biol 22:101–107

Ramirez F, Rifkin DB (2009) Extracellular microfibrils: contextual platforms for TGFbeta and BMP signaling. Curr Opin Cell Biol 21:616–622

Reed NI, Jo H, Chen C et al. (2015) The alphavbeta1 integrin plays a critical in vivo role in tissue fibrosis. Science translational medicine 7:288ra279

Rico MC, Manns JM, Driban JB et al (2008) Thrombospondin-1 and transforming growth factor beta are pro-inflammatory molecules in rheumatoid arthritis. Transl Res 152:95–98

Rico MC, Rough JJ, Del Carpio-Cano FE et al (2010) The axis of thrombospondin-1, transforming growth factor beta and connective tissue growth factor: an emerging therapeutic target in rheumatoid arthritis. Curr Vasc Pharmacol 8:338–343

Robinson PN, Neumann LM, Demuth S et al (2005) Shprintzen-Goldberg syndrome: fourteen new patients and a clinical analysis. Am J Med Genet A 135:251–262

Rodriguez-Vita J, Sanchez-Lopez E, Esteban V et al (2005) Angiotensin II activates the Smad pathway in vascular smooth muscle cells by a transforming growth factor-beta-independent mechanism. Circulation 111:2509–2517

Roman BL, Pham VN, Lawson ND et al (2002) Disruption of acvrl1 increases endothelial cell number in zebrafish cranial vessels. Development 129:3009–3019

Roth P, Silginer M, Goodman SL et al (2013) Integrin control of the transforming growth factor-beta pathway in glioblastoma. Brain 136:564–576

Seki T, Yun J, Oh SP (2003) Arterial endothelium-specific activin receptor-like kinase 1 expression suggests its role in arterialization and vascular remodeling. Circ Res 93:682–689

Selvamurugan N, Kwok S, Alliston T et al (2004) Transforming growth factor-beta 1 regulation of collagenase-3 expression in osteoblastic cells by cross-talk between the Smad and MAPK signaling pathways and their components, Smad2 and Runx2. J Biol Chem 279:19327–19334

Shi Y, Massague J (2003) Mechanisms of TGF-beta signaling from cell membrane to the nucleus. Cell 113:685–700

Shovlin CL, Guttmacher AE, Buscarini E et al (2000) Diagnostic criteria for hereditary hemorrhagic telangiectasia (Rendu-Osler-Weber syndrome). Am J Med Genet 91:66–67

Shovlin CL, Sulaiman NL, Govani FS et al (2007) Elevated factor VIII in hereditary haemorrhagic telangiectasia (HHT): association with venous thromboembolism. Thromb Haemost 98:1031–1039

Shovlin CL, Jackson JE, Bamford KB et al (2008) Primary determinants of ischaemic stroke/brain abscess risks are independent of severity of pulmonary arteriovenous malformations in hereditary haemorrhagic telangiectasia. Thorax 63:259–266

Shprintzen RJ, Goldberg RB (1982) A recurrent pattern syndrome of craniosynostosis associated with arachnodactyly and abdominal hernias. J Craniofac Genet Dev Biol 2:65–74

Sood S, Eldadah ZA, Krause WL et al (1996) Mutation in fibrillin-1 and the Marfanoid-craniosynostosis (Shprintzen-Goldberg) syndrome. Nat Genet 12:209–211

Sorrentino A, Thakur N, Grimsby S et al (2008) The type I TGF-beta receptor engages TRAF6 to activate TAK1 in a receptor kinase-independent manner. Nat Cell Biol 10:1199–1207

Stander M, Naumann U, Wick W et al (1999) Transforming growth factor-beta and p-21: multiple molecular targets of decorin-mediated suppression of neoplastic growth. Cell Tissue Res 296:221–227

Stefansson S, Petitclerc E, Wong MK et al (2001) Inhibition of angiogenesis in vivo by plasminogen activator inhibitor-1. J Biol Chem 276:8135–8141

Stheneur C, Collod-Beroud G, Faivre L et al (2008) Identification of 23 TGFBR2 and 6 TGFBR1 gene mutations and genotype-phenotype investigations in 457 patients with Marfan syndrome type I and II, Loeys-Dietz syndrome and related disorders. Hum Mutat 29:E284–E295

Superti-Furga A, Gugler E, Gitzelmann R et al (1988) Ehlers-Danlos syndrome type IV: a multi-exon deletion in one of the two COL3A1 alleles affecting structure, stability, and processing of type III procollagen. J Biol Chem 263:6226–6232

Sylvan J, Greenberg RK, Mastracci TM et al (2013) Treatment of a patient with vertebral and subclavian aneurysms in the setting of a TGFBR2 mutation. J Vasc Surg 57:1116–1119

Sys L, Van Den Hoogen FJ (2005) Rendu-Osler-Weber disease. Ned Tijdschr Tandheelkd 112:336–339

Trembath RC, Thomson JR, Machado RD et al (2001) Clinical and molecular genetic features of pulmonary hypertension in patients with hereditary hemorrhagic telangiectasia. N Engl J Med 345:325–334

Van De Laar IM, Oldenburg RA, Pals G et al (2011) Mutations in SMAD3 cause a syndromic form of aortic aneurysms and dissections with early-onset osteoarthritis. Nat Genet 43:121–126

Van De Laar IM, Van Der Linde D, Oei EH et al (2012) Phenotypic spectrum of the SMAD3-related aneurysms-osteoarthritis syndrome. J Med Genet 49:47–57

Van Steensel MA, Van Geel M, Parren LJ et al (2008) Shprintzen-Goldberg syndrome associated with a novel missense mutation in TGFBR2. Exp Dermatol 17:362–365

Von Kodolitsch Y, Nienaber CA, Dieckmann C et al (2004) Chest radiography for the diagnosis of acute aortic syndrome. Am J Med 116:73–77

Wahl SM, Chen W (2005) Transforming growth factor-beta-induced regulatory T cells referee inflammatory and autoimmune diseases. Arthritis Res Ther 7:62–68

Wallace GM, Shovlin CL (2000) A hereditary haemorrhagic telangiectasia family with pulmonary involvement is unlinked to the known HHT genes, endoglin and ALK-1. Thorax 55:685–690

Werner S, Alzheimer C (2006) Roles of activin in tissue repair, fibrosis, and inflammatory disease. Cytokine Growth Factor Rev 17:157–171

West J, Fagan K, Steudel W et al (2004) Pulmonary hypertension in transgenic mice expressing a dominant-negative BMPRII gene in smooth muscle. Circ Res 94:1109–1114

Wicks SJ, Grocott T, Haros K et al (2006) Reversible ubiquitination regulates the Smad/TGF-beta signalling pathway. Biochem Soc Trans 34:761–763

Wilkes MC, Mitchell H, Penheiter SG et al (2005) Transforming growth factor-beta activation of phosphatidylinositol 3-kinase is independent of Smad2 and Smad3 and regulates fibroblast responses via p21-activated kinase-2. Cancer Res 65:10431–10440

Wolf G, Ziyadeh FN, Stahl RA (1999) Angiotensin II stimulates expression of transforming growth factor beta receptor type II in cultured mouse proximal tubular cells. J Mol Med 77:556–564

Wrana JL, Attisano L, Carcamo J et al (1992) TGF beta signals through a heteromeric protein kinase receptor complex. Cell 71:1003–1014

Xiong W, Knispel RA, Dietz HC et al (2008) Doxycycline delays aneurysm rupture in a mouse model of Marfan syndrome. J Vasc Surg 47:166–172. discussion 172

Xiong W, Meisinger T, Knispel R et al (2012) MMP-2 regulates Erk1/2 phosphorylation and aortic dilatation in Marfan syndrome. Circ Res 110:e92–e101

Yamashita M, Fatyol K, Jin C et al (2008) TRAF6 mediates Smad-independent activation of JNK and p38 by TGF-beta. Mol Cell 31:918–924

Yan C, Boyd DD (2007) Regulation of matrix metalloproteinase gene expression. J Cell Physiol 211:19–26

You HJ, Bruinsma MW, How T et al (2007) The type III TGF-beta receptor signals through both Smad3 and the p38 MAP kinase pathways to contribute to inhibition of cell proliferation. Carcinogenesis 28:2491–2500

Yuan X, Downing AK, Knott V et al (1997) Solution structure of the transforming growth factor beta-binding protein-like module, a domain associated with matrix fibrils. EMBO J 16:6659–6666

Zeyer KA, Reinhardt DP (2015) Fibrillin-containing microfibrils are key signal relay stations for cell function. J Cell Commun Signal 9:309–325

Pathophysiology and Pathogenesis of Marfan Syndrome

8

Sanford M. Zeigler, Brandon Sloan, and Jeffrey A. Jones

Abstract

Marfan syndrome (MFS) is a systemic connective tissue disorder that is inherited in an autosomal dominant pattern with variable penetrance. While clinically this disease manifests in many different ways, the most life-threatening manifestations are related to cardiovascular complications including mitral valve prolapse, aortic insufficiency, dilatation of the aortic root, and aortic dissection. In the past 30 years, research efforts have not only identified the genetic locus responsible but have begun to elucidate the molecular pathogenesis underlying this disorder, allowing for the development of seemingly rational therapeutic strategies for treating affected individuals. In spite of these advancements, the cardiovascular complications still remain as the most life-threatening clinical manifestations. The present chapter will focus on the pathophysiology and clinical treatment of Marfan syndrome, providing an updated overview of the recent advancements in molecular genetics research and clinical trials, with an emphasis on how this information can focus future efforts toward finding betters ways to detect, diagnose, and treat this devastating condition.

Keywords

Aorta · Aneurysm · Extracellular matrix · Collagen · Metalloproteinase · Thoracic aortic aneurysm and dissection syndrome · Marfan syndrome (MFS) · Loeys-Dietz syndrome (LDS) · Transforming growth factor β (TGF-β) · Endoglin · Mitral valve · SMAD · TGF-β receptor · Mitogen-activated protein kinase (MAPK) · Extracellular signal related kinase (ERK) · Fibrillin · Genetic testing · Losartan · Ghent nosology · β-Blockers · Angiotensin receptor blockers (ARBs) · Fbn1C1039G/+ · mgR/mgR

Abbreviations

AT1R	Ang-II type-I specific receptor
ARB	Angiotensin receptor blocker
AIMS	Aortic Irbesartan Marfan Study

S. M. Zeigler · B. Sloan
Division of Cardiothoracic Surgery, Department of Surgery, Medical University of South Carolina, Charleston, SC, USA
e-mail: zeigle@musc.edu; sloanbr@musc.edu

J. A. Jones (✉)
Division of Cardiothoracic Surgery, Medical University of South Carolina and Ralph H. Johnson Veterans Affairs Medical Center, Charleston, SC, USA
e-mail: jonesja@musc.edu

J. Halper (ed.), *Progress in Heritable Soft Connective Tissue Diseases*, Advances in Experimental Medicine and Biology 1348, https://doi.org/10.1007/978-3-030-80614-9_8

AHT	Arterial hypertension
cbEGF	Calcium-binding EGF
COPD	Chronic obstructive pulmonary disease
CVG	Composite valved-graft
CCA	Congenital contractural arachnodactyly
CNV	Copy number variant
EGF	Epidermal growth factor
ECM	Extracellular matrix
EKR	Extracellular signal related kinase
FBN1	Fibrillin-1
HTAD	Heritable thoracic aortic disease
LLC	Large latent complex
LAP	Latency associated peptide
LTBP	Latent transforming growth factor binding protein
LDS	Loeys-Dietz syndrome
MFS	Marfan syndrome
MVPS	Mitral valve prolapse syndrome
MAPK	Mitogen-activated protein kinase
NGS	Next generation sequencing
PHN	Pediatric heart network
PEARS	Personalized external aortic root support
PTC	Premature termination codon
SLC	Small latent complex
TEVAR	Thoracic endovascular aortic repair
TGF-β	Transforming growth factor-β
VSARR	Valve-sparing root replacement
VUS	Variant of unknown significance
WES	Whole-exome sequencing
WGS	Whole-genome sequencing

8.1 Introduction

8.1.1 History and Overview

Approximately 125 years ago (1896), French pediatrician Antoine Bernard-Jean Marfan published a case report describing a 5½-year-old girl (Gabrielle P.) with extraordinary musculoskeletal abnormalities that included severe scoliosis and fibrous contractures of the fingers, in the absence of ocular and cardiac abnormalities (Marfan 1896). While much later, based on current diagnostic criteria (Beal and Hecht 1971), it was suggested that Gabrielle may have actually had Congenital Contractural Arachnodactyly (CCA) in place of MFS, this early observation nonetheless, laid the foundation for the description of a genetic syndrome that has come to be known as Marfan Syndrome (MFS; OMIM #154700). Interestingly, the first description of this disorder was likely made 20-years earlier by E. Williams, an Ophthalmologist from Cincinnati, OH. Williams published a case report in the Transactions of the American Ophthalmological Society, describing a brother-sister pair with ectopia lentis who both were extraordinarily tall and "loose-jointed" from birth (Williams 1876). It has been suggested that his keen observation went largely unnoticed because he did not practice at an academic center. Since that time, key discoveries have added to the symptomology and clinical description of this multisystem connective tissue disorder, providing better discrimination between similar heritable conditions. It wasn't until 1914 that Boerger, officially added ectopia lentis to the growing list of other MFS-related symptoms (Boerger 1914). The first cardiovascular manifestations were described by Salle in 1912. He reported the results of a necropsy following the death of a 2½ month old infant diagnosed with MFS, that displayed "*striking changes in the mitral valve*" (Salle 1912). A literature review written by Rados in 1942, supported this early observation by Salle, describing the presence and prevalence of mitral valve regurgitation ("*click-murmur syndrome*") in the MFS patient population (Rados 1942). The first descriptions of aortic aneurysm and dissection were recorded in 1943 by Etter and Glover (Etter and Glover 1943), and Baer and colleagues (Baer et al. 1943). This was followed by key advancements in 1955 by Victor McKusick, defining the heritable nature of MFS and its sequelae, along with the new cardiovascular aspects, including aortic root dilatation and aortic regurgitation (McKusick 1955a, b). Subsequently, this was followed by a postmortem study completed by Goyette and Palmer in 1963, in which they confirmed that evidence of cystic medial necrosis in the ascending aortic aneurysm, aortic insufficiency, and aortic dissection were prevalent in MFS patients (Goyette and Palmer 1963). In 1975, Brown and coworkers reported the first study examining MFS-related cardiovascular manifestations using cardiac ultrasound to record noninvasive quantitation of aortic size and mitral valve function (Brown et al. 1975). This pushed the limits of echocardiographic studies during a time when standardized criteria for measuring the aortic root had yet to be established, resulting in a wide "normal" range, suggesting that mild aortic root dilatation may have gone undiagnosed in many MFS patients, particularly children. It wasn't until the early 1990's that the genetic locus responsible for MFS was identified (Dietz et al. 1991a, b, 1993). Work by Dietz and Pyeritz localized key mutations to the gene encoding a large extracellular matrix structural protein, called fibrillin-1 (FBN1) (Dietz and Pyeritz 1995). For more than a century, many dedicated geneticists, physicians, and scientists have contributed toward understanding the mechanistic and clinical pathogenesis of this multisystem connective tissue disorder. We now know that MFS is inherited in an autosomal dominant manner with variable penetrance, affecting 1 in 3–5000 people without preference for ethnicity or geographic location (Pyeritz 2016). This connective tissue disorder affects multiple organ systems including the skeletal, pulmonary, ocular, central nervous, and cardiovascular systems of the body (Cook and Ramirez 2014; Pyeritz 2016).

8.1.2 Identification of the Primary Genetic Defect

The primary gene defect in MFS lies on chromosome 15q21.1, within the coding region of the fibrillin-1 gene (FBN1) (Dietz et al. 2005). Fibrillin-1 is a large 350 kDa glycoprotein, produced and secreted by fibroblasts, and incorporated into the extracellular matrix (ECM) as insoluble microfibrils (Kielty et al. 2005). These microfibrils serve as a scaffold for the deposition of elastin and are necessary to build proper elastic architecture to provide elasticity to dynamic connective tissues (Kielty et al. 2002). Structurally, each Fibrillin-1 monomer contains multiple epidermal growth factor (EGF)-like motifs that are arranged in tandem orientation. There is a total of 47 EGF-like motifs in all, 43 of which contain specialized calcium binding sequences, termed calcium-binding EGF (cbEGF) repeats (Yuan et al. 1997). These cbEGF modules sequester extracellular calcium in order to protect against fibrillin proteolysis, promote interactions between fibrillin monomers and other cellular components such as the integrins (e.g. $\alpha v\beta 3$), and serve to stabilize the structure of the microfibrils, which favors lateral packing (Ramirez and Pereira 1999; Ramirez et al. 2004). The tandem arrays of cbEGF motifs (1 to 12 repeats) are separated by seven 8-cys/TB modules. These modules have high homology to the latent transforming growth factor binding proteins (LTBPs), and are characterized by eight highly conserved cysteine residues, including a unique cysteine triplet (Ramirez and Pereira 1999; Yuan et al. 1997). Fibrillin monomers self-assemble into macroaggregates that form the basic structure on which mature elastin fibers are synthesized from tropoelastin subunits. Accordingly, it was originally hypothesized that mutations in fibrillin-1 resulted in weak and disordered elastic fiber formation, causing a disruption of the microfibril network connections to the adjacent interstitial cells (Pereira et al. 1999). In the aorta, it was suggested that this gave rise to a weakened vascular wall, prone to dilation and dissection; commonly observed in patients with MFS.

Contrary to the "weakened constitution" hypothesis, Dietz and colleagues suggested that MFS syndrome is caused by more than just a disordered microfibril matrix (Dietz et al. 2005). In addition to directing elastogenesis and providing structural integrity to the elastic lamellae, the

8-cys/TB domains of fibrillin are involved in the sequestration of latent transforming growth factor-beta (TGF-β). TGF-β is synthesized as a pre-pro-polypeptide that is cleaved in a post-Golgi compartment to yield a mature growth factor molecule complexed with an inactive cleavage fragment, termed the latency-associated peptide (LAP). Homodimers of the mature TGF-β and the LAP then uniquely combine and form a tight biologically inactive complex called the small latent complex (SLC). The SLC covalently binds to LTBP by forming disulfide bonds between cysteine residues in the LAP and a cysteine rich motif in the LTBP (Annes et al. 2003; Chaudhry et al. 2007; Ramirez et al. 2004). This large latent complex (LLC) is then secreted from the cell and functions to target latent-TGF-β to the ECM, specifically to fibrillin microfibrils, and the 8-cys/TB motifs they contain (Charbonneau et al. 2004; Isogai et al. 2003). This complex sequestration process serves a dual role in the regulation of TGF-β signaling. First, the LAP sequesters active TGF-β, rendering it inactive, preventing unregulated stimulation of the TGF-β signaling pathway. Second, interactions of the LLC with microfibrillar proteins, localizes and concentrates the inactive latent TGF-β at critical signaling sites within the ECM, making it available for rapid release and activation when required by the cell. As a result, when Dietz and coworkers localized multiple mutations within the FBN1 gene to sites that were associated with sequestering the LLC (Dietz and Pyeritz 1995), they hypothesized that fibrillin-1-dependent abnormalities observed in MFS may lead to impaired sequestration of latent TGF-β complexes and enhanced TGF-β signaling (Dietz et al. 2005).

In support of the Dietz hypothesis, evidence reported by Isogai and coworkers, demonstrated that fibrillin-1 and LTBP-1 were directly associated *in vivo*, and that the interaction was mediated by the cbEGF domains common to both proteins (Isogai et al. 2003). Chaudhry and coworkers identified a peptide of fibrillin-1 (encoded in exons 44–49) that was capable disrupting the interaction between fibrillin-1 and the C-terminal domain of LTBP-1 (Chaudhry et al. 2007). They demonstrated that the release of the LLC led to enhanced TGF-β signaling as determined by the phosphorylation Smad-2. While their work defined a physiological mechanism by which the LLC could be released from fibrillin-1, it did not address how TGF-β was released from the SLC. Studies by Ge et al. demonstrated that bone morphogenetic protein 1 (BMP1)-like metalloproteinases were capable of hydrolyzing LTBP-1 at two independent sites (Ge and Greenspan 2006). This resulted in the release of the SLC and subsequent activation of TGF-β via further cleavage of the LAP by non-BMP1-like proteinases. In similar fashion, work by Tatti *ei al.* demonstrated that LTBP-1 could be hydrolyzed by the action of membrane type-1 matrix metalloproteinase (Tatti et al. 2008). These studies suggested that proteolysis of LTBP-1 was sufficient to release the SLC, which was then subsequently acted upon by other extracellular proteases to release active TGF-β in a non-regulated fashion. In addition to proteolysis of the SLC, a number of other mechanisms have been shown to catalyze the release TGF-β or modify its association with the LAP by rendering it active; these include non-proteolytic dissociation by interaction with thrombospondin or integrin αvβ6, and release from the LAP by exposure to acidic or oxidative stress (reviewed in (Annes et al. 2003)).

Taken together, the missense mutations identified in the FBN1 gene, have been shown to alter the interaction between FBN1 and the TGF-β-LLC, resulting in aberrant sequestration of latent TGF-β. Upon further proteolytic action, LTBP releases the bound SLC, which subsequently releases activated homodimers of TGF-β that are capable of binding and activating the TGF-β signaling pathway. As a result of these biochemical studies, many of the clinical manifestations of MFS were looked at in a new light, focusing on how a weakened fibrillar elastic matrix could be exacerbated by enhanced TGF-β signaling.

8.2 Diagnosis and Clinical Presentation

8.2.1 Diagnostic Criteria

Diagnosis over the years often relied heavily on the treating physician to recognize the evidence of multiple clinical features scattered across multiple organ systems. Although these common clinical features do indeed appear in most patients, the variable penetrance of the genetic defect lead to inconsistencies in the time of onset, sometimes making a definitive diagnosis difficult. To improve diagnostic ability, an international clinical congress met in 1986 to formalize and categorize the common clinical diagnostic criteria; these criteria were thus termed the Berlin nosology (Beighton et al. 1988). Over time weaknesses were discovered in these criteria, which were further illuminated by the addition of novel molecular testing. As a result, the criteria were revised to include requirements for diagnosis of a first degree relative, skeletal involvement in 4 of 8 typical skeletal manifestations, any potential contribution from molecular diagnostics, and delineation of criteria for other heritable conditions with partial overlapping phenotypes (De Paepe et al. 1996). This set of diagnostic standards became known as the revised Ghent nosology. As information continues to be gathered about disease presentation, other heritable connective tissue disorders, and advanced molecular diagnostics, our capacity to discriminate similar but genetically and phenotypically unique diseases away from MFS has improved. The most recent iteration of the diagnostic criteria is known as the 2010 revised Ghent nosology (Table 8.1) (Loeys et al. 2010). This version of the diagnostic criteria relies heavily on the presence of aortic root aneurysm, ectopia lentis, family history, and genetics, but still requires a combination of two major criteria in different body systems, plus a minor criterion in a third body system. Furthermore, the revised criteria have advanced our ability to distinguish MFS from other heritable conditions, potentially avoiding unnecessary diagnostic testing, anxiety, and medical interventions. While the diagnosis of MFS in a patient that had a first degree relative who had previously been diagnosed with MFS was quite simple, patients lacking an affected first degree relative remain a bit more challenging. To facilitate the often-complex diagnostic process, a scoring system was developed (Table 8.2) (Loeys et al. 2010). Accordingly, the presence of an aortic root aneurysm, ectopia lentis, or a systemic score ≥ 7 provides sufficient confidence to make a positive diagnosis.

8.2.2 Clinical Presentation of MFS

Marfan syndrome, without appropriate and timely treatment, leads to a reduced lifespan with death usually in the third to fifth decades of life (Goldfinger et al. 2017). With appropriate medical and surgical care, however, patients can expect a lifespan approaching normal. Early diagnosis, expectant management, and multidisciplinary expertise throughout the patient's lifetime are key to achieving a near normal quantity of life for people with Marfan syndrome. Even so, there are significant detriments to quality of life based on repeat hospitalizations, a lifetime of medical and surgical therapies, as well as pain and disability from the musculoskeletal, ocular, neurologic, pulmonary, and cardiovascular sequelae (Goldfinger et al. 2017).

8.2.2.1 Musculoskeletal Manifestations

Musculoskeletal abnormalities are often the first finding that raises suspicion for MFS. The classic MFS phenotype includes exaggerated long bone growth accompanied by scoliosis, pectus deformities, and increased joint laxity. Aberrant TGF-β signaling, as a result of mutations in FBN1 during bone growth and mineralization, has been implicated in MFS-related osteoporosis and long bone overgrowth (Nistala et al. 2010). Dysregulated osteogenesis and osteoclastic activity lead to excessive growth in both long bones and the axial skeleton. Excessive rib length leads to anterior or posterior displacement of the sternum, causing pectus deformities. Dysregulated skeletal growth leads to other common musculoskeletal findings in MFS, including increased

Table 8.1 *Revised Ghent Nosology*. MASS: myopia, mitral valve prolapse, borderline aortic root dilation, striae, skeletal findings phenotype; MVPS: mitral valve prolapse syndrome

Aortic Root Dimension	Presence of Ectopia Lentis	Genetic Testing	Systemic Score	Other Factors	Diagnosis
Increased aortic dimensions					
Z-score ≥ 2	+	–	n/a		**MFS**
Z-score ≥2		Causal *FBN1* mutation	n/a		**MFS**
Z-score ≥ 2		–	≥ 7		**MFS**
Normal aortic dimensions					
Z-score < 2	+	Causal *FBN1* mutation	n/a		**MFS**
Z-score < 2	+	Non-MFS *FBN1* mutation **OR** No *FBN1* mutation	n/a		**Ectopia Lentis syndrome**
Z-score < 2	No	–	≥ 5		**MASS**
Z-score < 2	No	–	< 5	Mitral valve prolapse	**MVPS**

Note: *1*) aortic root z-score calculators for pediatric and adult populations can be found at (www.marfan.org/dx/home); 2) the presence of an aortic dissection can supplant a Z-Score cutoff for diagnosis of MFS

Table 8.2 *MFS Diagnostic Scoring System*. (www.marfan.org/dx/home). US/LS: Upper segment/lower segment ratio

Diagnostic Scoring System (A score ≥ 7 indicates systemic involvement)			
Features		**Points**	
Wrist AND thumb sign	Wrist **OR** thumb sign	3	1
Pectus carinatum deformity	Pectus excavatum **OR** chest asymmetry	2	1
Hindfoot deformity	Plain pes planus	2	1
Pneumothorax		2	
Dural ectasia		2	
Protrusio acetabuli		2	
Reduced US/LS AND increased arm/height AND no severe scoliosis		1	
Scoliosis or thoracolumbar kyphosis		1	
Reduced elbow extension		1	
Facial features (*3/5 required: Dolichocephaly, enophthalmos, down slanting palpebral fissures, malar hypoplasia, retrognathia*)		1	
Skin striae		1	
Myopia> 3 diopters		1	
Mitral valve prolapse (all types)		1	
Maximum		**20**	

arm span, pes planus, wrist sign, and the thumb sign. Patients with MFS also display a higher fracture rate, presumably due to osteopenia combined with joint instability and deformities that worsen with age (Trifiro et al. 2020). Scoliosis in MFS patients is quite common and can be both disabling and painful. Surgical repair of spinal alignment deformities, however, carries a higher risk for complications when compared to unaffected patients (Kurucan et al. 2019). While it is commonly held that MFS patients are tall and slender, this is not always true. Abnormal height should be evaluated relative to unaffected family members. Additionally, obesity and MFS can co-exist, which can serve to exacerbate joint and spinal deformities, and should likewise be carefully managed

8.2.2.2 Ocular and Craniofacial Manifestations

Visual disturbances are common in Marfan syndrome, related both to abnormal globe morphology and to laxity in the support structures of the lens and retina. The globe can be elongated with a flatter, thinner cornea, often showing posterior displacement within the orbit due to changes in the orbit volume (called endophthalmos), along with an increased distance between the eyes (hypertelorism) (Child 2017). One of the most common manifestations associated with MFS, and possibly the first identified, was thought to be caused by weakened structural filaments in the lens of the eye, likely due to the presence of mutated fibrillin-1 protein. This predisposed patients to ectopia lentis, a dislocation or displacement of the natural crystalline lens outside of the hyaloid fossa, free-floating in the vitreous, present in the anterior chamber, or lying directly on the retina (Clarke 1939). Around 30% of MFS patients will develop ectopia lentis, which is rarely seen outside of this disorder, but can result from facial trauma (Chandra et al. 2014; Gehle et al. 2017). Ectopia lentis is one of the cardinal features in the 2010 Ghent nosology. Myopia and astigmatism are also quite common and can be easily corrected with lenses. Retinal detachment occurs in about 15% of MFS patients (Chandra et al. 2014). Both ectopia lentis and retinal detachment may require ophthalmologic surgery to preserve vision. The majority of ophthalmologic problems manifest before age 30, and close follow up with an ophthalmologist starting in early childhood is recommended. Additional craniofacial findings may include a condition where the head is longer relative to its width (called dolichocephaly), down slanting palpebral fissures of the eye, malocclusion of the maxilla or mandible (retrognathia), a high arched palate, and underdevelopment of the upper jaw (malar hypoplasia)

8.2.2.3 Neurologic Manifestations

Dural Ectasia is dilation of the neural canal, usually in the lumbosacral region of the spinal cord. Though it can be found in a number of other connective tissue disorders, the presence of dural ectasia is another cardinal feature in the 2010 Ghent Nosology. It is commonly associated with MFS and found in more than 80% of patients with a prior diagnosis (Lundby et al. 2009). The dilation may lead to thinning of the cortex of the vertebral bodies and pedicles, and enlargement of the neural foramina, or anterior meningocele (De Paepe et al. 1996). These changes often lead to chronic back pain (Ahn et al. 2000), or to postural headaches related to intracranial hypotension.

8.2.2.4 Pulmonary Manifestations

Spontaneous pneumothorax is the chief pulmonary manifestation of Marfan syndrome, and blebs are commonly seen on radiographic imaging (Child 2017; Neuville et al. 2015). Chronic Obstructive Pulmonary Disease (COPD) can occur secondary to a weakened trachea (tracheomalacia), and has been associated with increased TGF-β signaling in the lung of MFS patients. Upper airway obstruction and obstructive sleep apnea related to abnormal oropharyngeal anatomy can also occur. Restrictive lung disease can be exacerbated in the setting of severe kyphosis or pectus excavatum, often associated with MFS.

8.2.2.5 Cutaneous Findings

Stretch marks on the skin (striae atrophicae) may occur in anyone as a result of rapid growth adolescence and pregnancy or with marked weight gain or loss. Patients with MFS, however, are prone to develop stretch marks, often at an early age and without weight change (Grahame and Pyeritz 1995). These striae, frequently seen on the trunks of MFS patients, tend to appear in near body parts subject to stress, such as the shoulders, hips, and lower back.

8.2.2.6 Cardiovascular Manifestations

Cardiovascular disease is the primary determinant of lifespan in MFS patients (van Karnebeek et al. 2001). The phenotype varies widely by age of presentation and severity of disease. The spectrum of MFS-related cardiovascular disease includes aneurysmal dilation of the aortic root, and proximal pulmonary arteries, thoracic aortic dissection and aneurysm, mitral and tricuspid valve prolapse, cardiomyopathy, and supraventricular arrhythmias (Figueiredo et al. 2001; Takeda et al. 2016). Of these, aortic aneurysm and dissection are the most immediately life-threatening (Groth et al. 2018). In the early 1970's, the outcomes of emergent surgical repair of the aortic valve and root in MFS patients were unacceptably poor. With the development of the Bentall procedure (Bentall and De Bono 1968), which was first performed on a patient with MFS, outcomes were dramatically improved and use of this composite graft (Bentall) procedure, rapidly became the international standard (Gott et al. 1999). Mortality related to prophylactic root replacement has proven to be safe and effective in MFS patients (Aalberts et al. 2008), and represents the single most effective intervention leading to a longer life span. In many cases, a valve-sparing procedure can be performed, which negates the need for long term anticoagulation and further improves long-term survival.

The classic aortic aneurysm associated with MFS is described as "*annulo-aortic ectasia*," highlighting the fact that both the aortic annulus and the sinuses of Valsalva are dilated. Uncorrected, this can lead to aortic insufficiency, aortic dissection, and congestive heart failure. In most instances, the ascending aorta and arch retain their normal dimensions, leading to an "*Erlenmeyer flask*" shaped aneurysm originating at the aortic valve annulus. As the aorta dilates, the risk for aortic rupture and dissection increases. Population studies have suggested that MFS patients are approximately 200 times more likely to die from aortic disease (Groth et al. 2018), and that half of patients <40 years old who suffer from aortic dissection likely have MFS (de Beaufort et al. 2017). It is unclear why the sinuses of Valsalva are more prone to early dilation than other aortic segments. However, recent biomechanical studies have demonstrated that aortic tissue from MFS patients demonstrated increased sinus stiffness with medial degeneration, both during aging and with aneurysmal growth (Yan et al. 2019). This suggests that the increased aortic stiffness present in in the aortic root, may significantly contribute to dilation and dissection early in disease pathogenesis. It is important to note, as MFS patients are now routinely surviving beyond their 30's following prophylactic aortic root repair, aortic dissection involving the aortic arch and/or descending thoracic aorta are becoming more common. Moreover, these patients experience elevated risk of thoracoabdominal aneurysms and dissections, as well as peripheral artery aneurysms (Yetman et al. 2011). Reports of the presence of myocardial fibrosis leading to distinct non-ischemic cardiomyopathy independent of valvular lesions, as well as an increased propensity for supraventricular tachycardias have also emerged, presumably due to the role of fibrillin-1 in maintaining normal structure and function of the myocardium (Cook et al. 2014; Judge and Dietz 2005; Sulejmani et al. 2017). In teens and young adults, aortic valve regurgitation and aortic root disease have become the dominant cardiac pathologies, however mitral valve prolapse and regurgitation are also commonly encountered.

Cardiovascular manifestations in neonates are less common, however with particularly aggressive forms of the MFS, they are affected by severe mitral and tricuspid valvular regurgitation, often leading to cardiac failure and death. This phenotype has commonly been called "*Neonatal Marfan Syndrome*" though some have suggested this term be retired in favor of "*Early Onset Marfan Syndrome*" or "*Rapidly Progressive Marfan Syndrome*" (Hennekam 2005; Yetman et al. 2011). Fortunately, aortic dissection in the pediatric MFS patient population remains rare.

Mitral valve dysfunction was the earliest cardiovascular manifestation of MFS syndrome identified (Salle 1912) and represents a key cardinal feature of the 2010 Ghent nosology. Myxomatous thickening and elongation of the mitral valve leaflets commonly occur in MFS patients in response to mutations in FBN1 (Judge et al. 2011). Prolapse of the anterior leaflet, or bileaflet prolapse, are more common in MFS than in the degenerative mitral valve regurgitation populations.

All patients with suspected or confirmed MFS should be followed by a cardiologist appropriate for the age group and diagnosis. Ideally, these patients should also be referred for genetic counseling, and have close access to a tertiary cardiovascular surgery program with experience in valve sparing aortic root operations, mitral valve repair, and thoracoabdominal aortic interventions.

8.3 Molecular Pathogenesis

8.3.1 Mouse Models of MFS

To enhance the ability to study the pathogenesis of MFS, small animal models designed to recapitulate the clinical phenotypes observed in MFS patients, were developed. Three primary strains emerged and have now been used for more than two decades to define mechanisms and disease sequalae. The first strain developed by Periera et al., designed to replicate the dominant-negative effects of fibrillin-1 mutations in MFS, targeted exons 19–24 for replacement with a neomycin resistant expression cassette (Pereira et al. 1997). The resulting mouse (designated $Fbn1^{mg\Delta/mg\Delta}$) resulted in a substantial reduction in fibrillin-1 protein (*almost complete*) but retained normal elastin levels. There was no fetal loss of transgenic pups, and no overt phenotypic abnormalities at birth. At approximately 3-weeks after birth however, they would all suddenly die from cardiovascular complications, showing evidence of hemothorax and large vessel catastrophe upon necropsy. The mgΔ/+ heterozygotes, on the other hand, were morphologically and histologically indistinguishable from wild-type animals, living a normal lifespan with normal fertility. The second MFS strain generated, was also developed by Periera et al., using the same targeting vector and scheme as the $Fbn1^{mg\Delta/mg\Delta}$ mouse. This strain, designated $Fbn1^{mgR/mgR}$, was the result of an aberrant recombination event that only replaced exons 20–24, retaining exon 19 (Pereira et al. 1999). Similar to the original strain, the Fbn1mgR/mgR were born with reduced fibrillin-1 protein levels. They contained on average, approximately 25% of the normal amount of fibrillin-1, with variability depending on secondary factors, which suggested there may be a minimum threshold, that once reached would be lethal embryologically. Importantly, the $Fbn1^{mgR/mgR}$ mice displayed classic MFS phenotypic features in the skeleton including progressive kyphosis, overgrowth of the ribs, and elongation of the long bones. The most severe manifestations were displayed in the cardiovascular system and included aortic aneurysm and dissection. The homozygous mgR mice typically survived to 3.8 months on average before dying (Pereira et al. 1999). Once again, the heterozygous mgR/+ mice were indistinguishable from their wild-type littermates. The third MFS mouse strain developed was based on a common missense mutation identified in the human MFS patient population (Judge et al. 2004). In this strain, mice heterozygous for a cysteine (C) to glycine (G) substitution in a calcium binding EGF-like domain (*designated* $Fbn1^{C1039G/+}$), resulted in a 50% reduction in fibrillin-1 protein at birth, and recapitulated many of the human MFS phenotypic features including the pulmonary, skeletal, and cardiovascular manifestations. Mice homozygous for the $Fbn1^{C1039G}$ mutation uniformly died in the perinatal period from aortic catastrophe. While a small number of other MFS mouse strains exist, these three strains were the primary tools utilized to define much of the molecular pathogenesis of MFS. These animals replicated clinical MFS features to different degrees, based on the severity of fibrillin-1 loss, and formed the basis for several key discoveries, as well as serving as a testbed for translational studies that ultimately resulted in novel clinical trials.

8.3.2 Identification of Disease Mechanisms

Several key laboratory studies were initiated to provide empirical support for the role of aberrant TGF-β signaling in the pathogenesis of MFS. First, Neptune and coworkers described that the lung abnormalities, evident in the immediate postnatal period in fibrillin-1 deficient mice ($Fbn1^{mg\Delta/mg\Delta}$), were related to enhanced TGF-β activation and signaling, and that the perinatal administration of a TGF-β neutralizing antibody could rescue the defect in alveolar septation (Neptune et al. 2003). The second study used mice carrying the hypomorphic allele of the fibrillin-1 missense mutation ($Fbn1^{C1039G/+}$) that effectively recapitulated most of the MFS clinical phenotypes. In this study, Ng and coworkers similarly demonstrated that treatment with a TGF-β neutralizing antibody was able to reverse myxomatous changes in the mitral valve, rescuing the prolapse defect (Ng et al. 2004). This was followed with a study by Habashi et al., using the same Fbn1$^{C1039G/+}$ mouse, which demonstrated that the treatment with the TGF-β neutralizing antibody was sufficient to attenuate the spontaneous aortic root dilatation, elastic fiber fragmentation, and activation of downstream TGF-β signaling intermediates (phospho-Smad2) (Habashi et al. 2006).

The use of β-blockers as a first line therapy for MFS patients was considered standard of care and offered to all patients with aortic root dilatation. As a class, β-blockers (e.g. atenolol) functioned by regulating systemic blood pressure as well as the cardiac pulse wave (dP/dt; rate-rise time). The underlying premise was that attenuation of the force of the cardiac pulse wave as it moved through the aorta, would decrease in systemic blood pressure and aortic wall tension and thereby must be advantageous towards the prevention of aneurysm rupture and continued root dilatation. With this in mind, Habashi and coworkers tried another class antihypertensives; the angiotensin-II receptor blockers (ARBs), specifically losartan (Habashi et al. 2006). There were several reasons supporting the use losartan in models of MFS. First, Daugherty et al., discovered that treatment with an Ang-II type-I specific receptor (AT1R) blocker (losartan) could attenuate AngII-induced abdominal aortic aneurysm formation, while treatment with an Ang-II type-II specific receptor (AT2R) inhibitor (PD123319) enhanced aortic pathology and aneurysm development. This suggested that activation of the AT1R pathway had deleterious effects on aneurysm development, while signaling through the AT2R pathway may have protective effects. Similarly, this drove the argument that AT1R selective antagonism may be better than the combined antagonism achieved with angiotensin converting enzyme (ACE) inhibitors (Daugherty et al. 2001). Secondly, many studies had previously demonstrated that losartan could attenuate TGF-β signaling by inhibiting AngII-dependent expression of TGF-β ligands and receptors (Everett et al. 1994; Fukuda et al. 2000; Naito et al. 2004; Wolf et al. 1999). These important discoveries led Habashi et al. to hypothesize that treating MFS mice with losartan may be advantageous both for its ability to lower systemic blood pressure and its ability to attenuate TGF-β signaling. Indeed, treating MFS mice (Fbn1$^{C1039G/+}$) with losartan attenuated aortic root dilatation, similar to treatment with the TGF-β neutralizing antibody (Habashi et al. 2006). These results indirectly implicated enhanced TGF-β signaling as the primary mechanism underlying FBN1 mutations. Subsequent studies by Carta et al. (Carta et al. 2009) and Rodriguez-Vita and coworkers (Rodriguez-Vita et al. 2005), demonstrated that AngII could stimulate the phosphorylation of Smad-2, independent of signaling through the type-I TGF-β receptor (ALK-5), resulting in enhanced production of connective tissue growth factor (CTGF). Importantly, losartan treatment was able to attenuate both the phosphorylation of Smad-2 and the accumulation of CTGF, suggesting a role for an as yet unknown downstream mediator in the AT1R pathway. Interestingly, AngII stimulation of Smad-2 phosphorylation was attenuated by treatment with a p38 MAPK selective inhibitor (SB203580), suggesting that AngII induced Smad-2 phosphorylation by a non-canonical pathway involving p38 MAPK that was independent of TGF-β receptor signaling (Carta et al. 2009). The importance of non-canonical TGF-β signaling in MFS has gained further support by a subsequent series of studies by Habashi (Habashi et al. 2011) and Holm (Holm et al. 2011) demonstrating that both ERK1/2 and JNK1 could be activated by AngII, and inhibition of either pathway could attenuate aortic disease in MFS mice (Fbn1$^{C1039G/+}$).

8.4 Genetic Testing and Patient Management

The classic ocular, skeletal, neurologic, and cardiovascular manifestations result from a combination of both fibrillin-1 structural weakness and abnormal TGF-β signaling, which leads to pleiomorphic changes depending on the local environment and the stage of tissue development. In fact, this deeper understanding of the role of TGF-β signaling has led to the appropriate delineation of similar but genetically unique inherited vascular connective disorders. The most notable of these, Loeys-Dietz Syndrome, is caused by an inherited defect in the TGF-β receptors and has a phenotype that is distinct from but commonly identified as MFS. To assist in the differentiation of heritable connective tissue disorders that have overlapping symptomology, genetic testing has taken on a larger role, and has been included as part of the criteria defined in the 2010 Ghent nosology.

8.4.1 Genetic Testing

Advances in genetic analysis have dramatically altered the landscape for diagnosis of heritable diseases, and as a result, the understanding of MFS and other related connective tissue disorders has grown rapidly since the early twenty-first century. Genetic analysis can be used both for diagnosis of a proband, and for screening of potentially affected relatives. On the other hand, with new information comes more questions.

Pathogenic variants causal for MFS are found on the FBN1 gene, on chromosome 15q21.1. However, not all variants in this locus are associated with MFS, and some of these may be associated with familial thoracic aortic aneurysms. These are often termed variants of unknown significance (VUS). Additionally, there are non-pathogenic variants that do not lead to any clinically relevant syndrome. In current practice, diagnosis of MFS and related diseases has shifted further and further away from clinical diagnosis to a genetic diagnosis (Loeys et al. 2010). However, correct diagnosis of MFS from genetic analysis is only possible when combined with clinical findings.

When heritable connective tissue or aortic disease is suspected, a consultation with a genetic counselor should be considered. However, in the appropriate clinical setting, genetic analysis is not necessary prior to prophylactic aortic root surgery or repair of disabling musculoskeletal or ocular features. The proband may be offered large panel sequencing, which examines a panel of different loci that have been implicated in hereditary aortopathies. If a causal variant is identified, then appropriate clinical follow up and surveillance imaging can be arranged, with a goal of providing prophylactic aortic surgery prior to aortic dissection or development of irreversible heart failure. If a causal variant, or a VUS is identified, but clinical criteria are not met, the diagnosis may be better described as the MASS phenotype (mitral valve prolapse (M), nonprogressive aortic root dilatation (A), musculoskeletal findings (S), and skin striae (S); OMIM# 604308), Shprintzen-Goldberg syndrome, the Weill-Marchesani syndrome, familial thoracic aortic aneurysm, or familial thoracic aortic dissection. Further differential diagnosis must then be based on clinical criteria (Arslan-Kirchner et al. 2008). It is important to note that our understanding of the role of isolated VUS and the interplay between VUS, bicuspid aortic valve disease, and less severe causal variants is incomplete.

Specific variants do impact severity of disease and prognosis. For instance, specific variants in the exon 24–32 region, colloquially known as the "*neonatal region*," are associated with a severe prognosis and death in the first few years of life (Faivre et al. 2007). Probands with premature termination codon (PTC), or non-sense variants had significantly reduced extracellular fibrillin deposition in addition to reduced fibrillin synthesis, whereas individuals with mis-sense cysteine substitutions had normal levels of fibrillin synthesis accompanied by reduced matrix deposition (Aubart et al. 2018; Faivre et al. 2007; Takeda et al. 2018). Genotype-phenotype correlations are further complicated by clinical heterogeneity among individuals with the same variant.

Genetic screening may also differentiate MFS from other related disorders arising from genetic variants remote from FBN1. These panels also screen for other potential pathogenic variants in a wide array of genes, such as TGFBR1 and TGFBR2, which may lead to Loeys-Dietz Syndrome; COL3A1 leading to vascular Ehlers-Danlos Syndrome; and ACTA2, the SMAD family of genes, and NOTCH genes. These panels may identify variants associated with both syndromic and non-syndromic hereditary aneurysm clusters (expertly reviewed by De Backer, et al.

(De Backer et al. 2019)). Once a specific variant is identified, the family group may choose to undergo focused cascade screening, which screens relatives only for the variant identified in the proband. Cascade screening is typically less expensive and can be done more quickly than large panel screening (Fig. 8.1).

Despite significant progress made in understanding the genetic and molecular basis of MFS, the diagnosis continues to depend heavily on clinical features described in the 2010 revised Ghent nosology, however, with the 2010 revision, genetic testing is playing a larger role in diagnosis (Ammash et al. 2008).

8.4.2 Management and Treatment of MFS in the Pediatric Population

All possible cases of MFS in the pediatric population should be regularly assessed by echocardiography, optometry, and skeletal survey as the child grows. Children often have an evolving phenotype and commonly need to be followed for several years before a MFS diagnosis can be confirmed (Stuart and Williams 2007). Early referral to a cardiologist and clinical geneticist is key. Repair of pectus deformity should be postponed until the adolescent growth spurt is completed unless there is cardiopulmonary compromise. Severe scoliosis may require surgical stabilization (Jones et al. 2002). These operations should be performed at centers with experience managing Marfan patients based on commonly used operative criteria.

All patients with MFS should undergo a yearly evaluation with a pediatric cardiologist to include physical exam and imaging of the aorta (Lindsay 2018). The preferred imaging modality is echocardiography in the majority of patients. In patients who are unable to be properly imaged with echocardiography due to either obesity or extreme pectus deformity, MRI or CT scanning can be used yearly for follow up; though MRI is preferred to avoid radiation exposure whenever practical (Lindsay 2018). In patients with an aortic root dimension above 4 cm, aortic root growth of more than 0.5 cm/year, or those involved in competitive sports, it is recommended to obtain aortic imaging every 6 months (Lindsay 2018).

Lifestyle modifications are another mainstay in the management of MFS. It is generally advised that children with MFS avoid isometric exercise, high impact sports, and competitive sports due to the risk of aortic dissection, long bone fractures, and retinal detachment. However, most patients should be encouraged to remain active with aerobic activities performed in moderation (Judge and Dietz 2005).

The commonly prescribed use of β-blockers for MFS patients is intended to reduce hemodynamic stress on the aortic wall and is considered by many to be the standard of care (Salim et al. 1994). Data supporting the therapeutic benefits of this class of drugs is conflicting, and controversy remains regarding the true benefit of β-blockade (Selamet Tierney et al. 2007). If the decision is made to use β-blockers it must be titrated to clinical guidelines. While studies using murine models of MFS, demonstrated the utility of Losartan, an angiotensin receptor blocker, in slowing aortic growth and preventing dissection, multiple randomized controlled human trials have been unable to consistently reproduce this effect (Forteza et al. 2016; Lacro et al. 2014; Milleron et al. 2015). When comparing losartan monotherapy to combination therapy with β-blockade, two studies have shown slower rates of aneurysm progression in the combination therapy (Forteza et al. 2016; Mullen et al. 2019). A recent meta-analysis concluded that ARB therapy and in combination with β-blockade, was capable of slowing the rate of aortic root dilation but did not change the rate of aortic complications and surgery (Al-Abcha et al. 2020). Accordingly, the use of ARBs may also favorably affect myocardial fibrosis, aortic stiffening, and circulating TGF-β levels (Karur et al. 2018). We recommend ARB as first line therapy, with the addition of β-blockade as tolerated.

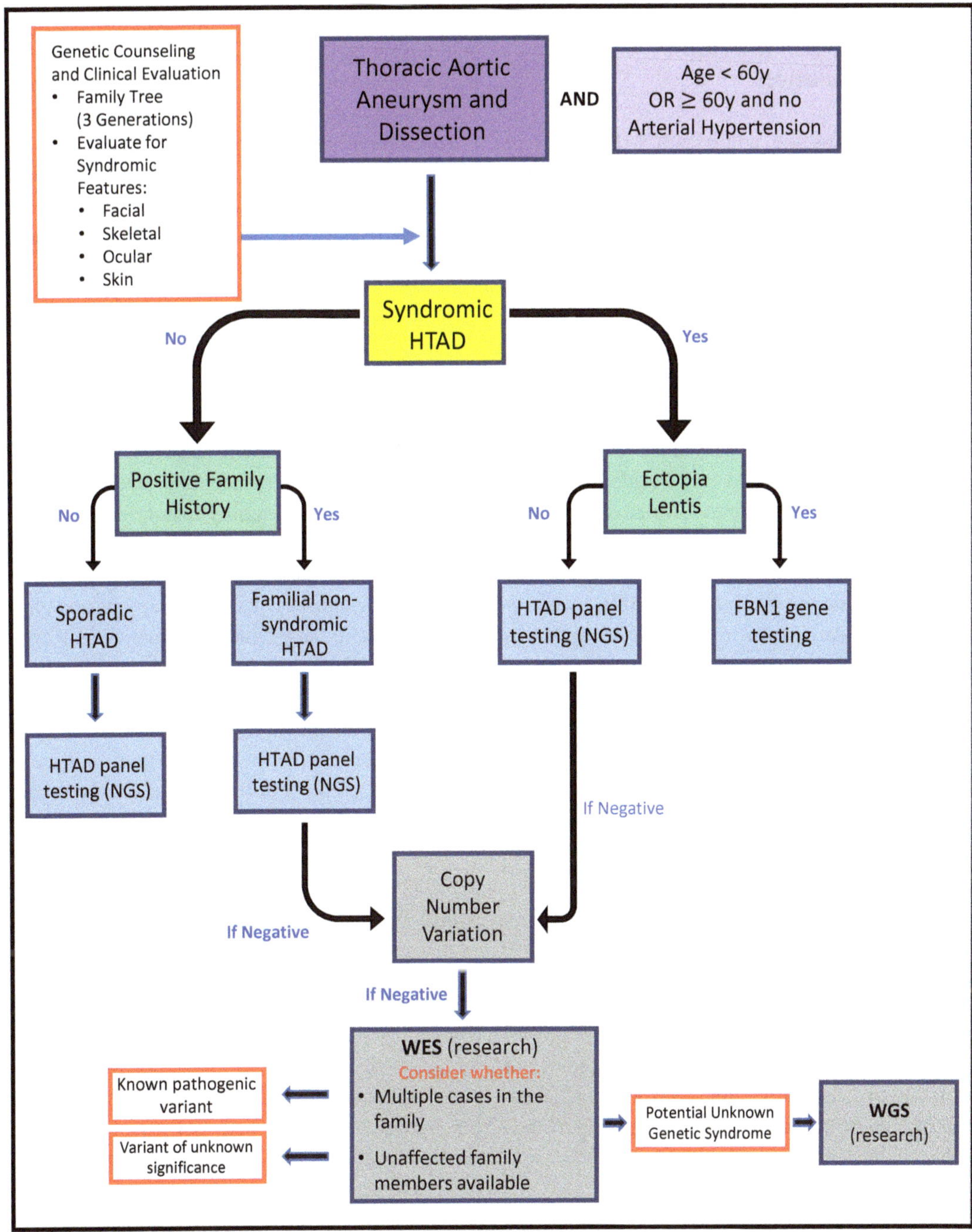

Fig. 8.1 *Genetic Testing Flow Chart*. (Adapted from De Backer, J. et al.; Curr Opin Cardiol 34:585–593; 2019 with permission from the publisher). Flowchart for the process of genetic evaluation in heritable thoracic aortic disease. HTAD, heritable thoracic aortic disease; NGS, next generation sequencing; WES, whole-exome sequencing; WGS, whole-genome sequencing

8.4.3 Management and Treatment of MFS in the Adult Population

Management of the adult with MFS revolves around medical drug therapy, lifestyle management, and appropriate surveillance of aortic and cardiac disease. The relative merits of β-blockade versus ARBs are as controversial in adults as in children, based on the same trial data. Hypertension should be avoided, with goal resting blood pressure < 130/80. Most cardiologists and aortic surgeons initiate prophylactic ARB or β-blocker therapy as tolerated, regardless of blood pressure. Both hydralazine and calcium channel blockers should be avoided in MFS patients, as these may be associated with higher rates of aortic dissection and death (Doyle et al. 2015). Furthermore, fluoroquinolones should be strongly avoided in patients with MFS, as multiple studies have demonstrated these medications are associated with *de novo* aneurysm formation, rapid aneurysmal dilation, and higher rates of dissection (Lee et al. 2015).

Lifestyle recommendations for adults are similar to those for children. High impact and isometric exercise, as well as competitive athletic pursuits should be avoided, though moderate aerobic exercise is generally recommended. Activities with dynamic atmospheric pressures, i.e. scuba and extreme altitude, should be avoided given the risk of spontaneous pneumothorax (Keane and Pyeritz 2008). It should be noted, there is little real-world data to support these lifestyle recommendations, however, they are unlikely to cause harm. Other cardiovascular risk factors, such as smoking, diabetes, hyperlipidemia, should be managed accordingly.

8.4.4 Cardiovascular Management

8.4.4.1 Imaging Surveillance

Imaging of the aortic root with echocardiography is recommended at the time of diagnosis, and at 6 month intervals after diagnosis in order to quantify baseline aortic root dimensions and initial rate of growth. Echocardiography will also determine the presence of any mitral valve pathology. Bi-annual transthoracic echocardiography should be considered if the maximal aortic dimension is greater than 4.5 cm.

It is important to note that size recommendations for intervention are based on cross-sectional imaging, not echocardiographic dimensions. CT measurements report the external dimension and tend to be a few millimeters larger than echocardiographic measurements (Taylor et al. 2010). Gated CT angiography should be performed if surgical repair is considered.

Surgical referral for prophylactic root replacement should be sought when the maximal aortic dimension reaches 5.0 cm. Other indications for prophylactic root replacement include rapid growth (>0.5 cm/yr), family history of dissection at smaller dimensions, or concomitantly when the primary indication is severe aortic or mitral valve regurgitation. For shorter patients and children, an aortic-height index greater than 10 has been associated with higher risk of dissection (Milewicz et al. 2005).

Often, a diagnosis of MFS is not made until after emergency treatment of a Stanford Type A aortic dissection. Following dissection repair, cross-sectional imaging should be obtained at 1, 3, 6, and 12 months, and annually to watch for aneurysmal dilation of the remaining aortic segments.

8.4.4.2 Surgical Management

The historic gold standard for treating aortic root aneurysm with or without aortic valve regurgitation in a patient with MFS is root replacement with a composite valved-graft (CVG), which incorporates a mechanical aortic valve into an aortic prosthesis (the Bentall procedure). This approach allows for resection of the aortic root and ascending aorta, as well as a durable, competent aortic prosthesis (Bentall and De Bono 1968). This approach still exposes the patient to lifelong anticoagulation with warfarin, as well as a significant risk of thromboembolic events, life threatening bleeding, and prosthetic valve endocarditis. Important refinements in surgical technique have led to the David technique becoming the preferred operation for prophylactic root replacement in the young, including in MFS patients. This technique, also known as aortic valve reimplantation, or less precisely, valve-sparing root replacement (VSARR), preserves the native aortic valve leaflets and can often restore competence to even a severely regurgitant valve, provided that there is limited pathology to the leaflets (David and Feindel 1992). A recent meta-analysis comparing VSARR to CVG in Marfan patients demonstrated very significant reductions in late complications including bleeding, thromboembolism, endocarditis, and mortality, with no difference in the need for reoperation. Overall survival for the entire cohort was excellent: 97.8% at 1 year and 90.5% at 10 years (Flynn et al. 2017).

Another interesting approach involves a custom-made external support, which becomes incorporated into the aortic wall to prevent dilation. Personalized external aortic root support (PEARS) was developed by an engineer with MFS, who then became the first patient in 2004. Mid-term follow up of the first 24 patients was published in 2018 and showed stability of the aortic root and ascending aorta, though 2 patients had technical failure of the operation (Izgi et al. 2018).

Patients presenting with an acute Type A dissection are much more difficult to manage, and often there is no opportunity for patient transfer to a high-volume aortic center without risking patient death. However, there is clear evidence that a more aggressive index operation, including root replacement and extension of the repair to the arch, results in less reoperation (Rylski et al. 2014). Failure to replace the aortic root in the index operation led to a 40% reintervention rate in their series.

The management of Type B aortic dissection in Marfan patients remains controversial. Traditionally, endovascular techniques have been avoided due to poor outcomes in the early experience with the technology. In low-risk patients with chronic dissection related thoracoabdominal aortic disease, open repair remains the treatment of choice on account of its reproducibility and high rate of technical success compared with TEVAR.

Open repair in the acute setting leads to poor outcomes; this has been largely abandoned except in rare circumstances. Outside of connective tissue disease, the use of thoracic endovascular aortic repair (TEVAR), even in uncomplicated dissection, is becoming more common. While initial experience with TEVAR in patients with connective tissue disorders was promising, due to the young age of the patients at implantation, the instability of the landing zones, a high rate of reintervention, and late complications was not uncommon (Nordon et al. 2009). The advantage to early TEVAR would include favorable remodeling by covering the primary tear. The debate continues as to whether this actually lessens the need for open thoracic aortic repairs (Cooper et al. 2009).

8.4.4.3 Mitral Valve Repair

Mitral valve pathology may present anytime in the course of a MFS patient's life. As previously discussed, bileaflet pathology is common, including prolapse and chordal rupture. Annular dilatation is also common. The indications for repair or replacement are not different for Marfan patients, and surgical repair should be considered when severe mitral regurgitation is present, regardless of symptoms or ventricular pathology (Otto et al. 2021). The presence of left ventricular myocardial dysfunction, heart failure symptoms, or atrial fibrillation is a strong indication for surgery. Despite concerns about repair failure, MFS patients who undergo mitral valve repair, rather than replacement, have improved long term survival and this is the preferred strategy (Fuzellier et al. 1998; Gillinov et al. 1994; Helder et al. 2014). Similar concerns about long term bleeding, thromboembolism, and infection apply for mitral valve surgery.

8.4.4.4 Treatment of MFS in Pregnancy

Pregnancy with MFS requires special attention. There are important considerations regarding prophylactic surgery and the risk for Type B dissection. The negative effects of Warfarin on fetal development, and the risk for dissection associated with lactation following delivery are key concerns. These patients need to be managed by a high-risk obstetrician, preferably at a medical center with expertise in aortic surgery and management of MFS patients.

8.5 Clinical Trials

Results from the early animal studies demonstrating that treatment with losartan, an already FDA approved ARB, could attenuate many of the MFS disease sequelae, stimulated physicians worldwide to start prescribing losartan to patients with thoracic aortic aneurysms secondary to MFS. As noted by Hoffmann Bowman and colleagues, this occurred well before randomized clinical trials of losartan had begun, in many cases in place of β-blocker therapy (Hofmann Bowman et al. 2019). Nonetheless, these animal studies served as a foundation for designing and implementing a series of exciting clinical trials that had the potential to change the medical management of MFS forever. The first landmark study was a small clinical trial that specifically recruited pediatric MFS patients with severe aortic root enlargement, that had already received β-blocker therapy that failed to reduce aortic root enlargement (Brooke et al. 2008). These patients received ARB therapy and were followed for 12–47 months. The results demonstrated that losartan was able to delay aortic root dilatation from 3.54+/−2.87 mm per year during previous β-blocker therapy to 0.46+/−0.62 mm per year during ARB therapy ($P < 0.001$) (Brooke et al. 2008). This early indication of success continued to feed the excitement and led to the initiation of multiple randomized clinical trials world-wide.

The first prospective trial assessed 233 Dutch MFS patients that were older than 18 years of age and were given losartan in addition to standard of care therapy (COMPARE trial), which often included a β-blocker (Groenink et al. 2013). The results were reported after 3 years of follow-up and suggested that losartan significantly reduced aortic root dilatation as compared to standard therapy alone. It was also reported, however that more patients receiving losartan were taking a β-blocker than in the standard of care treatment group (Groenink et al. 2013). Moreover, a substudy of this trial reported that of 117 patients that underwent FBN1 mutation analysis, there was a statically significant response to losartan in those patients that had FBN1 haploinsufficiency (*reduced protein content, with normal protein function*) versus FBN1 missense mutations (*normal protein content, with aberrant protein function*) (Groenink et al. 2013).

The next clinical trial to report results was completed in the US and sponsored by the Pediatric Heart Network (PHN). This randomized trial enrolled 608 MFS patients with aortic root enlargement and compared losartan with atenolol directly in children and young adults. The primary outcome measure was the rate of aortic-root enlargement indexed to body-surface area, measured over a 3-year period (Lacro et al. 2014). Unfortunately, the results revealed that there no significant difference in aortic-root dilatation rate after three years, as measured by standardized echocardiography, between patients randomized to the two treatment groups.

Shortly thereafter, a small parallel, randomized, double-blind study was initiated in Spain with 140 patients, using the same study design as the PHN (Forteza et al. 2016). After 3 years of follow-up this study likewise showed no change in aortic root diameter between the two groups, demonstrating no benefit of losartan over atenolol. Furthermore, follow-up of 128 of these patients at 6.7 years after receiving initial treatment, still failed to show a difference in aortic root diameter between the treatment groups (Forteza et al. 2016).

Importantly, the PHN and Spanish trials did not indicate that losartan was ineffective, it suggested only that it was equivocal with atenolol treatment, showing no specific benefit over standard of care. Interestingly, in the COMPARE trial substudy in which losartan was given in combination with β-blocker therapy, results suggested that there was a possible benefit in combination therapy. Accordingly, the “Marfan-Sartan” study addressed this question directly (Milleron et al. 2015). This study was a double-blind, randomized, multi-center, placebo-controlled, add-on trial that compared losartan vs. placebo in MFS patients that were older than 10 years of age. After 3.5 years of follow-up the 303 patients enrolled showed no difference in aortic root diameter (Milleron et al. 2015). In contrast to these results, the Aortic Irbesartan Marfan Study (AIMS) examined 192 patients with MFS that were receiving standard of care (in which approximately 56% of the patients were receiving a β-blocker) and randomized them to receive either irbesartan or a placebo (Mullen et al. 2019). The mean baseline aortic root diameter was 34.4 mm both groups at the start of the study. After 5-years of follow-up, the mean rate of aortic root dilatation was 0.53 mm per year in the irbesartan group, versus 0.74 mm per year in the placebo group, representing a statistically significant reduction in aortic root dilatation of −0.22 mm per year (Mullen et al. 2019). Importantly, irbesartan was well tolerated in both treatment groups, even in children.

While most of these clinical trials showed no benefit of ARB (losartan) over β-blocker (atenolol) therapy, there are a couple of important conclusions that came out of this work. First, there may be some hope in combining ARB and β-blocker therapies. Two of the smaller studies demonstrated benefit with the drug combination. Furthermore, the use of different β-blockers or ARBs may contribute to improved responses. Second, as additional studies began to look at the specific associations between genotype and phenotype, there may be subsets of patients that respond better than others. For example, as indicated above in the COMPARE trial, MFS patients that were FBN1 haploinsufficient, responded better to losartan than those that had missense mutations (Franken et al. 2015). The authors hypothesized that haploinsufficient patients may have a weaker aortic wall due to reduced FBN1 content, leading to enhanced mechanotransduction as a result of aortic dilatation and enhanced activation of the mechanosensitive angiotensin receptor pathway. Clearly further studies are warranted to delineate this and other genotype-phenotype associations and may suggest that additional treatment strategies are needed for MFS patients with dominant negative FBN1 mutations.

8.6 Conclusions

While the recent equivocal clinical trial results comparing ARB therapy versus β-blockade, have not been able to perfectly replicate the early promising animal studies, new data continues to build on the suggestion that combination therapy may provide the best benefit to all patients for now. Nonetheless, current therapeutic standards (β-blocker therapy) and prophylactic aortic root replacement, have extended the life of MFS patients to near normal. They are living longer and are now having to deal with medical issues related to later onset symptoms including joint disease and aneurysm of the descending thoracic aorta. Moreover, as genetic testing becomes increasingly prevalent in diagnosis and classification, additional data examining the FBN1 haploinsufficient genotype-phenotype association with ARB therapy response will certainly provide new insights and potential treatment options. Importantly, this highlights the need to develop additional therapeutic strategies to address those patients with less favorable genotypes. Overall, few medical discoveries in history have been elucidated to the degree that MFS has seen. From the early indications of a heritable disorder, to the identification of the gene responsible, to discerning the complex pathogenesis, culminating in rationale clinical trial design, the degree of thought and investment from the countless physicians and scientists involved has facilitated the truly remarkable progress that has been made in understanding this disorder.

References

Aalberts JJ, Waterbolk TW, van Tintelen JP, Hillege HL, Boonstra PW, van den Berg MP (2008) Prophylactic aortic root surgery in patients with Marfan syndrome: 10 years' experience with a protocol based on body surface area. Eur J Cardiothorac Surg 34:589–594

Ahn UM, Ahn NU, Buchowski JM, Garrett ES, Sieber AN, Kostuik JP (2000) Cauda equina syndrome secondary to lumbar disc herniation: a meta-analysis of surgical outcomes. Spine 25:1515–1522

Al-Abcha A, Saleh Y, Mujer M, Boumegouas M, Herzallah K, Charles L, Elkhatib L, Abdelkarim O, Kehdi M, Abela GS (2020) Meta-analysis examining

the usefulness of angiotensin receptor blockers for the prevention of aortic root dilation in patients with the Marfan syndrome. Am J Cardiol 128:101–106
Ammash NM, Sundt TM, Connolly HM (2008) Marfan syndrome-diagnosis and management. Curr Probl Cardiol 33:7–39
Annes JP, Munger JS, Rifkin DB (2003) Making sense of latent TGFbeta activation. J Cell Sci 116:217–224
Arslan-Kirchner M, von Kodolitsch Y, Schmidtke J (2008) The importance of genetic testing in the clinical management of patients with Marfan syndrome and related disorders. Dtsch Arztebl Int 105:483–491
Aubart M, Gazal S, Arnaud P, Benarroch L, Gross MS, Buratti J, Boland A, Meyer V, Zouali H, Hanna N, Milleron O, Stheneur C, Bourgeron T, Desguerre I, Jacob MP, Gouya L, Genin E, Deleuze JF, Jondeau G, Boileau C (2018) Association of modifiers and other genetic factors explain Marfan syndrome clinical variability. Eur J Hum Genet 26:1759–1772
Baer RW, Taissig HB, Oppenheimer EH (1943) Congenital aneurysmal dilatation of the aorta associated with arachnodactyly. Bull Johns Hopkins Hosp 72:309
Beals RK, Hecht F (1971) Congenital contractural arachnodactyly. A heritable disorder of connective tissue. J Bone Joint Surg Am 53:987–993
Beighton P, de Paepe A, Danks D, Finidori G, Gedde-Dahl T, Goodman R, Hall JG, Hollister DW, Horton W, McKusick VA et al (1988) International nosology of heritable disorders of connective tissue, Berlin, 1986. Am J Med Genet 29:581–594
Bentall H, De Bono A (1968) A technique for complete replacement of the ascending aorta. Thorax 23:338–339
Boerger F (1914) Ueber zwei Fälle von Arachnodaktylie. Monatsschr Kinderheilk 13:335
Brooke BS, Habashi JP, Judge DP, Patel N, Loeys B, Dietz HC 3rd. (2008) Angiotensin II blockade and aortic-root dilation in Marfan's syndrome. N Engl J Med 358:2787–2795
Brown OR, DeMots H, Kloster FE, Roberts A, Menashe VD, Beals RK (1975) Aortic root dilatation and mitral valve prolapse in Marfan's syndrome: an ECHOCARDIOgraphic study. Circulation 52:651–657
Carta L, Smaldone S, Zilberberg L, Loch D, Dietz HC, Rifkin DB, Ramirez F (2009) p38 MAPK is an early determinant of promiscuous Smad2/3 signaling in the aortas of fibrillin-1 (Fbn1)-null mice. J Biol Chem 284:5630–5636
Chandra A, Ekwalla V, Child A, Charteris D (2014) Prevalence of ectopia lentis and retinal detachment in Marfan syndrome. Acta Ophthalmol 92:e82–e83
Charbonneau NL, Ono RN, Corson GM, Keene DR, Sakai LY (2004) Fine tuning of growth factor signals depends on fibrillin microfibril networks. Birth Defects Res C Embryo Today 72:37–50
Chaudhry SS, Cain SA, Morgan A, Dallas SL, Shuttleworth CA, Kielty CM (2007) Fibrillin-1 regulates the bioavailability of TGFbeta1. J Cell Biol 176:355–367
Child AH (2017) Non-cardiac manifestations of Marfan syndrome. Ann Cardiothorac Surg 6:599–609
Clarke CC (1939) Ectopia lentis: a pathologic and clinical study. Arch Ophthalmol 21(1):124–153
Cook JR, Ramirez F (2014) Clinical, diagnostic, and therapeutic aspects of the Marfan syndrome. Adv Exp Med Biol 802:77–94
Cook JR, Carta L, Benard L, Chemaly ER, Chiu E, Rao SK, Hampton TG, Yurchenco P, Gen TACRC, Costa KD, Hajjar RJ, Ramirez F (2014) Abnormal muscle mechanosignaling triggers cardiomyopathy in mice with Marfan syndrome. J Clin Invest 124:1329–1339
Cooper DG, Walsh SR, Sadat U, Hayes PD, Boyle JR (2009) Treating the thoracic aorta in Marfan syndrome: surgery or TEVAR? J Endovasc Ther 16:60–70
Daugherty A, Manning MW, Cassis LA (2001) Antagonism of AT2 receptors augments angiotensin II-induced abdominal aortic aneurysms and atherosclerosis. Br J Pharmacol 134:865–870
David TE, Feindel CM (1992) An aortic valve-sparing operation for patients with aortic incompetence and aneurysm of the ascending aorta. J Thorac Cardiovasc Surg 103:617–621.; discussion 622
De Backer J, Jondeau G, Boileau C (2019) Genetic testing for aortopathies: primer for the nongeneticist. Curr Opin Cardiol 34:585–593
de Beaufort HWL, Trimarchi S, Korach A, Di Eusanio M, Gilon D, Montgomery DG, Evangelista A, Braverman AC, Chen EP, Isselbacher EM, Gleason TG, De Vincentiis C, Sundt TM, Patel HJ, Eagle KA (2017) Aortic dissection in patients with Marfan syndrome based on the IRAD data. Ann Cardiothorac Surg 6:633–641
De Paepe A, Devereux RB, Dietz HC, Hennekam RC, Pyeritz RE (1996) Revised diagnostic criteria for the Marfan syndrome. Am J Med Genet 62:417–426
Dietz HC, Pyeritz RE (1995) Mutations in the human gene for fibrillin-1 (FBN1) in the Marfan syndrome and related disorders. Hum Mol Genet 4:1799–1809
Dietz HC, Cutting GR, Pyeritz RE, Maslen CL, Sakai LY, Corson GM, Puffenberger EG, Hamosh A, Nanthakumar EJ, Curristin SM et al (1991a) Marfan syndrome caused by a recurrent de novo missense mutation in the fibrillin gene. Nature 352:337–339
Dietz HC, Pyeritz RE, Hall BD, Cadle RG, Hamosh A, Schwartz J, Meyers DA, Francomano CA (1991b) The Marfan syndrome locus: confirmation of assignment to chromosome 15 and identification of tightly linked markers at 15q15-q21.3. Genomics 9:355–361
Dietz HC, McIntosh I, Sakai LY, Corson GM, Chalberg SC, Pyeritz RE, Francomano CA (1993) Four novel FBN1 mutations: significance for mutant transcript level and EGF-like domain calcium binding in the pathogenesis of Marfan syndrome. Genomics 17:468–475
Dietz HC, Loeys B, Carta L, Ramirez F (2005) Recent progress towards a molecular understanding of Marfan syndrome. Am J Med Genet C Semin Med Genet 139C:4–9

Doyle JJ, Doyle AJ, Wilson NK, Habashi JP, Bedja D, Whitworth RE, Lindsay ME, Schoenhoff F, Myers L, Huso N, Bachir S, Squires O, Rusholme B, Ehsan H, Huso D, Thomas CJ, Caulfield MJ, Van Eyk JE, Judge DP, Dietz HC, TACRC G, Consortium ML (2015) A deleterious gene-by-environment interaction imposed by calcium channel blockers in Marfan syndrome. elife 4

Etter LE, Glover LP (1943) Arachnodactyly complicated by dislocated Lens and death from dissecting aneurysm of the aorta. JAMA 123:88

Everett AD, Tufro-McReddie A, Fisher A, Gomez RA (1994) Angiotensin receptor regulates cardiac hypertrophy and transforming growth factor-beta 1 expression. Hypertension 23:587–592

Faivre L, Collod-Beroud G, Loeys BL, Child A, Binquet C, Gautier E, Callewaert B, Arbustini E, Mayer K, Arslan-Kirchner M, Kiotsekoglou A, Comeglio P, Marziliano N, Dietz HC, Halliday D, Beroud C, Bonithon-Kopp C, Claustres M, Muti C, Plauchu H, Robinson PN, Ades LC, Biggin A, Benetts B, Brett M, Holman KJ, De Backer J, Coucke P, Francke U, De Paepe A, Jondeau G, Boileau C (2007) Effect of mutation type and location on clinical outcome in 1,013 probands with Marfan syndrome or related phenotypes and FBN1 mutations: an international study. Am J Hum Genet 81:454–466

Figueiredo S, Martins E, Lima MR, Alvares S (2001) Cardiovascular manifestations in Marfan syndrome. Rev Port Cardiol 20:1203–1218

Flynn CD, Tian DH, Wilson-Smith A, David T, Matalanis G, Misfeld M, Mastrobuoni S, El Khoury G, Yan TD (2017) Systematic review and meta-analysis of surgical outcomes in Marfan patients undergoing aortic root surgery by composite-valve graft or valve sparing root replacement. Ann Cardiothorac Surg 6:570–581

Forteza A, Evangelista A, Sanchez V, Teixido-Tura G, Sanz P, Gutierrez L, Gracia T, Centeno J, Rodriguez-Palomares J, Rufilanchas JJ, Cortina J, Ferreira-Gonzalez I, Garcia-Dorado D (2016) Efficacy of losartan vs. atenolol for the prevention of aortic dilation in Marfan syndrome: a randomized clinical trial. Eur Heart J 37:978–985

Franken R, den Hartog AW, Radonic T, Micha D, Maugeri A, van Dijk FS, Meijers-Heijboer HE, Timmermans J, Scholte AJ, van den Berg MP, Groenink M, Mulder BJ, Zwinderman AH, de Waard V, Pals G (2015) Beneficial outcome of losartan therapy depends on type of FBN1 mutation in Marfan syndrome. Circ Cardiovasc Genet 8:383–388

Fukuda N, Hu WY, Kubo A, Kishioka H, Satoh C, Soma M, Izumi Y, Kanmatsuse K (2000) Angiotensin II upregulates transforming growth factor-beta type I receptor on rat vascular smooth muscle cells. Am J Hypertens 13:191–198

Fuzellier JF, Chauvaud SM, Fornes P, Berrebi AJ, Lajos PS, Bruneval P, Carpentier AF (1998) Surgical management of mitral regurgitation associated with Marfan's syndrome. Ann Thorac Surg 66:68–72

Ge G, Greenspan DS (2006) BMP1 controls TGFbeta1 activation via cleavage of latent TGFbeta-binding protein. J Cell Biol 175:111–120

Gehle P, Goergen B, Pilger D, Ruokonen P, Robinson PN, Salchow DJ (2017) Biometric and structural ocular manifestations of Marfan syndrome. PLoS One 12:e0183370

Gillinov AM, Hulyalkar A, Cameron DE, Cho PW, Greene PS, Reitz BA, Pyeritz RE, Gott VL (1994) Mitral valve operation in patients with the Marfan syndrome. J Thorac Cardiovasc Surg 107:724–731

Goldfinger JZ, Preiss LR, Devereux RB, Roman MJ, Hendershot TP, Kroner BL, Eagle KA, Gen TACRC (2017) Marfan syndrome and quality of life in the GenTAC registry. J Am Coll Cardiol 69:2821–2830

Gott VL, Greene PS, Alejo DE, Cameron DE, Naftel DC, Miller DC, Gillinov AM, Laschinger JC, Pyeritz RE (1999) Replacement of the aortic root in patients with Marfan's syndrome. N Engl J Med 340:1307–1313

Goyette EM, Palmer PW (1963) Cardiovascular lesions in arachnodactyly. Circulation 7:363

Grahame R, Pyeritz RE (1995) The Marfan syndrome: joint and skin manifestations are prevalent and correlated. Br J Rheumatol 34:126–131

Groenink M, den Hartog AW, Franken R, Radonic T, de Waard V, Timmermans J, Scholte AJ, van den Berg MP, Spijkerboer AM, Marquering HA, Zwinderman AH, Mulder BJ (2013) Losartan reduces aortic dilatation rate in adults with Marfan syndrome: a randomized controlled trial. Eur Heart J 34:3491–3500

Groth KA, Stochholm K, Hove H, Andersen NH, Gravholt CH (2018) Causes of mortality in the Marfan syndrome(from a Nationwide register study). Am J Cardiol 122:1231–1235

Habashi JP, Judge DP, Holm TM, Cohn RD, Loeys BL, Cooper TK, Myers L, Klein EC, Liu G, Calvi C, Podowski M, Neptune ER, Halushka MK, Bedja D, Gabrielson K, Rifkin DB, Carta L, Ramirez F, Huso DL, Dietz HC (2006) Losartan, an AT1 antagonist, prevents aortic aneurysm in a mouse model of Marfan syndrome. Science 312:117–121

Habashi JP, Doyle JJ, Holm TM, Aziz H, Schoenhoff F, Bedja D, Chen Y, Modiri AN, Judge DP, Dietz HC (2011) Angiotensin II type 2 receptor signaling attenuates aortic aneurysm in mice through ERK antagonism. Science 332:361–365

Helder MR, Schaff HV, Dearani JA, Li Z, Stulak JM, Suri RM, Connolly HM (2014) Management of mitral regurgitation in Marfan syndrome: outcomes of valve repair versus replacement and comparison with myxomatous mitral valve disease. J Thorac Cardiovasc Surg 148:1020–1024.; discussion 1024

Hennekam RC (2005) Severe infantile Marfan syndrome versus neonatal Marfan syndrome. Am J Med Genet A 139:1

Hofmann Bowman MA, Eagle KA, Milewicz DM (2019) Update on clinical trials of losartan with and without beta-blockers to block aneurysm growth in patients with Marfan syndrome: a review. JAMA Cardiol 4:702–707

Holm TM, Habashi JP, Doyle JJ, Bedja D, Chen Y, van Erp C, Lindsay ME, Kim D, Schoenhoff F, Cohn RD, Loeys BL, Thomas CJ, Patnaik S, Marugan JJ, Judge DP, Dietz HC (2011) Noncanonical TGFbeta signaling contributes to aortic aneurysm progression in Marfan syndrome mice. Science 332:358–361

Isogai Z, Ono RN, Ushiro S, Keene DR, Chen Y, Mazzieri R, Charbonneau NL, Reinhardt DP, Rifkin DB, Sakai LY (2003) Latent transforming growth factor beta-binding protein 1 interacts with fibrillin and is a microfibril-associated protein. J Biol Chem 278:2750–2757

Izgi C, Newsome S, Alpendurada F, Nyktari E, Boutsikou M, Pepper J, Treasure T, Mohiaddin R (2018) External aortic root support to prevent aortic dilatation in patients with Marfan syndrome. J Am Coll Cardiol 72:1095–1105

Jones KB, Erkula G, Sponseller PD, Dormans JP (2002) Spine deformity correction in Marfan syndrome. Spine 27:2003–2012

Judge DP, Dietz HC (2005) Marfan's syndrome. Lancet 366:1965–1976

Judge DP, Biery NJ, Keene DR, Geubtner J, Myers L, Huso DL, Sakai LY, Dietz HC (2004) Evidence for a critical contribution of haploinsufficiency in the complex pathogenesis of Marfan syndrome. J Clin Invest 114:172–181

Judge DP, Rouf R, Habashi J, Dietz HC (2011) Mitral valve disease in Marfan syndrome and related disorders. J Cardiovasc Transl Res 4:741–747

Karur GR, Pagano JJ, Bradley T, Lam CZ, Seed M, Yoo SJ, Grosse-Wortmann L (2018) Diffuse myocardial fibrosis in children and adolescents with Marfan syndrome and Loeys-Dietz syndrome. J Am Coll Cardiol 72:2279–2281

Keane MG, Pyeritz RE (2008) Medical management of Marfan syndrome. Circulation 117:2802–2813

Kielty CM, Baldock C, Lee D, Rock MJ, Ashworth JL, Shuttleworth CA (2002) Fibrillin: from microfibril assembly to biomechanical function. Philos Trans R Soc Lond Ser B Biol Sci 357:207–217

Kielty CM, Sherratt MJ, Marson A, Baldock C (2005) Fibrillin microfibrils. Adv Protein Chem 70:405–436

Kurucan E, Bernstein DN, Ying M, Li Y, Menga EN, Sponseller PD, Mesfin A (2019) Trends in spinal deformity surgery in Marfan syndrome. Spine J 19:1934–1940

Lacro RV, Dietz HC, Sleeper LA, Yetman AT, Bradley TJ, Colan SD, Pearson GD, Selamet Tierney ES, Levine JC, Atz AM, Benson DW, Braverman AC, Chen S, De Backer J, Gelb BD, Grossfeld PD, Klein GL, Lai WW, Liou A, Loeys BL, Markham LW, Olson AK, Paridon SM, Pemberton VL, Pierpont ME, Pyeritz RE, Radojewski E, Roman MJ, Sharkey AM, Stylianou MP, Wechsler SB, Young LT, Mahony L, Pediatric Heart Network I (2014) Atenolol versus losartan in children and young adults with Marfan's syndrome. N Engl J Med 371:2061–2071

Lee CC, Lee MT, Chen YS, Lee SH, Chen YS, Chen SC, Chang SC (2015) Risk of aortic dissection and aortic aneurysm in patients taking Oral fluoroquinolone. JAMA Intern Med 175:1839–1847

Lindsay ME (2018) Medical management of aortic disease in children with Marfan syndrome. Curr Opin Pediatr 30:639–644

Loeys BL, Dietz HC, Braverman AC, Callewaert BL, De Backer J, Devereux RB, Hilhorst-Hofstee Y, Jondeau G, Faivre L, Milewicz DM, Pyeritz RE, Sponseller PD, Wordsworth P, De Paepe AM (2010) The revised Ghent nosology for the Marfan syndrome. J Med Genet 47:476–485

Lundby R, Rand-Hendriksen S, Hald JK, Lilleas FG, Pripp AH, Skaar S, Paus B, Geiran O, Smith HJ (2009) Dural ectasia in Marfan syndrome: a case control study. AJNR Am J Neuroradiol 30:1534–1540

Marfan AB (1896) Un cas de deformation congenitale des quatre membres, plus prononcee aux extremitds, characteriske par l'allongement des os avec un certain degre d'amincissement. Bull et mom Soc med hop Paris 13:220

McKusick VA (1955a) The cardiovascular aspects of Marfan's syndrome: a heritable disorder of connective tissue. Circulation 11:321–342

McKusick VA (1955b) Heritable disorders of connective tissue. III. The Marfan syndrome. J Chronic Dis 2:609–644

Milewicz DM, Dietz HC, Miller DC (2005) Treatment of aortic disease in patients with Marfan syndrome. Circulation 111:e150–e157

Milleron O, Arnoult F, Ropers J, Aegerter P, Detaint D, Delorme G, Attias D, Tubach F, Dupuis-Girod S, Plauchu H, Barthelet M, Sassolas F, Pangaud N, Naudion S, Thomas-Chabaneix J, Dulac Y, Edouard T, Wolf JE, Faivre L, Odent S, Basquin A, Habib G, Collignon P, Boileau C, Jondeau G (2015) Marfan Sartan: a randomized, double-blind, placebo-controlled trial. Eur Heart J 36:2160–2166

Mullen M, Jin XY, Child A, Stuart AG, Dodd M, Aragon-Martin JA, Gaze D, Kiotsekoglou A, Yuan L, Hu J, Foley C, Van Dyck L, Knight R, Clayton T, Swan L, Thomson JDR, Erdem G, Crossman D, Flather M, Investigators A (2019) Irbesartan in Marfan syndrome (AIMS): a double-blind, placebo-controlled randomised trial. Lancet 394:2263–2270

Naito T, Masaki T, Nikolic-Paterson DJ, Tanji C, Yorioka N, Kohno N (2004) Angiotensin II induces thrombospondin-1 production in human mesangial cells via p38 MAPK and JNK: a mechanism for activation of latent TGF-beta1. Am J Physiol Renal Physiol 286:F278–F287

Neptune ER, Frischmeyer PA, Arking DE, Myers L, Bunton TE, Gayraud B, Ramirez F, Sakai LY, Dietz HC (2003) Dysregulation of TGF-beta activation contributes to pathogenesis in Marfan syndrome. Nat Genet 33:407–411

Neuville M, Jondeau G, Crestani B, Taille C (2015) Respiratory manifestations of Marfan's syndrome. Rev Mal Respir 32:173–181

Ng CM, Cheng A, Myers LA, Martinez-Murillo F, Jie C, Bedja D, Gabrielson KL, Hausladen JM, Mecham

RP, Judge DP, Dietz HC (2004) TGF-beta-dependent pathogenesis of mitral valve prolapse in a mouse model of Marfan syndrome. J Clin Invest 114:1586–1592

Nistala H, Lee-Arteaga S, Smaldone S, Siciliano G, Carta L, Ono RN, Sengle G, Arteaga-Solis E, Levasseur R, Ducy P, Sakai LY, Karsenty G, Ramirez F (2010) Fibrillin-1 and -2 differentially modulate endogenous TGF-beta and BMP bioavailability during bone formation. J Cell Biol 190:1107–1121

Nordon IM, Hinchliffe RJ, Holt PJ, Morgan R, Jahangiri M, Loftus IM, Thompson MM (2009) Endovascular management of chronic aortic dissection in patients with Marfan syndrome. J Vasc Surg 50:987–991

Otto CM, Nishimura RA, Bonow RO, Carabello BA, Erwin JP 3rd, Gentile F, Jneid H, Krieger EV, Mack M, McLeod C, O'Gara PT, Rigolin VH, Sundt TM 3rd, Thompson A, Toly C (2021) 2020 ACC/AHA guideline for the Management of Patients with Valvular Heart Disease: a report of the American College of Cardiology/American Heart Association joint committee on clinical practice guidelines. Circulation 143:e72–e227

Pereira L, Andrikopoulos K, Tian J, Lee SY, Keene DR, Ono R, Reinhardt DP, Sakai LY, Biery NJ, Bunton T, Dietz HC, Ramirez F (1997) Targetting of the gene encoding fibrillin-1 recapitulates the vascular aspect of Marfan syndrome. Nat Genet 17:218–222

Pereira L, Lee SY, Gayraud B, Andrikopoulos K, Shapiro SD, Bunton T, Biery NJ, Dietz HC, Sakai LY, Ramirez F (1999) Pathogenetic sequence for aneurysm revealed in mice underexpressing fibrillin-1. Proc Natl Acad Sci U S A 96:3819–3823

Pyeritz RE (2016) Recent progress in understanding the natural and clinical histories of the Marfan syndrome. Trends Cardiovasc Med 26:423–428

Rados A (1942) Marfan's syndrome (arachnodactyly coupled with dislocation of the lens). Arch Ophthamol (Chicago) 27:477

Ramirez F, Pereira L (1999) The fibrillins. Int J Biochem Cell Biol 31:255–259

Ramirez F, Sakai LY, Dietz HC, Rifkin DB (2004) Fibrillin microfibrils: multipurpose extracellular networks in organismal physiology. Physiol Genomics 19:151–154

Rodriguez-Vita J, Sanchez-Lopez E, Esteban V, Ruperez M, Egido J, Ruiz-Ortega M (2005) Angiotensin II activates the Smad pathway in vascular smooth muscle cells by a transforming growth factor-beta-independent mechanism. Circulation 111:2509–2517

Rylski B, Bavaria JE, Beyersdorf F, Branchetti E, Desai ND, Milewski RK, Szeto WY, Vallabhajosyula P, Siepe M, Kari FA (2014) Type a aortic dissection in Marfan syndrome: extent of initial surgery determines long-term outcome. Circulation 129:1381–1386

Salim MA, Alpert BS, Ward JC, Pyeritz RE (1994) Effect of beta-adrenergic blockade on aortic root rate of dilation in the Marfan syndrome. Am J Cardiol 74:629–633

Salle V (1912) Ueber einen Fall von angeborner Grösse der Extremitäten mit einen an Akronemeglia erinnerden Symptomenkomplex. Jahrb Kinderheilk 75:540–550

Selamet Tierney ES, Feingold B, Printz BF, Park SC, Graham D, Kleinman CS, Mahnke CB, Timchak DM, Neches WH, Gersony WM (2007) Beta-blocker therapy does not alter the rate of aortic root dilation in pediatric patients with Marfan syndrome. J Pediatr 150:77–82

Stuart AG, Williams A (2007) Marfan's syndrome and the heart. Arch Dis Child 92:351–356

Sulejmani F, Pokutta-Paskaleva A, Ziganshin B, Leshnower B, Iannucci G, Elefteriades J, Sun W (2017) Biomechanical properties of the thoracic aorta in Marfan patients. Ann Cardiothorac Surg 6:610–624

Takeda N, Yagi H, Hara H, Fujiwara T, Fujita D, Nawata K, Inuzuka R, Taniguchi Y, Harada M, Toko H, Akazawa H, Komuro I (2016) Pathophysiology and Management of Cardiovascular Manifestations in Marfan and Loeys-Dietz syndromes. Int Heart J 57:271–277

Takeda N, Inuzuka R, Maemura S, Morita H, Nawata K, Fujita D, Taniguchi Y, Yamauchi H, Yagi H, Kato M, Nishimura H, Hirata Y, Ikeda Y, Kumagai H, Amiya E, Hara H, Fujiwara T, Akazawa H, Suzuki JI, Imai Y, Nagai R, Takamoto S, Hirata Y, Ono M, Komuro I (2018) Impact of pathogenic FBN1 variant types on the progression of aortic disease in patients with Marfan syndrome. Circ Genom Precis Med 11:e002058

Tatti O, Vehvilainen P, Lehti K, Keski-Oja J (2008) MT1-MMP releases latent TGF-beta1 from endothelial cell extracellular matrix via proteolytic processing of LTBP-1. Exp Cell Res 314:2501–2514

Taylor AJ, Cerqueira M, Hodgson JM, Mark D, Min J, O'Gara P, Rubin GD, American College of Cardiology Foundation Appropriate Use Criteria Task F, Society of Cardiovascular Computed T, American College of R, American Heart A, American Society of E, American Society of Nuclear C, North American Society for Cardiovascular I, Society for Cardiovascular A, Interventions, Society for Cardiovascular Magnetic R, Kramer CM, Berman D, Brown A, Chaudhry FA, Cury RC, Desai MY, Einstein AJ, Gomes AS, Harrington R, Hoffmann U, Khare R, Lesser J, McGann C, Rosenberg A, Schwartz R, Shelton M, Smetana GW, Smith SC Jr (2010) ACCF/SCCT/ACR/AHA/ASE/ASNC/NASCI/SCAI/SCMR 2010 appropriate use criteria for cardiac computed tomography. A report of the American College of Cardiology Foundation appropriate use criteria task force, the Society of Cardiovascular Computed Tomography, the American College of Radiology, the American Heart Association, the American Society of Echocardiography, the American Society of Nuclear Cardiology, the north American Society for Cardiovascular Imaging, the Society for Cardiovascular Angiography and Interventions, and the Society for Cardiovascular Magnetic Resonance. J Am Coll Cardiol 56:1864–1894

Trifiro G, Mora S, Marelli S, Luzi L, Pini A (2020) Increased fracture rate in children and adolescents with Marfan syndrome. Bone 135:115333 van Karnebeek CD, Naeff MS, Mulder BJ, Hennekam RC, Offringa M (2001) Natural history of cardiovascular manifestations in Marfan syndrome. Arch Dis Child 84:129–137

Williams E (1876) Rare cases, with practical remarks. Trans Am Ophthalmol Soc 1873–1879 2:291

Wolf G, Ziyadeh FN, Stahl RA (1999) Angiotensin II stimulates expression of transforming growth factor beta receptor type II in cultured mouse proximal tubular cells. J Mol Med 77:556–564

Yan J, Lehsau AC, Sauer B, Pieper B (2019) Mohamed SA, and mechanistic interrogation of bicuspid aortic valve associated Aortopathy Leducq C. comparison of biomechanical properties in ascending aortic aneurysms of patients with congenital bicuspid aortic valve and Marfan syndrome. Int J Cardiol 278:65–69

Yetman AT, Roosevelt GE, Veit N, Everitt MD (2011) Distal aortic and peripheral arterial aneurysms in patients with Marfan syndrome. J Am Coll Cardiol 58:2544–2545

Yuan X, Downing AK, Knott V, Handford PA (1997) Solution structure of the transforming growth factor beta-binding protein-like module, a domain associated with matrix fibrils. EMBO J 16:6659–6666

9 Ehlers-Danlos Syndromes, Joint Hypermobility and Hypermobility Spectrum Disorders

Lucia Micale, Carmela Fusco, and Marco Castori

Abstract

Ehlers-Danlos syndrome is an umbrella term for a clinically and genetically heterogeneous group of hereditary soft connective tissue disorders mainly featuring abnormal cutaneous texture (doughy/velvety, soft, thin, and/or variably hyperextensible skin), easy bruising, and joint hypermobility. Currently, musculoskeletal manifestations related to joint hypermobility are perceived as the most prevalent determinants of the quality of life of affected individuals. The 2017 International Classification of Ehlers-Danlos syndromes and related disorders identifies 13 clinical types due to deleterious variants in 19 different genes. Recent publications point out the possibility of a wider spectrum of conditions that may be considered members of the Ehlers-Danlos syndrome community. Most Ehlers-Danlos syndromes are due to inherited abnormalities affecting the biogenesis of fibrillar collagens and other components of the extracellular matrix. The introduction of next-generation sequencing technologies in the diagnostic setting fastened patients' classification and improved our knowledge on the phenotypic variability of many Ehlers-Danlos syndromes. This is impacting significantly patients' management and family counseling. At the same time, most individuals presenting with joint hypermobility and associated musculoskeletal manifestations still remain without a firm diagnosis, due to a too vague clinical presentation and/or the lack of an identifiable molecular biomarker. These individuals are currently defined with the term "hypermobility spectrum disorders". Hence, in parallel with a continuous update of the International Classification of Ehlers-Danlos syndromes, the scientific community is investing efforts in offering a more efficient framework for classifying and, hopefully, managing individuals with joint hypermobility.

Keywords

Collagen · Diagnosis · Ehlers-Danlos Syndrome · Extracellular Matrix · Hypermobility Spectrum Disorders · 2017 International Classification · Joint Hypermobility · Management.

L. Micale · C. Fusco · M. Castori (✉)
Division of Medical Genetics, Fondazione IRCCS-Casa Sollievo della Sofferenza, San Giovanni Rotondo, Italy
e-mail: m.castori@operapadrepio.it

J. Halper (ed.), *Progress in Heritable Soft Connective Tissue Diseases*, Advances in Experimental Medicine and Biology 1348, https://doi.org/10.1007/978-3-030-80614-9_9

Abbreviations

ACLP	aortic carboxypeptidase-like protein
CAH	congenital adrenal hyperplasia
CGH	comparative genomic hybridization
CNV	copy number variation
DS	dermatan sulfate
EDS	Ehlers-Danlos syndrome
ER	endoplasmic reticulum
GAG	glycosaminoglycan
hEDS	hypermobile Ehlers-Danlos syndrome
HSD	hypermobility spectrum disorders
HT-EDS	hypermobility type, Ehlers-Danlos syndrome
JHM	joint hypermobility
JHS	joint hypermobility syndrome
LZT	LIV-1 subfamily of ZIP zinc transporter
MLPA	multiplex ligation probe amplification
NGS	next generation sequencing
ROM	range of motion
SNP	single nucleotide polymorphism
WES	whole exome sequencing
WGS	whole genome sequencing

9.1 Introduction

The Ehlers-Danlos syndromes (EDS) are a group of clinically varied and genetically heterogeneous hereditary soft connective tissue disorders which mainly feature (i) abnormal cutaneous texture manifesting with doughy/velvety, soft, thin and/or variably hyperextensible skin, (ii) easy bruising with capillary fragility, and (ii) joint hypermobility (JHM) of various degree. Additional features which aid the distinction among the most common clinical types include skin fragility, abnormal wound healing and scarring formation, congenital contortionism (as a form of extreme generalized JHM), propensity to acquired dislocations which manifest spontaneously or due to trivial traumas, and fragility of vessels and internal organs with ruptures occurring in the absence of an underlying disease. Almost all of these characterizing features are intended as primary manifestations of a (dys)-histogenetic abnormality of non-ossified connective tissues. Therefore, EDS are "dysplasias" accordingly to the currently accepted terminology in Clinical Genetics (Hennekam et al. 2013).

Individuals with EDS may also present a constellation of additional connective tissue and pleiotropic manifestations, and this more frequently and distinctively occurs in the rarer types. Among these findings, the most clinically relevant include corneal thinning and eye fragility, congenital hypotonia or myopathy, severe scoliosis, skeletal dysplasia, joint contractures, congenital dislocations, congenital hernias, progressive/severe heart valve disease, and early-onset periodontitis. Finally, there is a series of anthropometric and morphological traits which usually do not impact the health status of the affected individual but rather support the physician in recognizing the disease, and mostly comprise unusual skin signs, such as piezogenic papules and pretibial plaques, peculiar facial features, mild foot deformities, and habitus variability (e.g. arachnodactyly and Marfanoid habitus).

Most EDS types are due to deleterious variants in genes encoding collagens I, III, V and XII, and enzymes involved in the biogenesis of fibrillar collagens or the glycosaminoglycan (GAG) chains of proteoglycans. All these molecules have a key role in the composition and functions of the extracellular matrix (ECM). The hypermobile EDS remains the unique variant without an identified molecular cause. In the last two decades, the interests around EDS proliferated due to the increasing number of newly discovered EDS types, and the emerging attention that the scientific community and lay people put on JHM and its ramifications in human pathology. This chapter is dedicated to offer an updated overview on EDS, and to present the current framework for approaching and classifying individuals presenting JHM.

9.2 Historical Overview

9.2.1 Original Description and the Berlin Nosology (1986)

EDS is an evolving concept. The very first descriptions by drs Chernogubow (Chernogubow 1892), Ehlers (Ehlers 1901) and Danlos (Danlos 1908) emphasized the tetrad of JHM, skin hyperextensibility, easy brusing and skin fragility as the paradigm of the disease. Afterwards, the growing observations allowed a deeper understanding of the background phenotype and the identification of an increasing number of "variations of the theme". Since the second half of the past century, the phenotypic core and boundaries of EDS changed significantly. The Berlin nosology of heritable disorders of connective tissue first identified eleven clinical types of EDS even before the "molecular era" (Beighton et al. 1988). Also at that time, the tetrad emphasized by Chernogubow, Ehlers and Danlos was recognized as more characteristic of what we currently define "classical" EDS (types I and II in the Berlin nosology), while the other nine types were distinguished by the attenuation of one or more of the key features, the presence of additional, sometime life-threatening features, and/or inheritance patterns different from the more common autosomal dominant.

9.2.2 Villefranche Nosology (1997)

From the last decade of the twentieth century, the introduction of molecular genetic technologies in the discovery of genes responsible for Mendelian disorders improved dramatically our understanding of the biological blueprint underlying the EDS nosology. Five major types were associated with deleterious variants in specific genes encoding collagens or enzymes directly involved in collagen production before the end of the last century. An updated (Villefranche) nosology of EDS was proposed in 1997, in which six major types were delineated (Beighton et al. 1998). At that time, seven genes (i.e. *ADAMTSL2, COL1A1, COL1A2, COL3A1, COL5A1, COL5A2* and *PLOD1*) were known as causative of EDS and the hypermobility-type EDS (HT-EDS) was the unique "major" type with unknown molecular bases. The Villefranche nosology also included a series of "minor types", for which available clinical descriptions were insufficient for delineating diagnostic criteria and, similarly to HT-EDS, molecular signatures were lacking.

9.2.3 The Beginning of the Twenty-First Century

The first decade of the current century opened with the publication of the Brighton criteria for the JHM syndrome (JHS) (Grahame et al. 2000). The term JHS was used to revise and extend the concept of the hypermobility syndrome, originally proposed by Kirk et al. (1967) as the presumably non-casual association of JHM and chronic musculoskeletal symptoms. The (revised) Brighton criteria for JHS were not limited to JHM and related musculoskeletal manifestations, but also included a series of soft-tissue manifestations, such as abnormal scarring, Marfanoid habitus and prolapses. Although JHS was originally intended as separate from HT-EDS, clinical practice suggested to a group of clinical geneticists and rheumatologists that HT-EDS and JHS should be considered one and the same at the phenotypic level until molecular studies will identify distinct profiles (Tinkle et al. 2009). This feeling was also supported by a family study, which demonstrated that, at least in some cases, the diagnostic criteria for HT-EDS and JHS may coexist in different family members or even in the same individual (Castori et al. 2014). While such a commonality of conditions was not accepted by some researchers, who recognized in such a presumed continuity between JHS and HT-EDS the premise for a drift of the concept of EDS and the related clinical research, these observations introduced the need of revising the EDS nosology and stimulating a reframe for the concept of JHM. This point was first presented at the scientific community in 2015 with a monographic issue of the American Journal of Medical Genetics dedicated to EDS and JHM (Castori and Colombi 2015).

9.2.4 International Classification (2017)

In 2017, the efforts of the "neonate" Ehlers-Danlos Society, risen from the merger of the national patients' associations of America and Great Britain, culminated with the publication of the International Classification of EDS and related disorders (Malfait et al. 2017). The Ehlers-Danlos Society invited dozens of international experts in the field with different interests and academic profiles to contribute in updating the clinical and molecular knowledge of previously defined EDS types, present an updated nosology including all other new monogenic forms of EDS described after the Villefranche nosology and propose a novel definition for HT-EDS (now renamed hypermobile EDS – hEDS). The 2017 International Classification of EDS and related disorders also included a set of satellite papers dedicated to single prevalent EDS types (i.e. classical, hypermobile and vascular) and to relevant clinical problems involving people with JHM and the various EDS, mostly hEDS (Bloom et al. 2017). Finally, this classification included a novel framework for people with JHM but not affected by a monogenic form of EDS (or any other genetic disorder featuring JHM and known molecular bases). This heterogeneous and presumably common group of people include individuals with non-syndromic, asymptomatic JHM, individuals with hEDS (which remains the unique type of EDS without known molecular basis), and individuals with apparently non-syndromic JHM in combination with various musculoskeletal complaints. The International Classification recognized that these three broad phenotypes are not clear-cut separated, but rather represent a continuous spectrum (i.e. the "spectrum") with non-syndromic, asymptomatic JHM on one end and a combination of pleiotropic features also including JHM and corresponding to the current diagnostic criteria for hEDS on the other. In the middle there are a series of possible combinations of JHM and musculoskeletal manifestations (with or without single or a few additional morphological signs compatible with hEDS, but in combinations not sufficient for a diagnosis of hEDS), which were defined hypermobility spectrum disorders (HSD). By definition, HSD include all presentations of "painful" JHM which cannot be associated with a deleterious variant in a known disease-gene and do not respect the hEDS criteria. Therefore, the term HSD incorporates JHS when it does not correspond to hEDS and solves the confusion previously generated by the possibility of co-diagnoses (i.e. both JHS and HT-EDS in the same individual) or "mutualistic" diagnoses (i.e. JHS instead of HT-EDS and *vice versa*) (Castori et al. 2017).

9.2.5 The Present

The International Classification of EDS and related disorders included 13 types of EDS due to deleterious variants in 19 different genes (Malfait et al. 2017). In 2018, Blackburn et al. (2018) published biallelic variants in the *AEBP1* gene in a disorder reminiscent of classical EDS but transmitted in an autosomal recessive pattern. This condition was provisionally added as the fourteenth type of EDS to the current classification (Malfait et al. 2020). More recently, individuals ascertained for a suspect of EDS and carrying deleterious variants in *COL1A1* and *COL1A2* were described as separate from the other EDS types with specific variants in these genes. These subjects present a phenotype strongly resembling EDS (often hEDS) and sometimes show minor features of osteogenesis imperfecta. In the past, such a condition was variably defined as "osteogenesis imperfecta/Ehlers-Danlos syndrome overlap" and currently renamed as "*COL1*-related overlap disorder" (Morlino et al. 2020). Whether such a condition will be recognized as an additional EDS type or not remains pending (Table 9.1).

Table 9.1 Current nosology of Ehlers-Danlos syndromes

Type	Gene(s)	Encoded Protein(s)	Inheritance	Prevalence	Major Features*
Types included in the 2017 international classification of Ehlers-Danlos syndromes and related disorders					
Classical	*COL5A1*, *COL5A2*, *COL1A1* [p.(Arg312Cys)]	α1(V) procollagen, α2(V) procollagen, α1(I) procollagen	AD	~1/20.000	Atrophic scarring Joint hypermobility (generalized) Skin hyperextensibility
Classical-like (type 1?)	*TNXB*	Tenascin-X	AR	Unknown (rare or ultrarare)	Easy bruising Joint hypermobility (generalized) Skin hyperextensibility **Note:** Absence of atrophic scarring
Cardiac-valvular	*COL1A2* (null alleles)	α2(I) procollagen	AR	Unknown (rare or ultrarare)	Atrophic scarring Easy bruising Joint hypermobility (variable) Severe/progressive cardiac valvular disease Skin hyperextensibility or thin skin
Vascular	*COL3A1*, *COL1A1* [p.(Arg312Cys), p.(Arg574Cys), p.(Arg1093Cys)]	α1(III) procollagen, α1(I) procollagen	AD	1/50.000–200.000	Arterial rupture at a young age Carotid-cavernous sinus fistula in the absence of trauma Colonic rupture in the absence of an underlying disease Uterine rupture during third trimester in the absence of C-section or perineum tears
Hypermobile	Unknown	Unknown	AD (?)	<1/5.000	Joint hypermobility (generalized) Secondary musculoskeletal complications Mild systemic involvement **Note:** See Table 9.2
Arthrochalasia	*COL1A1* (skipping of exon 6), *COL1A2* (skipping of exon 6)	α1(I) procollagen, α2(I) procollagen	AD	Unknown (rare or ultrarare)	Congenital bilateral hip dislocation Joint hypermobility (generalized) Skin hyperextensibility

(continued)

Table 9.1 (continued)

Type	Gene(s)	Encoded Protein(s)	Inheritance	Prevalence	Major Features*
Dermatosparaxis	*ADAMTS2*	Procollagen I N-procollagen	AR	Unknown (rare or ultrarare)	Easy bruising (severe) Extreme, congenital skin fragility Facial features Postnatal growth retardation Redundant/lax skin and increased palmar wrinkling Short limbs/hands/feet Umbilical hernia
Kyphoscoliotic	*PLOD1*	Lysyl-hydroxylase 1	AR	Unknown (rare or ultrarare)	Congenital hypotonia Congenital/early-onset kyphoscoliosis Joint hypermobility (generalized)
Brittle cornea syndrome	*PRDM5, ZNF469*	PR-domain containing protein 5, zinc finger protein 469	AR	Unknown (rare or ultrarare)	Blue sclerae Easy-onset, progressive keratoconus/globus Thin cornea
Spondylodysplastic	*B3GALT6, B4GALT7, SLC39A13*	Galactosyltransferases I and II, ZIP13	AR	Unknown (rare or ultrarare)	Bowing of limbs Hypotonia Short stature
Musculocontractural	*CHST14, DSE*	Dermatan-4-O-sulfotransferase 1, dermatan sulfate epimerase 1	AR	Unknown (rare or ultrarare)	Atrophic scarring Distal arthrogryposis Easy bruising Facial features Skin hyperextensibility
Myopathic	*COL12A1*	α1(XII) procollagen	AD	Unknown (rare or ultrarare)	Congenital hypotonia Joint contractures (proximal) Joint hypermobility (distal)
Periodontal	*C1R, C1S*	Complement subcomponents 1r and 1 s		Unknown (rare or ultrarare)	Early-onset, severe periodontitis Lack of attached gingiva Pretibial plaques

(continued)

Table 9.1 (continued)

Type	Gene(s)	Encoded Protein(s)	Inheritance	Prevalence	Major Features*
Subsequently identified conditions which could deserve inclusion in future revisions					
Classical-like (type 2?)	*AEBP1*	AE-binding protein 1	AD	Unknown (rare or ultrarare)	Not yet available, presumably: Atrophic scars Early-onset osteopenia Foot deformities Joint hypermobility (generalized) Skin hyperextensibility
COL1-related overlap disorder	*COL1A1*, *COL1A2*	α1(I) procollagen, α2(II) procollagen	AD, AR (rare)	Unknown (rare or ultrarare)	Blue sclerae Foot deformities Joint hypermobility (generalized) Skin hyperextensibility Soft and doughy skin

AD autosomal dominant, *AR* autosomal recessive. *Derived from the major diagnostic criteria in the 2017 International Classification of Ehlers-Danlos syndromes and related disorders

9.3 Joint Hypermobility

The current classification of EDS and related disorders clearly separates EDS from JHM. EDS is a group of pleiotropic conditions manifesting with multiple developmental anomalies also comprising JHM. Conversely, JHM is a clinical sign which may or may not occur within clinically relevant phenotypes. Clinical conditions related to JHM are not limited to the rare EDS (and any other Mendelian disorders featuring JHM – see below) but also extend to common chronic disorders primarily restricted to the musculoskeletal system (i.e. HSD).

9.3.1 Definition

JHM is the ability that a joint has to move beyond its normal limits along physiological axes. The adjective "normal" refers to the mean range of motion (ROM) that can be observed in the general population for any given joint. Major anatomical contributors to joint mobility include wellness of the articular surfaces, underlying bone morphology, muscle tone, and integrity of the soft tissues around joints, including ligaments, tendons and entheses. Such a variability in ROM relates to the evolution-driven peculiarities of the human musculoskeletal system, and a series of influencing non-modifiable factors, including sex, age and ethnicity. A series of modifiable factors may also impact the ROM and include, for example, nutritional status, weight, physical/sport training, past traumas and surgery. In the medical literature, synonyms of JHM are "joint laxity" and "joint hyperlaxity".

9.3.2 Epidemiology

JHM is common in the general population and shows an excess in females, with a prevalence of 6–57%, compared to males, who present JHM in 2–35% of the cases. Although epidemiological data are heterogeneous due to the variability of assessment methods used by the various research groups, all works present similar results (Remvig et al. 2007). Rate of JHM is also affected by ethnicity. It is more common in Inuit, African and Asian people and less common in Europe and Australasia. Intermediate rates are registered in North, Middle and South America (Cheng et al. 1991; Seow et al. 1999; Remvig et al. 2007). Clinical practice also indicates an excess of JHM

Table 9.2 Current criteria for hypermobile Ehlers-Danlos syndrome

CRITERION 1 (generalized joint hypermobility)
Beighton score ≥6 for pre-pubertal children and adolescents
OR Beighton score ≥5 for pubertal men and women up to the age of 50
OR Beighton score ≥4 for those >50 years of age for hEDS
Note: The 5-point questionnaire is used for individuals with historical (generalized) joint hypermobility and acquired joint limitations (past surgery, wheelchair, amputations, etc.) affecting the Beighton score; a diagnosis of generalized joint hypermobility can be made if BS is 1 point below the age- and sex-specific cut-off AND the 5-point questionnaire is positive with at least 2 positive items.
CRITERION 2
Two or more among the following features (A, B and C) MUST be present (i.e. A&B, or A& C, or B&C, or A&B&C).
Note: In individuals with an acquired connective tissue disorder (e.g., lupus, rheumatoid arthritis, etc.), additional diagnosis of hypermobile Ehlers-Danlos syndrome requires meeting both features A and B of criterion 2. Feature C of criterion 2 (chronic pain and/or instability) cannot be counted.
Feature A
Systemic manifestations of a more generalized connective tissue disorder (≥5 must be present among):
1. Unusually soft or velvety skin
2. Mild skin hyperextensibility
3. Unexplained striae such as striae distensae or rubrae at the back, groins, thighs, breasts and/or abdomen in adolescents, men or prepubertal women without a history of significant gain or loss of body fat or weight
4. Bilateral piezogenic papules of the heel
5. Recurrent or multiple abdominal hernia(s) (e.g., umbilical, inguinal, crural)
6. Atrophic scarring involving at least two sites and without the formation of truly papyraceous and/or hemosideric scars as seen in classical Ehlers-Danlos syndrome
7. Pelvic floor, rectal, and/or uterine prolapse in children, men or nulliparous women without a history of morbid obesity or other known predisposing medical condition
8. Dental crowding and high or narrow palate
9. Arachnodactyly, as defined in one or more of the following: a. positive wrist sign on both sides; b. positive thumb sign on both sides
10. Arm span-to-height ≥ 1.05
11. Mitral valve prolapse mild or greater based on strict echocardiographic criteria
12. Aortic root dilatation with Z-score > +2
Feature B
Positive family history, with one or more first degree relatives meeting the current diagnostic criteria for hypermobile Ehlers-Danlos syndrome.
Feature C
Musculoskeletal complications (must have at least one among):
1. Musculoskeletal pain in two or more limbs, recurring daily for at least 3 months
2. Chronic, widespread pain for ≥3 months
3. Recurrent joint dislocations or frank joint instability, in the absence of trauma (a or b)
a. Three or more atraumatic dislocations in the same joint or two or more atraumatic dislocations in two different joints occurring at different times
b. Medical confirmation of joint instability at 2 or more sites not related to trauma
CRITERION 3
All the following prerequisites MUST be met:
1. Absence of unusual skin fragility, which should prompt consideration of other types of Ehlers-Danlos syndrome
2. Exclusion of other heritable and acquired connective tissue disorders, including autoimmune rheumatologic conditions
3. Exclusion of genetic syndromes with joint hypermobility other than heritable soft connective tissue disorders
According to 2017 international criteria, the clinical diagnosis of hypermobile Ehlers-Danlos syndrome requires the simultaneous presence of CRITERIA I AND 2 AND 3.

in children and adolescents compared to adults and elder.

9.3.3 Patterns

Pattern analysis of JHM identifies four categories including generalized JHM, peripheral JHM, localized JHM and historical JHM. Generalized JHM is the most extended form of JHM and affect multiple joints (or groups of joints) of the appendicular and axial skeleton. Generalized JHM is reasonably the clinical equivalent of the lay terms "contortionism" and "double-jointedness". Peripheral JHM is a form of JHM that affects the extremities (i.e. hand and foot joints) only. It is often bilateral and is quite common in children. Localized JHM affects a single or a few (≤2) joints (or group of joints). Historical JHM is a questioned form of JHM. Its presence is testified by a personal history positive for some forms of congenital contortionism, but physical examination is, by definition, negative for excessive ROM at all assessed joints.

JHM can be also classified in congenital and acquired. Congenital JHM is (presumably) present at birth and, therefore, it is assumed a genetic and/or constitutional trait. On the contrary, acquired JHM is due to postnatal events (e.g. modifiable factors) and it is assumed not due to a genetic predisposition. Compared to generalized and peripheral JHM, localized JHM is more commonly acquired.

9.3.4 Clinical Measurements

In a clinical setting, JHM should be carefully assessed in as a higher number of body sites as possible. The ROM of almost all joints can be established with the use of appropriate tools (e.g. orthopedic goniometer and flexible tape) and by comparing the values to available standards (for example, accessible at https://www.cdc.gov/ncbddd/jointrom/index.html). The Beighton score is currently recognized as the most suitable tool for assessing generalized JHM, although this method has major limitations due to the lack of appropriate corrections considering modifiable and non-modifiable factors. The Beighton score includes the evaluation of nine joints or group of joints. In presence of clinically appreciable JHM, 1 point is attributed to each maneuver, while 0 points are assigned in case of normal ROM. The items composing the Beighton score include: (a) passive apposition of the thumb to the flexor aspect of the forearm (one point for each hand), (b) passive dorsiflexion of the V finger beyond 90° (one point for each hand), (c) hyperextension of the elbow beyond 10° (one point for each arm), (d) hyperextension of the knees beyond 10° (one point for each leg), (e) forward flexion of the trunk with the knees extended and the palms resting flat on the floor. A score of 5 or more is accepted for the "diagnosis" of generalized JHM in adults, while this cut-off is fixed to 6 in children (Juul-Kristensen et al. 2017). Historical JHM may be explored by the administration of the 5-point questionnaire (Hakim and Grahame 2003): *(a) can you now (or could you ever) place your hands flat on the floor without bending your knees?; (b) can you now (or could you ever) bend your thumb to touch your forearm?; (c) as a child did you amuse your friends by contorting your body into strange shapes or could you do the splits?; (d) as a child or teenager did your shoulder or kneecap dislocate on more than one occasion?; (e) do you consider yourself double-jointed?*. Positive answers to ≥2 questions are compatible with past JHM not yet appreciable at physical examination.

9.3.5 Secondary Musculoskeletal Manifestations

JHM facilitates the onset and worsening of a series of symptoms affecting the musculoskeletal system. JHM-related secondary musculoskeletal manifestations separates asymptomatic JHM from symptomatic JHM. Currently, the most common form of symptomatic JHM is HSD. However, similar manifestations are observed in hEDS and are presumably common also in the other EDS types. Joint instability is probably one of the leading mechanisms underlying JHM-related secondary musculoskeletal manifestations. Joint instability is a non-physiological condition of a joint or group of joints which can move along non-physiological axes and facilitates occasional, recurrent, and voluntary joint subluxations and luxations. Dislocations are usually acquired. However, in the most severely affected individuals, dislocations may be congenital particularly affecting the hips. JHM also predisposes to soft tissue injuries, sprains and strains. Musculoskeletal pain is common in combination with JHM. Types of musculoskeletal pain not related to the above mentioned events include acute/recurrent arthralgias and myalgias, cramps and neuropathic pain. Chronic widespread musculoskeletal pain is the worst pain-related complication in JHM (Castori et al.

2013; Castori 2016). There is only a fragmented knowledge of the natural history of musculoskeletal pain in HSD and EDS. Cross-sectional observations in people at different ages with hEDS/HSD suggest a possible evolution of pain, from occasional oligoarticular pain in combination or not with dislocations in infancy/childhood, to recurrent polyarticular pain in adolescence/adulthood, to chronic widespread pain with features of neuropathic pain and/or central sensitization in predisposed individuals (Castori et al. 2013). Preliminary evidence also indicates an association between JHM and developmental coordination disorders, attention deficit/hyperactivity disorders and learning disabilities in children (Piedimonte et al. 2018). The reasons as to why JHM co-exists with these neurodevelopmental attributes are undefined and, perhaps, related to a defective proprioception during development (Ghibellini et al. 2015).

9.4 The "Spectrum" and Hypermobility Spectrum Disorders

EDS is the most common pleiotropic syndrome presenting with JHM. For this reason, a growing number of physicians tend to suspect EDS when they notice this sign in their patients. Nevertheless, many individuals with JHM are healthy and most of the remaining are not affected by a recognizable genetic syndrome.

In the past, the too fragmented knowledge of the molecular basis of EDS, the difficulties in performing extensive molecular testing in a timely fashion and the confusion generated by the partially overlapping criteria for JHS and HT-EDS facilitated an over-diagnosis of EDS, mostly hEDS, in specific settings (e.g. rheumatology) and/or different countries. Presumably, most individuals who received a clinical diagnosis of EDS (i.e. molecular data unavailable or negative) or hEDS (i.e. Villefranche and/or Brighton criteria met) in the past, were affected by a chronic and sometimes disabling condition with widespread musculoskeletal involvement and a series of satellite functional disorders (the "so-called" joint hypermobility-related co-morbidities according to the 2017 international classification of EDS and related disorders – see below). Although such a presentation remains common among subjects respecting the 2017 criteria for hEDS, it is equally common in people referred for unexplained chronic pain but lacking these criteria or not affected by a syndrome with a deleterious variant in a known gene. The term HSD was introduced to define this latter category of patients, which likely represents an emerging phenotype presenting similarities with EDS but still unconvincing for a stringent diagnosis of EDS, which is, by definition, a rare condition (Castori et al. 2017).

Considering people without a molecularly-proved condition with JHM, HSD are put in the middle between non-syndromic, asymptomatic JHM and hEDS. More specifically, the presence of musculoskeletal complaints separates individuals with HSD from non-syndromic, asymptomatic JHM, while the presence of the 2017 diagnostic criteria separates hEDS from the remaining phenotypes within the "spectrum" (Fig. 9.1 and Table 9.2).

The current criteria for hEDS and, in particular, feature 2A (i.e. systemic involvement), which must be met by all index cases (i.e. family history/feature 2B not applicable), comprise a set of objective features standing for the primary involvement of connective tissue derivatives other than joints. Most of these ancillary findings are intended as *bona fide* manifestations of pleiotropy. On the contrary, musculoskeletal manifestations related to JHM may be (perhaps, simplistically) considered "consequences" of a pre-existing (congenital?) ligamentous laxity. Moreover, criterion 2A is distinguished from JHM-related co-morbidities which are, in turn, functional disorders equally common in the general population and, hence, not easily put under the umbrella of pleiotropy or interpreted as direct effects of JHM. The identification of a causative variant in a gene associated with a Mendelian syndrome with JHM, such as the other EDS types, moves the affected individual outside the "spectrum". Hence, the "spectrum" is, by definition, a transitory nosology used to classify indi-

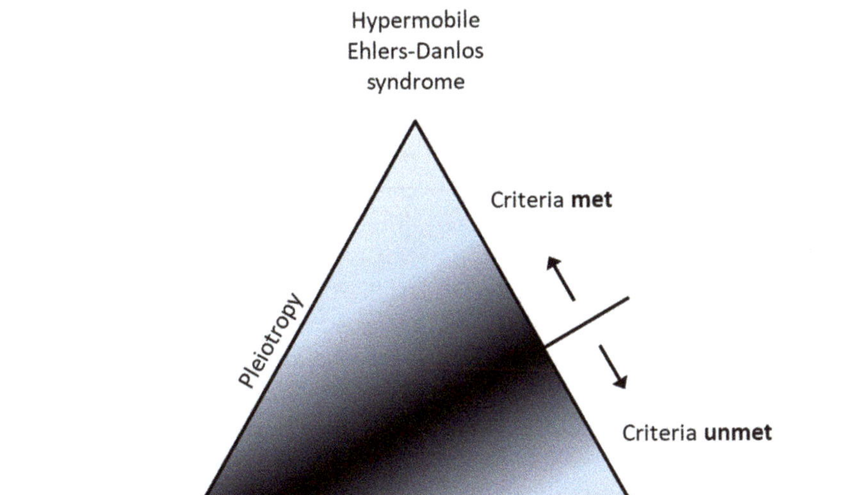

Fig. 9.1 Graphical representation of the relationships among joint hypermobility, hypermobility spectrum disorders and hypermobile Ehlers-Danlos syndrome. The presence of (secondary) musculoskeletal manifestations, with an increasing degree, separates (non-syndromic, **asymptomatic**) joint hypermobility from hypermobility spectrum disorders. On the other end, concurrent primary structural anomalies in other organs and tissues (i.e. pleiotropy) allow the distinction between (**non-syndromic,** asymptomatic) joint hypermobility and hypermobile Ehlers-Danlos syndrome. At the same time, it is accepted that secondary musculoskeletal manifestations are also common in hypermobile Ehlers-Danlos syndrome and that some attenuated/partial/incomplete pleiotropic features can be occasionally encountered in individuals with hypermobility spectrum disorders but lacking the criteria for other related diagnoses

viduals with JHM who cannot be diagnosed with more accurate methods, such as molecular testing.

At the time of this writing, HSD are considered exclusion diagnoses or descriptors for people with JHM (various phenotypes) and related musculoskeletal manifestations. No pathognomonic finding or set of diagnostic criteria are currently available for HSD, although some specifications have been put forward with an expert-based approach (Castori and Hakim 2017). HSD are provisionally classified mirroring the different patterns of JHM. Therefore, three different types of HSD are recognized and include generalized HSD (i.e. in presence of generalized JHM), peripheral HSD (i.e. in presence of peripheral JHM) and localized HSD (i.e. in presence of localized JHM). The existence of a fourth category of HSD for people with a history of (generalized) JHM but without evidence of JHM at examination (i.e. historical HSD) is questioned (Malfait et al. 2020). There are also concerns in attributing a diagnosis of HSD to all individuals with JHM and any musculoskeletal symptoms. Presumably, only a selected set of musculoskeletal manifestations may be intended as the symptomatic counterpart of a clinically demonstrable JHM and, therefore, eligible for a diagnosis of HSD. For example, musculoskeletal pain should be attributed to an underlying HSD only if (i) not associated with inflammatory/autoimmune disease, (ii) recurrent or chronic, and (iii) localized to joints with evidence or history of JHM. Dislocations could be considered only if (i) repeated (two or more events), (ii) not related to external forces sufficiently explaining the event, and (iii) affecting joint(s) with evidence or history of JHM. In addition, only a limited set of musculoskeletal physical traits (i.e. flexible flatfoot, scoliosis, *genua* and *cubita valga*), degenerative joint and bone disease (i.e. premature osteoarthritis and reduced bone mass), and neurodevelopmental attributes (i.e. benign hypotonia, simple motor delay, developmental coordination disorder, attention deficit/hyperactivity disorder) might be considered "extended" manifestations of an underlying HSD (Castori and Hakim 2017).

The diagnosis of HSD is established at physical examination including extensive evaluation of JHM, personal and family history, and exclusion of partially overlapping disorders (mostly, hEDS, the other syndromes featuring JHM and well-defined rheumatologic diseases causing pain). Hence, HSD is a nosologic category in which a patient falls if demonstrated as NOT affected by the above mentioned disorders, also including hEDS.

9.5 Joint Hypermobility-Related Co-Morbidities

In 2004, drs. Hakim and Grahame first put emphasis on a possible association between autonomic dysfunction and JHS (Hakim and Grahame 2004). In the ensuing years, this hypothesis attracted increasing attention by different research groups and an excess of various forms of orthostatic intolerance, mostly postural orthostatic tachycardia syndrome, is now accepted in people with JHM, HSD/JHS and hEDS/HT-EDS (Hakim et al. 2017; Roma et al. 2018). Intriguingly, hEDS and "isolated" generalized JHM are also common among individuals ascertained for postural orthostatic tachycardia syndrome (Miller et al. 2020). Such a bi-directional association is also observed for functional gastrointestinal disorders (Fikree et al. 2015; Lam et al. 2020), anxiety disorders (Bulbena et al. 2011; Bulbena et al. 2017), and pelvic prolapses (Veit-Rubin et al. 2016; Tinkle et al. 2017). Fatigue is also common in HSD and hEDS (Hakim et al. 2017b). The reasons as to why JHM (and related "spectrum" phenotypes) is so convincingly associated with orthostatic intolerance, functional gastrointestinal disorders, anxiety disorders, pelvic prolapses and fatigue are unknown. At the moment, pleiotropy is not an easy explanation as not any convincing molecular evidence is available. The more neutral term "co-morbidity" was preferred to describe this phenomenon, given the common occurrence of JHM and these associated disorders in the general population, and the complex etiopathogenesis presumed for all (Castori et al. 2017). In EDS and HSD, an underlying autonomic dysregulation could explain the combination of orthostatic intolerance and functional gastrointestinal disorders. If replicated in other populations, an increased rate of small fiber neuropathy demonstrated in Italian subjects with EDS might fill the gap between JHM and autonomic dysregulation (Cazzato et al. 2016).

9.6 Molecular Pathogenesis of Ehlers-Danlos Syndromes

ECM is a complex meshwork of insoluble fibrillar proteins and signaling factors interacting together to provide architectural and instructional cues to the surrounding cells. ECM exerts a plethora of biological functions. In addition to biomechanical support, ECM participates in cell communication and differentiation (Jover et al. 2018), and regulates development and homeostasis in all eukaryotic cell types. ECM is a reservoir of growth factors and bioactive molecules. The "core matrisome" comprises approximately 300 ECM constituents including collagens, proteoglycans, elastin, and cell-binding glycoproteins (Murphy-Ullrich and Helene 2014). ECM undergoes a continuous turnover either under physiological or in pathological circumstances, and its homeostasis is critical for connective tissue architecture and function (Karamanos 2019). Abnormal ECM remodeling is implicated in many pathological conditions such as cancer, osteoarthritis and fibrosis (Nallanthighal et al. 2019). Differential expression of genes encoding ECM components was reported as key driving factors in many connective tissue disorders and, among them, EDS (Chiarelli et al. 2016; Lim et al. 2019).

9.6.1 Collagen Biogenesis

Collagens represent the major ECM structural component and play a central role in providing the structural integrity of several connective tissue and various organs such as skin, lungs, and blood vessels. Collagens interact with other ECM components, as the integrins, a set of specific cell surface receptors that mediate the complex cell-matrix interactions. These bridging molecules, which are heterodimeric transmembrane receptors, connect ECM to cytoskeleton by interacting via their extracellular domain with collagens and other matrix molecules, thus mediating cell adhesion and motility (Campbell and Humphries 2011). Some integrins, the α5β1 and α2β1 are involved in ECM organization of fibronectin and fibrillar collagens, respectively and their expression results de-regulated in various types of EDS, reasonably as secondary effects of the primary collagen defect (Zoppi et al. 2004; Janecke et al. 2016).

Among vertebrates, there are 28 distinct collagen glycoproteins that are encoded by at least 45 genes. Deleterious variants in the genes encoding for (the various chains of) collagen I, III and V have been identified as molecular basis of many EDS types also including the most common ones, such as classical and vascular EDS. Each collagen gene encodes a unique alpha (α) chain. Each fibrillar collagen is usually made of three different polypeptides which are individually synthesized as procollagen chains (Fig. 9.2). These procollagens include the globular N- and C-propeptides flanking the prototypical collagen triple helix structures characterized by the Glycine-X-Y triple repeat, where X and Y can be any amino acid, more often proline and hydroxyproline, respectively. Once secreted, the procollagene polypeptides form trimers in the cytoplasm and the terminal propeptides are excised by N- and C-proteinase-mediated proteolytic cleavage. This action leads to the formation of the typical triple helix of the mature collagen molecule, which conserves short telopeptides at both ends. Procollagens are cleaved to form mature collagen molecules by disintegrin and metalloproteinase with thrombospondin motifs proteins (ADAMTS) and bone morphogenetic protein 1/tolloid-like proteinases.

The triple collagens fiber can be homotrimeric, composed of three identical α chains, or heterotrimeric, comprised of α chains encoded by distinct genes with possibly different proportions also according the expression sites (i.e. organs and structures). For instance, collagen V can be formed by one α1 and two α2 chains (encoded by *COL5A1* and *COL5A2*, respectively), by one α1, one α2 and one α3 chains (encoded by *COL5A1*, *COL5A2* and *COL5A3*), or three α1 chains (encoded by *COL5A1*). Collagen I is usually heterotrimeric and formed by two α1 and one α2 chains (encoded by *COL1A1* and *COL1A2*, respectively). Conversely, collagen III is always homotrimeric and formed by three α1 chains, encoded by *COL3A1*.

Since the Villefranche nosology, it was known that heterozygous variants in *COL5A1* and *COL5A2* cause classical EDS, while heterozygous variants in *COL3A1* cause vascular EDS (Beighton et al. 1998). At the moment, *COL5A3* variants have been never described in any EDS type. Not any strong genotype-phenotype correlation exists for classical and vascular EDS. Variants in *COL5A2* seem associated with a more severe cutaneous phenotype compared to *COL5A1* (Ritelli et al. 2013). In vascular EDS, in-frame exon skipping variants appear associated with the lowest median survival, followed by glycine substitutions with bulky residues and, then, by glycine substitutions with smaller residues (Pepin et al. 2014). Atypical missense variants in *COL3A1* leading to the substitution of a glutamic acid with lysine cause a cutaneous phenotype resembling classical EDS in people with vascular EDS (Ghali et al. 2019a). In rare families, an EDS phenotype with variable features of classical and vascular types has been described in association with the missense p.(Arg312Cys) in *COL1A1*. Vascular EDS has been also described in association with the non-glycine missense variants p.(Arg574Cys) and p.(Arg1093Cys) in *COL1A1*. While such rare heterozygous missense variants in *COL1A1* give rise to vascular and/or classical EDS genocopies, biallelic null alleles in *COL1A2* cause an ultrarare variant of EDS with

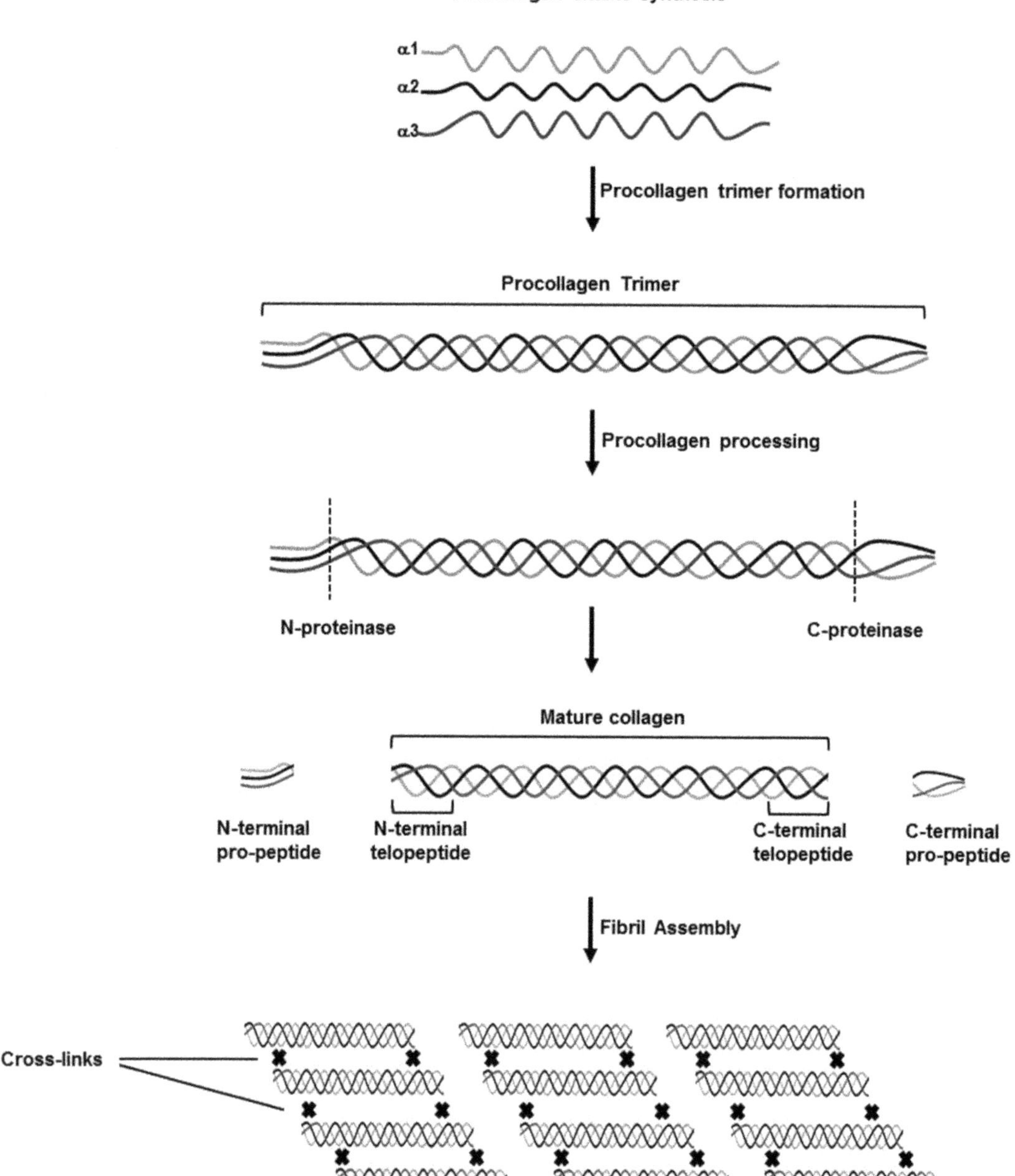

Fig. 9.2 Schematic representation of collagen fibrillogenesis by fibroblasts. Procollagen polypeptides are synthesized, and then, they form trimers in the fibroblasts' cytoplasm. Procollagens contain a central domain flanked by propeptide N- and C-terminal domains. Once secreted, the terminal propeptides are processed and this leaves the typical triple helix of the mature collagen molecule, which conserves short telopeptides at both ends. Collagen molecules assemble to form striated fibrils by covalent cross-links through these polipeptides

severe/progressive cardiac valve disease (i.e., cardiac-valvular EDS). In this variant of EDS, collagen I fibers are homotrimeric and always formed by α1 chains.

Heterozygous recurrent splicing variants causing loss of (the entire or part of) exon 6 of *COL1A1* and *COL1A2* are associated with arthrochalasia EDS. Exon 6 contains the N-terminal propeptide cleavage site in both genes. Such variants cause the loss of the substrate sequence for the procollagen N-proteinase, whose enzymatic activity is necessary for maturation of the collagen fiber, by skipping of the entire exon 6 or of the sequence corresponding to the N-proteinase cleavage site, when a cryptic exonic acceptor site is used (Chiodo et al. 1992). Similarly, biallelic loss-of-function variants in *ADAMTS2*, encoding the procollagen I N-proteinase, cause the autosomal recessive dermatosparaxis EDS. In this condition, there is the retention of the N-terminal propeptides in all collagen I chains and this generate fibrils that are completely distorted (Smith et al. 1992).

Structural integrity and stability of tissues and organs is also determined by the ability of mature collagen molecules to form covalent cross-links and then generate intermediate fibrils. Lysyl hydroxylase 1, encoded by *PLOD1*, is a critical enzyme involved in the formation of covalent cross-links and collagen glycosylation (Rautavuoma et al. 2004). This enzyme dependent collagen crosslinking stabilizes newly formed collagen fibers and enhances the stiffness of the matrix. Particularly, *PLOD1* regulates the hydroxylation of lysyl residues on collagen type V (Lim et al. 2019). Biallelic loss-of-function variants in *PLOD1* cause *PLOD1*-related kyphoscoliotic EDS and they explain the mechanical instability observed in the affected tissues, as well as the resulting clinical features (Giunta et al. 2005; Zahed-Cheikh et al. 2017).

Intermediate fibrils assemble into striated fibrils by the action of organizers, nucleators and regulators. At the plasma membrane, fibronectin and integrins act as organizers of fibril assembly. Type V collagen has an essential role during the early process of collagen fibril nucleation (Birk 2001; Wenstrup et al. 2004) inducing the addition of other collagen monomers. The intermediate fibrils are then deposited in the ECM and, subsequently, stabilized by regulators such as tenascin X, encoded by *TNXB*, and collagen type XII, encoded by *COL12A1*, which induce the covalent cross-links between telopeptides. These regulators influence the rate of assembly, size and structure of the collagen fibrils. Collagen XII and tenascin X form flexible bridges between collagen fibrils and other non-collagenous ECM molecules to regulate the organization and mechanical properties of collagen fibrils in several tissues. Heterozygous and biallelic variants in *COL12A1* have been found in individuals who presented with a phenotype that couples signs of EDS with myopathy (i.e. myopathic EDS). Recessive loss-of-function variants in *TNXB* are associated with classical-like EDS (type 1) (see "structure of the *TNXB*-surrounding genomic region and *TNXB* analysis").

9.6.2 Other Constituents of the Extracellular Matrix

GAGs form a considerable group of non-collagenous proteins abundant in connective tissue that display a variety of functions in ECM structure and remodeling. They are linear polysaccharides that form the side chain of proteoglycans. The structure of proteoglycans is composed of two fundamental molecules: core proteins and GAGs. Core protein represents a point of attachment to different GAGs such as heparan sulfate and/or dermatan sulfate (DS). Defects in the biosynthesis of GAGs are associated with rare forms of EDS, including spondylodysplastic and musculocontractural EDS. Spondylodysplastic EDS types 1 and 2 result from biallelic variants in *B4GALT7* and *B3GALT6* genes, which encode for galactosyltransferase I and galactosyltransferase II enzymes, respectively (Myllyharju and Kivirikko 2001; Nakajima et al. 2013). Both enzymes are involved in the biosynthesis of the proteoglycan core protein. Molecular studies on affected individuals' fibroblasts carrying *B4GALT7* or *B3GALT6* deleterious variants exhibited a reduced galactosyltransferase I and galactosyltransferase II enzyme activity, respectively, and showed an aberrant glycosylation of *B4GALT7* or *B3GALT6*-target proteins (Malfait et al. 2013; Ritelli et al. 2017).

A further form of spondylodysplastic EDS (type 3) is linked to defects in a protein involved in zinc transport, by deleterious variants in *SLC39A13* which encodes for ZIP13, a member of the LIV-1 subfamily of ZIP zinc transporters family (Bin et al. 2011). ZIP13 is important for connective tissue development, by monitoring the transport zinc and/or other metal ions from the extracellular space or from the organelles lumen into the cytoplasm. Deleterious variants in *SLC39A13* result in the disturbance of intracellular zinc homeostasis that appears to contribute to spondylodysplastic EDS pathogenesis. To date only three *SLC39A13* homozygous pathogenic variants have been identified: a 9-bp deletion, a missense and nonsense variants (Giunta et al. 2008; Fukada et al. 2008; Dusanic et al. 2018).

Musculocontractural EDS is caused by defects in the GAGs biosynthesis due to deleterious variants in *CHST14*, which encodes dermatan-4-O-sulfotransferase 1, and *DSE*, which encodes DS-epimerase 1. They encode two enzymes that are necessary for biosynthesis of DS, which is the core protein of GAGs (Miyake et al. 2010). Variants in *CHST14* and *DSE* lead to a reduced activity of DS that, in turn, compromises GAG chains and ECM organization. The reduced activity of DS has been demonstrated both in skin fibroblasts and urine from individuals harboring *CHST14* deleterious variants (Miyake et al. 2010; Mizumoto et al. 2017).

The ECM homeostasis is also assured by correct protein folding of mature fibrils of collagens in the endoplasmic reticulum (ER) lumen (Bateman et al. 2009). *FKBP14* encodes FKBP22, a peptidyl-prolyl cis-trans isomerase localized in the endoplasmic reticulum lumen that catalyzes procollagens folding. Biallelic variants in *FKBP14* cause a second form of kyphoscoliotic EDS. Inherited defects in this protein result in accumulation of collagen molecules in the ER, a fact that likely explains the enlargement of the ER cisterns that has been observed in *FKBP14*-related kyphoscoliotic EDS (Baumann et al. 2012).

The ECM equilibrium for the biosynthesis, processing and integrity of its constituents is maintained by the ECM protein degradation quality control system, which may be extended to the complement pathway. Kapferer-Seebacher and colleagues identified heterozygous variants in *C1R* or *C1S*, two contiguous genes encoding the complement subunits C1r and C1s as the genetic cause of periodontal EDS (Kapferer-Seebacher et al. 2016). These two proteins form a heterotetramer that combines with six C1q subunits and has a crucial role for targeted ECM proteins degradation.

In addition to ECM glycoproteins, many other proteins are matrix components required for the integrity of ECM. One of them is the aortic carboxypeptidase-like protein (ACLP) encoded by the *AEBP1* gene. ACLP binds fibrillary types I, III and V collagens, and assists in type I collagen polymerization (Blackburn et al. 2018). Recently, whole-exome sequencing study in individuals affected by an autosomal recessive variant of EDS reminiscent of classical EDS reported biallelic deleterious variants in *AEBP1* (Blackburn et al. 2018).

9.7 Diagnosis

In the field of JHM-related disorders, the "diagnosis" is a heterogeneous procedure mixing body measurements, review of systems for diagnostic criteria check listing and, in selected cases, molecular testing. Recognizing JHM is a physical procedure as previously reported (see "joint hypermobility"). Among individuals with a suspect of EDS, practitioners should take in their minds that many individuals with "symptomatic" JHM are usually affected by HSD or hEDS according to the current classification. Differently from the other "monogenic" EDS types whose recognition stands on positive molecular testing, the diagnosis of hEDS is uniquely established on the basis of clinical criteria (Table 9.2). Of relevance, criterion 3 of hEDS includes the exclusion of any other acquired and hereditary conditions which may represent an issue in specific cases. Therefore, although molecular testing is not indicated in all individuals with features of hEDS, it may be considered with a "case-by-case" approach. People with "symptomatic" JHM with specific musculoskeletal manifestations (see "the "spectrum" and hypermobility spectrum disorders") who do not respect the hEDS diagnostic criteria and have not received (or do not need) molecular confirmation of any other partially overlapping syndrome are labeled as HSD. Hence, molecular testing is currently indicated in (i) all individuals with JHM and a suspect of a genetic syndrome with known molecular basis, also including the monogenic EDS, and (ii) those with phenotypes compatible but not clinically convincing for a background diagnosis of hEDS or HSD. Molecular testing is not useful for distinguishing between HSD and hEDS.

The 2017 International Classification of EDS and related disorders highlights that the diagnosis of EDS should be confirmed by a "positive" molecular testing in all cases except for hEDS. If molecular testing is unavailable, a few exceptions exist in which the clinical diagnosis could be supported by other laboratory tests, including serum concentration measurement of tenascin-X for classical-type EDS (type 1) (Schalkwijk et al. 2001), and quantification of urinary deoxypyridinoline and pyridinoline crosslinks for *PLOD1*-related kyphoscoliotic EDS (Rohrbach et al. 2011) and, in a lesser extent, *SLC39A13*-related spondylodysplastic EDS (Giunta et al. 2008). However, these strategies are currently used for the study of variants of unknown significance detected at sequencing rather than as alternative first-tier analyses. Skin biopsy followed with electron microscopy is not efficient due to the very low predicting value of abnormal findings in most EDS types (Malfait et al. 2020). Therefore, molecular testing is currently accepted as the "gold standard" for the diagnosis of all EDS types (hEDS excluded).

Sanger sequencing has been widely used for several years for the molecular diagnosis of EDS (Ritelli et al. 2017). In the last decade, the advent of high-throughput next-generation sequencing (NGS) technologies has progressively changed the diagnostic approach in clinical laboratories by accelerating the rate of sequence generation and by reducing the costs. Based on laboratory setting, specificity of the phenotype and accuracy of the clinical diagnosis, first-tier analyses could be performed through three different approaches: (i) Sanger sequencing for selected genes or mutational hot-spots (e.g. arthrochalasia EDS or *COL5A1/COL5A2*-negative classical EDS), (ii) NGS analysis of custom targeted multigene NGS panels, or (iii) whole exome (WES) or whole genome (WGS) sequencing platforms. Sanger sequencing is usually limited to highly selected cases in which the causative variant is predicted falling in a single gene whose direct sequencing remains cost-effective compared to NGS according to the specific laboratory setting, or in case of stringent genotype-phenotype correlations linked to mutational hot-spots. The NGS multigene panel testing is probably the first-tier approach used in most specialized laboratories. Genes included in the panel could be restricted to those comprised in the current classification and those associated with emerging variants described after its publication in 2017 (i.e. *AEBP1*), or extended to partially overlapping conditions such as Marfan syndrome, Loeys-Dietz syndrome, hereditary cutis laxa syndromes, Stickler syndrome and myopathies with JHM. WES and WGS approaches are currently used with a virtual multigene panel strategy or with more open approaches based on the heterogeneity of the phenotype. In the latter case, the analysis is usually carried out on the proband and both parents simultaneously (trio analysis). Compared to WES, WGS data allow a higher predictive value for copy number variation (CNV) analysis (see below) and for the detection of balanced rearrangements.

In selected EDS types, the causative variant(s) is (are) single-exon, multi-exon or whole-gene deletion(s) or duplication(s). CNVs usually skip detection by standard NGS analyses carried out with short-reads NGS technologies. Selected laboratories implemented or are implementing *in silico* algorithms aimed at predicting CNVs, but this procedure has been currently integrated in the bioinformatics pipeline of a few laboratories only and its findings ideally need confirmation by a second technique for diagnostic purposes. CNVs could be investigated (or confirmed and refined) by second-tier technologies including locus-specific analyses (e.g. multiplex ligation probe amplification – MLPA) or high-resolution genomic hybridization arrays, including comparative genomic hybridization (CGH) and single nucleotide polymorphism (SNP) array (SNP-array). However, such approaches have a limited resolution at the exon level. Recently, a MLPA technology with extended CNV detection (i.e. digital MLPA) and SNP-array platforms with probes enriched for all coding exons have been commercialized.

An integrated analysis investigating point variants and CNVs, simultaneously or with a multi-tier approach, is recommended in most cases of molecular confirmation of EDS. In the current scenario, however, the analysis of *TNXB* remains a challenge due to the peculiar genomic structure in which *TNXB* maps. Therefore, the suspect of classical-like EDS (type 1) needs further molecular testing integrating NGS and exploring *TNXB/TNXA* chimeras and point variants in the *TNXB/TNXA* homologous regions (see "structure of the *TNXB*-surrounding genomic region and *TNXB* analysis"). In addition, intronic variants affecting canonical and non-canonical splice sites, and synonymous variants falling near to the exon-intron junctions detected at sequencing usually need demonstration of their effect in order to improve their clinical significance (unless this information is available in previous robust publications). In these cases, mRNA studies from peripheral blood or skin biopsy, or exon-trapping assays (in case of further sample unavailability, or genes not expressed in blood and skin) are indicated. Finally, several laboratories developed gel electrophoretic analyses of collagens I and III produced by cultured fibroblasts from affected individuals for the study of variants of unknown significance, also comprising splicing variants.

In the future, the integration of long reads and ultra-long reads NGS technologies, RNAseq and methylome studies, and more robust functional assays for variants of unknown significance in the diagnostics of EDS will certainly improve the detection rate and variant clinical interpretation in a more cost- and time-effective approach.

9.8 Structure of the *TNXB*-Surrounding Genomic Region and *TNXB* Analysis

Biallelic loss-of-function variants in *TNXB* result in autosomal recessive classical-like EDS (type 1). *TNXB* encodes for an ECM protein (i.e. tenascin X) 4267 amino acids long and with a mass of about 450 kDa. Tenascin X is a large ECM–forming glycoprotein characterized structurally by an N-terminal signal peptide followed by a series of repeats that resemble epidermal growth factor, a stretch of fibronectin type III repeats and a large C-terminal domain structurally related to fibrinogen (Bristow 1993). Tenascin X is a crucial component of connective tissue and is found in dermis, skeletal muscle, heart, and blood vessels and plays an important role in collagen fibrillogenesis and matrix maturation (Egging et al. 2007). It has been proposed to be essential for collagen I deposition by dermal fibroblasts and important for maintaining the distance between collagen fibrils by forming bridges through direct interactions with collagen fibrils (Lethias et al. 1996; Chiquet-Ehrismann et al. 2014; Bristow et al. 2005).

TNXB resides on chromosome 6p21.3, spans 68.2 kb and contains 44 exons. *TNXB* maps within the major histocompatibility complex locus of the human leukocyte antigen, which is characterized by highly homologous sequences between functional genes and their corresponding pseudogenes. *CYP21A2* encoding 21-hydroxylase, *RP1* encoding tryptophan/threonine kinase 19 (called "RP" in honor of the Nobel laureate Sir Rodney R. Porter; Shen et al. 1994), and *C4A* and *C4B* encoding the subcomplements C4a and C4b, are functional genes. *RP2*, *CYP21A1P* and *TNXA* are the nonfunctional forms of the *RP1*, *CYP21A2*, and *TNXB* genes, respectively. These genes are arranged in tandem in the RCCX module in the following order: *RP1–C4A-CYP21A1P–TNXA– RP2–C4B–CYP21A2–TNXB*. The *TNXA* and *TNXB* genes are located on the antisense strand, and their transcriptional direction is opposite that of the *C4s* and *CYP21s* genes (Lee 2005; Lee et al. 2004).

The last exon of *TNXA* and *TNXB* is located in the 3′-untranslated region of *CYP21A1P* and *CYP21A2*, respectively.

This genomic structure is prone to genic conversions and rearrangements, the latter resulting in the generation of chimeric *CYP21A1P/CYP21A2* and *TNXA/TNXB* genes combined with the deletion of the neighboring genes (Lee et al. 2004). When the resulting rearrangement cause haploinsufficiency of both *CYP21A2* and *TNXB* (by either chimera formation or large deletion), carrier individuals may be affected by both congenital adrenal hyperplasia (CAH) related to *CYP21A2*, and EDS. In this case, the term "CAH-X" syndrome is used (Burch et al. 1997). To date, three major types of *TNXA/TNXB* chimera have been identified to generate CAH-X rearrangements based on the junction site location (Miller and Merke 2018) (Fig. 9.3).

CAH-X chimera 1 (CH-1) and CAH-X chimera 2 (CH-2) have *TNXB* exons 35–44 and 40–44, respectively, replaced with *TNXA* (Burch et al. 1997; Merke et al. 2013; Morissette et al. 2015). CH-1 is characterized by a nonsense 120 base pairs deletion (c.11435_11524 + 30del) due to the substitution of *TNXB* exon 35 by *TNXA* that is the only well-documented discrepancy between *TNXB* and *TNXA* homologous sequence. CH-2 is characterized by the c.12174C > G, p.(Cys4058Trp) variant derived from the substitution of *TNXB* exon 40 by *TNXA*, which likely disrupts tenascin X function but not affect protein expression (Morissette et al. 2015; Chen et al. 2016). CAH-X chimera 3 (CH-3) has *TNXB* exons 41–44 substituted by *TNXA* and is characterized by a cluster of the three pseudogene variants: c.12218G > A p.(Arg4073His) in exon 41, and c.12514G > A, p.(Asp4172Asn) and c.12524G > A p.(Ser4175Asn) in exon 43.

Molecular testing of *TNXB* represents a challenging task for a diagnostic laboratory, due to the high homology between *TNXB* and *TNXA* fact that does not allow unequivocal assignment of sequence variants in this region to the gene or pseudogene. In fact, *TNXB* variants can be discarded and not be "called" in the short reads NGS data and *TNXA* variants can erroneously be assigned to *TNXB* (Mueller et al. 2013; Demirdas et al. 2017). Currently, effective mutational analysis of *TNXB* for a suspect of classical-like EDS (type 1) could include: (i) (short reads) NGS analysis of *TNXB* non-homologous region, (ii) multistep Sanger sequencing of the *TNXA/TNXB* homologous region, (iii) CNV analysis by MLPA kit or SNParray. Alternatively, the *TNXB* non-homologous region could be investigated by standard Sanger sequencing. The *TNXA/TNXB* homologous region study must be currently carried out by a combined long- and nested-polymerase chain reaction analysis.

9.9 Principles of Management

At the moment, there are no management, treatment and follow-up guidelines available for the various clinical manifestations related to JHM, HSD and EDS. Updated, expert-based recommendations are available in Malfait et al. (2020).

9.9.1 Mucocutaneous Manifestations and Tissue Fragility

While mucocutaneous manifestations are absent or minimal in HSD, doughy, soft, velvety and hyperextensible skin is one of the hallmark of EDS. Vascular EDS is the only consistent exception to this rule as it usually presents with thin and translucent skin. A limited subset of individuals with vascular EDS and atypical missense (glutamic acid to lysine) variants in *COL3A1* show significant cutaneous involvement with skin fragility and hyperextensibility (Ghali et al. 2019b). Delayed wound healing with propensity to atrophic scar formation is also common in the various EDS types. Skin fragility manifesting with propensity to lacerations and surgical complications is typical of EDS types with a more severe cutaneous involvement, such as classical EDS, kyphoscoliotic EDS, dermatosparaxis EDS and *AEBP1*-related classical-like EDS. In classical and *AEBP1*-related classical-like EDS types, defective tissue repair is marked with formation of extensively atrophic scars (i.e. papyraceous

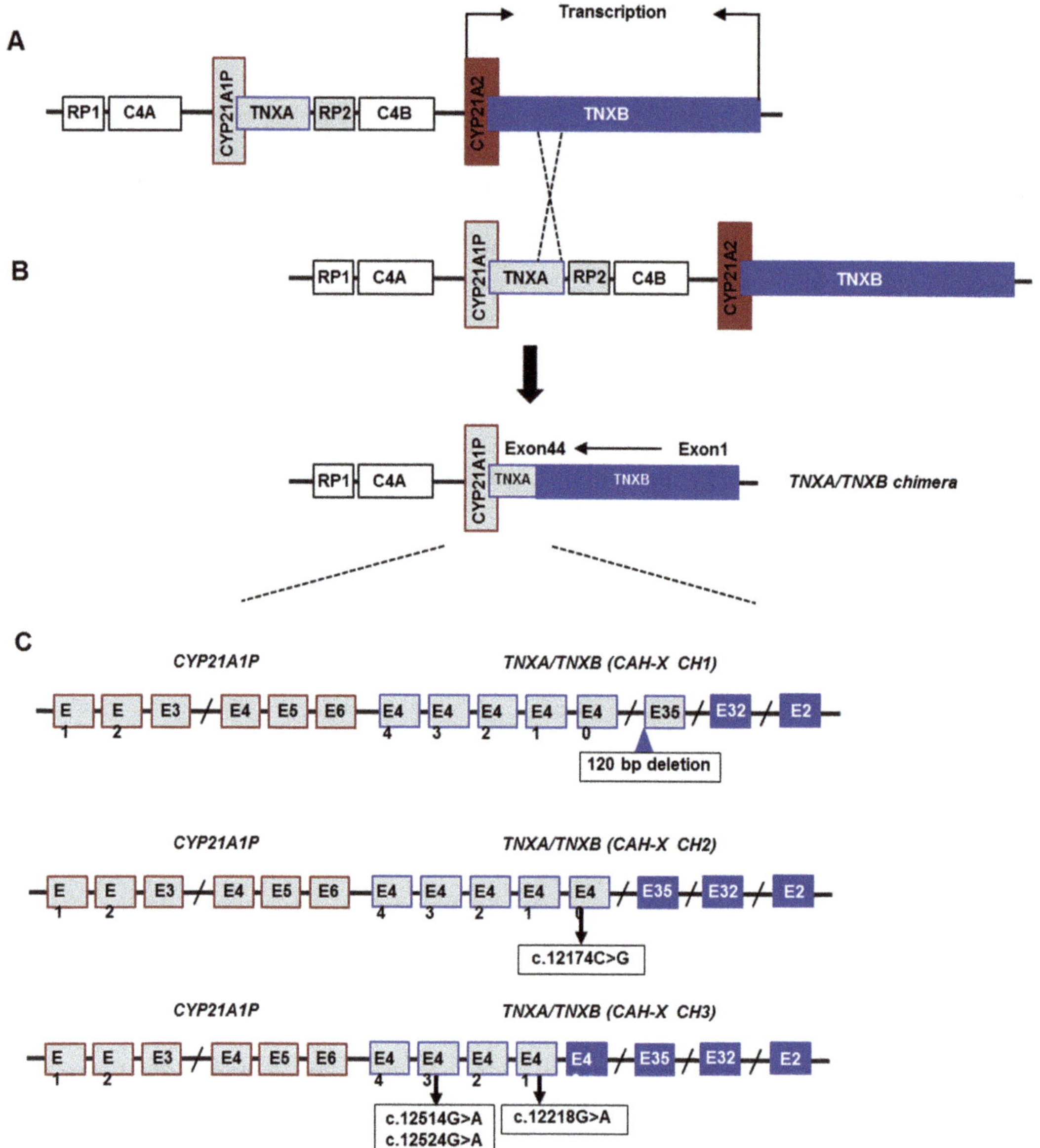

Fig. 9.3 Diagram of the *CYP21A2-TNXB* region and of the *TNXA/TNXB* chimeras CAH-X CH-1, CH-2 and CH-3. (A) The wild-type region is bimodular and includes two *RP-C4-CYP21-TNX* modules. (B) Unequal crossing over during meiosis generates the *TNXA/TNXB* chimeras, including CAH-X CH-1 (type 1), CH-2 (type 2) and CH-3 (type 3). (C) Representation of the three *TNXA/TNXB* chimeras. The solid triangle at the boundary of *TNXB* exon 35 and intron 35 denotes the 120 bp fragment absent in CAH-X CH-1. The lack of this 120 bp fragment is typical of CAH-X CH-1. The c.12174C > G pseudogene variant in exon 40 identifies CAH-X CH-2. CAH-X CH-3 is characterized by a cluster of 3 pseudogene variants: c.12218G > A in exon 41, and c.12514G > A and c.12524G > A in exon 43. Non-functional genes (pseudogenes) and their non-coding exons are indicated with grey boxes. *TNXB* and *CYP21A2* (and their exons are) ere indicated with blue and red boxes, respectively. The non-coding *TNXA* and *CYP21A1P* (and their non-coding exons) are marked with red and blue borders, respectively. The sizes of the exons and introns are not scaled

scars). Easy bruising is an additional characteristic cutaneous feature of EDS. In most EDS types, lesions are small- or medium-sized. In vascular EDS, easy bruising can manifest with extensive lesions involving entire limbs. Coagulation tests usually give normal or negative results.

People with EDS types with increased skin fragility could prevent lacerations, and delayed and defective scarring with protections especially during sport activities. This is particularly indicated in children. Education in preventing strategies and basic support after unexpected traumas is recommended in children and adults. Surgical repair should be carried out with care via sutures without tension. Stitches should be applied generously and left in place for a prolonged time (usually, twice as long as usual). Medications and tapes should be removed gently in order to prevent secondary skin lacerations. Easy bruising does not have any effective treatment. Anecdotal reports indicate amelioration by prolonged ascorbic acid oral assumption. People with EDS types with increased vascular fragility (e.g. vascular and kyphoscoliotic EDS) and those with other EDS types but a history of hemorrhagic events could benefit from an off-label use of 1-deamino-8-D-arginine vasopressin to prevent hemorrhages during major and minor surgery.

Fragility of the soft tissues is not limited to the dermis in EDS. People with vascular EDS have a risk of ~15% of perforation of the gastrointestinal tract, more commonly of the colon, in the absence of any underlying disease. Immediate surgery is the only indicated treatment and the most effective for reducing the risk of secondary infections. In case of colonic involvement, complete colectomy is the most commonly used approach. A study demonstrated that males with vascular EDS are more commonly affected and noted an increased risk of reperforation after resection of the rectosigmoid colon. Therefore, the authors proposed subtotal colectomy, particularly in males (Adham et al. 2019). Ruptures of the gut and perforation of colonic diverticula have been repeatedly described also in *CHST14*-related musculocontractural EDS and *TNXB*-related classical-like EDS (Kosho 2016; Demirdas et al. 2017; Green et al. 2020).

9.9.2 Secondary Musculoskeletal Manifestations

Management of secondary musculoskeletal manifestations is unrelated to the background phenotype (e.g. HSD, hEDS and EDS rare types). Occasional joint dislocations should be managed following standard procedures. People with JHM who experience a single dislocation should be instructed in the self-care of future events. Preventive strategies for activities at risk are indicated. Recurrent dislocations are often less painful, but may significantly impact quality of life. In people with JHM and recurrent dislocations, orthopedic surgery should be considered only after multidisciplinary consultation. Occasional and recurrent pain is often managed successfully by rest, short cycles of non-steroidal or steroidal anti-inflammatory drugs, and manual therapy. Recurrence of disabling events needs assessment by a specialist in physical medicine and rehabilitation for appropriate therapeutic and prevention strategies. People with chronic (widespread) musculoskeletal pain and neuropathic pain should be managed in a multidisciplinary setting also including the palliative care specialist for the appropriate use of opioids and drugs (neuropathic pain), specialist in physical therapy and rehabilitation medicine, and psychologist. Rest, cold or hot applications and non-steroidal anti-inflammatory drugs represent the routine clinical care of sprains, strains and soft-tissue injuries. Recurrence may be reduced by physical therapy programs improving proprioception, muscle tone and core stability. In children, neurodevelopmental attributes of JHM (i.e. developmental coordination disorders, learning difficulties, attention deficit/hyperactivity disorder) should be managed according standard procedures with focus on a likely underlying defect in proprioception.

9.9.3 Cardiovascular Manifestations

Cardiac valvular and arterial disease is absent or trivial in most EDS individuals and, in particular, hEDS. At the moment there is no evidence of progressive cardiovascular disease in HSD and non-syndromic, asymptomatic JHM. Conversely, progressive and potentially life-threatening cardiac valve disease has been considered the hallmark of cardiac valvular EDS due to biallelic null alleles in *COL1A2*. Vascular surgery is apparently the unique therapeutic resource for severe cardiac valvular disease in EDS and, ideally, is most effective if carried out as planned repair. Arterial/aortic aneurysms, spontaneous arterial rupture and dissections, arterial complications/fragility after surgery and trauma are typical of vascular EDS and are more common in kyphoscoliotic EDS. Elective surgical repair is the only available treatment for aneurysm. Although arterial complications has been registered in a wider spectrum of EDS types, the *a priori* risk seems relatively low for people with types other than vascular and kyphoscoliotic EDS (Malfait et al. 2017).

Two clinical trials have been published indicating a potential effect of celiprolol in reducing the risk of arterial ruptures in adults with vascular EDS (Ong et al. 2010; Frank et al. 2019). Given the rarity of the disorder, these studies have major limitations and concerns remain in the use of celiprolol in children with vascular EDS.

No specific recommendations are available for cardiovascular surveillance in EDS. Cardiac valves and thoracic aortic diameters could be assessed by heart ultrasound every 3–5 years or more frequently in presence of abnormal findings. The vascular tree could be assessed periodically (e.g. annually) in people with EDS types with increased vascular risk favoring non-invasive methods, such as head-to-pelvis magnetic resonance imaging angiography.

9.9.4 Reduced Bone Mass

Reduced bone mass is frequently observed in association with JHM, HSD and EDS (Guarnieri and Castori 2018). Recently, this feature is emerged as significantly impacting the overall clinical picture and, therefore, osteoporosis has been included among the JHM-related co-morbidities (Leganger et al. 2020). However, while documentation of a suboptimal bone composition is common in EDS, the risk of fractures is not well defined in the various JHM-related phenotypes (Guarnieri and Castori 2018; Basalom and Rauch 2020). Occasionally, (recurrent) fractures may occur in kyphoscoliotic and spondylodysplastic EDS (personal observation). There is also a subset of individuals with multiple features evocative of EDS in combination with minor manifestations, including mild increase of the pre-pubertal annual fracture rate, of osteogenesis imperfecta who share deleterious variants in *COL1A1* and *COL1A2*. Such a mixed phenotype has been recently defined COL1-related overlap disorder (Morlino et al. 2020). Recently, an increased prevalence of fractures has been also documented in women with hEDS/HSD compared to controls (Banica et al. 2020). Monitoring and treatment of reduced bone mass without evidence of an increased rate of fractures should be treated conservatively and following the recommendations available for the general population. People with EDS and a history of fractures, either spontaneous or after trivial/minor traumas, and those with COL1-related overlap disorder might benefit from the inclusion in an osteogenesis imperfecta management pathway.

9.9.5 Pregnancy and Anesthesiology

In the light of the current knowledge, pregnancy should not be an issue for fertile women with non-syndromic, asymptomatic JHM and HSD. Perhaps, those with HSD could have an increased risk of worsening of musculoskeletal pain and fatigue in the second half of the pregnancy and during puerperium. Concerning women with EDS, the associated risks are influenced by the type. Presumably in most EDS types but more commonly and severely in vascular EDS, pelvic floor laxity and membrane fragility may predispose to prematurity due to cervical incontinence or premature ruptures of the membranes. Women with vascular EDS also have an increased risk of maternal death due to uterine and vascular rupture in the third trimester, and of vaginal and cervical tears after vaginal delivery.

Recurrence risk remains currently undefined for non-syndromic, asymptomatic JHM and HSD. Couples with a member affected by hEDS has a theoretical risk of 50% of transmitting the disease to the offspring. However, the current uncertainties about the molecular basis of hEDS limit such an assumption. Couples with a member affected by any other Mendelian EDS type have a risk related to the inheritance pattern. If the causative genetic variant(s) is/are known, prenatal diagnosis is an option.

Tissue fragility rises concerns for anesthesia and procedures in people with EDS who need programmed and emergency surgical interventions. While HSD and hEDS seem to associate with minimal risks, this is not the same for the other Mendelian EDS types. Current literature is fragmented in single case reports and small case series and they are most commonly dedicated to vascular EDS. However, a series of (low) evidence-based recommendations are available in Wiesmann et al. (2014) and at https://www.orpha.net/data/patho/Ans/en/Ehlers-Danlos-Syndrome.pdf. Factors potentially influencing planning and outcome of anesthesiologic procedures may include (i) mucosal fragility, and temporomandibular joint and neck instability during intubation; (ii) blood volume instability during and after surgery due to unexpected arterial ruptures and/or defective tissue repair (major in vascular EDS); (iii) delayed wound healing; (iv) increased risk of orthostatic headache and cerebrospinal fluid leakage after spinal anesthesia; (v) resistance to local anesthesia; (vi) increased risk of pneumothorax during mechanical ventilation (vascular EDS); (vii) propensity to plexus and other compression neuropathies after prolonged uncomfortable positions.

9.9.6 Joint Hypermobility-Related Co-Morbidities

Comparably to secondary musculoskeletal manifestations, the management of JHM-related co-morbidities is unrelated to the background phenotype. At the moment, not any specific treatment is available for the wide range of JHM-related co-morbidities. Hence, their management should be based on the principles developed in the respective specialty setting. In people with evidence of fragility and/or softness of the tissues, surgical options should be considered in a multidisciplinary setting only.

References

Adham S, Zinzindohoué F, Jeunemaitre X, Frank M (2019) Natural history and surgical Management of Colonic Perforations in vascular Ehlers-Danlos syndrome: a retrospective review. Dis Colon Rectum 62:859–866

Banica T, Coussens M, Verroken C, Calders P, De Wandele I, Malfait F, Zmierczak HG, Goemaere S, Lapauw B, Rombaut L (2020) Higher fracture prevalence and smaller bone size in patients with hEDS/HSD-a prospective cohort study. Osteoporos Int 31:849–856

Basalom S, Rauch F (2020) Bone disease in patients with Ehlers-Danlos syndromes. Curr Osteoporos Rep 18:95–102

Bateman JF, Boot-Handford RP, Lamandé SR (2009) Genetic diseases of connective tissues: cellular and extracellular effects of ECM mutations. Nat Rev Genet 10:173–183

Baumann M, Giunta C, Krabichler B, Rüschendorf F, Zoppi N, Colombi M, Bittner RE, Quijano-Roy S, Muntoni F, Cirak S, Schreiber G, Zou Y, Hu Y, Romero NB, Carlier RY, Amberger A, Deutschmann A, Straub V, Rohrbach M, Steinmann B, Rostásy K, Karall D, Bönnemann CG, Zschocke J, Fauth C (2012) Mutations in FKBP14 cause a variant of Ehlers-Danlos syndrome with progressive kyphoscoliosis, myopathy, and hearing loss. Am J Hum Genet 90:201–216

Beighton P, de Paepe A, Danks D, Finidori G, Gedde-Dahl T, Goodman R, Hall JG, Hollister DW, Horton W, McKusick VA, Opitz JM, Pope FM, Pyeritz RE, Rimoin DL, Sillence D, Spranger JW, Thompson E, Tsipouras P, Viljoen D, Winship I, Young James I, Reynolds F (1988) International nosology of heritable disorders of connective tissue, Berlin, 1986. Am J Med Genet 29:581–594

Beighton P, De Paepe A, Steinmann B, Tsipouras P, Wenstrup RJ (1998) Ehlers-Danlos syndromes: revised nosology, Villefranche, 1997. Ehlers-Danlos National Foundation (USA) and Ehlers-Danlos support group (UK). Am J Med Genet 77:31–37

Bin BH, Fukada T, Hosaka T, Yamasaki S, Ohashi W, Hojyo S, Miyai T, Nishida K, Yokoyama S, Hirano T (2011) Biochemical characterization of human ZIP13 protein: a homo-dimerized zinc transporter involved in the spondylocheiro dysplastic Ehlers-Danlos syndrome. J Biol Chem 286:40255–40265

Birk DE (2001) Type V collagen: heterotypic type I/V collagen interactions in the regulation of fibril assembly. Micron 32:223–237

Blackburn PR, Xu Z, Tumelty KE, Zhao RW, Monis WJ, Harris KG, Gass JM, Cousin MA, Boczek NJ, Mitkov MV, Cappel MA, Francomano CA, Parisi JE, Klee EW, Faqeih E, Alkuraya FS, Layne MD, McDonnell NB, Atwal PS (2018) Bi-allelic alterations in AEBP1 Lead to defective collagen assembly and connective tissue structure resulting in a variant of Ehlers-Danlos syndrome. Am J Hum Genet 102:696–705

Bloom L, Byers P, Francomano C, Tinkle B, Malfait F, Steering Committee of The International Consortium on the Ehlers-Danlos Syndromes (2017) The international consortium on the Ehlers-Danlos syndromes. Am J Med Genet C Semin Med Genet 175:5–7

Bristow MR (1993) Changes in myocardial and vascular receptors in heart failure. J Am Coll Cardiol 22:61A–71A

Bristow J, Carey W, Egging D, Schalkwijk J (2005) Tenascin-X, collagen, elastin, and the Ehlers-Danlos syndrome. Am J Med Genet C Semin Med Genet 139:24–30

Bulbena A, Gago J, Pailhez G, Sperry L, Fullana MA, Vilarroya O (2011) Joint hypermobility syndrome is a risk factor trait for anxiety disorders: a 15-year follow-up cohort study. Gen Hosp Psychiatry 33:363–370

Bulbena A, Baeza-Velasco C, Bulbena-Cabré A, Pailhez G, Critchley H, Chopra P, Mallorquí-Bagué N, Frank C, Porges S (2017) Psychiatric and psychological aspects in the Ehlers-Danlos syndromes. Am J Med Genet C Semin Med Genet 175:237–245

Burch GH, Gong Y, Liu W, Dettman RW, Curry CJ, Smith L, Miller WL, Bristow J (1997) Tenascin-X deficiency is associated with Ehlers-Danlos syndrome. Nat Genet 17:104–108

Campbell ID, Humphries MJ (2011) Integrin structure, activation, and interactions. Cold Spring Harb Perspect Biol 3:a004994

Castori M (2016) Pain in Ehlers-Danlos syndromes: manifestations, therapeutic strategies and future perspectives. Exp Opin Orphan Drugs 4:1145–1158

Castori M, Colombi M (2015) Generalized joint hypermobility, joint hypermobility syndrome and Ehlers-Danlos syndrome, hypermobility type. Am J Med Genet C Semin Med Genet 169C:1–5

Castori M, Hakim A (2017) Contemporary approach to joint hypermobility and related disorders. Curr Opin Pediatr 29:640–649

Castori M, Morlino S, Celletti C, Ghibellini G, Bruschini M, Grammatico P, Blundo C, Camerota F (2013) Re-writing the natural history of pain and related symptoms in the joint hypermobility syndrome/Ehlers-Danlos syndrome, hypermobility type. Am J Med Genet A 161A:2989–3004

Castori M, Dordoni C, Valiante M, Sperduti I, Ritelli M, Morlino S, Chiarelli N, Celletti C, Venturini M, Camerota F, Calzavara-Pinton P, Grammatico P, Colombi M (2014) Nosology and inheritance pattern(s) of joint hypermobility syndrome and Ehlers-Danlos syndrome, hypermobility type: a study of intrafamilial and interfamilial variability in 23 Italian pedigrees. Am J Med Genet A 164A:3010–3020

Castori M, Tinkle B, Levy H, Grahame R, Malfait F, Hakim A (2017) A framework for the classification of joint hypermobility and related conditions. Am J Med Genet C Semin Med Genet 175:148–157

Cazzato D, Castori M, Lombardi R, Caravello F, Bella ED, Petrucci A, Grammatico P, Dordoni C, Colombi M, Lauria G (2016) Small fiber neuropathy is a common feature of Ehlers-Danlos syndromes. Neurology 87:155–159

Chen W, Perritt AF, Morissette R, Dreiling JL, Bohn MF, Mallappa A, Xu Z, Quezado M, Merke DP (2016) Ehlers-Danlos syndrome caused by Biallelic TNXB variants in patients with congenital adrenal hyperplasia. Hum Mutat 37:893–897

Cheng JC, Chan PS, Hui PW (1991) Joint laxity in children. J Pediatr Orthop 11:752–756

Chernogubow AN (1892) Uber einen Fall von Cutis Laxa [German]. Jahresber Ges Med 27:562

Chiarelli N, Carini G, Zoppi N, Dordoni C, Ritelli M, Venturini M, Castori M, Colombi M (2016) Transcriptome-wide expression profiling in skin fibroblasts of patients with joint hypermobility syndrome/Ehlers-Danlos syndrome hypermobility type. PLoS One 11:e0161347

Chiodo AA, Hockey A, Cole WG (1992) A base substitution at the splice acceptor site of intron 5 of the COL1A2 gene activates a cryptic splice site within exon 6 and generates abnormal type I procollagen in a patient with Ehlers-Danlos syndrome type VII. J Biol Chem 267:6361–6369

Chiquet-Ehrismann R, Orend G, Chiquet M, Tucker RP, Midwood KS (2014) Tenascins in stem cell niches. Matrix Biol 37:112–123

Danlos M (1908) Un cas de cutis laxa avec tumeurs par contusion chronique des coudes et des genoux (xanthome juvénile pseudo-diabétique de MM. Hallopeau et Macé de Lépinay) [French]. Bull Soc Fr Dermatol Syphiligr 19:70–72

Demirdas S, Dulfer E, Robert L, Kempers M, van Beek D, Micha D, van Engelen BG, Hamel B, Schalkwijk J, Loeys B, Maugeri A, Voermans NC (2017) Recognizing the tenascin-X deficient type of Ehlers-Danlos syndrome: a cross-sectional study in 17 patients. Clin Genet 91:411–425

Dusanic M, Dekomien G, Lücke T, Vorgerd M, Weis J, Epplen JT, Köhler C, Hoffjan S (2018) Novel nonsense mutation in SLC39A13 initially presenting as myopathy: case report and review of the literature. Mol Syndromol 9:100–109

Egging D, van Vlijmen-Willems I, van Tongeren T, Schalkwijk J, Peeters A (2007) Wound healing in tenascin-X deficient mice suggests that tenascin-X is involved in matrix maturation rather than matrix deposition. Connect Tissue Res 48:93–98

Ehlers E (1901) Cutis laxa, neigung zu haemorrhagien in der haut, lockerung meherer artikulationen [German]. Dermatol Z 8:173–174

Fikree A, Aktar R, Grahame R, Hakim AJ, Morris JK, Knowles CH, Aziz Q (2015) Functional gastrointestinal disorders are associated with the joint hypermobility syndrome in secondary care: a case-control study. Neurogastroenterol Motil 27:569–579

Frank M, Adham S, Seigle S, Legrand A, Mirault T, Henneton P, Albuisson J, Denarié N, Mazzella JM, Mousseaux E, Messas E, Boutouyrie P, Jeunemaitre X (2019) Vascular Ehlers-Danlos syndrome: long-term observational study. J Am Coll Cardiol 73:1948–1957

Fukada T, Civic N, Furuichi T, Shimoda S, Mishima K, Higashiyama H, Idaira Y, Asada Y, Kitamura H, Yamasaki S, Hojyo S, Nakayama M, Ohara O, Koseki H, Dos Santos HG, Bonafe L, Ha-Vinh R, Zankl A, Unger S, Kraenzlin ME, Beckmann JS, Saito I, Rivolta C, Ikegawa S, Superti-Furga A, Hirano T (2008) The zinc transporter SLC39A13/ZIP13 is required for connective tissue development; its involvement in BMP/TGF-beta signaling pathways. PLoS One 3:e3642

Ghali N, Baker D, Brady AF, Burrows N, Cervi E, Cilliers D, Frank M, Germain DP, Hulmes DJS, Jacquemont ML, Kannu P, Lefroy H, Legrand A, Pope FM, Robertson L, Vandersteen A, von Klemperer K, Warburton R, Whiteford M, van Dijk FS (2019a) Atypical COL3A1 variants (glutamic acid to lysine) cause vascular Ehlers-Danlos syndrome with a consistent phenotype of tissue fragility and skin hyperextensibility. Genet Med 21:2081–2091

Ghali N, Sobey G, Burrows N (2019b) Ehlers-Danlos syndromes. BMJ 366:l4966

Ghibellini G, Brancati F, Castori M (2015) Neurodevelopmental attributes of joint hypermobility syndrome/Ehlers-Danlos syndrome, hypermobility type: update and perspectives. Am J Med Genet C Semin Med Genet 169C:107–116

Giunta C, Randolph A, Steinmann B (2005) Mutation analysis of the PLOD1 gene: an efficient multistep approach to the molecular diagnosis of the kyphoscoliotic type of Ehlers-Danlos syndrome (EDS VIA). Mol Genet Metab 86:269–276

Giunta C, Elçioglu NH, Albrecht B, Eich G, Chambaz C, Janecke AR, Yeowell H, Weis M, Eyre DR, Kraenzlin M, Steinmann B (2008) Spondylocheiro dysplastic form of the Ehlers-Danlos syndrome--an autosomal-recessive entity caused by mutations in the zinc transporter gene SLC39A13. Am J Hum Genet 82:1290–1305

Grahame R, Bird HA, Child A (2000) The revised (Brighton 1998) criteria for the diagnosis of bening joint hypermobility syndrome (BJHS). J Rheumatol 27:1777–1779

Green C, Ghali N, Akilapa R, Angwin C, Baker D, Bartlett M, Bowen J, Brady AF, Brock J, Chamberlain E, Cheema H, McConnell V, Crookes R, Kazkaz H, Johnson D, Pope FM, Vandersteen A, Sobey G, van Dijk FS (2020) Classical-like Ehlers-Danlos syndrome: a clinical description of 20 newly identified individuals with evidence of tissue fragility. Genet Med 22:1576–1582

Guarnieri V, Castori M (2018) Clinical relevance of joint hypermobility and its impact on musculoskeletal pain and bone mass. Curr Osteoporos Rep 16:333–343

Hakim AJ, Grahame R (2003) A simple questionnaire to detect hypermobility: an adjunct to the assessment of patients with diffuse musculoskeletal pain. Int J Clin Pract v57:163–166

Hakim AJ, Grahame R (2004) Non-musculoskeletal symptoms in joint hypermobility syndrome. Indirect evidence for autonomic dysfunction? Rheumatology (Oxford) 43:1194–1195

Hakim A, O'Callaghan C, De Wandele I, Stiles L, Pocinki A, Rowe P (2017) Cardiovascular autonomic dysfunction in Ehlers-Danlos syndrome-hypermobile type. Am J Med Genet C Semin Med Genet 175:168–174

Hakim A, De Wandele I, O'Callaghan C, Pocinki A, Rowe P (2017b) Chronic fatigue in Ehlers-Danlos syndrome-hypermobile type. Am J Med Genet C Semin Med Genet 175:175–180

Hennekam RC, Biesecker LG, Allanson JE, Hall JG, Opitz JM, Temple IK, Carey JC, Elements of Morphology Consortium (2013) Elements of morphology: general terms for congenital anomalies. Am J Med Genet A 161A:2726–2733

Janecke AR, Li B, Boehm M, Krabichler B, Rohrbach M, Müller T, Fuchs I, Golas G, Katagiri Y, Ziegler SG, Gahl WA, Wilnai Y, Zoppi N, Geller HM, Giunta C, Slavotinek A, Steinmann B (2016) The phenotype of the musculocontractural type of Ehlers-Danlos syndrome due to CHST14 mutations. Am J Med Genet A 170A:103–115

Jover E, Silvente A, Marín F, Martínez-González J, Orriols M, Martinez CM, Puche CM, Valdés M, Rodriguez C, Hernández-Romero D (2018) Inhibition of enzymes involved in collagen cross-linking reduces vascular smooth muscle cell calcification. FASEB J 32:4459–4469

Juul-Kristensen B, Schmedling K, Rombaut L, Lund H, Engelbert RH (2017) Measurement properties of clinical assessment methods for classifying generalized joint hypermobility-a systematic review. Am J Med Genet C Semin Med Genet 175:116–147

Kapferer-Seebacher I, Pepin M, Werner R, Aitman TJ, Nordgren A, Stoiber H, Thielens N, Gaboriaud C, Amberger A, Schossig A, Gruber R, Giunta C, Bamshad M, Björck E, Chen C, Chitayat D, Dorschner M, Schmitt-Egenolf M, Hale CJ, Hanna D, Hennies HC, Heiss-Kisielewsky I, Lindstrand A, Lundberg P, Mitchell AL, Nickerson DA, Reinstein E, Rohrbach M, Romani N, Schmuth M, Silver R, Taylan F, Vandersteen A, Vandrovcova J, Weerakkody R, Yang M, Pope FM, Molecular Basis of Periodontal EDS Consortium, Byers PH, Zschocke J (2016) Periodontal Ehlers-Danlos syndrome is caused by mutations in C1R and C1S, which encode subcomponents C1r and C1s of complement. Am J Hum Genet 99:1005–1014

Karamanos NK (2019) Extracellular matrix: key structural and functional meshwork in health and disease. FEBS J 286:2826–2829

Kirk JA, Ansell BM, Bywaters EG (1967) The hypermobility syndrome. Musculoskeletal complaints associated with generalized joint hypermobility. Ann Rheum Dis 26:419–425

Kosho T (2016) CHST14/D4ST1 deficiency: new form of Ehlers-Danlos syndrome. Pediatr Int 58:88–99

Lam CY, Palsson OS, Whitehead WE, Sperber AD, Tornblom H, Simren M, Aziz I (2020) Rome IV functional gastrointestinal disorders and health impairment in subjects with hypermobility Spectrum disorders or hypermobile Ehlers-Danlos syndrome. Clin Gastroenterol Hepatol S1542-3565(20):30204–30204. (in press)

Lee HH (2005) Diversity of the CYP21P-like gene in CYP21 deficiency. DNA Cell Biol 24:1–9

Lee HH, Lee YJ, Lin CY (2004) PCR-based detection of the CYP21 deletion and TNXA/TNXB hybrid in the RCCX module. Genomics 83:944–950

Leganger J, Fonnes S, Kulas Søborg ML, Rosenberg J, Burcharth J (2020) The most common comorbidities in patients with Ehlers-Danlos syndrome: a 15-year nationwide population-based cohort study. Disabil Rehabil:1–5. (in press)

Lethias C, Labourdette L, Willems R, Comte J, Herbage D (1996) Composition and organization of the extracellular matrix of vein walls: collagen networks. Int Angiol 15:104–113

Lim PJ, Lindert U, Opitz L, Hausser I, Rohrbach M, Giunta C (2019) Transcriptome profiling of primary skin fibroblasts reveal distinct molecular features between PLOD1- and FKBP14-Kyphoscoliotic Ehlers-Danlos syndrome. Genes (Basel) 10:517

Malfait F (2018) Vascular aspects of the Ehlers-Danlos syndromes. Matrix Biol 71-72:380–395

Malfait F, Kariminejad A, Van Damme T, Gauche C, Syx D, Merhi-Soussi F, Gulberti S, Symoens S, Vanhauwaert S, Willaert A, Bozorgmehr B, Kariminejad MH, Ebrahimiadib N, Hausser I, Huysseune A, Fournel-Gigleux S, De Paepe A (2013) Defective initiation of glycosaminoglycan synthesis due to B3GALT6 mutations causes a pleiotropic Ehlers-Danlos-syndrome-like connective tissue disorder. Am J Hum Genet 92:935–945

Malfait F, Francomano C, Byers P, Belmont J, Berglund B, Black J, Bloom L, Bowen JM, Brady AF, Burrows NP, Castori M, Cohen H, Colombi M, Demirdas S, De Backer J, De Paepe A, Fournel-Gigleux S, Frank M, Ghali N, Giunta C, Grahame R, Hakim A, Jeunemaitre X, Johnson D, Juul-Kristensen B, Kapferer-Seebacher I, Kazkaz H, Kosho T, Lavallee ME, Levy H, Mendoza-Londono R, Pepin M, Pope FM, Reinstein E, Robert L, Rohrbach M, Sanders L, Sobey GJ, Van Damme T, Vandersteen A, van Mourik C, Voermans N, Wheeldon N, Zschocke J, Tinkle B (2017) The 2017 international classification of the Ehlers-Danlos syndromes. Am J Med Genet C Semin Med Genet 175:8–26

Malfait F, Castori M, Francomano CA, Giunta C, Kosho T, Byers PH (2020) The Ehlers-Danlos syndromes. Nat Rev Dis Primers 6:64

Merke DP, Chen W, Morissette R, Xu Z, Van Ryzin C, Sachdev V, Hannoush H, Shanbhag SM, Acevedo AT, Nishitani M, Arai AE, McDonnell NB (2013) Tenascin-X haploinsufficiency associated with Ehlers-Danlos syndrome in patients with congenital adrenal hyperplasia. J Clin Endocrinol Metab 98:E379–E387

Miller W, Merke DP (2018) Tenascin-X, congenital adrenal hyperplasia, and the CAH-X syndrome. Horm Res Paediatr 89:352–361

Miller AJ, Stiles LE, Sheehan T, Bascom R, Levy HP, Francomano CA, Arnold AC (2020) Prevalence of hypermobile Ehlers-Danlos syndrome in postural orthostatic tachycardia syndrome. Auton Neurosci 224:102637

Miyake N, Kosho T, Mizumoto S, Furuichi T, Hatamochi A, Nagashima Y, Arai E, Takahashi K, Kawamura R, Wakui K, Takahashi J, Kato H, Yasui H, Ishida T, Ohashi H, Nishimura G, Shiina M, Saitsu H, Tsurusaki Y, Doi H, Fukushima Y, Ikegawa S, Yamada S, Sugahara K, Matsumoto N (2010) Loss-of-function mutations of CHST14 in a new type of Ehlers-Danlos syndrome. Hum Mutat 31:966–974

Mizumoto S, Kosho T, Hatamochi A, Honda T, Yamaguchi T, Okamoto N, Miyake N, Yamada S, Sugahara K (2017) Defect in dermatan sulfate in urine of patients with Ehlers-Danlos syndrome caused by a CHST14/D4ST1 deficiency. Clin Biochem 50:670–677

Morissette R, Chen W, Perritt AF, Dreiling JL, Arai AE, Sachdev V, Hannoush H, Mallappa A, Xu Z, McDonnell NB, Quezado M, Merke DP (2015) Broadening the Spectrum of Ehlers Danlos syndrome in patients with congenital adrenal hyperplasia. J Clin Endocrinol Metab 100:E1143–E1152

Morlino S, Micale L, Ritelli M, Rohrbach M, Zoppi N, Vandersteen A, Mackay S, Agolini E, Cocciadiferro D, Sasaki E, Madeo A, Ferraris A, Reardon W, Di Rocco M, Novelli A, Grammatico P, Malfait F, Mazza T, Hakim A, Giunta C, Colombi M, Castori M (2020) COL1-related overlap disorder: a novel connective tissue disorder incorporating the osteogenesis imperfecta/Ehlers-Danlos syndrome overlap. Clin Genet 97:396–406

Mueller PW, Lyons J, Kerr G, Haase CP, Isett RB (2013) Standard enrichment methods for targeted next-generation sequencing in high-repeat genomic regions. Genet Med 15:910–911

Murphy-Ullrich J, Helene SE (2014) Revisiting the matricellular concept. Matrix Biol 37:1–14

Myllyharju J, Kivirikko KI (2001) Collagens and collagen-related diseases. Ann Med 33:7–21

Nakajima M, Mizumoto S, Miyake N, Kogawa R, Iida A, Ito H, Kitoh H, Hirayama A, Mitsubuchi H, Miyazaki O, Kosaki R, Horikawa R, Lai A, Mendoza-Londono R, Dupuis L, Chitayat D, Howard A, Leal GF, Cavalcanti D, Tsurusaki Y, Saitsu H, Watanabe S, Lausch E, Unger S, Bonafé L, Ohashi H, Superti-Furga A, Matsumoto N, Sugahara K, Nishimura G, Ikegawa S (2013) Mutations in B3GALT6, which encodes a glycosaminoglycan linker region enzyme, cause a spectrum of skeletal and connective tissue disorders. Am J Hum Genet 92:927–934

Nallanthighal S, Heiserman JP, Cheon DJ (2019) The role of the extracellular matrix in Cancer Stemness. Front Cell Dev Biol 7:86

Ong KT, Perdu J, De Backer J, Bozec E, Collignon P, Emmerich J, Fauret AL, Fiessinger JN, Germain DP, Georgesco G, Hulot JS, De Paepe A, Plauchu H, Jeunemaitre X, Laurent S, Boutouyrie P (2010) Effect of celiprolol on prevention of cardiovascular events in vascular Ehlers-Danlos syndrome: a prospective randomised, open, blinded-endpoints trial. Lancet 376:1476–1484

Pepin MG, Schwarze U, Rice KM, Liu M, Leistritz D, Byers PH (2014) Survival is affected by mutation type and molecular mechanism in vascular Ehlers-Danlos syndrome (EDS type IV). Genet Med 16:881–888

Piedimonte C, Penge R, Morlino S, Sperduti I, Terzani A, Giannini MT, Colombi M, Grammatico P, Cardona F, Castori M (2018) Exploring relationships between joint hypermobility and neurodevelopment in children (4-13 years) with hereditary connective tissue disorders and developmental coordination disorder. Am J Med Genet B Neuropsychiatr Genet 177:546–556

Rautavuoma K, Takaluoma K, Sormunen R, Myllyharju J, Kivirikko KI, Soininen R (2004) Premature aggregation of type IV collagen and early lethality in lysyl hydroxylase 3 null mice. Proc Natl Acad Sci U S A 101:14120–14125

Remvig L, Jensen DV, Ward RC (2007) Epidemiology of general joint hypermobility and basis for the proposed criteria for benign joint hypermobility syndrome: review of the literature. J Rheumatol 34:804–809

Ritelli M, Dordoni C, Venturini M, Chiarelli N, Quinzani S, Traversa M, Zoppi N, Vascellaro A, Wischmeijer A, Manfredini E, Garavelli L, Calzavara-Pinton P, Colombi M (2013) Clinical and molecular characterization of 40 patients with classic Ehlers-Danlos syndrome: identification of 18 COL5A1 and 2 COL5A2 novel mutations. Orphanet J Rare Dis 8:58

Ritelli M, Dordoni C, Cinquina V, Venturini M, Calzavara-Pinton P, Colombi M (2017) Expanding the clinical and mutational spectrum of B4GALT7-spondylodysplastic Ehlers-Danlos syndrome. Orphanet J Rare Dis 12:153

Rohrbach M, Vandersteen A, Yiş U, Serdaroglu G, Ataman E, Chopra M, Garcia S, Jones K, Kariminejad A, Kraenzlin M, Marcelis C, Baumgartner M, Giunta C (2011) Phenotypic variability of the kyphoscoliotic type of Ehlers-Danlos syndrome (EDS VIA): clinical, molecular and biochemical delineation. Orphanet J Rare Dis 6:46

Roma M, Marden CL, De Wandele I, Francomano CA, Rowe PC (2018) Postural tachycardia syndrome and other forms of orthostatic intolerance in Ehlers-Danlos syndrome. Auton Neurosci 215:89–96

Sasarman F, Maftei C, Campeau PM, Brunel-Guitton C, Mitchell GA, Allard P (2016) Biosynthesis of glycosaminoglycans: associated disorders and biochemical tests. J Inherit Metab Dis 39:173–188

Schalkwijk J, Zweers MC, Steijlen PM, Dean WB, Taylor G, van Vlijmen IM, van Haren B, Miller WL, Bristow J (2001) A recessive form of the Ehlers-Danlos syndrome caused by tenascin-X deficiency. N Engl J Med 345:1167–1175

Seow CC, Chow PK, Khong KS (1999) A study of joint mobility in a normal population. Ann Acad Med Singap 28:231–236

Shen L, Wu LC, Sanlioglu S, Chen R, Mendoza AR, Dangel AW, Carroll MC, Zipf WB, Yu CY (1994) Structure and genetics of the partially duplicated gene RP located immediately upstream of the complement C4A and the C4B genes in the HLA class III region. Molecular cloning, exon-intron structure, composite retroposon, and breakpoint of gene duplication. J Biol Chem 269:8466–8476

Smith LT, Wertelecki W, Milstone LM, Petty EM, Seashore MR, Braverman IM, Jenkins TG, Byers PH (1992) Human dermatosparaxis: a form of Ehlers-Danlos syndrome that results from failure to remove the amino-terminal propeptide of type I procollagen. Am J Hum Genet 51:235–244

Tinkle BT, Bird HA, Grahame R, Lavallee M, Levy HP, Sillence D (2009) The lack of clinical distinction between the hypermobility type of Ehlers-Danlos syndrome and the joint hypermobility syndrome (a.k.a. hypermobility syndrome). Am J Med Genet A 149A:2368–2370

Tinkle B, Castori M, Berglund B, Cohen H, Grahame R, Kazkaz H, Levy H (2017) Hypermobile Ehlers-Danlos syndrome (a.k.a. Ehlers-Danlos syndrome type III and Ehlers-Danlos syndrome hypermobility type): clinical description and natural history. Am J Med Genet C Semin Med Genet 175:48–69

Veit-Rubin N, Cartwright R, Singh AU, Digesu GA, Fernando R, Khullar V (2016) Association between joint hypermobility and pelvic organ prolapse in women: a systematic review and meta-analysis. Int Urogynecol J 27:1469–1478

Wenstrup RJ, Florer JB, Brunskill EW, Bell SM, Chervoneva I, Birk DE (2004) Type V collagen controls the initiation of collagen fibril assembly. J Biol Chem 279:53331–53337

Wiesmann T, Castori M, Malfait F, Wulf H (2014) Recommendations for anesthesia and perioperative management in patients with Ehlers-Danlos syndrome(s). Orphanet J Rare Dis 9:109

Zahed-Cheikh M, Tosello B, Coze S, Gire C (2017) Kyphoscolitic type of Ehlers-Danlos syndrome with prenatal stroke. Indian Pediatr 54:495–497

Zoppi N, Gardella R, De Paepe A, Barlati S, Colombi M (2004) Human fibroblasts with mutations in COL5A1 and COL3A1 genes do not organize collagens and fibronectin in the extracellular matrix, down-regulate alpha2beta1 integrin, and recruit alphavbeta3 instead of alpha5beta1 integrin. J Biol Chem 279:18157–18168

10 Ehlers Danlos Syndrome with Glycosaminoglycan Abnormalities

Noriko Miyake, Tomoki Kosho, and Naomichi Matsumoto

Abstract

Ehlers–Danlos syndrome (EDS) is a genetically and clinically heterogeneous group of connective tissue disorders that typically present with skin hyperextensibility, joint hypermobility, and tissue fragility. The major cause of EDS appears to be impaired biosynthesis and enzymatic modification of collagen. In this chapter, we discuss two types of EDS that are associated with proteoglycan abnormalities: spondylodysplastic EDS and musculocontractural EDS. Spondylodysplastic EDS is caused by pathogenic variants in *B4GALT7* or *B3GALT6*, both of which encode key enzymes that initiate glycosaminoglycan synthesis. Musculocontractural EDS is caused by mutations in *CHST14* or *DSE*, both of which encode enzymes responsible for the post-translational biosynthesis of dermatan sulfate. The clinical and molecular characteristics of both types of EDS are described in this chapter.

Keywords

Ehlers · Danlos syndrome (EDS) · Spondylodysplastic type · Musculocontractural type · *B4GALT7* · *B3GALT6* · *CHST14* · *DSE* · Proteoglycan · Glycosaminoglycan (GAG)

N. Miyake (✉)
Department of Human Genetics, National Center for Global Health and Medicine, Tokyo, Japan
e-mail: nomiyake@ri.ncgm.go.jp

T. Kosho
Department of Medical Genetics, Shinshu University School of Medicine, Matsumoto, Japan
e-mail: ktomoki@shinshu-u.ac.jp

N. Matsumoto
Department of Human Genetics, Yokohama City University Graduate School of Medicine, Yokohama, Japan
e-mail: naomat@yokohama-cu.ac.jp

Abbreviations

B4GALT7	Galactosyltransferase I
CHST14	Carbohydrate sulfotransferase 14
CS	Chondroitin sulfate
D4ST1	Dermatan 4-*O*-sulfotransferase 1
DS	Dermatan sulfate
DSE	Dermatan sulfate epimerase
EDS	Ehlers–Danlos syndrome
GAG	Glycosaminoglycan
Gal	Galactose
GalNAc	*N*-acetyl-D-galactosamine
GlcA	Glucuronic acid
IdoA	Iduronic acid
mcEDS	Musculocontractural EDS
PG	Proteoglycan
spEDS	Spondylodysplastic EDS
Xyl	Xylose

J. Halper (ed.), *Progress in Heritable Soft Connective Tissue Diseases*, Advances in Experimental Medicine and Biology 1348, https://doi.org/10.1007/978-3-030-80614-9_10

10.1 Introduction

Ehlers–Danlos syndrome (EDS) is a connective tissue disorder that affects as many as 1 in 5000 individuals. It is characterized by joint and skin laxity, and tissue fragility (Steinmann, and R.P.M., Superti-Furga A. 2002). EDS is clinically and genetically heterogeneous, and has been repeatedly reclassified upon the discovery of new genes and pathomechanism (Beighton et al. 1988, 1998). In the latest classification of EDS established in 2017, the syndrome is classified into 13 subtypes (Malfait et al. 2017) and it can also be classified into seven groups based on the molecular pathomechanism (Table 10.1). Among these, two subtypes, spondylodysplastic and musculocontractural EDS, are classified as "Group D, disorders of glycosaminoglycan (GAG) biosynthesis" (Malfait et al. 2017). Specifically, impaired GAG synthesis causes spondylodysplastic EDS (spEDS), while impaired biosynthesis of dermatan sulfate causes musculocontractural EDS (mcEDS). In this chapter, the clinical and molecular characteristics of both types of EDS are described.

Table 10.1 The latest classification of Ehlers–Danlos syndrome based on the pathomechanism

Group A: Disorders of collagen primary structure and collagen processing
Classical EDS (*COL5A1*, *COL5A2*, *COL1A1*; AD)
Vascular EDS (*COL3A1*, *COL1A1*; AD)
Arthrochalasia EDS (*COL1A1*, *COL1A2*; AD)
Dermatospraxis EDS (*ADAMTS2*; AR)
Cardiac-valvular EDS (COL1A2; AR)
Group B: Disorders of collagen folding and collagen cross-linking
Kyphoscoliotic EDS (*PLOD1*, *FKBB14*; AR)
Group C: Disorder of structure and function of myomatrix, the interface between muscle and ECM
Classical-like EDS (*TNXB*; AR)
Myopathic EDS (*COL12A1*; AD/AR)
Group D: Disorder of glycosaminoglycan biosynthesis
Spondylodysplastic EDS (*B4GALT7*, *B3GALT6*; AR)
Musculocontractural EDS (*CHST14*, *DSE*; AR)
Group E: Disorder of complementary pathway
Periodontal EDS (*C1R*, *C1S*; AD)
Group F: Disorders of intracellular processes
Spondylodysplastic EDS (*SLC39A13*; AR)
Brittle cornea syndrome (*ZNF469*, *PRDM5*; AR)
Unsolved forms of EDS
Hypermobile EDS (gene unknown; AD)

AD Autosomal dominant, *AR* autosomal recessive, *ECM* extracellular matrix

10.2 Background

Glycosylation is the addition of a sugar chain (a glycan) to a protein (generating a glycoprotein) or lipid (generating a glycolipid). Proteoglycans (PGs) are composed of core proteins and one or more glycans with modifications. PGs are present in the extracellular matrix and have important and diverse biological functions (Bulow and Hobert 2006). PG synthesis is initiated by the sequential addition of four monosaccharides [xylose (Xyl), two molecules of galactose (Gal), and glucuronic acid (GlcA)], known as a linker tetrasaccharide, to the serine residue of the core protein backbone (Fig. 10.1). In this process, xylose is transferred from UDP-xylose to a serine residue of the core protein by xylosyltransferases I and II encoded by *XYLT1* and *XYLT2*, respectively. Then, two molecules of galactose are serially added by beta-1,4-galactosyltransferase 7 encoded by *B4GALT7* and beta-1,3-galactosyltransferase 6 encoded by *B3GALT6*. Furthermore, a glucuronic acid is added by beta-1,3-glucuronyltransferase 3 encoded by *B3GAT3* (Fig. 10.1). Additional sugar chains are extended from the linker tetrasaccharide by the addition of repeated disaccharides (usually consisting of 50–150 disaccharides *in vivo*). Subsequently, some sugars are modified by a series of epimerases (epimerization) and sulfotransferases (sulfation).

GAGs are long unbranched polysaccharides consisting of repeating disaccharide units. They are highly negatively charged because of the acidic sugar residues and/or sulfation. Consequently, GAGs can change their conformation, attract cations, and bind water. Hydrated GAG gels enable joints and tissues to absorb large pressure changes, providing tissue elasticity. Post-translational modifications such as

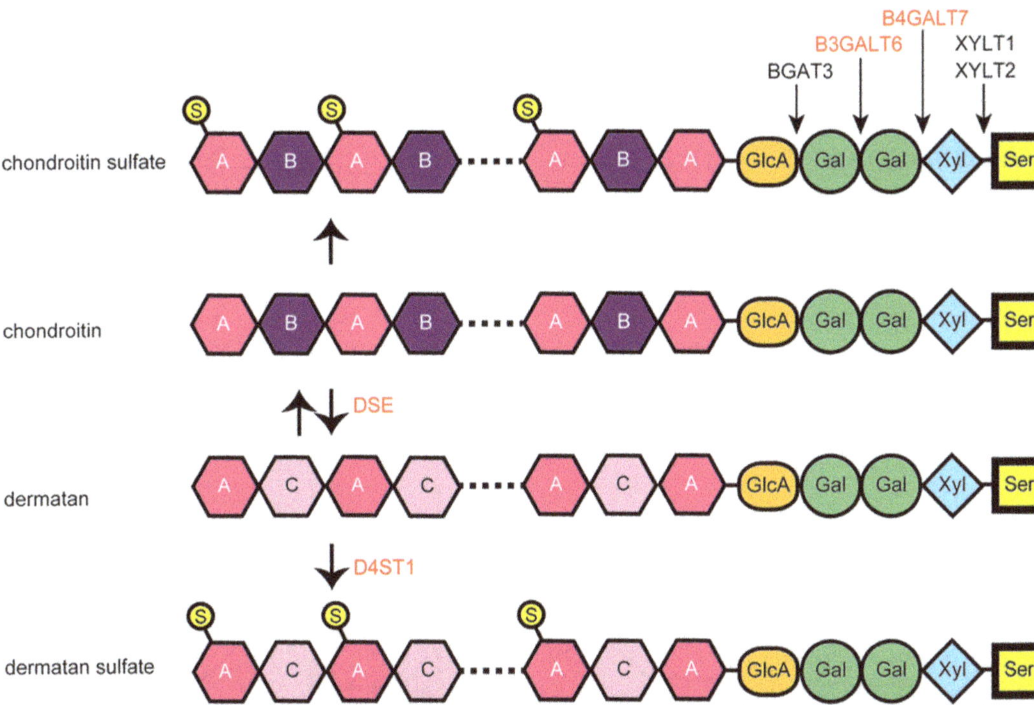

Fig. 10.1 Chondroitin sulfate and dermatan sulfate biosynthesis and related enzymes in two types of EDS associated with glycosaminoglycan abnormalities
The serine residue (Ser) of the core protein and the GAG chain are bound via a linker tetrasaccharide. In CS, the disaccharides are composed of *N*-acetylgalactosamine (GalNAc) [position A] and glucuronic acid (GlcA) [position B]. In DS, the disaccharides are composed of GalNAc [position A] and iduronic acid (IdoA) [position C]. B4GALT7 and B3GALT6 add the first and second galactoses (Gal), respectively, to the xylose of the linker tetrasaccharide. DSE alters GlcA to IdoA in the GAG chain in both directions. CHST14 adds the active sulfate (yellow circle with "S") to the 4-*O* position of GalNAc on dermatan. Four EDS genes associated with glycosaminoglycan abnormalities are shown in red. Gal: galactose; GlcA: glucuronic acid; S: active sulfate; Ser: serine; Xyl: xylose. (This figure has been reprinted from Miyake et al. (Miyake et al. 2014) with permission from the publisher.)

epimerization, sulfation, and acetylation/deacetylation result in the formation of diverse motifs in the GAG chains, which can bind to a large variety of ligands. Therefore, GAG chains play important roles in regulating growth factor signaling, cell adhesion, proliferation, differentiation, and motility (Bulow and Hobert 2006; Sugahara et al. 2003; Bishop et al. 2007).

GAGs can be divided into two groups: (1) galactosaminoglycans such as chondroitin sulfate (CS) and dermatan sulfate (DS), and (2) glucosaminoglycans such as hyaluronic acid, keratan sulfate, heparan sulfate, and heparin (Sisu et al. 2011). Two types of glycosylation are known: *O*-glycosylation and *N*-glycosylation (Fig. 10.2a). Most GAGs (except for keratan sulfate and hyaluronic acid) are *O*-glycans that bind to the glycan via an oxygen molecule in a serine or threonine residue of the core protein (Fig. 10.2a).

The CS and DS GAGs are produced via the same pathway (Fig. 10.1). After the linker tetrasaccharide attaches to the serine residue of the core protein, GalNAc (*N*-acetyl-D-galactosamine) transferase I elongates the glycan branch to create CS/DS. The enzyme C5-carboxy epimerase transforms glucuronic acid (GlcA) to iduronic acid (IdoA), which is specific for dermatan/DS. DS actually exists in a CS/DS hybrid state, containing GlcA–GalNAc

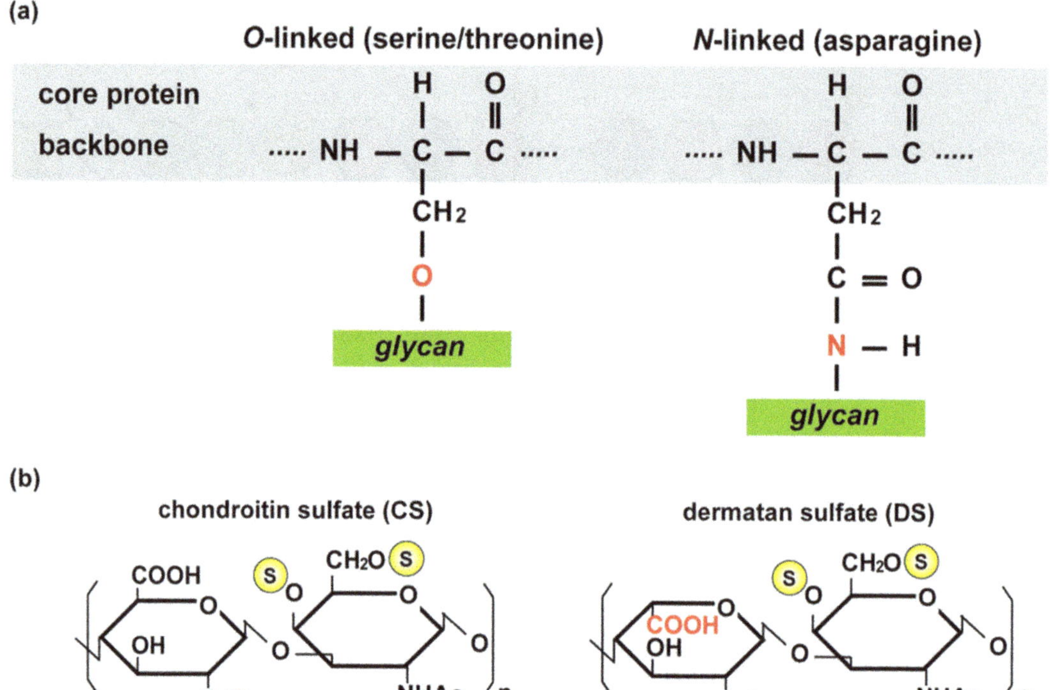

Fig. 10.2 Chemical structures of proteoglycan and disaccharides. (**a**) Chemical structures of *O*-linked and *N*-linked glycans. *O*-linked glycan can be linked via the *O*-element of serine or threonine. The diagram shows the linking for serine. (**b**) Chemical structures of the disaccharide units of CS (left) and DS (right). (This figure has been reprinted from Miyake et al. (Miyake et al. 2014) with permission from the publisher)

and IdoA–GalNAc disaccharides (Esko et al. 2009). Carbohydrate sulfotransferase 14 (CHST14, dermatan 4-*O*-sulfotransferase 1 (D4ST1) specifically transfers an active sulfate to the 4-*O* position on the GalNAc residue of dermatan. This transfer of an active sulfate is impaired in mcEDS type 1 (mcEDS-*CHST14*). DSE converts GlcA to IdoA residues in DS synthesis (Maccarana et al. 2006) (Figs. 10.1 and 10.2b). DSE defect leads to mcEDS type 2 (mcEDS-*DSE*).

10.3 Spondylodysplastic EDS (spEDS)

Mutations in at least three genes (*B4GALT7*, NM_007255.2, MIM*604327; *B3GALT6*, NM_080605.3, MIM*615291; *SLC39A13*, NM_152264.4, MIM*608735) are known to cause spEDS. Among them, *B4GALT7* and *B3GALT6* are associated with GAG abnormalities (Table 10.1). Here, we list the major and minor diagnostic criteria of spEDS according to the latest EDS classification (Malfait et al. 2017):

[Major criteria]

1. Short stature (progressive in childhood)
2. Muscle hypotonia (ranging from severe congenital, to mild later onset)
3. Bowing of limbs

[Minor criteria]

1. Skin hyperextensibility, soft, doughy skin, thin translucent skin
2. Pes planus

3. Delayed motor development
4. Osteopenia
5. Delayed cognitive development

The detailed information and additional minor diagnostic criteria of spEDS are presented in a gene-specific manner below.

10.3.1 EDS, Spondylodysplastic Type 1 (EDSSPD1, spEDS-*B4GALT7*, MIM#130070)

Disease gene: beta-1,4-galactosyltransferase 7 (*B4GALT7*)
Inheritance mode: autosomal recessive
Alternative names:
EDS, progeroid type 1, formerly
EDS with short stature and limb anomalies
Xylosylprotein 4-β-galactosyltransferase deficiency
Galactosyltransferase I deficiency

10.3.1.1 Clinical Manifestations

In addition to the diagnostic criteria common for spEDS, the following features are noted as *B4GALT7*-specific minor criteria (Malfait et al. 2017):

- Radioulnar synostosis
- Bilateral elbow contractures or limited elbow movement
- Generalized joint hypermobility
- Single transverse palmar crease
- Characteristic craniofacial features
- Characteristic radiographic findings
- Severe hypermetropia
- Clouded cornea

Craniofacial features include triangular face, wide-spaced eyes, proptosis, narrow mouth, low-set ears, sparse scalp hair, abnormal dentition, flat face, wide forehead, blue sclerae, and cleft palate/bifid uvula. Radioulnar synostosis, metaphyseal flaring, osteopenia, radial head subluxation/dislocation, and short clavicles with broad medial ends are the characteristic radiographic findings.

10.3.1.2 Genetic Information

In 1999, two different research groups (Almeida et al. 1999; Okajima et al. 1999a) identified compound heterozygous mutations of the gene encoding beta-1,4-galactosyltransferase 7 (*B4GALT7*, NM_007255.2), c.557C > A (p.Ala186Asp) and c.617 T > C (p.Leu206Pro), in the Danish patient originally reported by Kresse in 1987 (Kresse et al. 1987). In addition, two Qatari patients from a large consanguineous family were analyzed in 2004 (Faiyaz-Ul-Haque et al. 2004). Based on the autosomal recessive mode of inheritance, haplotype analysis using microsatellite markers for the limited candidate loci delineated a homozygous region from *D5S469* to *D5S2111*, which harbors *B4GALT7*, and a homozygous missense pathogenic variant (c.808C > T, p.Arg270Cys) in *B4GALT7* was identified (Faiyaz-Ul-Haque et al. 2004). Interestingly, the clinical phenotype of the Qatari patients was milder than that of the Danish one.

B4GALT7 was cloned by Okajima et al. (Okajima et al. 1999b). This gene consists of six coding exons with a 948-bp open reading frame. It encodes B4GALT7 (also known as galactosyltransferase I, and XGALT1), which is 327 amino acids long and has a molecular weight of 37.4 kDa. B4GALT7 is a type II transmembrane protein localized in the Golgi apparatus, and is a key initiator of GAG synthesis as it attaches the first galactose of the linker tetrasaccharide of PGs (Fig. 10.1). At least 11 pathogenic variants in *B4GALT7* have been reported to date, according to Human Genome Mutation Database (HGMD) Professional 2020.2 (https://portal.biobase-international.com/hgmd/pro/start.php).

10.3.1.3 Biochemical Characteristics

Kresse et al. reported that their patient's fibroblasts produced only PG chain-free core proteins (molecular weights: 46 and 44 kDa), whereas control fibroblasts produced normal PG chains (Kresse et al. 1987). Additionally, the GAG-free core proteins in that patient contained unsubstituted xylose residues.

Okajima et al. measured the enzyme activity of exogenously expressed proteins (wild type, p.Ala186Asp, p.Leu206Pro) in XGalT-1/

B4GALT7-deficient CHO cells (Okajima et al. 1999a). In total cell lysates, the enzyme activity of the p.Ala186Asp mutant was approximately 50% lower than that of the wild-type protein, whereas the activity of the p.Leu206Pro mutant was almost undetectable. Interestingly, the wild-type and p.Ala186Asp proteins were localized in the Golgi apparatus, whereas the p.Leu206Pro mutant existed in the cytoplasm. The α-helix disrupted by p.Leu206Pro may alter the protein's conformation, thus impairing intracellular trafficking and enzyme activity (Okajima et al. 1999a).

B4GALT7 activity in fibroblasts from another patient with a homozygous mutation, c.808C > T (p.Arg270Cys), was also lower than that of controls (Seidler et al. 2006). The extracellular matrix around the B4GALT7^{Arg270Cys} mutant fibroblasts was disorganized without banded fibrils. Furthermore, the proliferation of B4GALT7^{Arg270Cys} fibroblasts was significantly reduced to 45% of the level of control fibroblasts (Seidler et al. 2006).

Bui et al. measured the galactosyltransferase activity of B4GALT7 mutants expressed in CHO pgsB-618 cells using 4-methylumbelliferyl-β-D-xylopyranoside as an acceptor substrate. The enzyme activities of the p.Arg270Cys, p.Ala186Asp, and p.Leu206Pro mutants were decreased to 60%, 11%, and 0% (undetectable) of that of the wild-type enzyme (Bui et al. 2010). It has been reported that the clinical features of patients with homozygous p.Arg270Cys mutation appear to be milder than those of patients with compound heterozygous mutations, including p.Ala186Asp or p.Leu206Pro, supporting the different effects of these mutations.

10.3.2 EDS, Spondylodysplastic Type 2 (EDSSPD2, spEDS-*B3GALT6*, MIM#615349)

Disease gene: beta-1,3-galactosyltransferase 6 (*B3GALT6*)
Inheritance mode: autosomal recessive
Alternative name:
EDS, progeroid type 2, formerly

10.3.2.1 Clinical Manifestations

In addition to the major and minor diagnostic criteria of spEDS, the gene-specific minor criteria are as follows (Malfait et al. 2017):

- Kyphoscoliosis (congenital or early onset, progressive)
- Joint hypermobility, generalized or restricted to distal joints, with joint dislocations
- Joint contractures (congenital or progressive) (especially hands)
- Peculiar fingers (slender, tapered, arachnodactyly, spatulate, with broad distal phalanges)
- Talipes equinovarus
- Characteristic craniofacial features
- Tooth dislocation, dysplastic teeth
- Characteristic radiographic findings
- Osteoporosis with multiple spontaneous fractures
- Ascending aortic aneurysm
- Lung hypoplasia, restrictive lung disease

Craniofacial features are midfacial hypoplasia, frontal bossing, proptosis, or prominent eyes, blue sclerae, down-slanting palpebral fissures, depressed nasal bridge, long upper lip, low-set ears, micrognathia, abnormal dentition, cleft palate, and sparse hair (Malfait et al. 2017). Platyspondyly, anterior beak of vertebral body, short ilium, prominent lesser trochanter, acetabular dysplasia, metaphyseal flaring, metaphyseal dysplasia of femoral head, elbow malalignment, radial head dislocation, overtubulation, bowing of long bones, generalized osteoporosis, and healed fractures are the characteristic radiographic findings.

10.3.2.2 Genetic Information

In 2013, two independent groups, Nakajima et al. and Malfait et al., identified biallelic mutations of *B3GALT6* (NM_080605.3) in patients with the progeroid form of EDS (Nakajima et al. 2013) and in a pleiotropic EDS-like connective tissue disorder (Malfait et al. 2013). This intronless gene has a 990-bp open reading frame and encodes beta-1,3-galactosyltransferase 6 (also known as UDP-Gal:βGal β-1,3-galactosyltransferase polypeptide 6, galactosyltransferase-II: GalT-II), which is 329 amino acids long and has a molecular weight of 37.1 kDa. It is also a type II transmembrane protein localized in the Golgi apparatus, and it attaches the second galactose of the tetrasaccharide linker of PGs (Fig. 10.1). More than 40 pathogenic variants in *B3GALT6* have been reported so far, according to HGMD Professional 2020.2.

10.3.2.3 Biochemical Characteristics

Nakajima et al. measured the galactosyltransferase activity of *B3GALT6 in vitro* using soluble-FLAG-tagged proteins for WT and a mutant (p.Ser309Thr) found to be shared between two spEDS-*B3GALT6*-affected families. They revealed that the enzyme activity of the mutant protein was significantly decreased compared with that of the wild type (Nakajima et al. 2013). They also measured the amount of disaccharide of GAG chains from lymphoblastoid cells of the patients. The amount of disaccharide from heparan sulfate was significantly decreased and that from CS and DS chains was approximately five times higher than that in the control samples (Nakajima et al. 2013). In addition, Malfait et al. reported that the loss of B3GALT6 function impaired the formation of the tetrasaccharide linker region of PGs, produced immature decorin lacking its CS/DS chain, and decreased or abolished the heparan sulfate chains in patient-derived skin fibroblasts (Malfait et al. 2013).

10.3.2.4 Pathology

Malfait et al. observed the skin biopsy of a patient with a homozygous missense variant (c.649G > A, p.Gly217Ser) by transmission electron microscopy, and revealed loosely packed collagen fibrils of variable size and shape (Malfait et al. 2013). They also mentioned that these features are similar to those in tenascin-X-deficient patients.

10.4 Musculocontractural EDS (mcEDS)

Two causative genes (*CHST14*, NM_130468.3, MIM*608429; *DSE*, NM_ 013352.3, MIM*605942) of mcEDS have been identified. Here, we list the major and minor diagnostic criteria of mcEDS based on the latest classification of EDS (Malfait et al. 2017):

[Major criteria]

1. Congenital multiple contractures, characteristically adduction–flexion contractures and/or talipes equinovarus (clubfoot)
2. Characteristic craniofacial features, which are evident at birth or in early infancy
3. Characteristic cutaneous features including skin hyperextensibility, easy bruisability, skin fragility with atrophic scars, and increased palmar wrinkling

[Minor criteria]

1. Recurrent/chronic dislocations
2. Pectus deformities (flat, excavated)
3. Spinal deformities (scoliosis, kyphoscoliosis)
4. Peculiar fingers (tapering, slender, cylindrical)
5. Progressive talipes deformities (valgus, planus, cavum)
6. Large subcutaneous hematomas
7. Chronic constipation
8. Colonic diverticula
9. Pneumothorax/pneumohemothorax
10. Nephrolithiasis/cystolithiasis
11. Hydronephrosis
12. Cryptorchidism in males
13. Strabismus
14. Refractive errors (myopia, astigmatism)
15. Glaucoma/elevated intraocular pressure

Craniofacial features include large fontanelle, wide-spaced eyes, short and down-slanting palpebral fissures, blue sclerae, short nose with hypoplastic columella, low-set and rotated ears, high palate, long philtrum, thin upper lip vermilion, small mouth, and micrognathia.

10.4.1 EDS, Musculocontractural Type 1 (mcEDS-*CHST14*, MIM#601776)

Disease gene: carbohydrate sulfotransferase 14 (*CHST14*)
Inheritance mode: autosomal recessive
Alternative names:
D4ST1-deficient EDS
Ehlers–Danlos syndrome, type VIB, formerly
Adducted thumb–clubfoot syndrome
Adducted thumbs, clubfoot, and progressive joints and skin laxity syndrome
Arthrogryposis, distal, with peculiar faces and hydronephrosis

10.4.1.1 Clinical Manifestations

Originally, mcEDS-*CHST14* was recognized as kyphoscoliosis type without lysyl hydroxylase deficiency (formerly known as EDS type VIB) (Steinmann et al. 1975; Royce et al. 1989; Kosho et al. 2005). At birth, affected individuals show a characteristic facial features and multiple contractures (Fig. 10.3). The joint contractures become less evident with aging. The most unique and life-threatening complication is large subcutaneous hematoma formation, which sometimes requires intensive treatment. Intranasal administration of 1-desamino-8-D-arginine vasopressin was reported to prevent the development of large subcutaneous hematomas after trauma (Yasui et al. 2003). This syndrome is characterized by age-dependent phenotypes and a progressive clinical course.

10.4.1.2 Genetic Information

mcEDS-*CHST14* is considered to have an autosomal recessive mode of inheritance based on the presence of this syndrome in consanguineous families. Three independent groups performed homozygosity mapping and/or linkage analysis of independent mcEDS-*CHST14* families, and each showed that the gene carbohydrate sulfotransferase 14 (*CHST14*, NM_130468) was responsible for this syndrome (Dundar et al. 2009; Miyake et al. 2010; Malfait et al. 2010).

CHST14 was first cloned by Evers et al. (Evers et al. 2001). It contains one coding exon (1131-bp open reading frame) and is localized at 15q15.1. This gene encodes CHST14 (formerly D4ST1), a 376-amino-acid type II transmembrane protein (molecular weight: 43 kDa) that is localized in the Golgi membrane. It transfers a sulfate group from 3'-phosphoadenosine 5'-phosphosulfate to position 4 of the GalNAc residues in dermatan to generate DS (Fig. 10.1). Northern blotting revealed that *CHST14* is mainly expressed in the heart, placenta, liver, and pancreas, and is weakly expressed in the lung, skeletal muscle, and kidney (Evers et al. 2001).

At least 22 pathogenic variants in *CHST14* have been reported based on HGMD Professional 2020.2. Among them, 12 variants are missense and 10 are truncating. Therefore, these variants are likely to cause loss of function of CHST14.

10.4.1.3 Biochemical Information

Dündar et al. reported that DS-derived IdoA-GalNAc(4S) disaccharide was undetectable in fibroblasts derived from a patient with homozygous p.Arg213Pro mutation (Dundar et al. 2009). They also reported that GlcA-GalNAc(4S) content was greatly increased in the fibroblast extract and the culture media obtained from cultures of fibroblasts derived from this patient compared with the levels for control fibroblasts (Dundar et al. 2009). It was also found that the amount of nonsulfated disaccharides (GlcA-GalNAc and IdoA-GalNAc) was increased in the cell extract and its media from the patient's fibroblasts compared with those for normal control fibroblasts. From these results, Dündar et al. proposed that the epimerization of GlcA to IdoA by C5-carboxy epimerase is followed by the sulfation of the C4 hydroxyl on the adjacent GalNAc residue by CHST14. This process generates DS from dermatan and prevents back-epimerization from IdoA to GlcA (Dundar et al. 2009; Malmstrom 1984).

Miyake et al. measured the sulfotransferase activity of COS7 cells transfected with wild-type

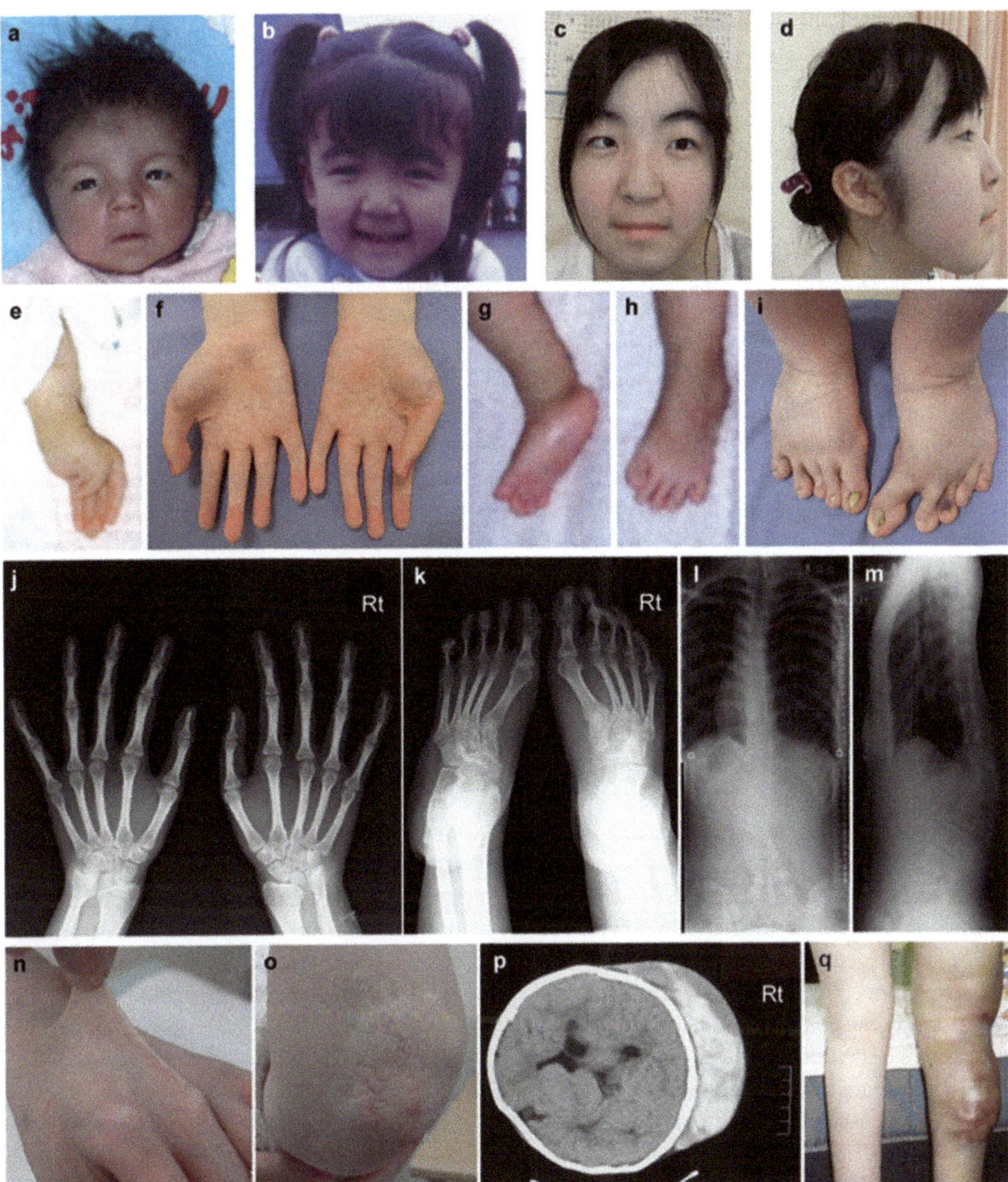

Fig. 10.3 Clinical photographs of patients with mcEDS-*CHST14*. (**a–d**) Facial features of a patient at 23 days (**a**), 3 years (**b**), and 16 years of age (**c**, **d**). (**e**, **f**) Images of the hand in a patient with an adducted thumb at 1 month of age (**e**) and cylindrical fingers at 19 years of age (**f**). (**g–i**) Images of the foot in a patient with bilateral clubfeet at 1 month of age (**g**, **h**) and progressive talipes deformities (planus and valgus) at 19 years of age (**i**). (j–m) Radiographs of a 16-year-old patient show diaphyseal narrowing of the phalanges and metacarpals (**j**, **k**) and kyphoscoliosis with tall vertebral bodies (**l**, **m**). (**n**, **o**) Cutaneous features of a 19-year-old patient with hyperextensibility (**n**), atrophic scars, and fistula formation (**o**). (**p**) A massive cranial subcutaneous hematoma in the head of a 6-year-old patient after falling onto the floor. (**q**) A subcutaneous hematoma in the leg of a 16-year-old patient. Modified from Kosho et al. [Am J Med Genet A. 2005 Oct 15;138A(3):282–7. doi: https://doi.org/10.1002/ajmg.a.30965; Am J Med Genet A. 2010 Jun;152A(6):1333–46. doi: https://doi.org/10.1002/ajmg.a.33498], and reprinted from Miyake et al. (Miyake et al. 2014) with permission from the publisher

and mutant CHST14 harboring the p.Lys69*, p.Pro281Leu, p.Cys289Ser, or p.Tyr293Cys mutation. The enzyme activity of the mutants was as low as that in mock-transfected cells, suggesting that these missense mutations result in loss of function (Miyake et al. 2010). The disaccharide composition of the decorin GAG chain isolated from the patient's fibroblasts consisted only of CS, without DS, while the chains isolated from normal fibroblasts consisted of CS/DS hybrid chains (Miyake et al. 2010). Furthermore, the level of nonsulfated dermatan was negligible in the patient's fibroblasts (Miyake et al. 2010). Thus, in this syndrome, the CS/DS chain is replaced by the CS chain, even though the core proteins are normal. Interestingly, DS was undetected in patients' urine while CS/DS and CS were detected in patients at levels comparable to those in sex- and age-matched healthy controls (Mizumoto et al. 2017). Thus, urine could be used for a noninvasive clinical test of mcEDS-*CHST14*.

10.4.1.4 Pathophysiology

Of the major DS PGs in skin, decorin was a focus of research about the pathomechanism of mcEDS-*CHST14* because it binds to collagen fibrils via its core protein and its GAG chains act as interfibrillar bridges (Scott 1996; Scott 2003). Three α-collagen chains are self-assembled to generate tropocollagen, in the form of a triple helix. Tropocollagen then self-assembles to form collagen fibrils via decorin (Fig. 10.4a). Collagen fibrils are assembled into a collagen fiber, known as the collagen bundle, via the antiparallel complex of the CS/DS hybrid GAG chains of decorin, which acts like a bridge to provide a space between individual fibrils and tighten the collagen fiber (Fig. 10.4b, c).

The GAGs span collagen fibrils in the extracellular matrix of skin and tendons, and the length of the GAG chain determines the width of the interfibrillar gap (Scott 1988; Scott 1992). Elasticity of the extracellular matrix is explained by the sliding filament model, which allows reversible longitudinal slippage between the antiparallel GAG chains (Fig. 10.4b) (Scott 2003). Because tissue stability and elasticity depend on the structure of the GAG bridges, irreversible damage can occur if the bridges are inelastic (Scott 2003).

Decorin is composed of a horseshoe-shaped core protein (molecular weight: ~45 kDa) and a single CS/DS hybrid chain on the N-terminal side (Fig. 10.4a) (Kobe and Deisenhofer 1993; Weber et al. 1996). Weber et al. reported that the model structure of decorin consists of an arch in which the inner concave surface is formed from a curved β-sheet and the outer convex surface is formed from α-helices. They also proposed that one tropocollagen fiber lies within the decorin convexity and another interacts with one arm of the arch (Weber et al. 1996). The IdoA:GlcA ratio in DS ranges from ~10% to >90% depending on the tissue type (Scott 2003). Importantly, L-IdoA residues in DS can easily undergo conformational changes, unlike GlcA in CS (Casu et al. 1988; Catlow et al. 2008). Thus, the IdoA:GlcA ratio should be higher in more flexible tissues (Scott 2003).

Light microscopic investigation of skin specimens from the patients showed that fine collagen fibers were predominant in the reticular to papillary dermis and the number of thick collagen

Fig. 10.4 (continued) The black lines represent unspecified GAGs. (**c**, **d**) In the normal state, the CS/DS chains can bend against the direction of mechanical compression and rebound to the original structure (**c**). Thus, the collagen bundles are refractory to compression stress (**d**). (**e**, **f**) In mcEDS-*CHST14*, the CS/DS chains are replaced with CS chains (red lines). These chains cannot resist mechanical compression, resulting in irreversible scattering of the collagen fvibrils. (**g**) The size and shape of the collagen fibrils are highly variable in decorin-deficient mice. (**h**) Composition of decorin core protein (green circle), GAG chain (blue line), and collagen fibrils (orange square) in the longitudinal view of a collagen fiber. (**i**, **j**) Scheme of the cross-sectional view of collagen fibrils (orange circle) with decorin core protein (green circle) and normal GAG (blue line in **i**) in a normal control or abnormal CS GAG chain (red lines in **j**) in a patient with mcEDS-*CHST14*. (This figure has been reprinted from Miyake et al. (Miyake et al. 2014) with permission from the publisher)

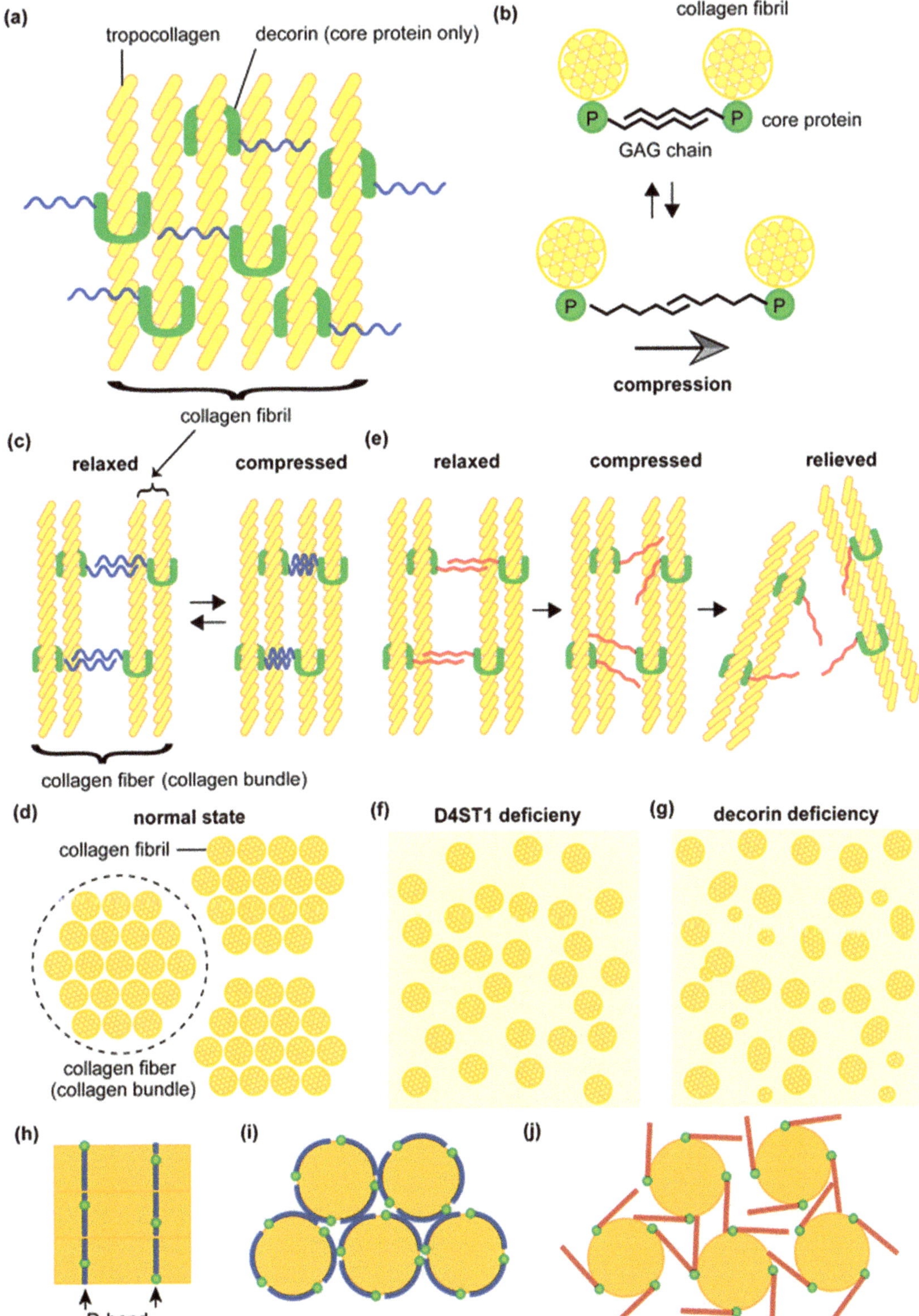

Fig. 10.4 Putative model of abnormal collagen bundle assembly in mcEDS-*CHST14*. (**a**) Tropocollagen directly binds to decorin and forms a collagen fibril. The blue lines represent the CS/DS hybrid chain. (**b**) Illustration of the sliding filament model showing reversible longitudinal slippage between the antiparallel GAG chains.

bundles was markedly reduced (Miyake et al. 2010; Mochida et al. 2016; Kono et al. 2016). Electron microscopic examination of the specimens showed that collagen fibrils were dispersed throughout the reticular dermis, whereas they were regularly and tightly assembled in control tissue (Miyake et al. 2010; Mochida et al. 2016; Kono et al. 2016; Janecke et al. 2016). Surprisingly, each collagen fibril was smooth and round, with little variation in size or shape, similar to the fibril in the control tissue (Fig. 10.4d, f) (Miyake et al. 2010). In addition, increased elastic fibers in the dermis were observed in the Elastica van Gieson-stained skin specimen of a patient (Kono et al. 2016).

The disaccharide composition of the decorin GAG chain from a patient's fibroblasts only consisted of CS, without DS disaccharide, whereas control fibroblasts consisted of a CS/DS hybrid (Miyake et al. 2010). The transition of decorin from the CS/DS hybrid chain to a CS chain probably decreases the flexibility of the GAG chain. The sliding filament model proposes that mechanical compression might also act in the CS chain of mcEDS-*CHST14* patients, but the inflexibility of the CS chain is unable to tolerate higher mechanical pressures or is too inelastic to maintain normal skin properties (Fig. 10.4e, f). This irreversible event could explain the progressive clinical course of this disease.

Interestingly, there were marked variations in the size and shape of dermal collagen fibrils in decorin-null mice (Fig. 10.4g) (Danielson et al. 1997). These findings suggest that the decorin core protein is important for collagen fibril formation, and that the CS/DS hybrid chain of decorin PG regulates the space between the collagen fibrils and the formation of collagen bundles, as previously reported (Scott 1995). These findings suggest that the main pathological basis of this disorder could be the insufficient assembly of collagen fibrils.

However, Dündar et al. reported that the light microscopic and electron microscopic findings of a patient's skin were unchanged relative to those of a normal control (Dundar et al. 2009). Malfait et al. reported that, in their mcEDS-*CHST14* patient, most collagen bundles had a small diameter, and some were composed of collagen fibrils of varying diameter that were separated by irregular interfibrillar spaces (Malfait et al. 2010). In addition, the fibroblasts exhibited an elongated and/or dilated endoplasmic reticulum.

Very recently, Hirose et al. successfully visualized the decorin localization and the shape of GAG in the skin specimens of patients with mcEDS-*CHST14* (Hirose et al. 2019). Immunohistochemistry revealed that decorin core proteins were colocalized with collagen bundles with a clear boundary in control samples, while they were stained like thin filaments without a clear boundary in the patients' skin specimens (Hirose et al. 2019). Transmission electron microscopy-based cupromeronic blue staining revealed that GAG chains attach perpendicularly to collagen fibrils and bind collagen fibrils at the D band in the longitudinal section of control and patients' samples (Fig. 10.4h). Interestingly, in the cross-sectional view, a majority of GAG chains appeared to be curved along with the surface of the collagen fibrils in normal control samples, while a majority of GAG chains appeared to be straight from the surface of the collagen fibrils in patients' samples (Fig. 10.4i and j).

10.4.2 EDS, Musculocontractural Type 2 (mcEDS-*DSE*, MIM#615539)

Disease gene: dermatan sulfate epimerase (*DSE*)
Inheritance mode: autosomal recessive

10.4.2.1 Clinical Manifestations

To date, only eight individuals affected by mcEDS-*DSE* have been reported (Müller et al. 2013; Syx et al. 2015; Schirwani et al. 2020; Lautrup et al. 2020). While the number of reported patients is limited in both types of mcEDS, their clinical features are quite similar (Lautrup et al. 2020). As no patients with mcEDS-*DSE* have shown life-threatening complications, it is possible that mcEDS-*DSE* is milder than mcEDS-*CHST14*. Further accumulation of patients is needed to confirm this.

10.4.2.2 Genetic Information

In 2013, Müller et al. reported that EDS, musculocontractural type 2, is caused by biallelic *DSE* variants (autosomal recessive inheritance) (Müller et al. 2013). To date, only six pathogenic variants in *DSE* have been reported (Müller et al. 2013; Syx et al. 2015; Schirwani et al. 2020; Lautrup et al. 2020). Three are missense and the other three are truncating. As in the case for mcEDS-*CHST14*, a loss-of-function mechanism may thus be involved in this condition.

10.4.2.3 Biochemical Information

DSE and dermatan sulfate epimerase like (DSEL) play an important role in the conversion from chondroitin to dermatan by converting glucuronic to iduronic acid of the chondroitin backbone; they can also perform conversion in the opposite direction (Fig. 10.1). Müller et al. measured the composition of CS/DS chains in fibroblasts from a patient harboring a homozygous missense variant (c.803C > T, p.S268L) and a control, and revealed that DS was significantly decreased in the cell fraction and conditioned media, while CS was increased in the cell fraction, but not in conditioned media, in the patient (Müller et al. 2013). This might indicate that decreased DS biosynthesis led to the increase of CS, which could not be converted to DS, in the patient (Müller et al. 2013). In addition, Lautrup et al. measured the disaccharide composition of DS and CS in the urine of a patient harboring a homozygous nonsense variant (c.960 T > A, p.Thr320*). DS was not detected in the patient, while CS was detected at a level similar to that in an age- and sex-matched control (Lautrup et al. 2020). Thus, urine could also be used for a noninvasive clinical test of mcEDS-*DSE*.

Acknowledgments We thank Edanz (https://en-author-services.edanzgroup.com/ac) for editing the English text of a draft of this manuscript.

Ethical Approval All procedures performed in studies involving human participants were in accordance with the ethical standards of the institutional and/or national research committee and with the 1964 Helsinki declaration and its later amendments or comparable ethical standards. Approval was granted by the Ethics Committee of the Shinshu University School of Medicine (# 610).

Informed Consent Informed consent for participation and publication was obtained from all individual participants or their parents/guardians in the case of minors included in the study.

References

Almeida R et al (1999) Cloning and expression of a proteoglycan UDP-galactose:beta-xylose beta1,4-galactosyltransferase I. A seventh member of the human beta4-galactosyltransferase gene family. J Biol Chem 274(37):26165–26171

Beighton P et al (1988) International nosology of heritable disorders of connective tissue, Berlin, 1986. Am J Med Genet 29(3):581–594

Beighton P et al (1998) Ehlers-Danlos syndromes: revised nosology, Villefranche, 1997. Ehlers-Danlos National Foundation (USA) and Ehlers-Danlos support group (UK). Am J Med Genet 77(1):31–37

Bishop JR et al (2007) Heparan sulphate proteoglycans fine-tune mammalian physiology. Nature 446(7139):1030–1037

Bui C et al (2010) Molecular characterization of beta1,4-galactosyltransferase 7 genetic mutations linked to the progeroid form of Ehlers-Danlos syndrome (EDS). FEBS Lett 584(18):3962–3968

Bulow HE, Hobert O (2006) The molecular diversity of glycosaminoglycans shapes animal development. Annu Rev Cell Dev Biol 22:375–407

Casu B et al (1988) Conformational flexibility: a new concept for explaining binding and biological properties of iduronic acid-containing glycosaminoglycans. Trends Biochem Sci 13(6):221–225

Catlow KR, Wei Z, Delehedde M, Fernig DG, Gherardi E, Gallagher JT, Pavao MS, Lyon M (2008) Interactions of hepatocyte growth factor/scatter factor with various glycosaminoglycans reveal an important interplay between the presence of iduronate and sulfate density. J Biol Chem 283:5235–5248

Danielson KG et al (1997) Targeted disruption of decorin leads to abnormal collagen fibril morphology and skin fragility. J Cell Biol 136(3):729–743

Dundar M et al (2009) Loss of dermatan-4-sulfotransferase 1 function results in adducted thumb-clubfoot syndrome. Am J Hum Genet 85(6):873–882

Esko, J.D. et al. (2009) Proteoglycans and sulfated Glycosaminoglycans. In essentials of Glycobiology (nd et al. eds)

Evers MR et al (2001) Molecular cloning and characterization of a dermatan-specific N-acetylgalactosamine 4-O-sulfotransferase. J Biol Chem 276(39):36344–36353

Faiyaz-Ul-Haque M et al (2004) A novel missense mutation in the galactosyltransferase-I (B4GALT7) gene in a family exhibiting facioskeletal anomalies and Ehlers-Danlos syndrome resembling the progeroid type. Am J Med Genet A 128A(1):39–45
Hirose T et al (2019) Structural alteration of glycosaminoglycan side chains and spatial disorganization of collagen networks in the skin of patients with mcEDS-CHST14. Biochim Biophys Acta Gen Subj 1863(3):623–631
Janecke AR et al (2016) The phenotype of the musculocontractural type of Ehlers-Danlos syndrome due to CHST14 mutations. Am J Med Genet A 170A(1):103–115
Kobe B, Deisenhofer J (1993) Crystal structure of porcine ribonuclease inhibitor, a protein with leucine-rich repeats. Nature 366(6457):751–756
Kono M et al (2016) A 45-year-old woman with Ehlers-Danlos syndrome caused by dermatan 4-O-sulfotransferase-1 deficiency: implications for early ageing. Acta Derm Venereol 96(6):830–831
Kosho T et al (2005) Ehlers-Danlos syndrome type VIB with characteristic facies, decreased curvatures of the spinal column, and joint contractures in two unrelated girls. Am J Med Genet A 138A(3):282–287
Kresse H et al (1987) Glycosaminoglycan-free small proteoglycan core protein is secreted by fibroblasts from a patient with a syndrome resembling progeroid. Am J Hum Genet 41(3):436–453
Lautrup CK et al (2020) Delineation of musculocontractural Ehlers-Danlos syndrome caused by dermatan sulfate epimerase deficiency. Mol Genet Genomic Med:e1197
Maccarana M et al (2006) Biosynthesis of dermatan sulfate: chondroitin-glucuronate C5-epimerase is identical to SART2. J Biol Chem 281(17):11560–11568
Malfait F et al (2010) Musculocontractural Ehlers-Danlos syndrome (former EDS type VIB) and adducted thumb clubfoot syndrome (ATCS) represent a single clinical entity caused by mutations in the dermatan-4-sulfotransferase 1 encoding CHST14 gene. Hum Mutat 31(11):1233–1239
Malfait F et al (2013) Defective initiation of glycosaminoglycan synthesis due to B3GALT6 mutations causes a pleiotropic Ehlers-Danlos-syndrome-like connective tissue disorder. Am J Hum Genet 92(6):935–945
Malfait F et al (2017) The 2017 international classification of the Ehlers-Danlos syndromes. Am J Med Genet C Semin Med Genet 175(1):8–26
Malmstrom A (1984) Biosynthesis of dermatan sulfate. II. Substrate specificity of the C-5 uronosyl epimerase. J Biol Chem 259(1):161–165
Miyake N et al (2010) Loss-of-function mutations of CHST14 in a new type of Ehlers-Danlos syndrome. Hum Mutat 31(8):966–974
Miyake N et al (2014) Ehlers-Danlos syndrome associated with glycosaminoglycan abnormalities. Adv Exp Med Biol 802:145–159
Mizumoto S et al (2017) Defect in dermatan sulfate in urine of patients with Ehlers-Danlos syndrome caused by a CHST14/D4ST1 deficiency. Clin Biochem 50(12):670–677
Mochida K et al (2016) Dermatan 4-O-sulfotransferase 1-deficient Ehlers-Danlos syndrome complicated by a large subcutaneous hematoma on the back. J Dermatol 43(7):832–833
Müller T et al (2013) Loss of dermatan sulfate epimerase (DSE) function results in musculocontractural Ehlers-Danlos syndrome. Hum Mol Genet 22(18):3761–3772
Nakajima M et al (2013) Mutations in B3GALT6, which encodes a glycosaminoglycan linker region enzyme, cause a spectrum of skeletal and connective tissue disorders. Am J Hum Genet 92(6):927–934
Okajima T et al (1999a) Molecular basis for the progeroid variant of Ehlers-Danlos syndrome. Identification and characterization of two mutations in galactosyltransferase I gene. J Biol Chem 274(41):28841–28844
Okajima T et al (1999b) Human homolog of Caenorhabditis elegans sqv-3 gene is galactosyltransferase I involved in the biosynthesis of the glycosaminoglycan-protein linkage region of proteoglycans. J Biol Chem 274(33):22915–22918
Royce PM et al (1989) Ehlers-Danlos syndrome type VI with normal lysyl hydroxylase activity cannot be explained by a defect in cellular uptake of ascorbic acid. Matrix 9(2):147–149
Schirwani S et al (2020) DSE associated musculocontractural EDS, a milder phenotype or phenotypic variability. Eur J Med Genet 63(4):103798
Scott JE (1988) Proteoglycan-fibrillar collagen interactions. Biochem J 252(2):313–323
Scott JE (1992) Morphometry of cupromeronic blue-stained proteoglycan molecules in animal corneas, versus that of purified proteoglycans stained in vitro, implies that tertiary structures contribute to corneal ultrastructure. J Anat 180(Pt 1):155–164
Scott JE (1995) Extracellular matrix, supramolecular organisation and shape. J Anat 187(Pt 2):259–269
Scott JE (1996) Proteodermatan and proteokeratan sulfate (decorin, lumican/fibromodulin) proteins are horseshoe shaped. Implications for their interactions with collagen. Biochemistry 35(27):8795–8799
Scott JE (2003) Elasticity in extracellular matrix 'shape modules' of tendon, cartilage, etc. A sliding proteoglycan-filament model. J Physiol 553(Pt 2):335–343
Seidler DG et al (2006) Defective glycosylation of decorin and biglycan, altered collagen structure, and abnormal phenotype of the skin fibroblasts of an Ehlers-Danlos syndrome patient carrying the novel Arg270Cys substitution in galactosyltransferase I (beta4GalT-7). J Mol Med (Berl) 84(7):583–594
Sisu E et al (2011) Modern developments in mass spectrometry of chondroitin and dermatan sulfate glycosaminoglycans. Amino Acids 41(2):235–256

Steinmann B, R.P.M., Superti-Furga A. (2002) The Ehlers - Danlos syndrome. Connective tissue and heritable disorders:431–523

Steinmann B et al (1975) Ehlers-Danlos syndrome in two siblings with deficient lysyl hydroxylase activity in cultured skin fibroblasts but only mild hydroxylysine deficit in skin. Helv Paediatr Acta 30(3):255–274

Sugahara K et al (2003) Recent advances in the structural biology of chondroitin sulfate and dermatan sulfate. Curr Opin Struct Biol 13(5):612–620

Syx D et al (2015) Genetic heterogeneity and clinical variability in musculocontractural Ehlers-Danlos syndrome caused by impaired dermatan sulfate biosynthesis. Hum Mutat 36(5):535–547

Weber IT et al (1996) Model structure of decorin and implications for collagen fibrillogenesis. J Biol Chem 271(50):31767–31770

Yasui H et al (2003) Combination therapy of DDAVP and conjugated estrogens for a recurrent large subcutaneous hematoma in Ehlers-Danlos syndrome. Am J Hematol 72(1):71–72

Loeys-Dietz Syndrome

11

Joe D. Velchev, Lut Van Laer, Ilse Luyckx, Harry Dietz, and Bart Loeys

Abstract

Loeys-Dietz syndrome is an autosomal dominant aortic aneurysm syndrome characterized by multisystemic involvement. The most typical clinical triad includes hypertelorism, bifid uvula or cleft palate and aortic aneurysm with tortuosity. Natural history is significant for aortic dissection at smaller aortic diameter and arterial aneurysms throughout the arterial tree. The genetic cause is heterogeneous and includes mutations in genes encoding for components of the transforming growth factor beta (TGFβ) signalling pathway: *TGFBR1, TGFBR2, SMAD2, SMAD3, TGFB2* and *TGFB3*. Despite the loss of function nature of these mutations, the patient-derived aortic tissues show evidence of increased (rather than decreased) TGFβ signalling. These insights offer new options for therapeutic interventions.

Keywords

Loeys-Dietz syndrome · Hypertelorism · Bifid uvula/cleft palate · Craniosynostosis · Dilatation of aortic root · Aortic aneurysm with tortuosity · Aortic dissection · Mutations in *TGFBR1* · *TGFBR2* · *SMAD3* or *TGFB2* · Increased TGFβ signalling · Overlap with Marfan and Ehlers-Danlos syndrome · Pectus excavatum or pectus carinatum · Scoliosis · Joint laxity · Arachnodactyly · Talipes equinovarus · Cervical spine malformation · Spondylolisthesis · Acetabular protrusion · Pes planus · Osteoporosis · Retrognathia · Strabismus · Blue sclerae · Myopia · Amblyopia · Translucent skin · Easy bruising · Dystrophic scars · Spontaneous bowel rupture · Peripartal uterine rupture · Aneurysm- osteoarthritis syndrome · Mutations in *SKI* · Shprintzen-Goldberg syndrome · Cutis laxa · Familial thoracic aortic aneurysm · Mutations in *ACTA2* · *MYH11* and *MYLK*

J. D. Velchev · L. Van Laer
Center for Medical Genetics, Antwerp University Hospital/University of Antwerp, Antwerp, Belgium
e-mail: joedavis.velchev@uantwerpen.be; lut.vanlaer@uantwerpen.be

I. Luyckx · B. Loeys (✉)
Center for Medical Genetics, Antwerp University Hospital/University of Antwerp, Antwerp, Belgium

Department of Human Genetics, Radboud University Medical Center, Nijmegen, The Netherlands
e-mail: ilse.luyckx@uantwerpen.be; bart.loeys@uantwerpen.be; http://www.genetica-antwerpen.be

H. Dietz
Harry Dietz, McKusick-Nathans Institute of Genetic Medicine, Johns Hopkins University School of Medicine, Baltimore, MD, USA
e-mail: hdietz@jhmi.edu

J. Halper (ed.), *Progress in Heritable Soft Connective Tissue Diseases*, Advances in Experimental Medicine and Biology 1348, https://doi.org/10.1007/978-3-030-80614-9_11

Abbreviations

AOS	Aneurysm-osteoarthritis syndrome
ARCL1	Autosomal recessive cutis laxa type 1
ASD/VSD	Atrial and ventricular septum defects
ATS	Arterial tortuosity syndrome
BAV	Bicuspid aortic valve
BMP	Bone morphogenetic protein
EDS	Ehlers-Danlos syndrome
LDS	Loeys-Dietz syndrome
MFS	Marfan syndrome
PDA	Patent ductus arteriosus
PLOD1	Procollagen-lysine, 2-oxoglutarate 5-dioxygenase 1
SGS	Shprintzen-Goldberg syndrome
TAAD	Thoracic aortic aneurysms and dissections
TGFβ	Transforming growth factor β
TGFβR	Transforming growth factor β receptor

11.1 Introduction

The Loeys-Dietz syndrome (LDS, MIM#609192) was first described by Loeys and Dietz in 2005. The initial paper presented ten probands with a novel aortic aneurysm syndrome characterized by the clinical triad of hypertelorism, bifid uvula/cleft palate and aortic/arterial aneurysms and tortuosity (Loeys et al. 2005). Although these presented the most typical characteristics, a widespread involvement of different organ systems was also recognized. These included craniofacial (e.g., craniosynostosis), skeletal (joint laxity and contractures), integumental (skin hyperextensibility, dural ectasia) and ocular findings (e.g., strabismus). Although LDS shows clinical overlap with Marfan syndrome (MFS), it can be clinically distinguished from the latter. Shared features include aortic root aneurysm, pectus deformities, scoliosis and arachnodactyly. Distinguishing findings are craniosynostosis, hypertelorism, cleft palate or bifid uvula, cervical spine instability, club feet, and most importantly widespread arterial aneurysms with tortuosity and early aortic rupture. Since the initial description of LDS, families with aortic aneurysms without significant outward features have also been described (Pannu et al. 2005; Tran-Fadulu et al. 2009).

11.2 Inheritance and Mutational Spectrum

LDS is an autosomal dominant disorder. About two-thirds of cases are the consequence of *de novo* mutations, whereas the other one-third are familial. In general, the more severe cases with marked craniofacial and skeletal findings are the consequence of a *de novo* mutation, whereas the milder cases tend to be familial. Both non-penetrance (Loeys et al. 2006) and mosaicism (Watanabe et al. 2008) have been reported.

Two major genes have been initially associated with LDS. These genes encode the transforming growth factor β receptors 1 and 2, *TGFBR1* and *TGFBR2*. *TGFBR1* is located on chromosome 9q and contains 9 exons, whereas *TGFBR2* is positioned on chromosome 3p and contains 8 exons. Mutations in the gene encoding *SMAD3* have been associated with a condition called aneurysm-osteoarthritis syndrome, showing significant clinical overlap with LDS (van de Laar et al. 2011). Finally, also mutations in *SMAD2, TGFB2* and *TGFB3* have been identified in patients with LDS-like presentations (Lindsay et al. 2012; Cannaerts et al. 2019; Bertoli-Avella et al. 2015).

TGFBR1/2 mutations are primarily located in the serine-threonine kinase domain, the intracellular part of the TGFβ receptor. Although occasional nonsense mutations or small intragenic deletions have been described in *TGFBR2*, these were all predicted to escape nonsense-mediated-mRNA decay (Loeys et al. 2006). Deletions involving *TGFBR2* lead to an LDS-like phenotype but patients lack significant aortic disease (Campbell et al. 2011). Moreover, *TGFBR1* nonsense mutations or mutations predicted to cause a complete loss-of-function have been shown to lead to a skin cancer phenotype, multiple self-healing squamous epitheliomas (Goudie et al. 2011). Haploinsufficiency and loss-of-function were suggested as mutational mechanisms for both *SMAD2/3* and *TGFB2/3* mutations. All findings hitherto suggest that although the *TGFBR*-mutations in LDS are also predicted to lead to loss-of-function (Cardoso et al. 2012), some residual protein activity seems to be required to cause the LDS phenotype (see pathogenesis).

11.3 Signs and Symptoms

An overview of the clinical features of LDS is given in Table 11.1. LDS is characterized by four major groups of clinical findings, affecting the vascular, craniofacial, skeletal and cutaneous system (Loeys et al. 2005). Although some clinical overlap with MFS exists, highly prevalent distinguishing features in LDS are cleft palate/bifid uvula, hypertelorism and arterial tortuosity. Interestingly, in some patients, the bifid uvula is the only visible marker to identify people at risk for aortic aneurysms.

11.3.1 Cardiovascular Manifestations

In the vascular system, the most common and prominent finding is the dilatation of the aortic root at the sinuses of Valsalva, which if undetected, leads to aortic dissection and rupture. These dissections have been described in patients as young as 6 months of age. Moreover, dissections have occurred at smaller diameters than those generally accepted at risk in MFS (Williams et al. 2007). In addition to the aortic root aneurysms, aneurysms throughout the arterial tree have been described, most prominently in the side branches of the aorta and the cerebral circulation. Finally, another striking finding is the presence of arterial tortuosity, which is usually most prominent in the head and neck vessels. Vertebral and carotid artery dissection and cerebral bleeding have been described; however, isolated carotid artery dissection in the absence of aortic root involvement has not been observed (Loeys et al. 2005, 2006; Eckman et al. 2009).

Table 11.1 Clinical features at initial diagnosis of LDS

Craniofacial features:	
Craniosynostosis	15%
Hypertelorism	48%
Cleft palate/uvula	72%
Exo/esotropia	18%
Blue sclerae	23%
Skeletal features:	
Pectus deformity	51%
Joint contractures	23%
Joint hypermobility	50%
Arachnodactyly	56%
Club feet	22%
Pes planus	51%
Scoliosis	70%
Cervical spine abnormality	39%
Skin features:	
Thin, translucent	33%
Smooth, velvety	23%
Easy bruising	24%
Delayed wound healing	12%
Herniae	25%
Vascular findings	
Arterial tortuosity	92%
Most common in head and neck vessels	
Carotids (55%)	
Vertebral (56%)	
Intracranial (37%)	
Ascending aorta (5%), aortic arch (10%)	
Descending thoracic (4%) or abdominal (7%) Ao, also other vessels (e.g. iliacs)	
Aneurysms	
Aorta	
Root	87%
Ascending	27%
Arch	10%
Desc thoracic	15%
Abdominal	12%
Vessel beyond Ao	30%

11.3.2 Skeletal Manifestations

Marfanoid skeletal features can be observed, although the actual overgrowth tends to be milder in LDS patients compared to MFS patients. Most typical LDS skeletal findings include pectus excavatum or pectus carinatum, scoliosis, joint laxity, arachnodactyly, talipes equinovarus and cervical spine malformation and/or instability. Arachnodactyly is present in some, but true dolichostenomelia (leading to an increase in the arm span-to-height ratio and a decrease in the upper-to-lower segment ratio) is less common in LDS than in MFS. The combined thumb and wrist signs are present in circa one-third of individuals with LDS. Joint hypermobility is very common and does include congenital hip dislocation and recurrent joint subluxations. Paradoxically, some individuals can show reduced joint mobility, especially of the hands (camptodactyly) and feet (club feet). Other recurrent skeletal findings include spondylolisthesis, acetabular protrusion and pes planus (Loeys et al. 2005, 2006). Preliminary evidence suggests that individuals with LDS have an increased incidence of osteoporosis with increased fracture incidence and delayed bone healing (Kirmani et al. 2010).

11.3.3 (Cranio)Facial Manifestations

Most typical craniofacial features consist of ocular hypertelorism and the presence of a cleft palate, or its mildest presentation, a bifid uvula. Sometimes the uvula is not bifid but has an unusual broad appearance with or without a midline raphe. Another common presenting feature in the more severely affected patients is craniosynostosis. In the latter all sutures can be involved: most commonly the sagittal suture (resulting in dolichocephaly), but also the coronal suture (resulting in brachycephaly) and metopic suture (resulting in trigonocephaly). Other common craniofacial characteristics are malar flattening and retrognathia. Besides the hypertelorism, ocular manifestations include strabismus, blue sclerae and myopia, but the latter is less frequent and less severe than in MFS. Significant refractive errors can lead to amblyopia. Retinal detachment has been reported rarely (Loeys et al. 2005, 2006). In our experience, ectopia lentis is not observed, although in the literature minimal lens(sub) luxation has been reported (Mizuguchi et al. 2004). Less common associated findings requiring further exploration include submandibular branchial cysts and defective tooth enamel (Loeys et al. 2006).

11.3.4 Cutaneous Manifestations

In persons without craniofacial features, important cutaneous findings can provide the clue towards diagnosis. These skin findings show significant overlap with those observed in Ehlers-Danlos syndrome (EDS) and include velvety, thin, translucent skin, easy bruising (other than the lower legs) and dystrophic scars. Comparable to the vascular type of EDS, life-threatening complications, such as spontaneous bowel rupture and peripartal uterine rupture have been reported (Loeys et al. 2006; Gutman et al. 2009). Although in the past, type I and II LDS have been described based on the presence of these vascular EDS-like findings, we now believe these are the representation of a continuum within the LDS spectrum of disease.

11.3.5 Other Findings

Finally, a minority of affected individuals present developmental delay. When present, developmental delay is most often associated with craniosynostosis and/or hydrocephalus, suggesting that learning disability is an extremely rare primary manifestation of LDS. Common neuroradiological findings are dural ectasia and Arnold-Chiari type I malformation (Rodrigues et al. 2009). The precise incidence of those two findings is unknown.

Other recurrent findings that need further documentation include muscle hypoplasia, dental problems with enamel dysplasia, allergic disease with seasonal allergies, asthma/sinusitis, eczema and important gastro-intestinal problems: food allergy, eosinophilic esophagitis, inflammatory bowel disease.

11.4 The Expanding Spectrum of LDS and Closely Related Disease

Van de Laar et al. described another autosomal dominant variant of LDS, also called aneurysm-osteoarthritis syndrome (AOS) (van de Laar et al.

2011). AOS is characterized by aneurysms, dissections and tortuosity throughout the arterial tree in addition to craniofacial (including hypertelorism and abnormal palate/uvula), skeletal (including arachnodactyly and scoliosis) and cutaneous (including striae and velvety skin) symptoms and thus perfectly fits in the phenotypic spectrum of LDS. A distinguishing feature, however, might be the presence of early-onset osteoarthritis. In the initially published series, about 50% of the patients present with osteochondritis dissecans and about 90% of patients have vertebral disc degeneration, suggesting that these findings are very common in *SMAD3* associated type of LDS (van de Laar et al. 2011). Since the initial publication, however, it has become clear that not all *SMAD3* mutation-positive patients do present with osteoarthritis (Wischmeijer et al. 2013; Regalado et al. 2011). The cardiovascular severity of AOS is similar to classical LDS with early-onset dissections at smaller diameters and marked tortuosity (van de Laar et al. 2012, 2012). As such, AOS is now classified as LDS type 3 (MIM#613795).

Subsequently, patients with mutations in the *TGFB2* gene, also present with an autosomal dominant disorder with many systemic features of both MFS and LDS, and are classified as LDS type 4 (MIM#614816). Features shared with MFS and LDS include aortic aneurysm, pectus deformity, arachnodactyly, scoliosis and skin striae. Features shared with LDS but not with MFS, consist of hypertelorism, bifid uvula, bicuspid aortic valve (BAV), arterial tortuosity, club feet and thin skin with easy bruising. Ectopia lentis was not observed (Lindsay et al. 2012). Heterozygous mutations in *TGFB3* lead to the mildest form of LDS, LDS type 5 (MIM#615582) (Bertoli-Avella et al. 2015). Although typical LDS features are observed non-penetrance is also very common. The most recently discovered gene for LDS involves *SMAD2* (LDS type 6) (Cannaerts et al. 2019). So far, few mutation-positive have been reported, so its precise position in the phenotypical spectrum still needs to be determined.

Interestingly, mutations in *SKI*, a functional repressor of TGFβ signalling, were identified as the cause of Shprintzen-Goldberg syndrome (SGS) (Doyle et al. 2012). SGS is characterized by craniosynostosis, distinctive craniofacial features with dolichocephaly, retrognathia, high arched palate, marfanoid skeletal changes including dolichostenomelia, arachnodactyly, camptodactyly, pes planus, pectus excavatum or carinatum, scoliosis, joint hypermobility, and contractures. Cardiovascular anomalies with mitral valve prolapse, mitral regurgitation, and aortic regurgitation may occur, but aortic root dilatation is usually mild. Minimal subcutaneous fat, abdominal wall defects, cryptorchidism in males, and myopia are also characteristic findings. Nearly all SGS patients present with developmental delay, a finding that is rare in LDS. Molecular analysis of a series of individuals with typical SGS did not reveal mutations in the *TGFBR1* or *TGFBR2* (Loeys et al. 2005).

The major clinical findings of MFS, LDS subtypes and SGS are summarized in a comparative table (Table 11.2).

11.5 Diagnostic Criteria for LDS

Although no formal diagnostic criteria have been developed, *LDS gene* testing (*TGFBR1/2, SMAD2/3, TGFB2/3*) should be considered in the following scenarios:

1. Patients with the typical clinical triad of hypertelorism, cleft palate/bifid uvula and arterial tortuosity/aneurysm
2. Early-onset aortic aneurysm with variable combination of other features including arachnodactyly, camptodactyly, club feet, craniosynostosis (all types), blue sclerae, thin skin with atrophic scars, easy bruising, joint hypermobility, BAV and patent ductus arteriosus (PDA), atrial and ventricular septum defects (ASD/VSD)
3. Patients with a MFS-like phenotype, especially those without ectopia lentis, but with

Table 11.2 Differential diagnostic features of MFS, LDS and SGS

	MFS	LDS			SGS
Clinical feature	*FBN1*	*TGFBR1/TGFBR2*	*SMAD2/3*	*TGFB2/3*	*SKI*
Ectopia lentis	+++	–	–	–	–
Cleft palate/bifid uvula	–	++	+	+	+
Hypertelorism	–	++	+	+	++
Craniosynostosis	–	++	+	–	+++
Tall stature	+++	+	+	++	
Arachnodactyly	+++	++	+	+	++
Pectus deformity	++	++	++	++	++
Club foot	–	++	+	++	+
Osteoarthritis	+	+	+++	+	–
Aortic root aneurysm	+++	++	++	++	+
Arterial aneurysm	–	++	+	+	+
Arterial tortuosity	–	++	++	+	+
Early dissection	+	+++	++	+	–
Bicuspid aortic valve	–	++	+	+	+
Mitral valve insufficiency	++	+	+	++	+
Striae	++	+	+	+	+
Dural ectasia	+	+	+	+	+
Developmental delay	–	–	–	–	++

aortic and skeletal features not fulfilling the MFS diagnostic criteria (Loeys et al. 2010)

4. Families with autosomal dominant thoracic aortic aneurysms, especially those families with precocious aortic/arterial dissection, aortic disease beyond the aortic root (including cerebral arteries)
5. Patients with a clinical tableau reminiscent of vascular EDS (thin skin with atrophic scars, easy bruising, joint hypermobility) and normal type III collagen biochemistry
6. Isolated young probands with aortic root dilatation/dissection

If patients present with premature onset of osteoarthritis in addition to any of the above clinical scenarios, *SMAD3* may be prioritized as the causal gene. If the clinical presentation is rather mild, mutation in *TGFB2* or *TGFB3* may also be considered. Although it should be stressed that the clinical overlap is so large, that it is impossible to predict the correct causal gene based on the clinical signs only. If craniosynostosis and intellectual disability are associated features, *SKI* might be the first gene to be analysed.

11.6 Differential Diagnosis

11.6.1 Syndromic Differential Diagnosis

11.6.1.1 Ehlers-Danlos Syndrome

EDS is a clinically and molecularly heterogeneous disorder (Beighton et al. 1998). Amongst the different subtypes, the vascular, valvular (Schwarze et al. 2004) and kyphoscoliosis type can present with significant cardiovascular complications.

The most typical clinical manifestations of vascular EDS include thin, translucent skin, characteristic facial appearance, vascular fragility demonstrated by extensive bruising and easy bleeding and spontaneous arterial/intestinal/uterine ruptures (Beighton et al. 1998). An abnormal type III collagen biochemistry confirms the diagnosis, but ultimate confirmation of the diagnosis lies in the identification of mutations in the *COL3A1* gene, encoding for the type III collagen α-chain 1. Interestingly, in a cohort of 40 patients displaying a vascular EDS-like phenotype but normal collagen III biochemistry, 30% carried *TGFBR1/2* mutations (Loeys et al. 2006), suggesting on the one hand that vascular EDS closely resembles LDS but on the other hand that *TGFBR* mutations may cause a broad spectrum of diseases associated with aortic aneurysms. Finally, arginine-to-cysteine mutations in *COL1A1* have been identified in a subset of affected individuals who typically present with aneurysms of the abdominal aorta and iliac arteries reminiscent of vascular EDS. For these cases, distinct abnormalities on collagen electrophoresis have been observed (Malfait et al. 2007).

The valvular type of EDS is a rare form of EDS with early-onset cardiac valvular dysfunction. This autosomal recessive condition is caused by nonsense mutations in *COL1A2*. Other recurrent findings include joint hypermobility and skin hyperextensibility (Schwarze et al. 2004).

Finally, aortic aneurysm and arterial rupture can also occur in the kyphoscoliotic form of EDS (the former type VI or Ocular-Scoliotic type). This generalized connective tissue disorder is characterized by kyphoscoliosis, joint laxity, muscle hypotonia, and, in some individuals, ocular problems. This autosomal recessive form of EDS is caused by deficient activity of the enzyme procollagen-lysine, 2-oxoglutarate 5-dioxygenase 1 (PLOD1, also called lysyl hydroxylase 1). The diagnosis of EDS, kyphoscoliotic type relies on the demonstration of an increased ratio of deoxypyridinoline to pyridinoline crosslinks in urine. Alternatively, an assay of lysyl hydroxylase enzyme activity in skin fibroblasts is diagnostic. Mutations in *PLOD1*, the gene encoding the enzyme lysyl hydroxylase 1, are causative (Pinnell et al. 1972).

11.6.1.2 Arterial Tortuosity Syndrome and Autosomal Recessive Cutis Laxa Type 1

Two other autosomal recessive connective tissue disorders present arterial tortuosity and aortic aneurysm as key findings.

The arterial tortuosity syndrome (ATS) is characterized by generalized tortuosity, elongation, stenosis and aneurysm formation in the major arteries. Patients often die at a young age due to cardiopulmonary complications. Features in common with LDS include arachnodactyly, hypertelorism, cleft palate and/or bifid uvula, joint laxity or contractions and micro/retrognathia. ATS is caused by loss-of-function mutations in *SLC2A10*, encoding GLUT10, which belongs to the glucose transporter family but its precise function remains unknown (Coucke et al. 2006).

Autosomal recessive cutis laxa type 1 (ARCL1) is another connective tissue disorder characterized by vascular anomalies, lung emphysema and diverticulae of the urinary and gastrointestinal tract aside from the cutaneous symptoms. As in ATS, prognosis can be severely compromised by cardiopulmonary complications. Mutations in two fibulin genes, *EFEMP2* (also called *FBLN4)* or *FBLN5*, are responsible for ARCL1 (Hucthagowder et al. 2006; Loeys et al. 2002). Arterial aneurysms and tortuosity are very prominent in patients with *FBLN4* mutations, while the cutaneous manifestations in *FBLN4* patients are limited and vascular stenosis is more pronounced in *FBLN5* patients. As such, ARCL1 caused by *FBLN4* mutations can be categorized within the LDS spectrum.

11.6.1.3 Meester-Loeys Syndrome

Loss-of-function mutations in *BGN* were identified in patients presenting with Meester-Loeys syndrome (MLS), a condition characterized by early-onset aortic dissection and LDS-associated features (i.e., hypertelorism, bifid uvula, and joint hypermobility and contractures) (Meester et al. 2016). Although most prominent in males as expected for an X-linked condition, a subset of mutation-carrying females also experience aortic aneurysm and even dissection. In other families, females were completely asymptomatic. Surprisingly skewed X-inactivation was shown not to underly this striking variability in females. *BGN* encodes the small leucine-rich proteoglycan biglycan that interacts with extracellular matrix components such as collagen I, III or elastin. Apart from this structural role, biglycan is also known to regulate cytokine activity (e.g. TGFβ and bone morphogenetic protein (BMP)).

11.6.2 Non-syndromic Differential Diagnosis

Non-syndromic types of thoracic aortic aneurysms and dissections (TAAD), or types in which only minor additional symptoms are present, exist as well. Occasionally, mutations in *FBN1* (Milewicz et al. 1996) and in *TGFBR1/2* (Pannu et al. 2005; Tran-Fadulu et al. 2009) causing TAAD have been described, perhaps representing the mildest end of the MFS/LDS phenotypic spectrum. Up to now, three genes, coding for components of the vascular smooth muscle contractile apparatus have been associated with familial thoracic aortic aneurysm: *ACTA2*, coding for vascular smooth cell specific α-actin, *MYH11* (β-myosin heavy chain 11) and *MYLK* (myosin light chain kinase) (Guo et al. 2007; Wang et al. 2010; Zhu et al. 2006). *ACTA2* -mutations have been identified in 14% of TAAD patients (Guo et al. 2007), while *MYH11* mutations have been found in TAAD patients with persistent ductus arteriosus (Zhu et al. 2006). Additional symptoms that can be found in *ACTA2* mutation positive patients include persistent ductus arteriosus, bicuspid aortic valve, iris flocculi, cerebrovascular accidents, Moya-Moya disease and coronary artery disease (Guo et al. 2009). Most recently, mutations in *MYLK, PRKG1, FOXE3* have been shown to account for a small subset of familial aortic aneurysmal disease (Wang et al. 2010; Guo et al. 2013; Kuang et al. 2016).

11.7 Pathology

Histologic examination of aortic tissue from LDS patients reveals elastic fibre fragmentation, loss of elastin content, a marked excess of collagen and accumulation of amorphous matrix components in the aortic media. Electron microscopy shows loss of the intimate spatial association between elastin deposits and vascular smooth muscle cells (Loeys et al. 2005). These findings have been reported already in very young children undergoing early aortic surgery and do occur in the absence of inflammation, suggesting a severe defect in elastogenesis rather than secondary elastic fibre destruction. LDS aortic samples had significantly more diffuse medial degeneration compared with MFS and control samples, but the changes are not specific for LDS (Loeys et al. 2005).

11.8 Biochemical Defects and Pathogenesis

For a long time, Marfan syndrome has been used as the sole paradigm for the pathogenetic study of thoracic aortic aneurysm. The study of Marfan mouse models has shifted our understanding of the pathogenetic mechanisms underlying this condition. In the past, it was believed that the structural deficiency of fibrillin-1 was responsible for many of the phenotypic characteristics, but recent work has also evoked a significant role for altered TGFβ signalling. It is now believed that deficient microfibrils fail to sequester TGFβ in an inactive state and that overactivation of the TGFβ signalling pathway contributes significantly to the disease pathogenesis. The discovery of the genetic basis of LDS has deepened our insights into the role of TGFβ in aortic aneurysm formation.

LDS is most frequently caused by mutations in the genes encoding the transforming growth factor beta (TGFβ) receptor subunits, TβRI and TβRII. The majority of LDS mutations are missense mutations positioned within the intracellular kinase domain, impairing kinase activity but not altering receptor expression or trafficking (Loeys et al. 2005, 2006). These mutations are predicted to cause loss-of-function of TβRI and TβRII. Interestingly, a recent report describes a cutaneous neoplastic phenotype without aortic or systemic involvement in people with heterozygous mutations that confer haploinsufficiency for *TGFBR1* (Goudie et al. 2011). In LDS, it was hypothesized that loss-of-function of the TGFβ receptors could lead to a paradoxical upregulation of TGFβ signalling. Indeed, aortic tissue-derived fibroblast studies documented that heterozygous patient cells show full preservation of the acute-phase response to TGFβ, and that patient-derived tissues show evidence of increased (rather than decreased) TGFβ signalling (Loeys et al. 2005, 2006). While this finding intuitively corroborates the essential role of TGFβ in the pathogenesis of aortic aneurysm, it was not clear how a loss-of-function of the TGFβ receptors could lead to the same upregulation of TGFβ activity as seen in the Marfan mouse models.

The current data suggest that expression of a receptor with impaired kinase activity is necessary to generate the LDS phenotype and would be compatible with either a dominant-negative or complex gain-of-function mechanism of disease. On the one hand, at least two studies, either using heterozygous patient cells or co-transfection experiments, could not find evidence for dominant-negative activity (Loeys et al. 2005; Mizuguchi et al. 2004). On the other hand, a third study provided a somewhat complicated argument for a dominant-negative mechanism despite evidence that co-transfection of equal amounts (both 1X) of DNA encoding wild-type and mutant receptor subunits did not result in less than half the signalling activity seen upon transfection of a 2X complement of wild-type DNA (as expected for a dominant-negative mechanism)(Horbelt et al. 2010). Given that the TGFβ receptor complex involves association between two TβRI and two TβRII subunits, one might argue that dominant-negative activity is both intuitive and inevitable. However, recent evidence suggests that the individual TβRI:TβRII dimers within this tetrameric complex bind ligand and signal independently, yielding a dominant-negative mechanism untenable (Huang et al. 2011). When considered in combination with the repetitive observation of paradoxically increased TGFβ activity in LDS patient tissues, hypotheses have focused on the prospect of excessive and nonproductive compensatory mechanisms, most likely proximately induced by an imbalance of the various signalling functions (eg canonical versus noncanonical) supported by TGFβ receptors in a given cell type or an imbalance of TGFβ signalling in general between distinct but neighbouring cell populations (Lindsay and Dietz 2011). This hypothesis was first supported by accentuation of the aneurysm phenotype in *Fbn1*$^{C1039G/+}$ mice after the introduction of *Smad4* haploinsufficiency in the context of maintained high levels of Smad-dependent signalling and enhanced Smad-independent signalling (Holm et al. 2011). Furthermore, loss-of-function mutations in *SMAD3* or *TGFB2/3* were also associated with an overall increased TGFβ signature in the aortic wall (van de Laar et al. 2011; Lindsay et al. 2012).

Together, these findings indicate that TGFβ signalling is under the control of multiple feedback regulatory pathways. While adding to the complexity, the data support the contentions that many features of microfibril disorders likely manifest failure of proper regulation of TFGβ function, and that consideration of both primary and secondary events will be required to attain full mechanistic insight. Overall, the observations confirm the central role of TGFβ in the final common pathway leading to aortic aneurysms in different syndromes.

11.9 Treatment and Management

11.9.1 Natural History

Comparison of the natural history of Marfan syndrome and Loeys-Dietz syndrome has lead to two important lessons. First, in the most severe cases of LDS (with more outward features of LDS), aortic dissections at smaller diameters as in MFS have been observed, leading to the need for earlier surgical intervention (see below). Secondly, it has been observed that the aortic disease is far more widespread in LDS with aortic disease beyond the aortic root and prominent involvement of aortic sidebranches, necessitating a complete imaging of the arterial tree from head to pelvis.

11.9.2 Medical Treatment

Many of the treatment strategies in LDS are derived from knowledge derived from MFS patient management. The current treatment for aortic aneurysms in MFS is not causal and purely symptomatic. Preventive treatment with beta-blockers is believed to slow down the aortic root growth but in general this does not prevent aortic surgery at later age. Based on initial experiments that demonstrated the rescue of the lung phenotype in Marfan mouse models through the use of TGFβ neutralizing antibodies (Neptune et al. 2003), it was hypothesized that similar treatments may be beneficial for the aortic phenotype in MFS patients. Proof-of-principle was obtained from a Marfan mouse trial (Habashi et al. 2006). The intraperitoneal injection of TGFβ neutralizing antibody blocked aortic root growth in these mice. Subsequently, similar results were obtained using an angiotensin II type 1 receptor blocker, losartan. Losartan does not only have an effect on the renin-angiotensin-aldosterone axis but has also an effect on TGFβ signalling. It is believed to reduce both the total and active amount of TGFβ in the extracellular matrix, probably through effects on thrombospondin, a TGFβ activator. In a placebo-controlled trial on Marfan mice, losartan resulted in significantly reduced aortic growth compared to atenolol, despite the similar hemodynamic effect. In addition to a major effect on the aortic growth, the histology of elastic fibres in the aortic wall of the losartan treated MFS mice was also indistinguishable from wild type mice (Habashi et al. 2006).

The beneficial effect of angiotensin receptor blocker treatment on aortic growth was confirmed in a preliminary observational study in severely affected pediatric MFS patients. Similar to the MFS mice, a significant decrease in rate of change of aortic root dimension after starting angiotensin receptor blocker therapy was observed. Again, as there was no difference in the effect of hemodynamic parameters, the data suggest that these achieved protective effects were likely to be attributed to TGFβ antagonism (Brooke et al. 2008). This study has provided the first evidence for a significant benefit of angiotensin receptor blocking agents over current therapies in reducing aortic root dilation in severe pediatric MFS patients.

Based on the mouse data and the preliminary human study a large, randomized clinical trial in MFS patients has been initiated. This trial, supported by the Pediatric Heart Network through the U.S. National Heart, Lung and Blood Institute (NHLBI), compares atenolol with losartan treatment in more than 600 patients for a three-year treatment (Lacro et al. 2007). In addition, a dozen other trials with different designs and inclusion criteria have been initiated in Belgium, France, Italy, The Netherlands, Taiwan and the United Kingdom (Detaint et al. 2010; Gambarin et al. 2009; Moberg et al. 2012; Radonic et al. 2010). A meta-analysis suggests that losartan is associated with a slower progression of aortic root dilation when compared with placebo and as an addition to beta-blocker therapy (Pyeritz and Loeys 2011; Al-Abcha et al. 2020; Elbadawi et al. 2019). Nevertheless, the precise mode of action of losartan is still unclear and might express its effect independently of the targeted angiotensin receptor, as a recent study in MFS murine model suggested (Sellers et al. 2018).

Taken together, in recent years, major steps have been made in elucidating the pathogenesis of thoracic aneurysm diseases and their clinical treatment. However, until now, there are only limited therapeutic prospects to completely halt its progression. This points out the need for further clinical trials including more patients and longer follow up periods in order to delineate even the minute effects that current and future treatments might express.

11.9.3 Surgical Treatment

Given the safety and the increasing availability of the valve-sparing procedure, the following recommendations have been issued for aortic surgery in LDS (Patel et al. 2011). First, for young children with severe systemic findings of LDS, surgical repair of the ascending aorta should be considered once the maximal dimension exceeds three standard deviations and the aortic annulus exceeds 1.8 cm, allowing the placement of a graft of sufficient size to accommodate growth. Second, for adolescents and adults, surgical repair of the ascending aorta should be considered once the maximal dimension approaches 4.0–4.5 cm. This advice is based on both numerous examples of documented aortic dissection in adults with aortic root dimensions at or below 4.5 cm and the excellent outcome of prophylactic surgery. An extensive family history of larger aortic dimension without dissection could alter this practice for individual patients (Augoustides et al. 2009).

11.10 Genetic Counselling

LDS is inherited in an autosomal dominant manner. About one-quarter of LDS patients has an affected parent whereas approximately three-quarters of probands have LDS as the result of a *de novo* mutation. If the parent is affected, each child has a 50% chance of inheriting the mutation and thus the disorder. Prenatal diagnosis for pregnancies at increased risk for LDS is possible if the disease-causing mutation in the family is known.

References

Al-Abcha A, Saleh Y, Mujer M, Boumegouas M, Herzallah K, Charles L, Elkhatib L, Abdelkarim O, Kehdi M, Abela GS (2020) Meta-analysis examining the usefulness of angiotensin receptor blockers for the prevention of aortic root dilation in patients with the Marfan syndrome. Am J Cardiol 128:101–106

Augoustides JG, Plappert T, Bavaria JE (2009) Aortic decision-making in the Loeys-Dietz syndrome: aortic root aneurysm and a normal caliber ascending aorta and aortic arch. J Thorac Cardiovasc Surg 138:502–503

Beighton P, De Paepe A, Steinmann B, Tsipouras P, Wenstrup RJ (1998) Ehlers-Danlos syndromes: revised nosology, Villefranche, 1997. Ehlers-Danlos National Foundation (USA) and Ehlers-Danlos support group (UK). Am J Med Genet 77:31–37

Bertoli-Avella AM, Gillis E, Morisaki H, Verhagen JMA, de Graaf BM, van de Beek G, Gallo E, Kruithof BPT, Venselaar H, Myers LA, Laga S, Doyle AJ, Oswald G, van Cappellen GWA, Yamanaka I, van der Helm RM, Beverloo B, de Klein A, Pardo L, Lammens M, Evers C, Devriendt K, Dumoulein M, Timmermans J, Bruggenwirth HT, Verheijen F, Rodrigus I, Baynam G, Kempers M, Saenen J, Van Craenenbroeck EM, Minatoya K, Matsukawa R, Tsukube T, Kubo N, Hofstra R, Goumans MJ, Bekkers JA, Roos-Hesselink JW, van de Laar I, Dietz HC, Van Laer L, Morisaki T, Wessels MW, Loeys BL (2015) Mutations in a TGF-beta ligand, TGFB3, cause syndromic aortic aneurysms and dissections. J Am Coll Cardiol 65:1324–1336

Brooke BS, Habashi JP, Judge DP, Patel N, Loeys B, Dietz HC 3rd. (2008) Angiotensin II blockade and aortic-root dilation in Marfan's syndrome. N Engl J Med 358:2787–2795

Campbell IM, Kolodziejska KE, Quach MM, Wolf VL, Cheung SW, Lalani SR, Ramocki MB, Stankiewicz P (2011) TGFBR2 deletion in a 20-month-old female with developmental delay and microcephaly. Am J Med Genet A 155A:1442–1447

Cannaerts E, Kempers M, Maugeri A, Marcelis C, Gardeitchik T, Richer J, Micha D, Beauchesne L, Timmermans J, Vermeersch P, Meyten N, Chenier S, van de Beek G, Peeters N, Alaerts M, Schepers D, Van Laer L, Verstraeten A, Loeys B (2019) Novel pathogenic SMAD2 variants in five families with arterial aneurysm and dissection: further delineation of the phenotype. J Med Genet 56:220–227

Cardoso S, Robertson SP, Daniel PB (2012) TGFBR1 mutations associated with Loeys-Dietz syndrome are inactivating. J Recept Signal Transduct Res 32:150–155

Coucke PJ, Willaert A, Wessels MW, Callewaert B, Zoppi N, De Backer J, Fox JE, Mancini GM, Kambouris M, Gardella R, Facchetti F, Willems PJ, Forsyth R, Dietz HC, Barlati S, Colombi M, Loeys B, De Paepe A (2006) Mutations in the facilitative glucose transporter GLUT10 alter angiogenesis and cause arterial tortuosity syndrome. Nat Genet 38:452–457

Detaint D, Aegerter P, Tubach F, Hoffman I, Plauchu H, Dulac Y, Faivre LO, Delrue MA, Collignon P, Odent S, Tchitchinadze M, Bouffard C, Arnoult F, Gautier M, Boileau C, Jondeau G (2010) Rationale and design of a randomized clinical trial (Marfan Sartan) of angiotensin II receptor blocker therapy versus placebo in individuals with Marfan syndrome. Arch Cardiovasc Dis 103:317–325

Doyle AJ, Doyle JJ, Bessling SL, Maragh S, Lindsay ME, Schepers D, Gillis E, Mortier G, Homfray T, Sauls K, Norris RA, Huso ND, Leahy D, Mohr DW, Caulfield MJ, Scott AF, Destree A, Hennekam RC, Arn PH, Curry CJ, Van Laer L, McCallion AS, Loeys BL, Dietz HC (2012) Mutations in the TGF-beta repressor SKI cause Shprintzen-Goldberg syndrome with aortic aneurysm. Nat Genet

Eckman PM, Hsich E, Rodriguez ER, Gonzalez-Stawinski GV, Moran R, Taylor DO (2009) Impaired systolic function in Loeys-Dietz syndrome: a novel cardiomyopathy? Circ Heart Fail 2:707–708

Elbadawi A, Omer MA, Elgendy IY, Abuzaid A, Mohamed AH, Rai D, Saad M, Mentias A, Rezq A, Kamal D, Khalife W, London B, Morsy M (2019) Losartan for preventing aortic root dilatation in patients with Marfan syndrome: a meta-analysis of randomized trials. Cardiol Ther 8:365–372

Gambarin FI, Favalli V, Serio A, Regazzi M, Pasotti M, Klersy C, Dore R, Mannarino S, Vigano M, Odero A, Amato S, Tavazzi L, Arbustini E (2009) Rationale and design of a trial evaluating the effects of losartan vs. nebivolol vs. the association of both on the progression of aortic root dilation in Marfan syndrome with FBN1 gene mutations. J Cardiovasc Med (Hagerstown) 10:354–362

Goudie DR, D'Alessandro M, Merriman B, Lee H, Szeverenyi I, Avery S, O'Connor BD, Nelson SF, Coats SE, Stewart A, Christie L, Pichert G, Friedel J, Hayes I, Burrows N, Whittaker S, Gerdes AM, Broesby-Olsen S, Ferguson-Smith MA, Verma C, Lunny DP, Reversade B, Lane EB (2011) Multiple self-healing squamous epithelioma is caused by a disease-specific spectrum of mutations in TGFBR1. Nat Genet 43:365–369

Guo DC, Pannu H, Tran-Fadulu V, Papke CL, Yu RK, Avidan N, Bourgeois S, Estrera AL, Safi HJ, Sparks E, Amor D, Ades L, McConnell V, Willoughby CE, Abuelo D, Willing M, Lewis RA, Kim DH, Scherer S, Tung PP, Ahn C, Buja LM, Raman CS, Shete SS, Milewicz DM (2007) Mutations in smooth muscle alpha-actin (ACTA2) lead to thoracic aortic aneurysms and dissections. Nat Genet 39:1488–1493

Guo DC, Papke CL, Tran-Fadulu V, Regalado ES, Avidan N, Johnson RJ, Kim DH, Pannu H, Willing MC, Sparks E, Pyeritz RE, Singh MN, Dalman RL, Grotta JC, Marian AJ, Boerwinkle EA, Frazier LQ, LeMaire SA, Coselli JS, Estrera AL, Safi HJ, Veeraraghavan S, Muzny DM, Wheeler DA, Willerson JT, Yu RK, Shete SS, Scherer SE, Raman CS, Buja LM, Milewicz DM (2009) Mutations in smooth muscle alpha-actin (ACTA2) cause coronary artery disease, stroke, and Moyamoya disease, along with thoracic aortic disease. Am J Hum Genet 84:617–627

Guo DC, Regalado E, Casteel DE, Santos-Cortez RL, Gong L, Kim JJ, Dyack S, Horne SG, Chang G, Jondeau G, Boileau C, Coselli JS, Li Z, Leal SM, Shendure J, Rieder MJ, Bamshad MJ, Nickerson DA, Gen TACRC, National Heart L, Blood Institute Grand Opportunity Exome Sequencing, P, Kim C, Milewicz DM (2013) Recurrent gain-of-function mutation in PRKG1 causes thoracic aortic aneurysms and acute aortic dissections. Am J Hum Genet 93:398–404

Gutman G, Baris HN, Hirsch R, Mandel D, Yaron Y, Lessing JB, Kuperminc MJ (2009) Loeys-Dietz syndrome in pregnancy: a case description and report of a novel mutation. Fetal Diagn Ther 26:35–37

Habashi JP, Judge DP, Holm TM, Cohn RD, Loeys BL, Cooper TK, Myers L, Klein EC, Liu G, Calvi C, Podowski M, Neptune ER, Halushka MK, Bedja D, Gabrielson K, Rifkin DB, Carta L, Ramirez F, Huso DL, Dietz HC (2006) Losartan, an AT1 antagonist, prevents aortic aneurysm in a mouse model of Marfan syndrome. Science 312:117–121

Holm TM, Habashi JP, Doyle JJ, Bedja D, Chen Y, van Erp C, Lindsay ME, Kim D, Schoenhoff F, Cohn RD, Loeys BL, Thomas CJ, Patnaik S, Marugan JJ, Judge DP, Dietz HC (2011) Noncanonical TGFbeta signaling contributes to aortic aneurysm progression in Marfan syndrome mice. Science (New York, NY) 332:358–361

Horbelt D, Guo G, Robinson PN, Knaus P (2010) Quantitative analysis of TGFBR2 mutations in Marfan-syndrome-related disorders suggests a correlation between phenotypic severity and Smad signaling activity. J Cell Sci 123:4340–4350

Huang T, David L, Mendoza V, Yang Y, Villarreal M, De K, Sun L, Fang X, Lopez-Casillas F, Wrana JL, Hinck AP (2011) TGF-beta signalling is mediated by two autonomously functioning TbetaRI:TbetaRII pairs. EMBO J 30:1263–1276

Hucthagowder V, Sausgruber N, Kim KH, Angle B, Marmorstein LY, Urban Z (2006) Fibulin-4: a novel gene for an autosomal recessive cutis laxa syndrome. Am J Hum Genet 78:1075–1080

Kirmani S, Tebben PJ, Lteif AN, Gordon D, Clarke BL, Hefferan TE, Yaszemski MJ, McGrann PS, Lindor NM, Ellison JW (2010) Germline TGF-beta receptor mutations and skeletal fragility: a report on two patients with Loeys-Dietz syndrome. Am J Med Genet A 152A:1016–1019

Kuang SQ, Medina-Martinez O, Guo DC, Gong L, Regalado ES, Reynolds CL, Boileau C, Jondeau G, Prakash SK, Kwartler CS, Zhu LY, Peters AM, Duan XY, Bamshad MJ, Shendure J, Nickerson DA, Santos-Cortez RL, Dong X, Leal SM, Majesky MW, Swindell EC, Jamrich M, Milewicz DM (2016) FOXE3 mutations predispose to thoracic aortic aneurysms and dissections. J Clin Invest 126:948–961

Lacro RV, Dietz HC, Wruck LM, Bradley TJ, Colan SD, Devereux RB, Klein GL, Li JS, Minich LL, Paridon SM, Pearson GD, Printz BF, Pyeritz RE, Radojewski E, Roman MJ, Saul JP, Stylianou MP, Mahony L (2007) Rationale and design of a randomized clinical trial of beta-blocker therapy (atenolol) versus angiotensin II receptor blocker therapy (losartan) in individuals with Marfan syndrome. Am Heart J 154:624–631

Lindsay ME, Dietz HC (2011) Lessons on the pathogenesis of aneurysm from heritable conditions. Nature 473:308–316

Lindsay ME, Schepers D, Bolar NA, Doyle JJ, Gallo E, Fert-Bober J, Kempers MJ, Fishman EK, Chen Y, Myers L, Bjeda D, Oswald G, Elias AF, Levy HP, Anderlid BM, Yang MH, Bongers EM, Timmermans J, Braverman AC, Canham N, Mortier GR, Brunner HG, Byers PH, Van Eyk J, Van Laer L, Dietz HC, Loeys BL (2012) Loss-of-function mutations in TGFB2 cause a syndromic presentation of thoracic aortic aneurysm. Nat Genet 44:922–927

Loeys B, Van Maldergem L, Mortier G, Coucke P, Gerniers S, Naeyaert JM, De Paepe A (2002) Homozygosity for a missense mutation in fibulin-5 (FBLN5) results in a severe form of cutis laxa. Hum Mol Genet 11:2113–2118

Loeys BL, Chen J, Neptune ER, Judge DP, Podowski M, Holm T, Meyers J, Leitch CC, Katsanis N, Sharifi N, Xu FL, Myers LA, Spevak PJ, Cameron DE, De Backer J, Hellemans J, Chen Y, Davis EC, Webb CL, Kress W, Coucke P, Rifkin DB, De Paepe AM, Dietz HC (2005) A syndrome of altered cardiovascular, craniofacial, neurocognitive and skeletal development caused by mutations in TGFBR1 or TGFBR2. Nat Genet 37:275–281

Loeys BL, Schwarze U, Holm T, Callewaert BL, Thomas GH, Pannu H, De Backer JF, Oswald GL, Symoens S, Manouvrier S, Roberts AE, Faravelli F, Greco MA, Pyeritz RE, Milewicz DM, Coucke PJ, Cameron DE, Braverman AC, Byers PH, De Paepe AM, Dietz HC (2006) Aneurysm syndromes caused by mutations in the TGF-beta receptor. N Engl J Med 355:788–798

Loeys BL, Dietz HC, Braverman AC, Callewaert BL, De Backer J, Devereux RB, Hilhorst-Hofstee Y, Jondeau G, Faivre L, Milewicz DM, Pyeritz RE, Sponseller PD, Wordsworth P, De Paepe AM (2010) The revised Ghent nosology for the Marfan syndrome. J Med Genet 47:476–485

Malfait F, Symoens S, De Backer J, Hermanns-Le T, Sakalihasan N, Lapiere CM, Coucke P, De Paepe A (2007) Three arginine to cysteine substitutions in the pro-alpha (I)-collagen chain cause Ehlers-Danlos syndrome with a propensity to arterial rupture in early adulthood. Hum Mutat 28:387–395

Meester JA, Vandeweyer G, Pintelon I, Lammens M, Van Hoorick L, De Belder S, Waitzman K, Young L, Markham LW, Vogt J, Richer J, Beauchesne LM, Unger S, Superti-Furga A, Prsa M, Dhillon R, Reyniers E, Dietz HC, Wuyts W, Mortier G, Verstraeten A, Van Laer L, Loeys BL (2016) Loss-of-function mutations in the X-linked biglycan gene cause a severe syndromic form of thoracic aortic aneurysms and dissections. Genet Med

Milewicz DM, Michael K, Fisher N, Coselli JS, Markello T, Biddinger A (1996) Fibrillin-1 (FBN1) mutations in patients with thoracic aortic aneurysms. Circulation 94:2708–2711

Mizuguchi T, Collod-Beroud G, Akiyama T, Abifadel M, Harada N, Morisaki T, Allard D, Varret M, Claustres M, Morisaki H, Ihara M, Kinoshita A, Yoshiura K, Junien C, Kajii T, Jondeau G, Ohta T, Kishino T, Furukawa Y, Nakamura Y, Niikawa N, Boileau C, Matsumoto N (2004) Heterozygous TGFBR2 mutations in Marfan syndrome. Nat Genet 36:855–860

Moberg K, De Nobele S, Devos D, Goetghebeur E, Segers P, Trachet B, Vervaet C, Renard M, Coucke P, Loeys B, De Paepe A, De Backer J (2012) The Ghent Marfan trial – a randomized, double-blind placebo controlled trial with losartan in Marfan patients treated with beta-blockers. Int J Cardiol 157:354–358

Neptune ER, Frischmeyer PA, Arking DE, Myers L, Bunton TE, Gayraud B, Ramirez F, Sakai LY, Dietz HC (2003) Dysregulation of TGF-beta activation contributes to pathogenesis in Marfan syndrome. Nat Genet 33:407–411

Pannu H, Fadulu VT, Chang J, Lafont A, Hasham SN, Sparks E, Giampietro PF, Zaleski C, Estrera AL, Safi HJ, Shete S, Willing MC, Raman CS, Milewicz DM (2005) Mutations in transforming growth factor-beta receptor type II cause familial thoracic aortic aneurysms and dissections. Circulation 112:513–520

Patel ND, Arnaoutakis GJ, George TJ, Allen JG, Alejo DE, Dietz HC, Cameron DE, Vricella LA (2011) Valve-sparing aortic root replacement in Loeys-Dietz syndrome. Ann Thorac Surg 92:556–560. discussion 560-551

Pinnell SR, Krane SM, Kenzora JE, Glimcher MJ (1972) A heritable disorder of connective tissue. Hydroxylysine-deficient collagen disease. N Engl J Med 286:1013–1020

Pyeritz RE, Loeys B (2011) The 8th international research symposium on the Marfan syndrome and related conditions. Am J Med Genet A

Radonic T, de Witte P, Baars MJ, Zwinderman AH, Mulder BJ, Groenink M (2010) Losartan therapy in adults with Marfan syndrome: study protocol of the multi-center randomized controlled COMPARE trial. Trials 11:3

Regalado ES, Guo DC, Villamizar C, Avidan N, Gilchrist D, McGillivray B, Clarke L, Bernier F, Santos-Cortez RL, Leal SM, Bertoli-Avella AM, Shendure J, Rieder MJ, Nickerson DA, Milewicz DM (2011) Exome sequencing identifies SMAD3 mutations as a cause of familial thoracic aortic aneurysm and dissection with intracranial and other arterial aneurysms. Circ Res 109:680–686

Rodrigues VJ, Elsayed S, Loeys BL, Dietz HC, Yousem DM (2009) Neuroradiologic manifestations of Loeys-Dietz syndrome type 1. AJNR Am J Neuroradiol 30:1614–1619

Schwarze U, Hata R, McKusick VA, Shinkai H, Hoyme HE, Pyeritz RE, Byers PH (2004) Rare autosomal recessive cardiac valvular form of Ehlers-Danlos syndrome results from mutations in the COL1A2 gene that activate the nonsense-mediated RNA decay pathway. Am J Hum Genet 74:917–930

Sellers SL, Milad N, Chan R, Mielnik M, Jermilova U, Huang PL, de Crom R, Hirota JA, Hogg JC, Sandor GG, Van Breemen C, Esfandiarei M, Seidman MA, Bernatchez P (2018) Inhibition of Marfan syndrome aortic root dilation by losartan: role of angiotensin II receptor type 1-independent activation of endothelial function. Am J Pathol 188:574–585

Tran-Fadulu V, Pannu H, Kim DH, Vick GW 3rd, Lonsford CM, Lafont AL, Boccalandro C, Smart S, Peterson KL, Hain JZ, Willing MC, Coselli JS, LeMaire SA, Ahn C, Byers PH, Milewicz DM (2009) Analysis of multigenerational families with thoracic aortic aneurysms and dissections due to TGFBR1 or TGFBR2 mutations. J Med Genet 46:607–613

van de Laar IM, Oldenburg RA, Pals G, Roos-Hesselink JW, de Graaf BM, Verhagen JM, Hoedemaekers YM, Willemsen R, Severijnen LA, Venselaar H, Vriend G, Pattynama PM, Collee M, Majoor-Krakauer D, Poldermans D, Frohn-Mulder IM, Micha D, Timmermans J, Hilhorst-Hofstee Y, Bierma-Zeinstra SM, Willems PJ, Kros JM, Oei EH, Oostra BA, Wessels MW, Bertoli-Avella AM (2011) Mutations in SMAD3 cause a syndromic form of aortic aneurysms and dissections with early-onset osteoarthritis. Nat Genet 43:121–126

van de Laar IM, van der Linde D, Oei EH, Bos PK, Bessems JH, Bierma-Zeinstra SM, van Meer BL, Pals

G, Oldenburg RA, Bekkers JA, Moelker A, de Graaf BM, Matyas G, Frohn-Mulder IM, Timmermans J, Hilhorst-Hofstee Y, Cobben JM, Bruggenwirth HT, van Laer L, Loeys B, De Backer J, Coucke PJ, Dietz HC, Willems PJ, Oostra BA, De Paepe A, Roos-Hesselink JW, Bertoli-Avella AM, Wessels MW (2012) Phenotypic spectrum of the SMAD3-related aneurysms-osteoarthritis syndrome. J Med Genet 49:47–57

van der Linde D, van de Laar IMBM, Bertoli-Avella AM, Oldenburg RA, Bekkers JA, Mattace-Raso FUS, van den Meiracker AH, Moelker A, Tanghe HLJ, van Kooten F, Frohn IME, Timmermans J, Moltzer E, Cobben JM, Van Laer L, Loeys B, De Backer J, Coucke PJ, De Paepe A, Wessels MW, Roos-Hesselink JW (2012) Cardiovascular phenotype of the recently discovered aneurysms-osteoarthritis syndrome (AOS) caused by SMAD3 mutations. J Am Coll Cardiol. (In press)

Wang L, Guo DC, Cao J, Gong L, Kamm KE, Regalado E, Li L, Shete S, He WQ, Zhu MS, Offermanns S, Gilchrist D, Elefteriades J, Stull JT, Milewicz DM (2010) Mutations in myosin light chain kinase cause familial aortic dissections. Am J Hum Genet 87:701–707

Watanabe Y, Sakai H, Nishimura A, Miyake N, Saitsu H, Mizuguchi T, Matsumoto N (2008) Paternal somatic mosaicism of a TGFBR2 mutation transmitting to an affected son with Loeys-Dietz syndrome. Am J Med Genet A 146A:3070–3074

Williams JA, Loeys BL, Nwakanma LU, Dietz HC, Spevak PJ, Patel ND, Francois K, DeBacker J, Gott VL, Vricella LA, Cameron DE (2007) Early surgical experience with Loeys-Dietz: a new syndrome of aggressive thoracic aortic aneurysm disease. Ann Thorac Surg 83:S757–S763. discussion S785-790

Wischmeijer A, Van Laer L, Tortora G, Ajit Bolar N, Van Camp G, Fransen E, Peeters N, di Bartolomeo R, Pacini D, Gargiulo G, Turci S, Bonvicini M, Mariucci E, Lovato L, Brusori S, Ritelli M, Colombi M, Garavelli L, Seri M, Loeys BL (2013) First report of thoracic aortic aneurysm in infancy in a family with aneurysms-osteoarthritis syndrome due to a novel SMAD3 mutation: further delineation of the clinical phenotype. Am J Med Genet A. Minor revision

Zhu L, Vranckx R, Khau Van Kien P, Lalande A, Boisset N, Mathieu F, Wegman M, Glancy L, Gasc JM, Brunotte F, Bruneval P, Wolf JE, Michel JB, Jeunemaitre X (2006) Mutations in myosin heavy chain 11 cause a syndrome associating thoracic aortic aneurysm/aortic dissection and patent ductus arteriosus. Nat Genet 38:343–349

12 Meester-Loeys Syndrome

Josephina A. N. Meester, Pauline De Kinderen,
Aline Verstraeten, and Bart Loeys

Abstract

Meester-Loeys syndrome is an X-linked form of syndromic thoracic aortic aneurysm, characterized by the involvement of multiple organ systems. More specifically, the cardiovascular, skeletal, craniofacial, cutaneous and neurological systems are affected. Clear clinical overlap with Marfan syndrome and Loeys-Dietz syndrome is observed. Aortic dissections occur typically at young ages and are most often observed in males. Meester-Loeys syndrome is caused by loss-of-function mutations in *BGN*, encoding the small leucine-rich proteoglycan biglycan. Although functional consequences of these mutations remain largely elusive, increased TGF-β signaling has been observed. Novel insights will provide opportunities for preventive therapeutic interventions.

Keywords

Meester-Loeys syndrome · Loss-of-function mutations · Biglycan · Thoracic aortic aneurysm · Aortic dissection · Marfan syndrome · Loeys-Dietz syndrome · Connective tissue disorder

J. A. N. Meester · P. De Kinderen · A. Verstraeten
Center for Medical Genetics, Antwerp University Hospital/University of Antwerp, Antwerp, Belgium
e-mail: josephina.meester@uantwerpen.be; pauline.dekinderen@uantwerpen.be; aline.verstraeten@uantwerpen.be

B. Loeys (✉)
Center for Medical Genetics, Antwerp University Hospital/University of Antwerp, Antwerp, Belgium

Department of Human Genetics, Radboud University Medical Center, Nijmegen, The Netherlands
e-mail: bart.loeys@uantwerpen.be; http://www.genetica-antwerpen.be

Abbreviations

BGN	Biglycan
ECM	Extracellular matrix
GAG	Glycosaminoglycan
LDS	Loeys-Dietz syndrome
MFS	Marfan syndrome
MRLS	Meester-Loeys syndrome
NMD	Nonsense mediated RNA decay
SEMDX	Spondylo-epi-metaphyseal dysplasia
TAA	Thoracic aortic aneurysm

12.1 Introduction

Meester-Loeys syndrome (MRLS, OMIM #300989) was first described in 2017 and is caused by loss-of-function mutations in the *BGN* gene (Meester et al. 2017). This initial report

J. Halper (ed.), *Progress in Heritable Soft Connective Tissue Diseases*, Advances in Experimental Medicine and Biology 1348, https://doi.org/10.1007/978-3-030-80614-9_12

described five probands, all of which suffered from cardiovascular complications (i.e. predominantly thoracic aortic aneurysm (TAA) but also mild valve insufficiency was seen in some families), in addition to other connective tissue findings affecting the skeleton (e.g., joint hypermobility and contractures, and flat feet), skin (e.g., striae and hypertrichosis), neurological system (e.g., dilated cerebral ventricles and relative macrocephaly) and craniofacial structures (e.g., hypertelorism, down-slanting eyes, proptosis, malar hypoplasia, frontal bossing and dolichocephaly). A clear clinical overlap is observed with Loeys-Dietz syndrome (LDS, OMIM #609192, #610168, #613795, #614816, #615582, and #601366) (Loeys et al. 2005; Bertoli-Avella et al. 2015; Boileau et al. 2012; Lindsay et al. 2012; van de Laar et al. 2011; Micha et al. 2015) and Marfan syndrome (MFS, OMIM #154700), (Dietz et al. 1991) evidenced by the presence of TAA, early onset aortic dissections, hypertelorism, bifid uvula, arterial tortuosity, joint hypermobility and joint contractures. Unique features for MRLS include ventricular dilatation on brain imaging, hypertrichosis, gingival hypertrophy and relative macrocephaly.

12.2 Genetics

Although familial TAA is most often inherited in an autosomal dominant manner, rare X-linked families have been described before (Coady et al. 1999). Prior to the publication of loss-of-function mutations in *BGN* as the cause for MRLS, *FLNA* (encoding for filamin A) was the only X-linked gene associated with aortic aneurysm formation. However, only 18% of *FLNA* mutation carriers suffers from TAA (29% males and 71% females), and dissections are rarely observed, (Chen et al. 2018) whereas all male patients (N = 9) and 67% of female patients (N = 6) with a *BGN* loss-of-function mutation were reported to suffer from TAAs (Meester et al. 2017). Due to its X-chromosomal location (Xq28), *BGN* shows different phenotypes in mutation carriers across genders and could play an important role in sexual dimorphism of TAA in general.

To date, five causal mutations for MRLS have been described. More specifically, these include two large deletions (respectively 21 kb and 28 kb in size), deleting all coding exons of *BGN* and a portion of downstream DNA (including the 5′ untranslated region of several isoforms of *ATP2B3*), one nonsense mutation at the very N-terminal part of the protein (p.(Trp2*)), and two missense mutations (p.(Gly80Ser) and p.(Gln303Pro)), which are both located at the exon-intron boundaries. These missense mutations (are predicted to) lead to nonsense mediated RNA decay (NMD), due to splicing aberrations. All *BGN* mutations were inherited, so *de novo* mutations have not yet been described for this syndrome (Meester et al. 2017).

Interestingly, a more pronounced skeletal phenotype was described in male patients with a deletion of the (coding part of the) *BGN* gene. This skeletal phenotype involved brachydactyly, arachnodactyly, flat feet, short stature, joint hypermobility, and joint contractures. Unique skeletal features observed in these patients include spatulous fingers, joint dislocation and C1/2 spine malformation (Meester et al. 2017).

12.3 Clinical Features

Table 12.1 lists the clinical features of MRLS with their observed frequencies in the total cohort of mutation carriers as well as in males and females separately.

12.3.1 Cardiovascular Manifestations

The cardiovascular manifestations in MRLS already manifest at a very young age and are particularly severe in males. All male carriers suffer from (severe) aortic root dilatation and aortic dissection was reported in two male *BGN* deletion carriers. In two other hemizygous carriers, TAA was observed both at the level of the ascending aorta and aortic root (Meester et al. 2017).

In contrast to what has been observed in males, females have a very variable degree of cardiovascular severity. One female died from

Table 12.1 Clinical features of MRLS

Organ system	Clinical feature	Frequency total	% total	Frequency males	% males	Frequency females	% females
Cardiovascular	Aortic root dilatation	13/15	87%	9/9	100%	4/6	67%
	Aortic dissection	3/14	21%	2/9	22%	1/5	20%
	Dilated ascending aorta	2/5	40%	2/4	50%	0/1	0%
Skeletal	Pectus deformity	2/11	18%	2/8	25%	0/3	0%
	Arachnodactyly	4/10	40%	4/8	50%	0/2	0%
	Brachydactyly	3/7	43%	2/4	50%	1/3	33%
	Spatulous fingers	3/6	50%	3/4	75%	0/2	0%
	Flat feet	5/9	56%	4/7	57%	1/2	50%
	Joint hypermobility	8/10	80%	7/8	88%	1/2	50%
	Joint dislocation	3/7	43%	2/5	40%	1/2	50%
	Joint contracture	4/7	57%	4/5	80%	0/2	0%
	Short stature	5/11	45%	4/8	50%	1/3	33%
Craniofacial	Dolichocephaly	4/7	57%	4/5	80%	0/2	0%
	Frontal bossing	5/7	71%	5/5	100%	0/2	0%
	Hypertelorism	8/11	73%	5/8	63%	3/3	100%
	Down-slanting eyes	6/7	86%	4/5	80%	2/2	100%
	Proptosis	5/8	63%	3/5	60%	2/3	67%
	Malar hypoplasia	7/8	88%	6/6	100%	1/2	50%
	Bifid uvula	1/7	14%	1/6	17%	0/1	0%
	Highly arched palate	2/5	40%	2/4	50%	0/1	0%
	Gingival hypertrophy	2/3	67%	1/1	100%	1/2	50%
Cutaneous	Striae	4/9	44%	2/6	33%	2/3	67%
	Hypertrichosis	2/3	67%	2/2	100%	0/1	0%
	Delayed wound healing	1/5	20%	0/3	0%	1/2	50%
	Easy bruising	1/5	20%	0/3	0%	1/2	50%
	Umbilical hernia	1/6	17%	1/4	25%	0/2	0%
Neurological	Mild learning problems	1/5	20%	1/4	25%	0/1	0%
	Dilated cerebral ventricles	3/3	100%	3/3	100%	0/0	0%
	Relative macrocephaly	3/3	100%	3/3	100%	0/0	0%

Some features were only investigated in a subset of individuals. Therefore, percentages should be interpreted accordingly aortic dissection and several others suffered from aortic root dilatation, while some do not show any aortic dilatation (even at old age). Skewed X inactivation was investigated, but could not provide a solid explanation to the observed differences in phenotypes (Meester et al. 2017).

Furthermore, aneurysms are not restricted to the aorta, as cerebral, patent ductus arteriosus, and pulmonary artery aneurysms were also observed. Complete imaging from head to pelvis is therefore recommended in the management of this syndrome. Other cardiovascular features include atrial septal defect, aortic valve regurgitation, arterial tortuosity, and mitral valve prolapse, but these features have only been observed once (Meester et al. 2017).

12.3.2 Skeletal Manifestations

MRLS presents with a variety of skeletal manifestations, some of which are also observed in MFS and LDS. Joint hypermobility was observed most frequently, but this feature is present in many connective tissue disorders and can therefore not be used to distinguish MRLS from other TAA syndromes (Meester et al. 2017).

In LDS and especially MFS, skeletal overgrowth underlies most observed skeletal features. Clinically overlapping features have the following frequencies in MRLS: pectus deformity (18%, 2/11), arachnodactyly (40%, 4/10), flat feet (56%, 5/9), joint hypermobility (80%, 8/10), joint contractures (57%, 4/7), and joint dislocation (43%, 3/7). However, in MRLS, the opposite phenotype is also observed, as 43% (3/7) of patients had brachydactyly and 45% (5/11) had a rather short stature (Table 12.1) (Meester et al. 2017).

In both MRLS and LDS, cervical spine malformation has been observed. Unique features of MRLS patients are the presence of spatulous fingers, skeletal dysplasia with hip dislocation, platyspondyly, phalangeal dysplasia and dysplastic epiphysis of the long bones. However, these unique features have only been observed in males with a deletion of *BGN* (Meester et al. 2017).

12.3.3 Craniofacial Manifestations

Most typical craniofacial features of MRLS include, dolichocephaly, frontal bossing, downslanting palpebral fissures, proptosis and malar hypoplasia. Bifid uvula, a cardinal feature of LDS, has been observed, in addition to its milder equivalent, a broad uvula. A highly arched palate has also been observed occasionally. Although rarely observed in MFS, a large proportion of LDS patients has hypertelorism, which is also the case for MRLS. A specific feature for MRLS patients is gingival hypertrophy, observed in two patients with a deletion of *BGN*. Interestingly, no specific ocular features have been associated to this syndrome yet (Meester et al. 2017).

12.3.4 Cutaneous Manifestations

Striae have been reported in 44% of cases. Two brothers were also reported to have hypertrichosis, a feature that has not been described for LDS or MFS. Other occasional skin findings, which are frequently observed in other connective tissue disorders, include delayed wound healing, easy bruising, and umbilical hernia (Meester et al. 2017).

12.3.5 Neurological Manifestations

Unique features to MRLS include dilated cerebral ventricles and relative macrocephaly. However, they were only observed in males with a *BGN* deletion. It remains to be determined if these features are specific for *BGN* deletion carriers, or whether they can also be observed in other MRLS patients. Mild learning problems were described in one male, but it is unknown if this feature is part of the clinical MRLS spectrum or whether this was an incidental finding (Meester et al. 2017).

12.4 Diagnostic Criteria for MRLS

Because MRLS was not described until 2017, the number of reported patients is very limited. Therefore, no formal diagnostic criteria have been proposed yet. These criteria can be established when a larger number of cases have been reported. There are a number of recurrent features which could be considered for the diagnostic criteria, including aortic aneurysms and dissections, joint hypermobility, and joint contractures. Furthermore, unique features of MRLS should also be included in these criteria, as they allow to differentiate between other connective tissue disorders. Ventricular dilatation on brain imaging, hypertrichosis, gingival hypertrophy, relative macrocephaly and several skeletal findings are the currently suggested discriminating features. However, only a limited number of *BGN* patients have been described so far and they potentially represent the most severe end of the spectrum. Therefore, it remains to be determined what the entire clinical spectrum encompasses. Finally, proof of loss-of-function for the identified *BGN* variant should be included in the future diagnostic criteria (Meester et al. 2017).

12.5 Differential Diagnosis

Due to significant clinical overlap, both MFS and LDS should be considered in diagnostic decision making. Although these conditions are caused by mutations in different genes, all patients have overlapping connective tissue findings, e.g. TAA and/or dissection, joint hypermobility, pectus deformity, and joint contractures. Other overlapping features between MRLS and LDS include hypertelorism and a bifid (or broad) uvula.

The skeletal features that can be observed in specific MRLS families (i.e. in families with a *BGN* deletion) show also significant clinical overlap with Melnick-Needles syndrome. Both are X-linked disorders, but Melnick-Needles syndrome is caused by gain-of-function mutations in *FLNA* and is often lethal in males (Robertson et al. 2003; Foley et al. 2010). Clinical overlap is evidenced by the presence of proptosis, frontal bossing, short stature, and phalangeal dysplasia (Foley et al. 2010).

Connective tissue findings, other than TAA and aortic dissections, can also be mild in MRLS patients. Therefore, the syndromic presentation is not always visible at a first glance. For this reason, non-syndromic TAA should also part of the differential diagnosis. Several causal genes have been described for familial TAA, including *ACTA2*, *MYH11*, *PRKG1*, *MFAP5*, *LOX*, and *MYLK*. (Guo et al. 2007; Wang et al. 2010; Luyckx et al. 2017; Zhu et al. 2006; Guo et al. 2013; Barbier et al. 2014; Guo et al. 2016)

12.6 Biglycan

BGN (Xq28) encodes **biglycan**, which belongs to the class I small leucine-rich proteoglycans and is mainly involved in maintenance and assembly of the extracellular matrix (ECM) (Halper 2014). The small protein core contains 10 leucine-rich repeats, to which two tissue-specific chondroitin or dermatan-sulfate glycosaminoglycan (GAG) chains are attached (Roughley and White 1989). Through these GAG chains and its core region, biglycan interacts with many other ECM proteins, including collagen type I, II, III, and VI and elastin (Douglas et al. 2006; Wiberg et al. 2002; Reinboth et al. 2002). Apart from being a mechanical link between matrix components, biglycan is also involved in regulation of signalling pathways by growth factor binding, including TGF-β (Hildebrand et al. 1994). *BGN* expression has been found in various tissues (e.g. bone, cartilage, skin, heart, lung, artery) and specialized cell types (e.g. endothelial cells, skeletal myocytes, differentiating keratinocytes) (Bianco et al. 1990; Yeo et al. 1995).

12.7 Pathogenesis

In contrast to what has been described for LDS and MFS, where fragmentation of elastic fibers is typically present, elastic fibers appeared normal in the aortic walls of two patients with MRLS (Meester et al. 2017). This is similar to what has been described for a *Bgn* knock-out (KO) mouse model, where elastic fibers in the aorta also appeared normal (Heegaard et al. 2007). Furthermore, an increase in collagen content is typical for LDS and MFS patients, while a low to normal collagen content was observed in patients with MRLS (Meester et al. 2017). This is also in line with what has been described in this Bgn KO mouse model, where no change in collagen content was observed, but clear structural collagen abnormalities, evidenced by variable sizes of collagen fibril diameters, were apparent (Heegaard et al. 2007). This suggests that biglycan, as a structural component of the extracellular matrix, plays an important role in the regulation of collagen fibril diameters, rather than in the regulation of overall collagen content (Meester et al. 2017).

The TGF-β signaling pathway is well known to play a key role in the pathogenesis of MFS and LDS. MFS is caused by mutations in the ECM protein fibrillin-1, while LDS is caused by mutations in genes encoding key signaling components of the TGF-β pathway, including *TGFBR1/2*, *TGFB2/3*, and *SMAD2/3*. Similar to fibrillin-1, biglycan is described to play a role in growth factor and cytokine binding, including TGF-β (Meester et al. 2017; Hildebrand et al. 1994; Neptune et al. 2003). It is suggested that biglycan, similar to fibrillin-1, forms a reservoir for inactive TGF-β. Because of clinical similarities to these syndromes and the suggested role of biglycan in growth factor regulation, this pathway has also been investigated in the frame of MRLS. Indeed, an upregulation of the TGF-β pathway, evidenced by an increase in pSMAD2 positive nuclei, has been observed in patient aortic walls (Meester et al. 2017). This observation can be related to the regulatory role that biglycan performs. It remains to be investigated what exact downstream effects of this increased TGF-β signaling is, as no clear effects on extracellular matrix organization and content have been observed yet. Furthermore, pathomechanistic insights into the differences in (skeletal) phenotypes of *BGN* mutation carriers are lacking and provide interesting avenues for further investigation.

12.8 Treatment and Management

Many lessons can be learned from MFS and LDS regarding the management and treatment of MRLS patients. Similar to LDS, aortic dissections tend to occur at younger ages and aneurysms are not restricted to the aorta. Therefore, complete imaging from head to pelvis is necessary to identify any potential life-threatening aneurysms at locations beyond the ascending aorta. As in LDS, yearly echocardiographical follow-up of the aorta is recommended with intermittent CT-angio or MR-angio of the complete arterial tree.

Both a medical and surgical treatment are used to treat (syndromic) TAA patients. Due to the limited number of identified *BGN* mutation carriers, no firm recommendations regarding treatment can be given yet, but treatment can be guided by experience in LDS. Preventive treatment with losartan or β-blockers is believed to slow down aortic growth progression, but this does not prevent any surgical treatment that is required at a later age. Isometric exercises and competitive sports should be avoided. Until more MRLS specific experience is available, thresholds for aortic surgery defined in LDS can be applied in MRLS (MacCarrick et al. 2014).

12.9 Genetic Counseling

MRLS is inherited in an X-linked manner, which differs from all other identified TAAD genes (except for *FLNA*). To date, it is unknown what proportion of *BGN* variants are the result of a *de novo* mutation, as none have been described yet. A child has a 50% chance of inheriting the mutation from a mutation carrying mother. An affected father on the other hand, cannot pass the mutated allele to his son, but will pass it on to his daughters.

12.10 SEMDX

In parallel to what has been observed in MRLS, two missense mutations (p.(Lys147Gly) and p.(Gly259Val)) in *BGN*, were described as the cause of an X-linked spondylo-epi-metaphyseal dysplasia (SEMDX) in three families. These mutations were not predicted to affect splicing and no cardiovascular anomalies were identified in these *BGN* mutation carriers. SEMD refers to a heterogeneous group of disorders with different modes of inheritance and are associated with vertebral, epiphyseal, and metaphyseal anomalies. To date, *BGN* is the only known X-linked SEMD gene. *BGN*-related SEMD cases have vertebral anomalies and show epi- and metaphyseal changes in the long bones, leading to short stature, brachydactyly and osteoarthritis. The initially proposed disease mechanism was either impaired binding of TGF-β to the concave binding site of biglycan or decreased biglycan stability (and as such also LOF), but no functional evidence was provided for these hypotheses (Cho et al. 2016). Pathomechanistic insights are required to determine how different *BGN* mutations lead to two seemingly very different phenotypes.

References

Barbier M, Gross MS, Aubart M et al (2014) MFAP5 loss-of-function mutations underscore the involvement of matrix alteration in the pathogenesis of familial thoracic aortic aneurysms and dissections. Am J Hum Genet 95(6):736–743

Bertoli-Avella AM, Gillis E, Morisaki H et al (2015) Mutations in a TGF-beta ligand, TGFB3, cause syndromic aortic aneurysms and dissections. J Am Coll Cardiol 65(13):1324–1336

Bianco P, Fisher LW, Young MF, Termine JD, Robey PG (1990) Expression and localization of the two small proteoglycans biglycan and decorin in developing human skeletal and non-skeletal tissues. J Histochem Cytochem 38(11):1549–1563

Boileau C, Guo DC, Hanna N et al (2012) TGFB2 mutations cause familial thoracic aortic aneurysms and dissections associated with mild systemic features of Marfan syndrome. Nat Genet 44(8):916–921

Chen MH, Choudhury S, Hirata M, Khalsa S, Chang B, Walsh CA (2018) Thoracic aortic aneurysm in patients with loss of function Filamin a mutations: clinical characterization, genetics, and recommendations. Am J Med Genet A 176(2):337–350

Cho SY, Bae JS, Kim NKD et al (2016) BGN mutations in X-linked Spondyloepimetaphyseal dysplasia. Am J Hum Genet 98(6):1243–1248

Coady MA, Davies RR, Roberts M et al (1999) Familial patterns of thoracic aortic aneurysms. Arch Surg 134(4):361–367

Dietz HC, Cutting GR, Pyeritz RE et al (1991) Marfan syndrome caused by a recurrent de novo missense mutation in the fibrillin gene. Nature 352(6333):337–339

Douglas T, Heinemann S, Bierbaum S, Scharnweber D, Worch H (2006) Fibrillogenesis of collagen types I, II, and III with small leucine-rich proteoglycans decorin and biglycan. Biomacromolecules 7(8):2388–2393

Foley C, Roberts K, Tchrakian N et al (2010) Expansion of the Spectrum of FLNA mutations associated with Melnick-needles syndrome. Molecular syndromology 1(3):121–126

Guo DC, Pannu H, Tran-Fadulu V et al (2007) Mutations in smooth muscle alpha-actin (ACTA2) lead to thoracic aortic aneurysms and dissections. Nat Genet 39(12):1488–1493

Guo DC, Regalado E, Casteel DE et al (2013) Recurrent gain-of-function mutation in PRKG1 causes thoracic aortic aneurysms and acute aortic dissections. Am J Hum Genet 93(2):398–404

Guo DC, Regalado ES, Gong L et al (2016) LOX mutations predispose to thoracic aortic aneurysms and dissections. Circ Res 118(6):928–934

Halper J (2014) Proteoglycans and diseases of soft tissues. Adv Exp Med Biol 802:49–58

Heegaard AM, Corsi A, Danielsen CC et al (2007) Biglycan deficiency causes spontaneous aortic dissection and rupture in mice. Circulation 115(21):2731–2738

Hildebrand A, Romaris M, Rasmussen LM et al (1994) Interaction of the small interstitial proteoglycans biglycan, decorin and fibromodulin with transforming growth factor beta. Biochem J 302(Pt 2):527–534

Lindsay ME, Schepers D, Bolar NA et al (2012) Loss-of-function mutations in TGFB2 cause a syndromic presentation of thoracic aortic aneurysm. Nat Genet 44(8):922–927

Loeys BL, Chen J, Neptune ER et al (2005) A syndrome of altered cardiovascular, craniofacial, neurocognitive and skeletal development caused by mutations in TGFBR1 or TGFBR2. Nat Genet 37(3):275–281

Luyckx I, Proost D, Hendriks JMH et al (2017) Two novel MYLK nonsense mutations causing thoracic aortic aneurysms/dissections in patients without apparent family history. Clin Genet 92(4):444–446

MacCarrick G, Black JH, Bowdin S et al (2014) Loeys-Dietz syndrome: a primer for diagnosis and management. Genet Med 16(8):576–587

Meester JA, Vandeweyer G, Pintelon I et al (2017) Loss-of-function mutations in the X-linked biglycan gene cause a severe syndromic form of thoracic aortic aneurysms and dissections. Genet Med 19(4):386–395

Micha D, Guo DC, Hilhorst-Hofstee Y et al (2015) SMAD2 mutations are associated with arterial aneurysms and dissections. Hum Mutat 36(12):1145–1149
Neptune ER, Frischmeyer PA, Arking DE et al (2003) Dysregulation of TGF-beta activation contributes to pathogenesis in Marfan syndrome. Nat Genet 33(3):407–411
Reinboth B, Hanssen E, Cleary EG, Gibson MA (2002) Molecular interactions of biglycan and decorin with elastic fiber components: biglycan forms a ternary complex with tropoelastin and microfibril-associated glycoprotein 1. J Biol Chem 277(6):3950–3957
Robertson SP, Twigg SR, Sutherland-Smith AJ et al (2003) Localized mutations in the gene encoding the cytoskeletal protein filamin a cause diverse malformations in humans. Nat Genet 33(4):487–491
Roughley PJ, White RJ (1989) Dermatan sulphate proteoglycans of human articular cartilage. The properties of dermatan sulphate proteoglycans I and II. Biochem J 262(3):823–827
van de Laar IM, Oldenburg RA, Pals G et al (2011) Mutations in SMAD3 cause a syndromic form of aortic aneurysms and dissections with early-onset osteoarthritis. Nat Genet 43(2):121–126
Wang L, Guo DC, Cao J et al (2010) Mutations in myosin light chain kinase cause familial aortic dissections. Am J Hum Genet 87(5):701–707
Wiberg C, Heinegard D, Wenglen C, Timpl R, Morgelin M (2002) Biglycan organizes collagen VI into hexagonal-like networks resembling tissue structures. J Biol Chem 277(51):49120–49126
Yeo TK, Torok MA, Kraus HL, Evans SA, Zhou Y, Marcum JA (1995) Distribution of biglycan and its propeptide form in rat and bovine aortic tissue. J Vasc Res 32(3):175–182
Zhu L, Vranckx R, Khau Van Kien P et al (2006) Mutations in myosin heavy chain 11 cause a syndrome associating thoracic aortic aneurysm/aortic dissection and patent ductus arteriosus. Nat Genet 38(3):343–349

Clinical and Molecular Delineation of Cutis Laxa Syndromes: Paradigms for Elastic Fiber Homeostasis

13

Aude Beyens, Lore Pottie, Patrick Sips, and Bert Callewaert

Abstract

Cutis laxa (CL) syndromes are a large and heterogeneous group of rare connective tissue disorders that share loose redundant skin as a hallmark clinical feature, which reflects dermal elastic fiber fragmentation. Both acquired and congenital-Mendelian- forms exist. Acquired forms are progressive and often preceded by inflammatory triggers in the skin, but may show systemic elastolysis. Mendelian forms are often pleiotropic in nature and classified upon systemic manifestations and mode of inheritance. Though impaired elastogenesis is a common denominator in all Mendelian forms of CL, the underlying gene defects are diverse and affect structural components of the elastic fiber or impair metabolic pathways interfering with cellular trafficking, proline synthesis, or mitochondrial functioning. In this chapter we provide a detailed overview of the clinical and molecular characteristics of the different cutis laxa types and review the latest insights on elastic fiber assembly and homeostasis from both human and animal studies.

Keywords

Cutis Laxa · De Barsy syndrome (DBS) · Elastic Fiber · Extracellular matrix · Glycosylation · Proline synthesis · Krebs cycle · *ELN* · *FBLN4* · *SLC2A10* · *FBLN5* · *LTBP4* · *ATP6V0A2* · *ATP6V1E1* · *ATP6V1A* · *ALDH18A1* · *PYCR1* · ARCL type 1 (ARCL1) · ARCL type 2 (ARCL2) · Urban-Rifkin-Davis syndrome · Debré-type · ARCL type 3 (ARCL3)

The original version of this chapter was revised: The book was inadvertently published with incorrect Chapter title which has been corrected now. The correction to this chapter is available at https://doi.org/10.1007/978-3-030-80614-9_16

A. Beyens
Center for Medical Genetics Ghent, Department of Dermatology, Department of Biomolecular Medicine, Ghent University Hospital, Ghent University, Ghent, Belgium
e-mail: aude.beyens@ugent.be

L. Pottie · P. Sips
Center for Medical Genetics Ghent, Ghent University Hospital, Ghent, Belgium

Department of Biomolecular Medicine, Ghent University, Ghent, Belgium
e-mail: lore.pottie@ugent.be; patrick.sips@ugent.be

B. Callewaert (✉)
Center for Medical Genetics Ghent, Department of Biomolecular Medicine, Ghent University Hospital, Ghent University, Ghent, Belgium
e-mail: bert.callewaert@ugent.be

J. Halper (ed.), *Progress in Heritable Soft Connective Tissue Diseases*, Advances in Experimental Medicine and Biology 1348, https://doi.org/10.1007/978-3-030-80614-9_13

Abbreviations

AA	Ascorbic acid
ADAMTS	a Disintegrin and metalloproteinase with thrombospondin motifs
ADCL	Autosomal dominant cutis laxa
ALDH	Aldehyde dehydrogenase
Apo C-III	Apolipoprotein CIII
ARCL	Autosomal recessive cutis laxa
ARCL1	ARCL Type 1
ARCL2	ARCL Type 2
ARCL3	ARCL Type 3
ARD	Aortic root dilatatio
ATS	Arterial tortuosity syndrome
CDG	Congenital disorder of glycosylation
CFC	Cardiofaciocutaneous
CHST14	Carbohydrate sulfotransferase 14
CL	Cutis laxa
CRISPR	Clustered regularly interspaced short palindromic repeats
DBS	De Barsy syndrome
DSE	Dermatan sulfate epimerase
EBP	Elastin binding proteins
ECM	Extracellular Matrix
EDS	Ehlers Danlos syndrome
EGR1	Early growth response 1
ELN	Elastin
ER	Endoplasmic reticulum
FLN	Fibulin
GLUT10	Glucose transporter 10
GO	Geroderma osteodysplastica
HGPS	Hutchinson-Gilford progeria syndrome
IUGR	Intra-uterine growth retardation
KD	Knockdown
KI	Knockin
KO	Knockout
LAP	Latency-associated protein
LOX	Lysyl oxidase
LTBP	Latent transforming growth factor β
MACS	Macrocephaly, alopecia, cutis laxa and scoliosis
MD	Menkes disease
MFS	Marfan syndrome
MGUS	Monoclonal gammopathy of undetermined significance
MO	Morpholino
NGS	Next-generation sequencing
OHS	Occipital horn syndrome
P5CSΔ1	Pyrroline-5-carboxylate synthase enzyme
PAR1	Protease activating factor–1
PLOD1	Procollagen-Lysine,2-Oxoglutarate 5-Dioxygenase
PYCR1	Pyrroline-5-carboxylate reductase 1
SCARF	Skeletal abnormalities, cutis laxa, craniostenosis, ambiguous genitalia, retardation and facial abnormalities
TALDO	Transaldolase
TEM	Transmission electron microscopy
TIEF	Transferrin-isoelectric focusing analysis
VSMC	Vascular smooth muscle cell
VUR	Vesicourethral reflux
VUS	Variants of unknown significance
UPR	Unfolded protein response
WRS	Wiedeman-Rautenstrauch syndrome
WSS	Wrinkly skin syndrome
XRCL	X-linked recessive cutis laxa

13.1 Introduction

Cutis laxa (CL) syndromes are multisystem connective tissue disorders that share loose redundant skin folds as a hallmark clinical feature that reflects severe elastic fiber fragmentation in light microscopy of skin biopsies. Both acquired and inherited subtypes exist, that vary significantly in severity and spectrum of the associated clinical manifestations.

Historically, the classification of cutis laxa disorders is based on the mode of inheritance and systemic involvement. Autosomal dominant (ADCL; MIM 123700, MIM 614434, MIM 616603), autosomal recessive (ARCL; MIM 219100, MIM 614437, MIM 613177, MIM 219200, MIM 612940, MIM 617402, MIM 617403, MIM 219150, MIM 614438, MIM 231070, MIM 612075) and X-linked recessive patterns (XRCL; MIM 304150) are considered (Berk et al. 2012). However, the clinical use of this classification is hampered by the extensive overlap between different phenotypes, the limited clinical characterization of rare subtypes, the presence of related entities, and ongoing identification of new subtypes. Hence, the diagnostic work-up and management of CL patients is often challenging.

The molecular defects underlying congenital forms of cutis laxa affect different steps in the synthesis and/or association of elastic fiber associated extracellular matrix (ECM) proteins. Initial research focused on the extracellular matrix, identifying defects in elastin, latent transforming growth factor β proteins and fibulins (*ELN*, *LTBP4*, *FBLN4*, *FBLN5*) (Zhang et al. 1999; Urban et al. 2009; Nampoothiri et al. 2010; Loeys et al. 2002). However, more recent insights in the pathogenesis also include defects in cellular trafficking (*ATP6V0A2*, *ATP6V1E1*, *ATP6V1A*, *SCYL1BP1*, *RIN2*, and *ATP7A*) and metabolism (*ALDH18A1*, *PYCR1*, *SLC2A10*), implicating disturbed intracellular transport, lysosomal dysfunction, mitochondrial dysfunction, and proline synthesis (Kornak et al. 2008a; Van Damme et al. 2017; Kaler et al. 1994; Reversade et al. 2009; Bicknell et al. 2008).

CL entities due to defects in ECM proteins include the type 1 dominant (caused by *ELN* mutations) and type 1 recessive forms (caused by mutations in *FBLN4*, *FBLN5* and *LTBP4*). Patients with autosomal dominant CL (MIM 123700) show generalized, but variable skin involvement and a risk for developing aortic root dilatation and emphysema (Callewaert et al. 2011; Hadj-Rabia et al. 2013). Autosomal recessive cutis laxa type 1 is associated with severe cardiopulmonary complications and historically consists of 3 subtypes (ARCL1A, -1B and -1C). *FBLN5*- and *LTBP4*-related CL (ARCL1A, MIM 604850; ARCL1C, Urban-Rifkin-Davis syndrome, MIM 604710) are disorders with extensive clinical overlap presenting with life-threatening pulmonary emphysema, arterial stenoses and diverticula of the gastrointestinal and urinary tract (Urban et al. 2009; Loeys et al. 2002). *FBLN4/EFEMP2*-related CL (ARCL1B, MIM 603633) is characterized by widespread arterial tortuosity, aneurysms, and stenoses (Kappanayil et al. 2012). Arterial tortuosity syndrome (ATS, MIM 208050) is closely related to type 1B recessive CL with severe tortuosity of the arterial bed and a propensity for aneurysm formation. ATS is caused by mutations in *SLC2A10*, encoding the glucose transporter 10 (GLUT10) that has been suggested to transport dehydroascorbic acid into different intracellular compartments (Coucke et al. 2006; Boel et al. 2019a).

X-linked cutis laxa or Occipital Horn syndrome (OHS, MIM 304150) presents with variable neurological involvement, bony exostoses [most prominently on the occiput ('occipital horn')], and connective tissue abnormalities including hernias, joint laxity, and bladder diverticula. Being allelic with Menkes disease (MD, OMIM 309400), OHS is considered to be situated at the milder end of the phenotypic spectrum resulting from pathogenic variants in *ATP7A*, encoding a copper transporting P-type ATPase (Kaler et al. 1994). Copper is a cofactor of multiple enzymes, including lysyl oxidase that cross-links elastin and collagens.

Defects in cellular trafficking (*ATP6V0A2*, *ATP6V1E1*, *ATP6V1A*), metabolism (*de novo* proline synthesis), and mitochondrial function (*PYCR1*, *ALDH18A1*) result in metabolic cutis laxa syndromes with skeletal and central nervous system abnormalities. *ATP6V0A2*-related CL (ARCL type 2a, MIM 219200) comprises the phenotypic spectrum of both Debré type CL and the milder Wrinkly Skin Syndrome (WSS) (Kornak et al. 2008b) with neurological impairment, lipodystrophy and variable skeletal manifestations. The more recently described *ATP6V1E1*- and *ATP6V1A*-related CL (MIM 617402 and 617403) combine marfanoid habitus, lipodystrophy, hypotonia, cardiovascular abnormalities, and pneumothorax in a related congenital disorder of glycosylation (CDG) entity (Van Damme et al. 2017). *PYCR1* mutations (ARCL2B, MIM 612940) cause a variable phenotype with intrauterine growth retardation, skeletal problems, central nervous abnormalities, with or without cataract or corneal clouding (Reversade et al. 2009). If the latter symptom is present, the disorder is known as De Barsy syndrome (DBS) or autosomal recessive cutis laxa type 3B (ARCL3B, OMIM 614438). *ALDH18A1*-related CL (ARCL3A, OMIM 219150) reveals a more severe phenotype within the same spectrum (Bicknell et al. 2008).

ADCL3 (MIM 616603) is caused by *de novo* recurrent pathogenic variants in *ALDH18A1* and results in a milder form of DBS compared to its recessive counterparts. Patients show intrauterine growth retardation, mild intellectual disability, corneal clouding, cataract, and intracranial arterial tortuosity.

13.2 Clinical Delineation of Cutis Laxa Syndromes

13.2.1 CL Entities Due to Defects in Extracellular Matrix Proteins

13.2.1.1 *ELN*-related Autosomal Dominant Cutis Laxa

Autosomal dominant cutis laxa (ADCL) presents with generalized, but variable skin involvement, ranging from mild skin hyperextensibility to loose redundant skin folds, most pronounced in the axillar, inguinal, and lower facial regions (Hadj-Rabia et al. 2013). Typical facial characteristics include a high forehead, sagging cheeks, prominent nasolabial folds, large ears, a long philtrum, and a convex nasal ridge (Fig. 13.1) (Callewaert et al. 2011; Hadj-Rabia et al. 2013). A hoarse voice, inguinal or umbilical hernias, and joint hypermobility are usually present.

Although previously considered a mild disease restricted to the dermatological manifestations, several reports indicate a risk for major complications including aortic root dilatation with a risk for dissection and pulmonary emphysema (Callewaert et al. 2011; Hadj-Rabia et al. 2013; Szabo et al. 2006), occurring in 55 and 35% of patients, respectively. Some patients may show bicuspid aortic valves with a dilatation more distally on the ascending aorta (Callewaert et al. 2011). Frequent respiratory infections and bronchiectasis may cause morbidity, both early and later in life (Callewaert et al. 2011; Vodo et al. 2015; Graul-Neumann et al. 2008). As these pulmonary and cardiovascular complications can lead to early demise, a close follow-up of the aortic root diameter with echocardiography and pulmonary function with spirometry is recommended (Szabo et al. 2006; Urban et al. 2005).

13.2.1.2 *FBLN4/EFEMP2* -related Cutis Laxa

Mutations in *FBLN4* (or *EFEMP2)*, cause ARCL type 1B associated with severe cardiovascular abnormalities including aortic dilatation, aortic and arterial tortuosity, focal aortic narrowing at the isthmus, and dilatation or stenosis of the pulmonary arteries (Kappanayil et al. 2012; Loeys et al. 1993). A high risk for cardiovascular complications including stroke, dissection, and cardiopulmonary failure results in a poor prognosis (Kappanayil et al. 2012; Hucthagowder et al. 2006; Renard et al. 2010). However, clinical variability is wide (personal observation). The phenotype in *FBLN4*-related CL is closely related to ATS (see 2.1.3.), that however has a far better prognosis despite even more prominent arterial tortuosity.

Successful vascular surgery, mainly replacement of the ascending aorta (Al-Hassnan et al. 2012), has been applied in multiple cases of ARCL1b. Skin features are usually mild with a rather hyperextensible or velvety skin, or some redundant skin folds in the axillae and groins (Fig. 13.2). Rarely, inguinal or diaphragmatic hernias can be associated (Renard et al. 2010). In contrast to *FBLN5*-related cutis laxa, developmental pulmonary emphysema is uncommon, and reported in only one individual, who also presented diaphragmatic hypoplasia (Hoyer et al. 2009). Nonetheless, aneurysmatic compression of the trachea and bronchi may cause respiratory complications such as atelectasis and respiratory distress (Kappanayil et al. 2012; Deshpande et al. 2017). Musculoskeletal features include arachnodactyly, joint hypermobility, pectus deformities,

Fig. 13.1 (continued) individual), sagging cheeks, and large ears. *FBLN5*-related CL: 'droopy' face, convex nasal ridge, large ears, and micrognathia. *LTBP4*-related CL: periorbital fullness, depressed nasal bridge, and anteverted nares. OHS: long face, large ears, and sagging cheeks. *ATP6V0A2*-related CL: long face, downslanted palpebral fissures, broad nasal bridge, marked nasolabial folds and coarse hair. *ATP6V1E1/A*-related CL: 'mask'-like triangular face, short forehead, entropion, and pointed chin. *PYCR1*- and *ALDH18A1*(dom)-related CL: triangular face, short pinched nose, thin lips, and long philtrum

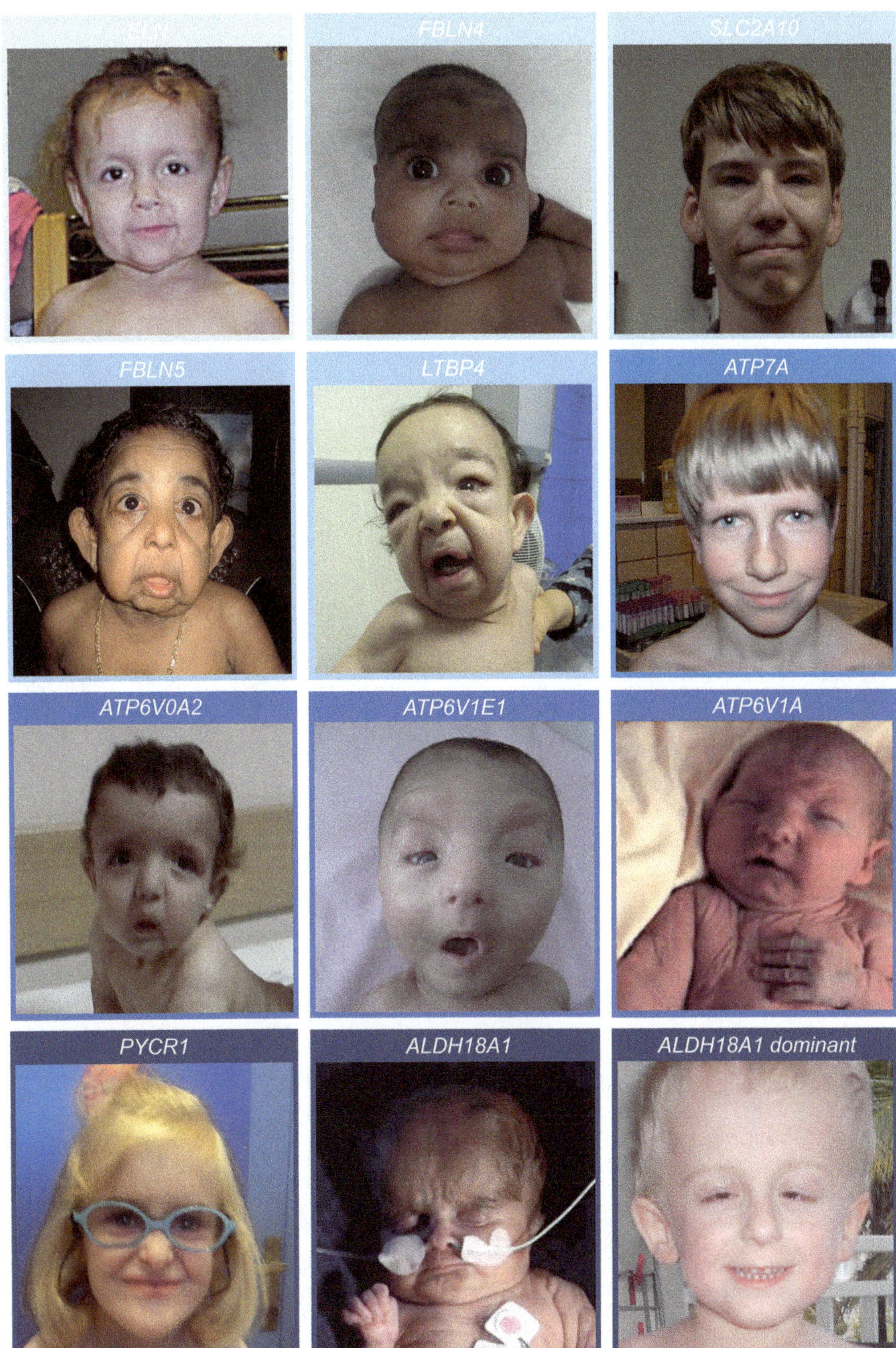

Fig. 13.1 Craniofacial characteristics in cutis laxa syndromes. *ELN*-related CL: long face, sagging cheeks, long philtrum, prominent nasolabial folds, and large ears. *FBLN4*-related CL: hypertelorism, retrognathia, long philtrum, and thin upper vermillion. ATS: long face, downslanting palpebral fissures (not present in this

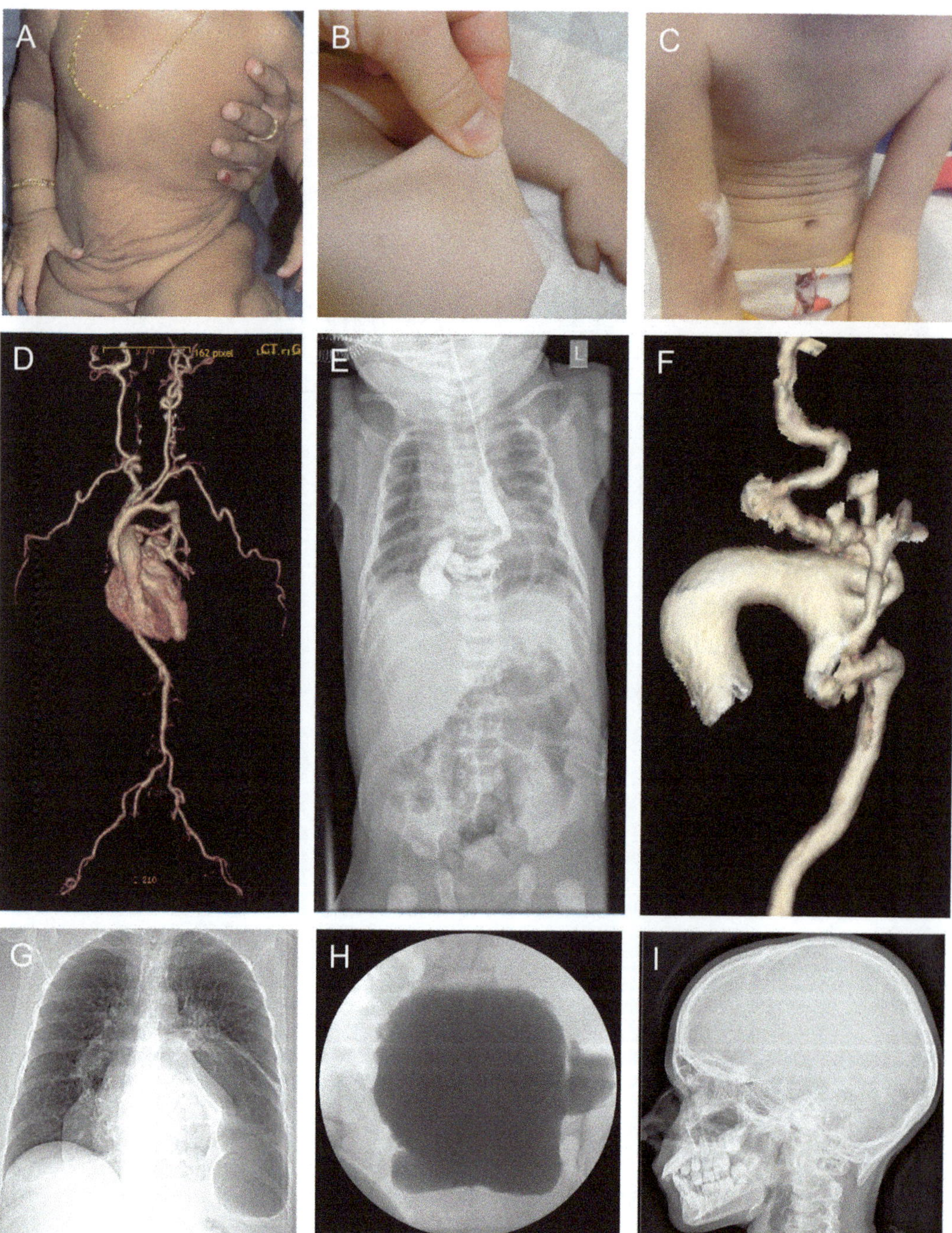

Fig. 13.2 Clinical features of cutis laxa syndromes. **a–c**: Skin features include loose redundant skin folds in *ELN*-, *FBLN5*- and *LTBP4*-related CL (**a**), mild CL with predominantly hyperextensible skin in *FBLN4*-related CL and ATS (**b**) and skin wrinkling on the abdomen and dorsum of hands and feet in *ATP6V0A2*-, *PYCR1*- and *ALDH18A1*-related CL). **d-f**: Cardiovascular abnormalities are common in *FBLN4*-related CL and ATS and include generalized arterial tortuosity, aortic root dilatation and stenosis. Image (**e**) also shows the presence of a hiatal hernia. **g**: Severe emphysema is the main feature of *FBLN5*-and *LTBP4*-related CL. **h**: Bladder diverticulae are common in *FBLN5*-related and *LTBP4*-related CL, but also a main feature of OHS. **i**: Pathognomic occipital horns in OHS

and osteoporosis with easy fracturing (Hoyer et al. 2009; Dasouki et al. 2007). Despite hypotonia, neurodevelopment and intelligence are normal. Affected individuals may show specific craniofacial features such as hypertelorism, a long philtrum with thin upper vermilion, a highly arched palate, and retrognathia. (Fig. 13.1).

13.2.1.3 Arterial Tortuosity Syndrome

Arterial tortuosity syndrome (ATS) is characterized by the elongation and tortuosity of the large and medium-sized arteries with a propensity for aneurysm formation, dissection, and ischemic events. Most patients present in early infancy with cardiovascular manifestations, including a cardiac murmur, coarctation, or narrowing of the descending aortae. Other reported initial presentations are infant respiratory distress syndrome, skin laxity, failure to thrive, and pyloric stenosis. In some cases, prenatal manifestations such as intra-uterine growth retardation (IUGR) and oligohydramnios were observed (Beyens et al. 2018a).

Facial features suggestive for ATS constitute of a long face, hypertelorism, downslanting palpebral fissures, epicanthal folds, sagging cheeks, large ears, and a highly arched palate (Fig. 13.1). Patients can display variable connective tissue manifestations including increased skin laxity, hernias (diaphragmatic, inguinal and umbilical), joint hypermobility, and muscular hypoplasia. Arachnodactyly and pectus deformities are frequently observed. Hypotonia may interfere with early motor development, but normalizes later in life (Beyens et al. 2018a; Callewaert et al. 2008a). Ocular problems, including corneal thinning and pellucid corneas have been reported in several ATS patients, and result in irregular astigmatism, keratoconus, and keratoglobus (Hardin et al. 2018).

Tortuosity of the aorta and/or middle-sized arteries is the main cardiovascular manifestation in ATS and is invariably present (Fig. 13.2). The head, neck, and pulmonary arteries are most frequently affected. Arterial narrowing can be associated, mainly located in the aorta and pulmonary arteries. The latter often leads to pulmonary hypertension and secondary feeding difficulties. A risk for aortic or renal artery stenosis warrants caution when starting agents interfering with intraglomerular pressure (angiotensin-converting enzyme inhibitors, angiotensin II receptor antagonists and nonsteroidal anti-inflammatory agents). The occurrence of early aortic root dilatation (ARD) at young age warrants initial echocardiographic follow-up every 3 months until the age of 5 years (Beyens et al. 2018a). So far, no arterial dissections have been unequivocally recorded in ATS (even with large diameters), which suggests that an aggressive surgical approach may not be of first choice in cases of slowly progressive ARD (Callewaert et al. 2008a). As such, management guidelines similar to those being used for Marfan syndrome (MFS) might be appropriate. Ischemic cerebrovascular events are one of the main concerns in ATS, as well as intestinal perforations secondary to local ischemia (Beyens et al. 2018a).

Taking in mind the generalized arterial anomalies, ATS was thought to have a poor prognosis with reported mortality rates up to 40% before the age of 5 years (Wessels et al. 2004). However, early studies may have been biased by severity and more recent studies indicate a milder course with better survival (Beyens et al. 2018a; Callewaert et al. 2008a).

13.2.1.4 *FBLN5*-related Cutis Laxa

FBLN5-related CL is characterized by severe, early-onset pulmonary emphysema, aortic and pulmonary artery stenosis, and diverticula of the urinary tract. Patients show severe cutis laxa with loose redundant skin folds over the whole body, but most obviously in the neck, axillae and groin. The generalized skin sagging results in a 'droopy' or prematurely aged facial appearance with sagging cheeks, long philtrum, downslanted palpebral fissures, eyelid ptosis, and downturned mouth corners. Other craniofacial features include a convex nasal ridge, large ear lobules, and micrognathia (Fig. 13.1) (Loeys et al. 2002; Kantaputra et al. 2014; Malakan Rad et al. 2016; Van Maldergem and Loeys 1993). *FBLN5*-related CL has a poor prognosis with childhood mortality rates between 30–50%, mainly attributable to cardiopulmonary failure and rupture of hollow visceral diverticula (Loeys et al. 2002; Elahi et al. 2006; de Schepper et al. 2003; Van Maldergem et al. 1988). In most cases, panlobular emphysema develops during early childhood, but it can already be present in the neonatal period (Van Maldergem et al. 1988). Recurrent lower respiratory infections resulting in pneumonia are reported in most patients (Loeys et al. 2002; Kantaputra et al. 2014; Malakan Rad et al. 2016; Elahi et al. 2006; de Schepper et al. 2003). Supravalvular aortic stenosis and peripheral pulmonary artery stenosis are the most common cardiovascular abnormalities. Peripheral pulmonary artery stenosis often leads to right ventricular dilatation and progressive heart failure. Tricuspid valve dysplasia, including stenosis and regurgitation may be present (Callewaert et al. 2013). Apart from severe cutis laxa, patients show signs of a generalized connective disorder. Inguinal hernias are frequent, while umbilical hernias are rare, and diaphragmatic hernias have not been observed. Bladder diverticula, vesicourethral reflux (VUR) and hydrourethronephrosis can cause recurrent upper and lower urinary infections and even lead to end-stage kidney failure (Loeys et al. 2002; Van Maldergem et al. 1988; Callewaert et al. 2013). Neuromotor and skeletal development is normal in *FBLN5*-related CL (Van Maldergem et al. 1988; Callewaert et al. 2013).

13.2.1.5 *LTBP4*-related Cutis Laxa (Urban-Rifkin-Davis Syndrome)

The clinical features of *LTBP4*-related CL (ARCL type 1C), eponymously named Urban-Rifkin-Davis syndrome, are variable and highly similar to *FBLN5*-related CL. It is a severe disorder characterized by early childhood pulmonary emphysema, peripheral pulmonary artery stenosis, and signs of a generalized connective tissue disorder (Urban et al. 2009).

Obvious at birth in most patients, the cutis laxa is generalized and resembles the skin features seen in *FBLN5*-related CL (Fig. 13.2). Craniofacial features include a 'droopy' facial appearance (as described in *FBLN5*-related CL), periorbital fullness, depressed nasal bridge, anteverted nares, and micrognathia (Fig. 13.1). Severe pulmonary emphysema usually becomes obvious in the first months of life. However, in some patients who survived beyond childhood, emphysema can be mild or even absent, even though lung function tests shows obstructive lung disease (Fig. 13.2) (Urban et al. 2009; Su et al. 2015). Precipitating factors to respiratory failure include recurrent respiratory infections, pulmonary hypertension, tracheomalacia, and positive pressure ventilation (Urban et al. 2009; Callewaert et al. 2013; Su et al. 2015). Bladder diverticula can be found in about 50% of patients, appear and enlarge progressively, and, in addition to urethral weakness and prolapse, contribute to voiding problems. VUR and fragility of the collective system can induce hydronephrosis. Patients are at risk of upper and lower urinary tract infections (Urban et al. 2009; Callewaert et al. 2013). Gastrointestinal abnormalities are a discernable feature of *LTBP4*-related CL rarely found in other CL subtypes (Urban et al. 2009). Newborns at risk for (early onset) pyloric stenosis. Diverticula, elongation and tortuosity of the gastrointestinal tract predispose to intestinal wall fragility and possible rupture. Rectal prolapse may occur and cause obstipation. Diaphragmatic involvement is present in half of reported patients and includes congenital, sliding, or hiatal hernia, and diaphragmatic eventration, frequently aggravating respiratory problems. Gastroesophageal reflux frequently occurs. Nissen fundoplication has been successfully applied in *LTBP4*-related CL (Urban et al. 2009).

Cardiovascular features primarily include stenosis of the peripheral pulmonary arteries, frequently complicated by pulmonary hypertension. Septal defects and valvular dysfunction are less common. Cognition is likely normal, but hypotonia commonly results in delayed motor milestones. Joint hyperlaxity and pectus deformities are present in a minority of patients. Overall prognosis is poor, with mortality rates up to 50% during childhood and thus similar to *FBLN5*-related cutis laxa (Urban et al. 2009; Callewaert et al. 2013). The distinction between *FBLN5*- and *LTBP4*-related cutis laxa can be challenging purely based on clinical grounds. In *LTBP4*-

related CL, supravalvular aortic stenosis is generally not observed and the gastrointestinal tract is more commonly affected than in *FBLN5*-related CL. Pulmonary hypertension appears to be a feature specific for *LTBP4*-related CL, despite the presence of (peripheral) pulmonary artery stenosis in both entities (Urban et al. 2009; Callewaert et al. 2013; Callewaert and Urban 1993).

13.2.2 Neurometabolic Cutis Laxa Syndromes

13.2.2.1 CL Entities Due to Defects in Cellular Trafficking

(a) *ATP6V0A2*-related Cutis Laxa

Debré type CL and the milder WSS are comprised in the phenotypic spectrum of *ATP6V0A2*-related CL (ARCL2A, MIM 219200). The clinical features include typical facial characteristics, variable skin involvement, skeletal and neurological abnormalities. Patients usually present with normal growth parameters and generalized cutis laxa at birth (Morava et al. 2008). This ranges from confined areas with wrinkly skin to marked CL with loose redundant skin folds. In general, skin features seem to improve over time and shift from wrinkly to more overfolded skin (Van Maldergem et al. 1993; Van Maldergem et al. 2008). Facial features include a long face, downslanting palpebral fissures, a convex nasal ridge, a broad nasal root, sagging cheeks, marked nasolabial folds, and coarse hair (Fig. 13.1) (Van Maldergem and Loeys 1993; Fischer et al. 2012). The fontanelles are invariably large and show delayed closure (with an anterior fontanel over 6x6cm in a newborn and over 3x3cm in babies aged 1 year) (Hucthagowder et al. 2009). Skeletal abnormalities include congenital hip dislocation, microcephaly, short stature, joint hypermobility, and the occurrence of bony exostoses in some older individuals (Fischer et al. 2012).

Delayed neuromotor development is common, but not present in all patients. Intellectual disability varies in presence and severity, but is usually mild to moderate. Seizures (generalized or partial complex) and neurologic regression (spasticity and cerebellar signs) can occur by the end of the first decade and are often associated with cortical and cerebellar malformations, such as 'cobblestone' brain dysgenesis reflecting lissencephaly, polymicrogyria, and cerebellar vermis hypoplasia or a classic Dandy-Walker malformation (Van Maldergem et al. 2008; Gardeitchik et al. 2014). In contrast to *PYCR1*- and *ALDH18A1*- related CL, the corpus callosum is normal in all but one reported patient (Beyens et al. 2018b). Sensorineural hearing loss can be found in some patients (Morava et al. 2008). Patients may display different ophthalmological problems, which can consist of strabismus, high myopia (<−5 diopters) or amblyopia with hypermetropia. Cardiovascular features are rare, but there is an increased risk for aortic root dilatation and aortic coarctation (Morava et al. 2008; Greally et al. 2014). Though lung emphysema was not regarded to be part of the *ATP6V0A2* spectrum, a recent report mentioned one young adult with emphysema in the absence of risk factors (Beyens et al. 2018b). Bleeding diathesis linked to coagulation factor deficiencies, including Van Willebrand disease type 2, may lead to prolonged bleeding time with easy bruising, gingival bleeding and menorrhagia (Beyens et al. 2018b). All patients have a combined disorder of glycosylation with type II pattern on plasma transferrin-isoelectric focusing analysis (TIEF; detection of *N*-glycosylation) and abnormal isoelectric focusing of plasma apolipoprotein CIII (Apo C-III) consistent with defects in the core 1 mucin type *O*-glycan biosynthesis (Morava et al. 2008).

(b) *ATP6V1E1*/*ATP6V1A*-related CL

ATP6V1E1- and *ATP6V1A*-related CL are two recently described cutis laxa entities within the of CDG disease group. In infancy, both diseases show considerable overlap with generalized cutis laxa, lipodystrophy, striking facial dysmorphology, hypotonia, congenital hip dysplasia, and joint contractures. Craniofacial features include a 'mask'-like triangular face, short forehead, hypertelorism, entropion, a convex nasal ride, narrow nostrils, short pointed chin, and low-set ears with misfolded helices (Fig. 13.1). However, whereas skin features in *ATP6V1E1*-related CL consist of generalized skin wrinkling, the *ATP6V1A* subtype rather shows loose redundant skin folds and abnormal fat distribution (mainly on the buttocks) that tends to improve over time. Hypotonia is severe and disabling in all patients, but *ATP6V1A* mutations cause an additional neurological phenotype with intellectual disability, seizures, and brain abnormalities (enlarged ventricles with white matter involvement, polymicrogyria, and thin corpus callosum), so far not observed in *ATP6V1E1*-related CL. Less frequently observed features include marfanoid habitus, congenital hip dysplasia, congenital contractures, kyphoscoliosis, inguinal hernia, and cryptorchidism (Van Damme et al. 2017; Alazami et al. 2016).

Both disorders are associated with an increased risk for potentially life-threatening cardiopulmonary complications, such as pneumothorax, aortic root dilation, cardiomyopathy, and congenital heart defects (septal defects, right hypoplastic heart syndrome, and cardiac valve abnormalities). These features allow to differentiate both entities from the more prevalent *ATP6V0A2* subtype, as these patients rarely show any cardiopulmonary involvement. In addition, the cardiopulmonary features are also discernable from those observed in ARCL type 1 (*FBLN4*, *FBLN5* and *LTBP4*-related CL), where patients present with emphysema and arterial tortuosity. Finally, variable glycosylation abnormalities can be found. Despite the limited number of reported patients, the prognosis *ad vitam* seems rather poor (mortality *ATP6V1E1*: 2/6 (33%), *ATP6V1A*: 1/3 (33%) (Van Damme et al. 2017; Alazami et al. 2016).

(c) X-linked cutis laxa/Occipital horn syndrome and Menkes disease

Occipital horn syndrome (OHS, MIM 304150) and Menkes disease (MD, MIM 309400) are allelic X-linked recessive disorders caused by pathogenic variants in the *ATP7A* gene. *ATP7A* encodes a copper transporter. Most clinical features in both disorders are related to the malfunctioning of different cupro-enzymes, such as lysyl oxidase (LOX), dopamine ß-hydroxylase, tyrosinase and cytochrome C oxidase (Byers et al. 1980a; Petris et al. 2000; Kodama et al. 1989; Royce et al. 1980). OHS, previously known as Ehlers-Danlos syndrome (EDS) type IX or X-linked CL, is characterized by the formation of bony exostoses, and variable generalized connective tissue abnormalities including cutis laxa, hernias, joint hypermobility, and bladder diverticula. Downward pointing exostoses situated in the tendinous insertions of the sternocleidomastoid and trapezius muscles, known as 'occipital horns' appear in childhood and are pathognomonic (Fig. 13.2) (Kaler et al. 1994; Beyens et al. 2019; Kaler 2013).

MD is a lethal multisystemic disorder of copper metabolism characterized by progressive neurodegeneration, growth retardation, vascular abnormalities, kinky hair (pili torti and trichorrhexis nodosa), dysautonomia, and different connective tissue problems (Kaler 2013; Tümer and Møller 2010).

With only a limited number of reported patients, OHS is far less frequent than MD, but has a better prognosis and survival rate with limited childhood mortality. However, significant mortality can still be present in young adults and be related to gastrointestinal, respiratory bleeding complications, such as intestinal perforations due to gastric ulcers or (post-surgery) apnea (Beyens et al. 2019). Patients may display variable mild craniofacial features, including a long face, large ears, sagging cheeks, and hair abnormalities (coarse hair, pili torti). The ubiquitous connective tissue manifestations include cutis laxa and inguinal hernia in most patients. Skeletal abnormalities are widespread in OHS and often disabling, requiring repeated surgical correction. The age-dependent pathognomonic occipital horns are reported in all patients but one (Tang et al. 2006; Tsukahara et al. 1994; Lazoff et al. 1975). Less specific for OHS are hammer-shaped clavicula and radial/tibial exostoses. Other skeletal problems include scoliosis, pectus deformities, coxa and genua valga, joint hypermobility, and joint luxations. Urological complications are of utter concern in OHS, with most patients presenting giant bladder diverticula and VUR leading to complications such as urinary tract infections, urinary retention, bladder rupture, and ultimately VUR-related renal failure (Fig. 13.2). Since surgical diverticulectomy has a high failure rate, self-catherisation or vesicostomy are often indicated to maintain proper voiding. The neurological symptoms in OHS are significantly milder than the reported neurodegeneration in MD, but still consist of developmental delay, mild intellectual disability and seizures in up to two-thirds of patients (Beyens et al. 2019). Hypotonia can also be present. Symptoms of dysautonomia can be disabling and consist of chronic diarrhea, hypothermia and orthostatic hypotension (Kaler et al. 1994; Kaler 2011). Tortuosity of the intracranial arteries is reported in two thirds of patients and to lesser extent in the cervical, splenic and mesenteric arteries. Importantly, aneurysm formation can complicate pre-existing tortuosity and impose risks for aneurysm rupture and bleeding (Beyens et al. 2019).

13.2.2.2 Mitochondrial CL Disorders

ARCL type 3, caused by *PYCR1* and *ALDH18A1* mutations, comprises progeroid disorders overlapping with DBS. DBS can be situated at the severe phenotypic end of ARCL-2B. *ALDH18A1* and *PYCR1* encode the mitochondrial enzymes Δ1-pyrroline-5-carboxylate synthase enzyme (P5CS) and pyrroline-5-carboxylate reductase 1 involved in glutamate and proline biosynthesis, respectively (Reversade et al. 2009; Bicknell et al. 2008; Vanakker et al. 2015; Mohamed et al. 2011).

(a) *PYCR1*-related CL

PYCR1-related CL consists of a variable phenotype with severe intrauterine growth retardation, neurological, and skeletal abnormalities. Patients present with a typical progeroid facial gestalt, including triangular face, short pinched nose, thin lips, long philtrum, and large ears (Fig. 13.1). Cutis laxa is mainly characterized by thin, translucent, and wrinkled skin, most notable on the dorsum of hands and feet (Fig. 13.2) (Yildirim et al. 2011). Prominent cutaneous venous patterning is often present on the scalp and chest. The type and extend of neurological involvement, such as psychomotor delay, intellectual disability, hypotonia, seizures, and pachygyria, is variable and not discriminative from other neurometabolic cutis laxa syndromes. Choreoathethosis and corpus callosum agenesis are more specific for this type of cutis laxa, but their absence does not rule out the diagnosis (Dimopoulou et al. 2013; Morava et al. 2009). Skeletal abnormalities include congenital hip dislocation, short stature, microcephaly, joint hypermobility, joint contractures (adducted thumbs), and osteopenia. Corneal clouding, resulting from the rupture of the Descemet's membrane, and cataract are specific for ARCL type 3, but less frequently present in *PYCR1*-related CL compared to *ALDH18A1*-related CL. Cardiopulmonary and urogenital abnormalities are usually not associated. Nonetheless, mild aortic root dilatation was described in two patients (Lessel et al. 2018; Lin et al. 2011a, b). Inguinal hernia is present in a few patients. Despite some patients being severely disabled due to profound psychomotor retardation and ataxia, the prognosis is better compared to *ALDH18A1*-related CL with no reported mortality (Table 13.1).

(b) *ALDH18A1*-related CL

Biallelic *ALDH18A1* mutations cause a more severe phenotype within the spectrum of DBS-ARCL3A. With fewer reported cases than *PYCR1*-related CL, it is considered to be the most severe disease of the ARCL spectrum. Patients with P5CS deficiency share most features with *PYCR1*-related CL, including the previously described IUGR, typical facial gestalt, skeletal, neurological, and ophthalmological abnormalities (Fig. 13.1) (Mohamed et al. 2011). Thin, translucent skin with prominent venous patterning is the main skin feature. The neurological phenotype is characterized by severe intellectual disability, structural brain abnormalities (cortical thinning, cerebellar hypoplasia, enlarged ventricles, corpus callosum agenesis), progressive neurodegeneration, choreo-athetosis, and seizures (Handley et al. 2014; Zampatti et al. 2012; Martinelli et al. 2010). Severe hypotonia

Table 13.1 Clinical characteristics of the different CL syndromes

Cutis laxa entity	ELN	FBLN4	SLC2A10	FBLN5	LTBP4	ATP6V0A2	ATP6V1E1 ATP6V1A	PYCR1	ALDH18A1 AR	ALDH18A1 AD	ATP7A OHS
Connective tissue											
Loose redundant skin folds	+	+/–	+/–	+	+	+	+	–	–	–	+
Thin skin	–	–	–	–	–	–	–	+	+	+	–
Hyperextensible skin	+/–	+	+	–	–	–	–	–	–	–	+
Wrinkly skin	–	–	–	–	–	+	+	+	+	+	+
Inguinal hernia	+	+	+	+	+	+	+	+	+	+	+
Diaphragmatic hernia	–	+	+	–	+	–	–	–	–	–	–
Musculoskeletal											
Arachnodactyly	–	+	+	–	–	–	–	–	–	–	–
Pectus deformity	–	+	+	–	+	–	–	–	–	–	+
Congenital hip dislocation	–	–	–	–	–	+	+	+	+	+	–
Contractures	–	–	–	–	–	–	+	+	+	+	+/–
Joint hypermobility	+	+	+	–	+	+	–	+	+	+	+
Osteopenia/ fractures	–	+	–	–	–	–	–	+	+	+	–
Muscle hypotonia	–	–	+	–	+	+	+	+	+	+	+
Neurological											
Developmental delay	–	–	–	–	–	+	–	+	+	+	+
Intellectual disability	–	–	–	–	–	+	–	+	+	+	+/–
Seizures	–	–	–	–	–	+	+	+	+	+	–
Choreoathetosis	–	–	–	–	–	–	–	+	+	–	–
Structural brain abnormalities	–	–	–	–	–	+	+	+	+	–	–
Stroke	–	+	+	–	–	–	–	–	–	–	–
Cardiovascular											
Aortic aneurysm	+	+	+	–	–	–	+	–	–	–	–
Aortic tortuosity	–	+	+	–	–	–	–	–	–	–	–
Intracranial tortuosity	–	+	+	–	–	–	–	–	+	+	+
Arterial dissections	+	+	–	–	–	–	–	–	–	–	–
Aortic stenosis	–	+	+	+	–	–	–	–	–	–	–
Pulmonary artery stenosis	–	–	+	+	+	–	–	–	–	–	–

Pulmonary										
Respiratory distress	–	+	+	+	+	–	+	–	–	–	–
Pulmonary emphysema	+	–	–	+	+	–	–	–	–	–	–
Pneumothorax	–	–	–	–	–	–	+	–	–	–	–
Apnea	–	–	–	–	–	–	–	–	–	–	+
Other											
Cataract/ corneal clouding	–	–	–	–	–	–	–	+	+	+	–
Urological diverticula	–	–	–	+	+	+	–	–	–	–	+
Hearing loss	–	–	–	–	–	+	–	–	–	–	–
Bleeding diathesis	–	–	–	–	–	+	–	–	–	–	–
Gastrointestinal diverticula	–	–	–	–	+	–	–	–	–	–	–
IUGR	–	–	–	–	–	–	–	+	+	+	–

and absence of language development can worsen the neurological prognosis (Bicknell et al. 2008). Bilateral subcapsular cataract or corneal clouding is present in most patients (Mohamed et al. 2011). In addition to the congenital hip dislocation, microcephaly, joint hypermobility, short stature, and finger contractures, patients may show kyphoscoliosis and abnormally shaped skeletal elements (metacarpal bowing, rib hypoplasia) (Fischer et al. 2014; Wolthuis et al. 2014). Arterial tortuosity limited to the intracranial vessels is specific for *ALDH18A1*-related CL and may impose a risk for stroke with fatal outcome (Fischer et al. 2014; Dutta et al. 2016; Skidmore et al. 2011; Martinelli et al. 2012). In some severe cases, congenital heart malformations, such as cardiac septal defects and patent ductus arteriosus, were found (Alazami et al. 2016; Fischer et al. 2014; Skidmore et al. 2011). The overall prognosis is poor with several reports of early childhood mortality due to progressive failure-to-thrive and stroke (Fischer et al. 2014; Skidmore et al. 2011).

In addition, *de novo* heterozygous mutations of a highly conserved Arginine at p.138 in *ALDH18A1* cause an autosomal dominant cutis laxa disorder with progeroid features. The phenotype is milder compared to ARCL3 with IUGR, a wrinkled and translucent skin, mild to moderate intellectual disability, congenital hip dislocation, postnatal growth retardation, cataract, and intracranial arterial tortuosity (Sinnige et al. 2017). This phenotype more faithfully resembles the initial description of the De Barsy syndrome (Fischer-Zirnsak et al. 2015).

13.2.3 Acquired Cutis Laxa

The acquired form of cutis laxa is associated with skin inflammation inducing elastolysis and skin wrinkling. Two subtypes have been described in acquired cutis laxa. Type 1 acquired CL often starts in young adulthood and can be generalized or localized. Clinically, the inflammatory phase shows an erythema, often starting at the neck or trunk, expanding to the rest of the body and relatively sparing the limbs. Different triggers have been described to initiate the disease, including neoplastic disorders [monoclonal gammopathy of undetermined significance (MGUS) or multiple myeloma], infections (viral, atypical bacterial or parasitic), hypersensitivity reactions to drugs (penicillin, penicillamin or isoniazide), or insect bites (Paulsen et al. 2014). This process starts in early adulthood and may continue for several years. In some cases, the process can spread to other internal organs, causing aortic root dilatation or pulmonary emphysema. Type 2 acquired cutis laxa or Marshall syndrome is characterized by signs of an inflammatory dermatitis with neutrophilic component (mostly urticarial or papular eruptions). Skin elasticity occurs over the inflammatory area without underlying systemic involvement (Fontenelle et al. 2013; Nabatanzi et al. 2020). In both types, the pathogenesis remains incompletely understood.

13.3 Diagnosis of Cutis Laxa Syndromes

13.3.1 Histological and Ultrastructural Findings in Cutis Laxa Syndromes

Light microscopy in affected skin demonstrates reduced, fragmented elastic fibers in the reticular dermis. All types of CL show elastic fiber abnormalities, but no findings are discriminative for individual subtypes. Moreover, mild abnormalities of elastic fibers are difficult to detect upon histochemical analysis and the absence of major elastic fiber abnormalities does not exclude the diagnosis of CL.

Transmission Electron Microscopy (TEM) has showed specific abnormalities in several cutis laxa subtypes and overrules light microscopy when diagnosing the type of CL. Nevertheless, molecular genetic testing has become the golden standard to diagnose the correct cutis laxa type. However, the increased use of next-generation sequencing-based techniques (NGS), including multiple gene panels, exome or genome sequencing, has led to the identification of variants of unknown significance (VUS), which may hamper a straightforward diagnosis. As such, additional functional studies to deliver supporting evidence of pathogenicity are needed and may be provided by performing TEM on the dermis of CL patients. In the next paragraph, we give an overview of the ultrastructural abnormalities of the elastic fibers reported in specific CL subtypes.

In *ELN*-related CL, TEM reveals a reduced number of elastic fibers with disorganized elastin deposition onto microfibrillar bundles. The amount of amorphous elastic material is reduced, shows extensive branching and fragmentation, and is not properly associated with microfibrils. The elastin is organized in separate globules with an increasing electron density from the inner to outer elastic fiber (Callewaert et al. 2011). *LTBP4*-related CL is associated with distinctive elastic fiber anomalies including large, globular deposits of elastin poorly integrated with the microfibrils. As in *ELN*-related CL, separate elastic globules are found spread over a normally appearing microfibrillin scaffold (Urban et al. 2009; Callewaert et al. 2013). Comparably, patients with *FBLN5* mutations show rounded and poorly integrated elastin deposits, but these are smaller and more uniform compared to the observations in *LTBP4*-related CL (Callewaert et al. 2013). In addition, microfibrils are reduced in size and abnormally shaped. No electron microscopy studies have been performed in *FBLN4* patients. In ATS, patient dermis shows fragmentation of the periphery of the elastic fiber, as the elastin core is disrupted by the presence of microfibrils and the peripheral mantle of microfibrils is extensive and random directionality exists (Beyens et al. 2018a).

Neurometabolic CL syndromes due to impaired vesicular trafficking, including *ATP6V0A2*- and *ATP6V1E1*-related CL, associate with severe elastic fiber anomalies. The elastic fibers are sparse and show disorganized microfibrillar structures with complete disruption of the elastic fiber core. The deposited elastin shows a frayed and moth-eaten aspect with a network of fragmented elastin clumps (Van Damme et al. 2017; Beyens et al. 2018b). In mitochondrial CL syndromes, such as *PYCR1*- and *ALDH18A1*-related CL, the elastic fibers are sparse and thin, show minimal peripheral fragmentation, and are surrounded by an excessive mantle of microfibrils (Fischer et al. 2014; Kretz et al. 2011). Interestingly, in X-linked or *ATP7A*-related CL, similar elastic fiber abnormalities are reported with a debris extending outside the elastic fiber and loss of the microfibrillar organization with, in addition, a complete disruption of the elastin core (Beyens et al. 2019).

13.3.2 Differential Diagnosis of Related Entities Presenting with CL-like Features (Table 13.2)

13.3.2.1 Related Entities

(a) Geroderma Osteodysplasticum

Geroderma Osteodysplastica (GO) refers to the old aspect of the skin and the bone abnormalities typically associated with this syndrome. Facial characteristics include a prematurely aged appearance, sagging cheeks, maxillary hypoplasia, and oblique furrowing extending from the outer canthus to the lateral border of the supraorbital ridge (Kariminejad et al. 2017). Osteoporosis or severe osteopenia and spontaneous fractures are present in most patients, often affecting the vertebra. Additional musculoskeletal abnormalities include congenital hip dislocation, short stature, and hypotonia. Intelligence is normal (Kariminejad et al. 2017; Al-Dosari and Alkuraya 2009; Hunter 1988). GO is caused by mutations in the *SCYLB1/GORAB* gene, a Rab-6 interacting golgin and COP1 scaffolding factor at the trans-Golgi (Hennies et al. 2008; Witkos et al. 2019).

(b) MACS syndrome

MACS syndrome is an acronym for macrocephaly, alopecia, cutis laxa, and scoliosis. Patients further display a coarse and swollen facial appearance affecting the eyelids, lips, and cheeks. There are no ocular, neurological, or respiratory abnormalities (Basel-Vanagaite et al. 2009). The causative gene is *RIN2*, which plays a role in endocytic trafficking and as a guanine

Table 13.2 Differential diagnosis of cutis laxa syndromes

Disorder	Gene	Dermatological abnormalities	Additional features
Geroderma Osteodysplaticum	*SCYLBP1* (*GORAB*)	WS, mainly on dorsum of hands, feet and abdomen	Osteopenia, fractures, microcephaly, hip dislocation, SS
MACS syndrome	*RIN2*	CL, hyperextensible skin, alopecia	Macrocephaly, scoliosis, JH
Ehlers-Danlos syndromes (EDS)			
Classical EDS	*COL5A1* *COL5A2*	Hyperextensible skin, atrophic scarring, easy bruising	JH, luxations, aortic aneurysms
Vascular EDS	*COL3A1*	Thin, translucent, hyperextensible skin, easy bruising	Arterial disease, gastrointestinal perforation, uterine rupture
Hypermobile EDS	Unknown	Mild hyperextensible, soft skin	JH, (sub)luxations and pain
Kyphoscoliotic EDS	*PLOD1* *FKBP14*	Thin, hyperextensible, soft skin, easy bruising	Kyphoscoliosis, ocular fragility (*PLOD1*), hearing loss (*FKBP14*)
Dermatosparaxis EDS	*ADAMTS2*	Sagging, soft skin, easy bruising	Delayed closure of fontanelles, SS, brachydactyly, blue sclerae
Arthrochalastic EDS	*COL1A1* *COL1A2*	Thin, hyperextensible, soft skin, atrophic scars, easy bruising	Congenital hip dislocation, JH, hypotonia, SS
Musculocontractural EDS	*CHST14* *DES*		Adducted thumbs, kyphoscoliosis, congenital talipes
Brittle cornea syndrome	*ZNF469* *PRDM5*	WS, mainly on palms and soles, atrophic scars	Corneal thinning, keratoconus and -globus
Pseudoxanthoma elasticum	*ABCC6*	WS, yellowish papules and plaques, mainly in flexural areas	Retinal abnormalities, peripheral vessel disease
Transaldolase defiency	*TALDO1*	WS	Hepatosplenomegaly, cardiac, renal and pulmonary features
Cantu syndrome	*ABCC9* *KCNJ8*	CL	Osteochondrodysplasia, hypertrichosis, cardiomegaly
GAPO syndrome	*ANTXR1*	CL, alopecia, prominent scalp veins	Growth retardation, pseudoanodontia, ophthalmological features
Sotos syndrome	*NSD1 NFIX*	CL, mainly on dorsum of hands and feet	Overgrowth, advanced bone age, brain anomalies
RASopathies			
Noonan syndrome	*PTPN11*	Dry WS, hyperpigmentation, hypotrichosis, palmar creases	CHD, cardiomyopathy, SS, NMD
Cardiofaciocutaneous syndrome	*BRAF*		CHD, cardiomyopathy, ID
Costello syndrome	*HRAS*		Prenatal overgrowth, SS, FTT, CHD, cardiomyopathy NMD
Juvenile progeroid syndromes			
Hutchinson Gilford progeria	*LMNA*	WS, thin skin, prominent veins, hypotrichosis	Aged appearance, SS, lipodystrophy
Wiedemann-Rautenstrauch syndrome	*POL3RA*		IUGR, failure to thrive, lipodystrophy, NMD
Lenz-Majewski hyperostotic dwarfism	*PTDS11*	WS, thin skin, prominent veins	Sclerosing bone dysplasia, dental anomalies, severe brachydactyly
SCARF syndrome	Unknown	CL	Craniostenosis, ambiguous genitalia, ID
Barber-say syndrome	*TWIST2*	CL, hypertrichosis	Macrostomia, ambiguous genitalia, hypertrichosis

SS short stature, *JH* joint hypermobility, *NMD* neuromotor delay, *CHD* congenital heart disease, *ID* intellectual disability mental retardation, *FTT* failure to thrive, *WS* wrinkly skin exchange factor for Rab5, a guanosine triphosphate involved in membrane and protein trafficking (Syx et al. 2010; Kimura et al. 2006).

13.3.2.2 Congenital Malformation Syndromes Associated with Cutis Laxa

The Ehlers Danlos syndromes (EDS) are characterized by joint hypermobility, skin hyperextensibility, and tissue fragility. Skin hyperextensibility has to be distinguished from cutis laxa, as it reduces quickly to normal shape after stretching and does not show the typical redundancy. In addition, EDS is often associated with abnormal wound healing resulting in atrophic scars. The 2017 revised classification provides 13 subtypes and is based on the main clinical symptoms (Malfait et al. 2017). The differential diagnosis between EDS and CL is often challenging, as the dermatological features are not always discernable and overlap exists between the involved organ systems in both entities. Classical, vascular and hypermobile EDS account for the most frequent subtypes within the group of disorders (Ghali et al. 2019).

Classical EDS is caused by monoallelic mutations in the *COL5A1* and *COL5A2* genes. Patients show features of skin hyperextensibility and tissue fragility resulting in widened atrophic scars. Easy bruising is commonly seen. Joint hypermobility is pronounced and patients are prone to repetitive joint luxations and subluxations. Aortic aneurysms and dissections occur in less than 3% of the patients (Ghali et al. 2019).

The diagnosis of vascular EDS, caused by dominant *COL3A1* mutations, is usually suspected based on the clinical history of arterial aneurysms, rupture or dissection, gastrointestinal perforation, or pregnancy complications such as uterine rupture (Byers et al. 2017). In up to 25% of cases, complications occur before the age of 20 (Pepin et al. 2000). Clinical features on examination may be rather subtle and include a thin, translucent skin, easy bruising, typical facial features (thin nose, prominent eyes, lobeless ears), joint hypermobility, and congenital talipes (Ghali et al. 2019).

The hypermobility type of EDS has no identified underlying genetic etiology and is diagnosed in patients with predominant musculoskeletal complaints including joint hypermobility, joint (sub)luxations, and pain, as well as some skin and soft tissue problems (Tinkle et al. 2017).

Other rarer forms include kyphoscoliotic, dermatosparaxis, arthrolochalastic, and musculocontractural EDS, and brittle cornea syndrome. Kyphoscoliotic EDS is an autosomal recessive disorder caused by *PLOD1* or *FKBP14* mutations and characterized by progressive kyphoscoliosis, hypotonia and gross motor delay. Ocular and scleral fragility is specific for *PLOD1* mutations, whereas hearing loss is an additional feature for *FKBP14* (Ghali et al. 2019; Rohrbach et al. 2011). Dermatosparaxis EDS may resemble cutis laxa since redundant, sagging skin can be present, as well as delayed closure of the fontanelles, short stature, brachydactyly, and blue sclerae. The underlying genetic defect consists of biallelic mutations in *ADAMTS2*, a gene for a disintegrin and metalloproteinase with thrombospondin motifs 2 (ADAMTS2) (Colige et al. 1999). Arthrochalastic EDS is mainly characterized by congenital hip dislocation, severe skin features, prominent joint hypermobility, and hypotonia. Causative genes include *COL1A1* and *COL1A2* (Cole et al. 1986; Byers et al. 1997). Adducted thumbs, progressive kyphoscoliosis, arachnodactyly, and congenital talipes are the key clinical features of musculocontractural EDS, caused by *CHST14* and *DSE* mutations. Finally, brittle cornea syndrome presents with thinning of the cornea and predisposition to corneal rupture, as well as early onset keratoconus and keratoglobus (Al-Hussain et al. 2004).

Pseudoxanthoma elasticum is characterized by the triad of dermatological, ocular, and cardiovascular symptoms. Skin features include skin wrinkling and yellowish papules and plaques, often in flexural areas. Patients develop ophthalmological abnormalities upon slid lap examination with angioid streaks, peau d'orange, and calcium deposits, which may result in neovascularization with a risk of retinal bleeding. Furthermore, patients are at risk for peripheral artery disease. This recessive disorder is caused by *ABCC6* mutations, an ATP transporter expressed in the liver and kidneys (Vanakker et al. 2015; Le Saux et al. 2001). In addition, a related phenotype shows progressive cutis laxa, mainly on the belly, with leathery skin folds, in addition to vitamin K dependent clotting deficiency. Ocular findings are usually less severe in this subtype due to pathogenic variants in *GGCX* (Vanakker et al. 2007).

Transaldolase deficiency is characterized by a wide clinical phenotype including fetal hydrops, hepatosplenomegaly, thrombocytopenia, anemia, renal, pulmonary, and cardiac abnormalities. Patients can have cutis laxa and dysmorphic features such as a triangular face, prominent, short philtrum, and low-set ears (Verhoeven et al. 2001; Samland and Sprenger 2009). Transaldolase deficiency is an inborn error of the non-oxidative phase of the pentose phosphate pathway and is caused by biallelic *TALDO1* mutations (Verhoeven et al. 2001).

Cantu syndrome is characterized by the triad of osteochondrodysplasia, congenital hypertrichosis, and cardiomegaly. Distinctive coarse facial features (broad nasal bridge, epicanthal folds, wide mouth with full lips) are present in association with cutis laxa in a number of patients. Molecular diagnosis is made based on the detection heterozygous pathogenic variant in *ABCC9* or *KCNJ8* (van Bon et al. 2012; Cooper et al. 2014).

GAPO syndrome, acronymic for growth retardation, alopecia, pseudoanodontia, and ophthalmological abnormalities, is phenotypically similar to MACS syndrome as in both disorders cutis laxa can be present limited to the face. Biallelic *ANTXR1* mutations cause GAPO syndrome (Stranecky et al. 2013).

The main clinical features of Sotos syndrome include overgrowth from the prenatal period through childhood, advanced bone age, acromegalic features, and brain anomalies resulting in impaired intellectual development. A limited degree of CL can be present, mainly on the dorsum of hands and feet. Sotos is caused by mutations is *NSD1* and *NFIX* (Robertson and Bankier 1999; Kurotaki et al. 2002; Malan et al. 2010).

Several entities belonging to the group of RASopathies or developmental disorders of the Ras/MAPK pathway, including Noonan, cardiofaciocutaneous, and Costello syndrome, have been reported to present with dry skin that can be wrinkled, mainly in the flexural regions. The clinical features of Noonan syndrome include congenital cardiac abnormalities, short stature, variable neuromotor delay, and typical facial features (Tartaglia et al. 2002). In more than 50% of patients, a *PTPN11* mutation is found (Tartaglia et al. 2001). Likewise, cardiofaciocutaneous (CFC) syndrome is characterized by congenital heart defects, intellectual disability, and distinct facial features, including high forehead with bitemporal narrowing, hypoplastic supraorbital ridges, a depressed nasal bridge, and posteriorly rotated ears. Heterozygous *BRAF* mutations usually underlie CFC syndrome (Niihori et al. 2006). Costello syndrome shows phenotypic overlap with Noonan and CFC syndrome, as it is associated with characteristic coarse facial features, prenatal overgrowth, short stature, failure to thrive, cardiac anomalies, and neuromotor delay. Additional skin features include hyperpigmentation, perioral papillomatosis, hypotrichosis, and deep palmar creases. Most patients carry *de novo* heterozygous mutations in *HRAS* (Kerr et al. 2006; Aoki et al. 2005).

Juvenile progeroid syndromes, including Hutchinson-Gilford progeria syndrome (HGPS, caused by *LMNA* mutations) and Wiedeman-Rautenstrauch syndrome (WRS, caused by *POL3RA* mutations), can be difficult to discern from neurometabolic cutis laxa syndromes (Lessel et al. 2018). In addition to the presence of loose, wrinkly skin, they can share several other progeroid features such growth retardation, sparse hair and lipodystrophy. The onset of HGPS is usually within the first year and clinical features consist of an aged appearance, short stature, thin skin, lipodystrophy, scleroderma, early hair loss, and decreased joint mobility (Hennekam 2006). WRS constitutes a syndrome characterized by IUGR, progeroid facial features with a pinched nose, failure to thrive, sparse hair, lipodystrophy and developmental delay (Toriello 1990).

A similarly progeroid, thin and translucent skin can be observed in Lenz-Majewski hyperostotic dwarfism. Additional features include sclerosing bone dysplasia leading to severe growth retardation, distinct craniofacial features, dental abnormalities, severe brachydactyly, and symphalangism. Lenz-Majewski syndrome is caused by *de novo* autosomal dominant *PTDSS1* mutations (Sousa et al. 2014). SCARF syndrome (skeletal abnormalities, cutis laxa, craniostenosis, ambiguous genitalia, (mental) retardation, and facial abnormalities) is a very rare disorder with some features suggestive for Lenz-Majewski syndrome but without brachydactyly and hitherto unknown molecular cause (Koppe et al. 1989).

Finally, Barber-Say syndrome, easily recognizable by the combination of hypertrichosis and macrostomia, can further present with a lax and redundant skin, ambiguous genitalia, and facial dysmorphisms including a low frontal hairline telecanthus, eyelid anomalies, and a bulbous nasal tip with hypoplastic alae nasi (Roche et al. 2010). Barber Say syndrome is caused by heterozygous *TWIST2* mutations (Marchegiani et al. 2015).

13.4 Animal Models for Cutis Laxa Syndromes

Historically, animal modeling for CL syndromes has focused on mouse models. Different knockout (KO) and knockin (KI) mouse models were established for genes encoding ECM proteins including *FBLN4*, *FBLN5,* and *LTBP4*. However, the lack of suitable *in vivo* models often hampers the investigation of the pathogenic mechanisms in recently discovered CL syndromes. Therefore, we will briefly touch on the benefits of implementing zebrafish models for future CL syndrome studies in a second paragraph. In addition, we summarized the major phenotypic characteristics of the CL animal models described in this book chapter in Table 13.3.

13.4.1 Mouse Models for CL Syndromes Caused by Defects in Extracellular matrix Proteins

Complete abolishment of *Fbln4* in mice induces perinatal lethality, vascular and lung defects, including tortuous arteries, hemorrhages, and emphysema (McLaughlin et al. 2006; Huang et al. 2010; Horiguchi et al. 2009). To circumvent early lethality, vascular smooth muscle cell -specific *Fbln4* conditional KO mice, hypomorphic $Fbln4^{R/R}$, and $Fbln4^{E57K/E57K}$ KI mice, all harboring residual expression of the *Fbln4* gene, were generated (Huang et al. 2010; Horiguchi et al. 2009; Hanada et al. 2007; Igoucheva et al. 2015). These mice survive until adulthood and show ascending aortic aneurysms, aortic valve insufficiency and ventricular hypotrophy. Aortic tissue obtained from these models shows reduced mature hydroxyproline collagen cross-links, irregular and increased collagen fibril diameters, and reduced elastin-specific (isodesmosine and desmosine) cross-links pointing toward impaired LOX function. Indeed, fibulin-4 is relevant for maturation of pro-LOX and for copper ion transfer from ATP7A to LOX (Noda et al. 2020). *Fbln4* deficiency further leads to abnormal mechanosensing upregulating the transcription factor, early growth response 1 (EGR1), illustrated by the prevention of aneurysm formation in vascular smooth muscle cell-specific *Fbln4* conditional KO mice by EGR1 knock-out. Furthermore, mechanical stimuli, matrix metalloproteinase-9 and thrombin seem to activate protease activating factor–1 (PAR1) that acts upstream of EGR1. Alterations in mitochondrial protein composition leading to changes in *in vitro* oxygen consumption of $Fbln4^{R/R}$ thoracic aortas and the implication of metabolic pathways add further complexity to the pathogenesis in *FBLN4*-related CL (Papke et al. 2015; Amin et al. 2012; van der Pluijm et al. 2018; Bultmann-Mellin et al. 2016).

$Fbln5^{-/-}$ mice survive into adulthood, develop progressive elastic fiber defects including lax skin, sagging jowls, senescent appearance, and excess folds of abdominal skin, but also tortuous aorta, emphysematous lungs, and genital pro-

Table 13.3 Animal models for cutis laxa syndromes.

Gene	CL subtype	Disease model	Abnormal elastogenesis	Mortality	Main phenotypic observations	References
Mouse models						
FBLN5	ARCL1A	*Fbln5*$^{-/-}$	+	Adulthood	Loose skin	Yanagisawa et al. (2002)
					Vascular abnormalities	
					Emphysema	
		Fbln5$^{RGE/RGE}$	–	Adulthood	Normal elastic fiber formation	Budatha et al. (2011)
		Fbln5$^{-/-}$	+	Adulthood	Aortic tortuosity	Nakamura et al. (2002)
					Emphysema Loose skin	
FBLN4	ARCL1B	*Fbln4*$^{\text{vascular smooth muscle cell KO}}$	NA	Adulthood	Ascending aortic aneurysm	Huang et al. (2010) and Horiguchi et al. (2009)
					Aortic valve insufficiency	
					Ventricular hypotrophy	
		Fbln4$^{\text{germinal KO}}$	–	Perinatal	Emphysema	Huang et al. (2010)
					Arterial tortuosity	
		Conditional *Fbln4*$^{-/-}$	+	Perinatal	Diaphragmatic hernia	Horiguchi et al. (2009)
					Arterial tortuosity	
					Impaired development of distal airways	
		Fbln4$^{-/-}$	–	Perinatal	Aortic rupture	McLaughlin et al. (2006)
					Emphysema	
					Arterial tortuosity	
		Fbln4$^{R/R}$	+	Adulthood	Aortic aneurysm	Hanada et al. (2007)
					Elastic fiber fragmentation	
					Valvular leaflets defects	
		Fbln4$^{E57K/E57K}$	+	Adulthood	Loose skin	Igoucheva et al. (2015)
					Arterial tortuosity	
					Emphysema	
					Bent forelimbs	

LTBP4	ARCL1C	*Ltbp4S*$^{-/-}$	+	Adulthood	Emphysema	Dabovic et al. (2015)
					Alveolar septation defects	
					Dilated cardiomyopathy	
					Colorectal cancer	
		Ltbp4$^{-/-}$	+	Perinatal	Reduced body size	Sterner-Kock et al. (2002)
					Alveolar septation defects	
					Reduced dermal thickness	
					Increased thickness of the aortic wall	
					Right ventricular hypertrophy	
Zebrafish models						
PYCR1	ARCL3B	*pycr1*$^{-/-}$	ND	Adulthood (80% dies around 7 months)	Progeria-like	Gistelinck et al. (2016b)
					Reduced locomotion	
					Reduced social interaction/interest	
		pycr1 morphant in symbiosis with Xenopous	ND	ND	Growth retardation	Reversade et al. (2009)
					Ectodermal edema over the entire body	
					Skin wrinkling	
					Disorganized architecture of epidermis with enlarged cells, nuclei and cellular protrusions	
		pycr1 morphants	ND	ND	Reduced body length	Reversade et al. (2009)
					Bowed tail	
					Loosening of the epidermis around the yolk sac	
					Apoptotic cells	
SLC2A10	ATS	*slc2a10* morphants	ND	ND	Cardiac edema	Liang et al. (2019)
					Reduced heart rate and blood flow	
					Incomplete and irregular patterning of the vasculature	
					Blood pooling in the sinus venosus	

ARCL autosomal recessive cutis laxa, *ATS* arterial tortuosity syndrome, *LTBP4* latent transforming growth factor ß binding protein 4, *FBLN4* fibulin-4, *FBLN5* fibulin-5, *SLC2A10* solute carrier family 2 member 10, *URDS* Urban-Rifkin-Davis Syndrome, *PYCR1* Δ1-pyrroline-5-carboxylate reductase 1, *DBS* De Barsy Syndrome, *ND* not determined lapse (Yanagisawa et al. 2002; Nakamura et al. 2002). Subsequent studies performed detailed cardiovascular analysis focusing on biomechanical properties of the large arteries (Wan and Gleason Jr. 2013; Wan et al. 2010; Le et al. 2014; Budatha et al. 2011), but the underlying mechanisms of arterial tortuosity remain unclear.

Mice lacking the short isoform of *Ltbp-4*, *Ltbp4S*$^{-/-}$, survive until adulthood and have air-sac septation defects that deteriorate into massive pulmonary emphysema associated with dilated cardiomyopathy. Lack of *Ltbp4S* impairs the incorporation of elastin in the microfibrillar bundles (with the appearance of non-fibrillar elastin aggregates). However, this mouse model does not recapitulate the severe prognosis of this CL subtype (Dabovic et al. 2015; Sterner-Kock et al. 2002; Bultmann-Mellin et al. 2015). Moreover, *Ltbp4S*$^{-/-}$ mice lack the urological and gastrointestinal features reminiscent of *LTBP4*-related CL. All together, this suggests that the long isoform of Ltbp4, Ltbp4L, may partly compensate for the loss of Ltbp4S (Urban and Davis 2014). Complete ablation of both the short and long isoform of *Ltbp4* in mice, *Ltbp4*$^{-/-}$, results in early death, approximately 2 weeks after birth (Bultmann-Mellin et al. 2015). *Ltbp4*$^{-/-}$ mice have reduced dermal thickness, postnatal growth delay and right ventricular dilatation, probably secondary to the lung problems. Overall, these studies indicate that *Ltbp4L* is crucial for postnatal survival in mice and that *Ltbp4*$^{-/-}$ mice better recapitulate the clinical presentation of *LTBP4*-related CL (Bultmann-Mellin et al. 2016; Bultmann-Mellin et al. 2015).

Ltbp4S$^{-/-}$;*Fbln5*$^{-/-}$ mice have no additional phenotypic defects compared to *Ltbp4S*$^{-/-}$ or *Fbln5*$^{-/-}$ mice. Paradoxically, these mice have improved air-sac septation and elastic fiber architecture compared to *Ltbp4S*$^{-/-}$ mice and demonstrate intact elastic fibers in the subepithelial region of lung airways. This observation rendered the hypothesis that either an alternative elastogenesis pathway exists and/or that compensation mechanisms are triggered, most likely involving *Fbln4* and *Ltbp4L* (Bultmann-Mellin et al. 2016; Dabovic et al. 2015). To test the latter hypothesis, *Ltbp4S*$^{-/-}$;*Fbln4*$^{R/R}$ mice were created that expressed the long isoform of *Ltbp4* and had reduced *Fbln4* expression. Lifespan is significantly decreased in *Ltbp4S*$^{-/-}$;*Fbln4*$^{R/R}$ mice compared to *Ltbp4S*$^{-/-}$ mice. The pulmonary and cardiovascular abnormalities in Ltbp4S$^{-/-}$;Fbln4$^{R/R}$ mice were more severe, causing higher postnatal mortality, compared to *Ltbp4S*$^{-/-}$ or *Fbln4*$^{R/R}$ mice. Based on insights from these models, the current model for elastic fiber assembly states that both fibulin-4 and fibulin-5 bind coacervated elastin and require LTBP4 to subsequently deposit elastin on the fibrillin microfibrils. More specifically, fibulin-4 interacts with LTBP4L and fibulin-5 with LTBP4S (for a review, see (Shin and Yanagisawa 2019) and below).

13.4.2 Zebrafish: An Emerging Model System for CL Syndromes

In recent years, zebrafish (*danio rerio*) have emerged from pure developmental research purposes to efficient animal modeling for gene validation for novel clinical entities, evaluation of the pathogenic mechanisms, and investigation of therapeutic compounds. Zebrafish research benefits from low-cost husbandry, a large progeny, transparent ex-utero embryogenesis, and a large collection of background reporter lines expressing fluorescent proteins in different cell types for *in vivo* visualization (Teame et al. 2019). Approximately 70% of human genes have at least one obvious zebrafish orthologue compared to the human reference genome, suggesting that most human physiology and pathologies can be modeled in zebrafish (Zhang and Peterson 2020; Howe et al. 2013). Moreover, zebrafish are highly suitable for fast and efficient genetic manipulation. Most common methods are gene knockdown by using morpholinos, viral mutagenesis, ENU-induced mutagenesis, and site-directed mutagenesis using zinc finger nuclease, transcription activator-like effector nuclease, or clustered regularly interspaced short palindromic repeats (CRISPR)/Cas9 techniques (Hosen et al. 2013; Gagnon et al. 2014; Nasevicius and Ekker 2000; Hwang et al. 2014; Amsterdam et al. 2011;

Rohner et al. 2011; Sander et al. 2011; Woods and Schier 2008). Several zebrafish models have been generated and characterized for heritable connective tissue disorders including osteogenesis imperfecta, Bruck Syndrome, and EDS (Delbaere et al. 2019; Gistelinck et al. 2016a, b). At this moment, zebrafish models for CL syndromes are limited to *pycr1*- and *scl2a10*-related disease.

In recent years, antisense morpholino (MO)-based approaches to knockdown (KD) genes of interest became less popular due to the higher risk of off-target effects and wash-out of the morpholino reagents due to cell division. Nevertheless, KD of *pycr1* in frogs results in epidermal hypoplasia with enlarged cells and nuclei in the epithelium of the skin, stunted growth and increased apoptosis. In addition, they also show no circulating red blood cells. In order to exclude the effect of anemia on the skin phenotype, both morphant and control frogs were crossed with zebrafish, but a shared circulation did not improve the phenotype (Reversade et al. 2009). Recently, an adult *pycr1*$^{-/-}$ zebrafish model was described in which the larvae showed early aging, as evidenced by senescence-associated beta-galactosidase (SA-ß-gal) and TUNEL assays. Additional behavioral phenotyping showed that adult *pycr1*$^{-/-}$ zebrafish have reduced social interactions, with a predator avoidance test stating altered locomotion trajectories. At the molecular level, *pycr1*$^{-/-}$ zebrafish show reduced the levels of hydroxyproline, dermatan sulfate, chondroitin sulfate, keratin sulfate, and heparan sulfate (Liang et al. 2019).

Morpholino KD of *slc2a10* results in a dosage-sensitive phenotype. KD of *slc2a10* in zebrafish causes a wavy notochord and cardiovascular abnormalities with a reduced heart rate, and incomplete and irregular vascular patterning. Furthermore, maximal oxygen consumption rate was reduced, despite normal mitochondrial morphology. *Slc2a10* KD zebrafish phenocopy wild-type zebrafish treated with a small molecule inhibitor of TGFβ receptor type 1 (tgfbr1/alk5). In addition, the dosage-sensitive phenotype is partially rescued by co-injection of MOs targeting *smad7* (a TFGβ inhibitor) (Willaert et al. 2012). Microarray analysis of the transcriptome of zebrafish treated with tgfbr1 inhibition and *slc2a10* KD zebrafish both show consistent dysregulation of genes important for cardiovascular, cartilage and eye development and neurogenesis related to TGFβ function. Moreover, *alk5*$^{-/-}$ zebrafish show specific dilation of the outflow tract resembling the phenotype of a MO *slc2a10* KD. It was recently established that restoration of *alk5* expression or increasing *fbln5* expression compensates for Alk5 deficiency and could be a potential therapeutic target (Boezio et al. 2020). To investigate the disease mechanisms at later stages, KO and KI models for *slc2a10* by using CRISPR/cas9 mutagenesis could be generated.

13.5 Pathophysiology of Cutis Laxa Syndromes: State of the Art

Elastic fibers provide resilience and elasticity to the connective tissue, but also contribute to tissue homeostasis by regulating cytokine signaling and cell-matrix interactions. Elastic fiber assembly, or elastogenesis, is a complex multistep process that is precisely regulated in a spatiotemporal manner and depends on proper growth factor signaling and mechanosensing (Vanakker et al. 2015; Papke and Yanagisawa 2014). Mature elastic fibers consist of an inner core of amorphous elastin surrounded by a mantle of fibrillin microfibrils. These microfibrils consist of fibrillin monomers assembling in a head-to-tail configuration and the microfibrillar meshwork serves as a scaffold on which elastin is deposited and extensively cross-linked during elastic fiber formation (Baldwin et al. 2013; Czirok et al. 2006). Elastin is synthesized as a monomer, tropoelastin, and is chaperoned through the secretory pathway by elastin binding proteins (EBP). During secretion, tropoelastin molecules undergo coacervation, a temperature- and pH dependent process of self-aggregation into smaller globules, prior to cross-linking (Yeo et al. 2011). Coacervated tropoelastin binds to fibulins and latent transforming growth factor ß (TGFß) binding proteins (LTBPs). The latter guide globules to the microfibrillar scaffold, where they coascelence into larger aggregates. Further maturation into insoluble elastin occurs through crosslinking by transglutaminase and lysyl oxidases (Vanakker et al. 2015; Baldwin et al. 2013).

The underlying molecular defects in cutis laxa syndromes involve all steps in elastic fiber assembly and provide anchor points to elucidate this complex process (Fig. 13.3).

In this section we will give a detailed review of the pathophysiological mechanisms in different cutis laxa syndromes and their relation to ECM homeostasis, intracellular trafficking and mitochondrial function.

13.5.1 Mutations in Extracellular Matrix Proteins

Autosomal dominant cutis laxa is in most cases caused by frameshift mutations within the 5 last exons of the *ELN* gene, extending the reading frame at the 3′ terminus to include a part of the adjoining 3′ untranslated region (Callewaert et al. 2011; Hadj-Rabia et al. 2013; Urban et al. 2005). This stable mutant mRNA results in the replacement of the C-terminus of tropoelastin with a missense and extended peptide sequence. The mutant tropoelastin shows increased self-association properties (as evidenced by a lower coacervation temperature), resulting in enhanced globule formation. As such, maturation of tropoelastin into soluble elastin is impaired and precludes efficient globule deposition on microfibrils in a dominant negative manner (Callewaert et al. 2011). Elastic fiber formation is further mitigated by a decreased molecular interaction between tropoelastin and the microfibrillar components fibulin-5 and fibrillin-1 (Sato et al. 2006).

In addition, mutation specific induction of the unfolded protein response (UPR) was observed, depending on both the length and the primary structure of the missense amino acid sequence. This leads to increased apoptosis in patient fibroblasts and in a transgenic mouse model for *ELN*-related CL (Callewaert et al. 2011; Hu et al. 2010). In this mouse model, mutant tropoelastin impairs the mechanic function of lung elastic fibers. However, little to no effect on elastic fibers in skin and aorta was observed, pointing towards tissue-specific mechanisms (Hu et al. 2010). The specific C-terminal frameshift variants in ADCL contrast to the *ELN* variants causing supravalvular aortic stenosis and Williams-Beuren syndrome that induce loss-of-function and/or (functional) haploinsufficiency. This points in the direction of a variety of specific functions of the tropoelastin C-terminus including cell attachment through heparin sulfate proteoglycans and $\alpha_V\beta_3$-integrin binding (Urban 2012; Broekelmann et al. 2005; Bax et al. 2009).

Fibulin-4 and -5 are critical molecules in the assembly of elastic fibers, as both interact with various molecules in each step of elastogenesis and knockdown of either fibulin-4 or -5 abolishes elastin fiber formation in dermal fibroblasts (Horiguchi et al. 2009; Liu et al. 2004; Yamauchi et al. 2010). Both molecules have distinct binding affinities for a set of shared ligands, but different

Fig. 13.3 (continued) with various molecules in each step of elastogenesis. Fibulin-4 strongly binds LOX and tropoelastin, facilitating deposition on the microfibrillar scaffold and elastin cross-linking. Fibulin-5 inhibits maturation of coacervation and interacts with LTBP4, tropoelastin and LOXL1. LTBP4 has dual functions in elastogenesis: sequestration of TGFß and facilitation of elastic fiber assembly through guiding the deposition of the FBLN5-elastin-LOXL1 complex to the microfibrils. *B. Mutations affecting intracellular trafficking*. Defects in *ATP6V0A2*, *ATP6V1E1* and *ATP6V1A* interfere with intracellular trafficking as they encode for different subunits of the V-type H^+-ATPase, that creates the acidic environment in the secretory vesicular compartments and in membrane trafficking processes. Increased pH and slower trafficking of the secretory vesicles are supposed to result in increased elastin coarvation and aberrant deposition on the microfibrillar scaffold. *ATP7A* encodes a P-type ATPase involved in copper transport in the Golgi network. *ATP7A* loss-of-function results in decreased LOX activity as it uses copper as a co-factor C. *Mutations affecting metabolism and mitochondrial functioning*. *ALDH18A1* encodes the mitochondrial enzyme P5CS that catalyses the conversion of L-glutamate to Δ1-pyrroline-5-carboxylate. This is further converted to proline by PYCR1, which catalyses the final and obligatory step in *de novo* proline synthesis, or is converted to ornithine by ornithine aminotransferase and entered into the urea cycle. Figure reproduced from Beyens et al. (2021) using the Biorender software.

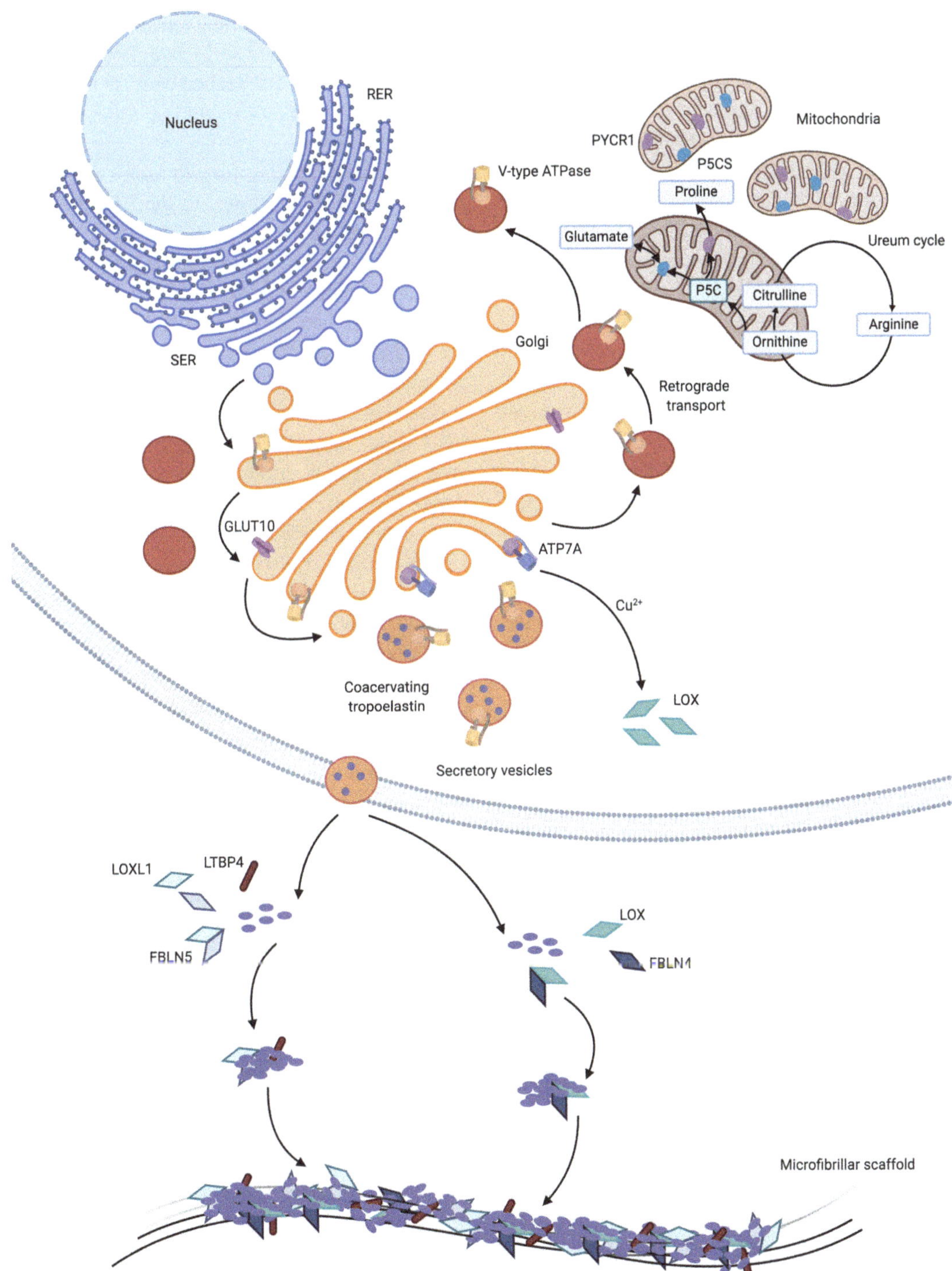

Fig. 13.3 The underlying molecular defects in CL impair different steps of elastogenesis. Elastin is synthesized as a monomer, tropoelastin, and undergoes coacervation prior to cross-linking. Coacervated tropoelastin binds to fibulins and LTBPs, which guide the globules to the microfibrillar scaffold. Further maturation to insoluble elastin is established through crosslinking by lysyl oxidases. *A. Mutations in extracellular matrix proteins.* C-terminal mutations in *ELN* result into a stable mutant protein with increased self-association properties, resulting in enhanced globule formation and disturbed deposition on the microfibrillar scaffold. Fibulin-4 and -5 both interact

spatiotemporal expression and additional unique binding agents, justifying the observed phenotypic differences with mainly a cardiovascular phenotype in *FBLN4*-related CL and severe CL with pulmonary emphysema in *FBLN5*-related CL. Fibulin-5 is expressed at much higher levels than fibulin-4 in major elastogenic organs, including the skin, aorta and lungs, and exhibits higher binding affinity to tropoelastin compared to fibulin-4 (Kobayashi et al. 2007; Choudhury et al. 2009), while fibulin-4 binds tropoelastin in a Ca^{2+}-dependent manner. Fibulin-5 is known to bind full-length tropoelastin, potentiates coacervation and lowers the coacervation temperature in a dose-dependent manner (Hirai et al. 2007; Wachi et al. 2008). Fibulin-5, and to lesser extend fibulin-4, were shown to inhibit maturation of the elastin coacervate (Cirulis et al. 2008). Fibulin-4 and -5 bind N-terminal fibrillin-1 (Freeman et al. 2005; Zheng et al. 2007). Consequent studies show that fibulin-5 binds tropoelastin to navigate onto a fibrillin-1 dominant microfibrillar scaffold while concomitantly mediating proper coacervation (Choudhury et al. 2009; El-Hallous et al. 2007). In contrast, fibulin-4 seems to play a relatively minor role in this process (Papke and Yanagisawa 2014). Oxidative modification of tropoelastin decreases fibulin-4/-5 binding and results in decreased cross-linking with the formation of large elastin positive aggregates instead of intact elastic fibers *in vitro* (Akhtar et al. 2010). These findings correlate with the ultrastructural abnormalities found in skin of patients with *FBLN5*-related cutis laxa and $Fbln5^{-/-}$ mice.

The next step in elastic fiber formation, which consists of elastin cross-linking, is mediated by the lysyl oxidase enzymes (LOX and LOXL-1) (Vallet and Ricard-Blum 2019). Both *in vitro* binding analysis and co-immunoprecipitation studies show that LOX strongly binds fibulin-4, which is mediated through the N-terminal region of fibulin-4 and the LOX propeptide (Choudhury et al. 2009). Moreover, fibulin-4 increases tropoelastin binding to LOX (Hirai et al. 2007). Counter to fibulin-4, fibulin-5 is involved in LOXL-1 binding, activation of preproLOXL-1 into active LOXL-1, and relocation to the elastic fibers (Liu et al. 2004; Choi et al. 2009). As such, interaction between LOX and fibulin-4 and LOXL-1 and fibulin-5, respectively, is important for elastic fiber crosslinking (Papke and Yanagisawa 2014).

Latent transforming growth factor ß binding protein-4 (LTBP4) belongs to the LTBP family, consisting of four members (LTBP1, −2, −3 and − 4). Three out of four isoforms (LTBP1, −3 and − 4) form latent complexes with TGFß by forming disulfide bonds with the TGFß propeptide (latency-associated protein or LAP) in the endoplasmic reticulum (Zilberberg et al. 2012). LAP is cleaved from the mature TGFß precursor in the trans-Golgi network, but stays strongly associated with TGFß through non-covalent interactions. Together, LAP, TGFß and LTBP form the large latent complex (Robertson et al. 2015). LTBP4 utilizes 3 alternative transcripts producing one small (LTBP4S) and 2 large (LTBP4L) isoforms. LTBP4 has a dual function: the aforementioned TGFß sequestration (with LTBP4 disruption resulting in increased TGFß signaling) and facilitation of elastic fiber assembly. As mentioned above, the $Ltbp4S^{-/-}$ mouse model recapitulates some of the elastic fiber abnormalities in humans with the observation of both emphysema and abnormal morphology of elastic fibers with large globular elastin deposits (Urban et al. 2009; Sterner-Kock et al. 2002). However, it lacks the urological and gastrointestinal features reminiscent of *LTBP4*-related cutis laxa, suggesting a role for LTBP4L in the development of these organ systems (Urban and Davis 2014). As LTPB4 binds both fibrillin-1 and fibulin-5, it guides fibulin-5 coated elastin globules to the microfibrilin scaffold (Vanakker et al. 2015; Dabovic et al. 2015). This mechanism sheds light on the extensive phenotypic overlap between *FBLN5*- and *LTBP4*-related CL.

Dysregulation of TGFß signaling is not limited to cutis laxa disorders caused by genes involved in TGFß sequestration, since increased TGFß signaling has been reported in *ELN*-, *FBLN4*-, and *ATP6V0A2*-related cutis laxa (Callewaert et al. 2011; Renard et al. 2010; Callewaert et al. 2013; Fischer et al. 2010). Elevated TGFß signaling in cutis laxa can point to the loss of a negative feedback signal due to impaired ECM assembly (Urban and Davis 2014). In support of this, TGFß signaling is upregulated in many other elastic fiber diseases, including MFS (caused by mutations in the *FBN1* gene encoding fibrillin-1) and Loeys-Dietz syndrome (caused by mutations in the transforming growth factor receptors 1 and 2 or their downstream transducers *SMAD2* and *SMAD3*). However, the exact spatiotemporal regulation remains a matter of debate. The pathogenesis of the arterial tortuosity syndrome (ATS), closely related to *FBLN4*-related cutis laxa, could shed more light on TGFß signaling as a common final pathway in vascular disorders. ATS is caused by biallelic mutations in *SLC2A10*, encoding the facilitative glucose transporter GLUT10 with hitherto unknown cellular substrate (Coucke et al. 2006). While initial studies show strongly increased TGFß signaling in arterial tissues as evidenced by immunostaining experiments for connective tissue growth factor and pSMAD, this could not be confirmed in later studies, possibly indicating variability in spatiotemporal TGFß signaling (Beyens et al. 2018a; Boel et al. 2019b). This could be further investigated in genome-edited animal models of ATS. Unfortunately, currently reported models lack a distinct phenotype (Callewaert et al. 2008b; Boel et al. 2020). However, morpholino-mediated *slc2a10* knockdown in a zebrafish model resulted in severe cardiovascular abnormalities and decreased TGFß signaling in early stages (Willaert et al. 2012), which underlines the variable spatiotemporal regulation of TGFß. Additional hypotheses in ATS depict GLUT10 as a transporter of dehydroxyascorbic acid present both over the mitochondrial membranes and in the endoplasmic reticulum (ER). In mitochondria, reduced dehydroxyascorbic (ascorbic acid or AA) acts a reactive oxygen species scavenger, which protects cells against oxidative stress. In the ER, AA functions as a hydroxylation cofactor for both prolyl and lysyl residues, which in its turn is a crucial reaction for both elastin maturation (Lee et al. 2010). Therefore, TGFß signaling could reside a delicate balance between oxidative stress and defective ECM, especially since it is not clear if perturbed TGFß signaling precedes ECM remodeling or is the consequence of altered ECM homeostasis (Boel et al. 2019b).

13.5.2 Mutations Affecting Intracellular Trafficking

A second group of causative genes in CL (*ATP6V0A2*, *ATP6V1E1*, *ATP6V1A*, *ATP7A*), involved in intracellular trafficking, reflect the importance of the secretory pathway in elastic fiber homeostasis. The V-type proton (H^+) ATPase is an ATP-dependent H^+-pump involved in the maintenance of the acidic environment of intracellular organelles, including secretory granules, endosomes and lysosomes, as well as extracellular compartments (Beyenbach and Wieczorek 2006). Within intracellular membranes, ATPases function in membrane trafficking processes such as receptor-mediated endocytosis and intracellular trafficking of lysosomal enzymes (Forgac 2007). Other functions include protein degradation, pH-dependent (e.g. Wnt and Notch) or – independent (e.g. mTORC1 and APMK) regulation of intracellular signaling, and different pathological processes such as drug resistance, virus and toxin entry and cancer cell survival, migration and invasion (Cotter et al. 2015). The V-type H^+-ATPase consists of two multi-subunit domains: the catalytic cytosolic V_1 domain and a H^+-pumping, membrane-embedded V_0 domain. Both domains comprise different smaller subunits and are connected by the a-subunit, composed of 8–9 transmembrane helices and a long cytoplasmic N-terminal domain. Four closely related paralogs exist for the a-subunit, that are ubiquitously expressed (except for a4) (Jefferies et al. 2008). Loss-of-function mutations in *ATP6V0A2* (encoding the a2-subunit of the V_0 complex) cause reduced acidification of trafficking vesicles in patient fibroblasts. Increased pH induces elastin coacervation and which further impairs elastin cross-linking and maturation. This is supported by the presence of swelling and fragmentation of the Golgi-apparatus, the accumulation of abnormal, tropoelastin-positive Golgi vesicles and decreased deposition of abnormal globular tropoelastin aggregates. Retrograde Golgi transport and membrane fusion of vesicles can be affected, illustrated by the delayed Golgi collapse following Brefeldin A treatment of *ATP6V0A2* mutant fibroblasts. In addition, N-glycosylation anomalies are present in all patients and probably due to pH-dependent defective functioning of glycosylation enzymes (Hucthagowder et al. 2009). In the more recently identified *ATP6V1E1*- and *ATP6V1A*-related CL, homozygous mutations affect highly conserved amino acids in the E1 and A subunits, resulting in overall assembly and structure of the entire V-ATPase complex in addition to reduced ATP-hydrolysis-coupled H^+-transport (Van Damme et al. 2017). Comparable to mutations in *ATP6V0A2*, mutations in *ATP6V1E1* and *ATP6V1A* affect protein glycosylation and affect retrograde Golgi trafficking. Some of the features in the *ATP6V1E1* and *ATP6V1A* entities, including marfanoid habitus, pneumothorax, and aortic root dilatation are reminiscent of MFS, caused by fibrillin-1 mutations. Fibrillin-1 is, unlike elastin and collagen, a highly glycosylated protein that co-assembles with elastin and other ECM glycoproteins to form elastic fibers. Therefore, disturbed glycosylation of proteins such as fibrillin-1 might explain the overlap with MFS and the observed elastic fiber abnormalities (Van Damme et al. 2017).

Occipital horn syndrome, or X-linked CL is caused by hemizygous mutations in *ATP7A*, which encodes a P-type ATPase, involved in copper transport into the Golgi network. OHS is associated with decreased LOX activity as copper acts a cofactor for LOX, essential for collagen fibril stabilization and the integrity and elasticity of mature elastin, and thus explaining the variable connective tissue manifestations (Vallet and Ricard-Blum 2019). Additional features including aberrant pigmentation and central and peripheral nerve system abnormalities can be explained by the malfunctioning of other cuproenzymes, including tyrosinase, dopamine ß-hydroxylase, and cytochrome C oxidase (Royce et al. 1980; Byers et al. 1980b; Petris et al. 1998; Kodama et al. 2005).

13.5.3 Mutations Affecting Metabolism and Mitochondrial Functioning

ALDH18A1-, dominant *ALDH18A1*- and *PYCR1*-related CL belong to the De Barsy spectrum and are caused by mutations involved in cellular metabolism, including mitochondrial functioning, Krebs cycle and proline synthesis (Reversade et al. 2009; Bicknell et al. 2008; Skidmore et al. 2011; Fischer-Zirnsak et al. 2015; Guernsey et al. 2009). *ALDH18A1* encodes the mitochondrial enzyme P5CS that catalyses the conversion of L-glutamate to Δ1-pyrroline-5-carboxylate. This is further converted to proline by pyrroline-5-carboxylate reductase 1 (PYCR1), which catalyses the final and obligatory step in *de novo* proline synthesis, or is converted to ornithine by ornithine aminotransferase and entered into the urea cycle. Recessive phenotypes likely result from loss-of-function mechanisms and the dominant type results from a mutation with dominant negative effect and reduced enzymatic function of the P5CS multimer (Fischer-Zirnsak et al. 2015). P5CS uses glutamate, the main excitatory neurotransmitter in the brain, as a substrate to initiate proline synthesis. Proline, apart from its established role in protein synthesis, is hypothesized to play a role in neurotransmission, as a proline transporter is present in the synaptic vesicles and axon terminals of glutamergic neurons (Mohamed et al. 2011; Shafqat et al. 1995). In addition, P5CS uses glutamate, the main excitatory neurotransmitter in the brain, as a substrate to initiate proline synthesis. Alternatively, glutamate is also the substrate for glutamate decarboxylase to synthetize gamma-aminobutyric acid (GABA), the main inhibitory neurotransmitter in the brain. Disequilibrium of these neurotransmitters may add to the neurological phenotype, which constitutes the main clinical feature in this entity. However, only a minority of patients with *ALDH18A1* mutations manifest abnormal levels of glutamate, proline, and intermediary products of the urea cycle, suggesting that other mechanisms also contribute to the pathogenesis

(Fischer-Zirnsak et al. 2015; Baumgartner et al. 2000). Similarly, *PYCR1* patient fibroblasts did not show evidence of low proline, but showed a decreased mitochondrial membrane potential and increase in cell death upon short exposure to H_2O_2. These findings suggest the involvement of *PYCR1* in the cell's protective response to oxidative stress, and might explain the observed progeroid changes in connective tissues. Likewise, knockdown of the orthologous genes in Xenopus and zebrafish led epidermal hypoplasia and blistering, accompanied by massive apoptosis (Reversade et al. 2009). Finally, a hypothesis can be made that reduced levels of proline would impair collagen and elastin synthesis due to their proline content, as reduced P5CS activity has been shown to reduce protein synthesis and slow down the cell cycle (Kardos et al. 2015).

13.6 Avenues in Cutis Laxa Research

Mendelian cutis laxa syndromes provide relevant models to study elastic fiber biology. The recent advances in the understanding of the different subtypes have thaught us that elastic fiber biology extends well beyond ECM assembly and relies on many cellular processes including growth factor signaling, metabolic pathways, intracellular transport, organelle acidification, and mitochondrial functioning. Nevertheless, in many cases of congenital cutis laxa, the underlying gene defect remains unknown (Callewaert, unpublished). However, the rarity of the cases related to novel genes represent a major burden in gene discovery. Therefore, international collaboration through databases (as proposed by the European Reference Networks (ERN-Skin, www.ern-skin.eu) and initiatives as GeneMatcher are pivotal to identify individuals with these rare entities.

It is tempting to speculate that many of these mechanisms may be relevant for common age-related diseases of the elastic fibers such as emphysema, aortic aneurysm, skin ageing, and pelvic organ prolapse (Saito et al. 2018; Cloonan et al. 2016; Ryter et al. 2018; Li et al. 2020; Cooper et al. 2020; Goumans and ten Dijke 2018). Animal models will provide opportunities to unravel all players in ECM biology. The surge of compound screening in zebrafish research will evidently become important for identification of compounds to identify novel targets for therapy (Farr 3rd et al. 2020; Bergen et al. 2019). Moreover, unbiased compound screens may even identify novel players and targets in ECM homeostasis.

Ethical Statement This work is supported by the Special Research Fund, Flanders of Ghent University (Belgium, grant 01N04516C to B.C.), the Research Foundation Flanders (project G035620N to B.C.) and the European Academy of Dermatovenereology (Switzerland, EADV project PPRC-2018-50 to B.C.). B.C. is a senior clinical investigator of the Research Foundation Flanders. The Ghent University Hospital is a member of the European Reference Network for skin disorders (ERN-skin). Clinical pictures are published after written informed consent of the patient or, in case of minors, their legal guardian. Previously published pictures are reprinted with permission. Clinical pictures were not masked due to scientific and educational reasons, allowing for better patient recognition and care.

References

Akhtar K, Broekelmann TJ, Miao M, Keeley FW, Starcher BC, Pierce RA et al (2010) Oxidative and nitrosative modifications of tropoelastin prevent elastic fiber assembly in vitro. J Biol Chem 285(48):37396–37404

Alazami AM, Al-Qattan SM, Faqeih E, Alhashem A, Alshammari M, Alzahrani F et al (2016) Expanding the clinical and genetic heterogeneity of hereditary disorders of connective tissue. Hum Genet 135(5):525–540

Al-Dosari M, Alkuraya FS (2009) A novel missense mutation in SCYL1BP1 produces geroderma osteodysplastica phenotype indistinguishable from that caused by nullimorphic mutations. Am J Med Genet A 149A(10):2093–2098

Al-Hussain H, Zeisberger SM, Huber PR, Giunta C, Steinmann B (2004) Brittle cornea syndrome and its delineation from the kyphoscoliotic type of Ehlers-Danlos syndrome (EDS VI): report on 23 patients and review of the literature. Am J Med Genet A 124A(1):28–34

Al-Hassnan ZN, Almesned AR, Tulbah S, Hakami A, Al-Omrani A, Al Sehly A et al (2012) Recessively inherited severe aortic aneurysm caused by mutated EFEMP2. Am J Cardiol 109(11):1677–80. https://doi.org/10.1016/j.amjcard.2012.01.394

Amin M, Le VP, Wagenseil JE (2012) Mechanical testing of mouse carotid arteries: from newborn to adult. J Vis Exp 60

Amsterdam A, Varshney GK, Burgess SM (2011) Retroviral-mediated insertional mutagenesis in zebrafish. Methods Cell Biol 104:59–82

Aoki Y, Niihori T, Kawame H, Kurosawa K, Ohashi H, Tanaka Y et al (2005) Germline mutations in HRAS proto-oncogene cause Costello syndrome. Nat Genet 37(10):1038–1040

Baldwin AK, Simpson A, Steer R, Cain SA, Kielty CM (2013) Elastic fibres in health and disease. Expert Rev Mol Med 15:e8

Basel-Vanagaite L, Sarig O, Hershkovitz D, Fuchs-Telem D, Rapaport D, Gat A et al (2009) RIN2 deficiency results in macrocephaly, alopecia, cutis laxa, and scoliosis: MACS syndrome. Am J Hum Genet 85(2):254–263

Baumgartner MR, Hu CAA, Almashanu S, Steel G, Obie C, Aral B et al (2000) Hyperammonemia with reduced ornithine, citrulline, arginine and proline: a new inborn error caused by a mutationin in the gene encoding Δ1-pyrroline-5-carboxylate synthase. Hum Mol Genet 9(19):2853–2858

Bax DV, Rodgers UR, Bilek MM, Weiss AS (2009) Cell adhesion to tropoelastin is mediated via the C-terminal GRKRK motif and integrin alphaVbeta3. J Biol Chem 284(42):28616–28623

Bergen DJM, Kague E, Hammond CL (2019) Zebrafish as an emerging model for osteoporosis: a primary testing platform for screening new Osteo-active compounds. Front Endocrinol 10:6

Berk DR, Bentley DD, Bayliss SJ, Lind A, Urban Z (2012) Cutis laxa: a review. J Am Acad Dermatol 66(5):842.e1-17

Beyenbach KW, Wieczorek H (2006) The V-type H+ ATPase: molecular structure and function, physiological roles and regulation. J Exp Biol 209(Pt 4):577–589

Beyens A, Albuisson J, Boel A, Al-Essa M, Al-Manea W, Bonnet D et al (2018a) Arterial tortuosity syndrome: 40 new families and literature review. Genet Med 20(10):1236–1245

Beyens A, Moreno-Artero E, Bodemer C, Cox H, Gezdirici A, Yilmaz Gulec E et al (2018b) ATP6V0A2-related cutis laxa in 10 novel patients: focus on clinical variability and expansion of the phenotype. Exp Dermatol

Beyens A, Van Meensel K, Pottie L, De Rycke R, De Bruyne M, Baeke F et al (2019) Defining the clinical, molecular and ultrastructural characteristics in occipital horn syndrome: two new cases and review of the literature. Genes (Basel) 10(7)

Beyens A, Boel A, Symoens S, Callewaert B (2021) Cutis laxa: a comprehensive overview of clinical characteristics and pathophysiology. Clin Genet 99(1):53–66

Bicknell LS, Pitt J, Aftimos S, Ramadas R, Maw MA, Robertson SP (2008) A missense mutation in ALDH18A1, encoding Delta1-pyrroline-5-carboxylate synthase (P5CS), causes an autosomal recessive neurocutaneous syndrome. Eur J Hum Genet 16(10):1176–1186

Boel A, Veszelyi K, Nemeth CE, Beyens A, Willaert A, Coucke P, et al. Arterial tortuosity syndrome: an ascorbate compartmentalization disorder? Antioxidants & redox signaling. 2019a

Boel A, Veszelyi K, Nemeth CE, Beyens A, Willaert A, Coucke P, et al. Arterial tortuosity syndrome: an ascorbate compartmentalization disorder? Antioxid redox signal. 2019b

Boel A, Burger J, Vanhomwegen M, Beyens A, Renard M, Barnhoorn S et al (2020) Slc2a10 knock-out mice deficient in ascorbic acid synthesis recapitulate aspects of arterial tortuosity syndrome and display mitochondrial respiration defects. Hum Mol Genet 29(9):1476–1488

Boezio GL, Bensimon-Brito A, Piesker J, Guenther S, Helker CS, Stainier DY (2020) Endothelial TGF-beta signaling instructs smooth muscle cell development in the cardiac outflow tract. elife 9

Broekelmann TJ, Kozel BA, Ishibashi H, Werneck CC, Keeley FW, Zhang L et al (2005) Tropoelastin interacts with cell-surface glycosaminoglycans via its COOH-terminal domain. J Biol Chem 280(49):40939–40947

Budatha M, Roshanravan S, Zheng Q, Weislander C, Chapman SL, Davis EC et al (2011) Extracellular matrix proteases contribute to progression of pelvic organ prolapse in mice and humans. J Clin Invest 121(5):2048–2059

Bultmann-Mellin I, Conradi A, Maul AC, Dinger K, Wempe F, Wohl AP et al (2015) Modeling autosomal recessive cutis laxa type 1C in mice reveals distinct functions for Ltbp-4 isoforms. Dis Model Mech 8(4):403–415

Bultmann-Mellin I, Essers J, van Heijingen PM, von Melchner H, Sengle G, Sterner-Kock A (2016) Function of Ltbp-4L and fibulin-4 in survival and elastogenesis in mice. Dis Model Mech 9(11):1367–1374

Byers PH, Siegel RC, Holbrook KA (1980a) X-linked cutis laxa. Defective cross-link formation in collagen due to decreased lysyl oxidase activity. N Engl J Med 303(2):61–65

Byers PH, Siegel RC, Holbrook KA, Narayanan AS, Bornstein P, Hall JG (1980b) X-linked cutis laxa: defective cross-link formation in collagen due to decreased lysyl oxidase activity. N Engl J Med 303(2):61–65

Byers PH, Duvic M, Atkinson M, Robinow M, Smith LT, Krane SM et al (1997) Ehlers-Danlos syndrome type VIIA and VIIB result from splice-junction mutations or genomic deletions that involve exon 6 in the COL1A1 and COL1A2 genes of type I collagen. Am J Med Genet 72(1):94–105

Byers PH, Belmont J, Black J, De Backer J, Frank M, Jeunemaitre X et al (2017) Diagnosis, natural history, and management in vascular Ehlers-Danlos syndrome. Am J Med Genet C Semin Med Genet 175(1):40–47

Callewaert BL, Urban Z (1993) LTBP4-Related Cutis Laxa. In: Adam MP, Ardinger HH, Pagon RA, Wallace SE, Bean LJH, Stephens K, et al., editors. GeneReviews((R)). Seattle (WA): University of Washington, Seattle University of Washington,

Seattle. GeneReviews is a registered trademark of the University of Washington, Seattle. All rights reserved
Callewaert BL, Willaert A, Kerstjens-Frederikse WS, De Backer J, Devriendt K, Albrecht B et al (2008a) Arterial tortuosity syndrome: clinical and molecular findings in 12 newly identified families. Hum Mutat 29(1):150–158
Callewaert BL, Loeys BL, Casteleyn C, Willaert A, Dewint P, De Backer J et al (2008b) Absence of arterial phenotype in mice with homozygous slc2A10 missense substitutions. Genesis 46(8):385–389
Callewaert B, Renard M, Hucthagowder V, Albrecht B, Hausser I, Blair E et al (2011) New insights into the pathogenesis of autosomal-dominant cutis laxa with report of five ELN mutations. Hum Mutat 32(4):445–455
Callewaert B, Su CT, Van Damme T, Vlummens P, Malfait F, Vanakker O et al (2013) Comprehensive clinical and molecular analysis of 12 families with type 1 recessive cutis laxa. Hum Mutat 34(1):111–121
Choi J, Bergdahl A, Zheng Q, Starcher B, Yanagisawa H, Davis EC (2009) Analysis of dermal elastic fibers in the absence of fibulin-5 reveals potential roles for fibulin-5 in elastic fiber assembly. Matrix Biol J Int Soc Matrix Biol 28(4):211–220
Choudhury R, McGovern A, Ridley C, Cain SA, Baldwin A, Wang MC et al (2009) Differential regulation of elastic fiber formation by fibulin-4 and -5. J Biol Chem 284(36):24553–24567
Cirulis JT, Bellingham CM, Davis EC, Hubmacher D, Reinhardt DP, Mecham RP et al (2008) Fibrillins, fibulins, and matrix-associated glycoprotein modulate the kinetics and morphology of in vitro self-assembly of a recombinant elastin-like polypeptide. Biochemistry 47(47):12601–12613
Cloonan SM, Glass K, Laucho-Contreras ME, Bhashyam AR, Cervo M, Pabon MA et al (2016) Mitochondrial iron chelation ameliorates cigarette smoke-induced bronchitis and emphysema in mice. Nat Med 22(2):163–174
Cole WG, Chan D, Chambers GW, Walker ID, Bateman JF (1986) Deletion of 24 amino acids from the pro-alpha 1(I) chain of type I procollagen in a patient with the Ehlers-Danlos syndrome type VII. J Biol Chem 261(12):5496–5503
Colige A, Sieron AL, Li SW, Schwarze U, Petty E, Wertelecki W et al (1999) Human Ehlers-Danlos syndrome type VII C and bovine dermatosparaxis are caused by mutations in the procollagen I N-proteinase gene. Am J Hum Genet 65(2):308–317
Cooper PE, Reutter H, Woelfle J, Engels H, Grange DK, van Haaften G et al (2014) Cantu syndrome resulting from activating mutation in the KCNJ8 gene. Hum Mutat 35(7):809–813
Cooper HA, Cicalese S, Preston KJ, Kawai T, Okuno K, Choi ET et al (2020) Targeting mitochondrial fission as a potential therapeutic for abdominal aortic aneurysm. Cardiovasc Res
Cotter K, Stransky L, McGuire C, Forgac M (2015) Recent insights into the structure, regulation, and function of the V-ATPases. Trends Biochem Sci 40(10):611–622
Coucke PJ, Willaert A, Wessels MW, Callewaert B, Zoppi N, De Backer J et al (2006) Mutations in the facilitative glucose transporter GLUT10 alter angiogenesis and cause arterial tortuosity syndrome. Nat Genet 38(4):452–457
Czirok A, Zach J, Kozel BA, Mecham RP, Davis EC, Rongish BJ (2006) Elastic fiber macro-assembly is a hierarchical, cell motion-mediated process. J Cell Physiol 207(1):97–106
Dabovic B, Robertson IB, Zilberberg L, Vassallo M, Davis EC, Rifkin DB (2015) Function of latent TGFbeta binding protein 4 and fibulin 5 in elastogenesis and lung development. J Cell Physiol 230(1):226–236
Dasouki M, Markova D, Garola R, Sasaki T, Charbonneau NL, Sakai LY et al (2007) Compound heterozygous mutations in fibulin-4 causing neonatal lethal pulmonary artery occlusion, aortic aneurysm, arachnodactyly, and mild cutis laxa. Am J Med Genet Part A 143(22):2635–2641
de Schepper S, Loeys B, de Paepe A, Lambert J, Naeyaert JM (2003) Cutis laxa of the autosomal recessive type in a consanguineous family. Eur J Dermatol: EJD 13(6):529–533
Delbaere S, Van Damme T, Syx D, Symoens S, Coucke P, Willaert A et al (2019) Hypomorphic zebrafish models mimic the musculoskeletal phenotype of beta4GalT7-deficient Ehlers-Danlos syndrome. Matrix Biol J Int Soc Matrix Biol
Deshpande S, Alazraki A, Khoshnam N, Kantarovich D, Hoseinzadeh P, Hebson C et al (2017) Four new cases of pediatric thoracic aortic aneurysm (TAA) with review of the molecular genetic basis, utilizing the newly published consensus nomenclature. Cardiovascul Pathol Off J Soc Cardiovascul Pathol 31:34–40
Dimopoulou A, Fischer B, Gardeitchik T, Schröter P, Kayserili H, Schlack C et al (2013) Genotype-phenotype spectrum of PYCR1-related autosomal recessive cutis laxa. Mol Genet Metab 110(3):352–361
Dutta A, Ghosh SK, Ghosh A, Roy S (2016) A 5-year journey with cutis Laxa in an Indian child: the De Barsy syndrome revisited. Indian J Dermatol 61(1):81–84
Elahi E, Kalhor R, Banihosseini SS, Torabi N, Pour-Jafari H, Houshmand M et al (2006) Homozygous missense mutation in fibulin-5 in an Iranian autosomal recessive cutis laxa pedigree and associated haplotype. J Invest Dermatol 126(7):1506–1509
El-Hallous E, Sasaki T, Hubmacher D, Getie M, Tiedemann K, Brinckmann J et al (2007) Fibrillin-1 interactions with fibulins depend on the first hybrid domain and provide an adaptor function to tropoelastin. J Biol Chem 282(12):8935–8946
Farr GH 3rd, Morris M, Gomez A, Pham T, Kilroy E, Parker EU et al (2020) A novel chemical-combination screen in zebrafish identifies epigenetic small

molecule candidates for the treatment of Duchenne muscular dystrophy. Skelet Muscle 10(1):29

Fischer B, Chan HWL, Dimopoulou A, Egerer J, Chan D, Mundlos S et al (2010) Loss of ATP6V0A2 impairs trafficking in several intracellular compartments and results in apoptotic cell death. Med Genet 22(1):151–152

Fischer B, Dimopoulou A, Egerer J, Gardeitchik T, Kidd A, Jost D et al (2012) Further characterization of ATP6V0A2-related autosomal recessive cutis laxa. Hum Genet 131(11):1761–1773

Fischer B, Callewaert B, Schroter P, Coucke PJ, Schlack C, Ott CE et al (2014) Severe congenital cutis laxa with cardiovascular manifestations due to homozygous deletions in ALDH18A1. Mol Genet Metab 112(4):310–316

Fischer-Zirnsak B, Escande-Beillard N, Ganesh J, Tan YX, Al Bughaili M, Lin AE et al (2015) Recurrent De novo mutations affecting residue Arg138 of Pyrroline-5-carboxylate synthase cause a Progeroid form of autosomal-dominant cutis Laxa. Am J Hum Genet 97(3):483–492

Fontenelle E, de Almeida APM, Souza GMAA (2013) Marshall's syndrome. An Bras Dermatol 88(2):279–282

Forgac M (2007) Vacuolar ATPases: rotary proton pumps in physiology and pathophysiology. Nat Rev Mol Cell Biol 8(11):917–929

Freeman LJ, Lomas A, Hodson N, Sherratt MJ, Mellody KT, Weiss AS et al (2005) Fibulin-5 interacts with fibrillin-1 molecules and microfibrils. Biochem J 388(Pt 1):1–5

Gagnon JA, Valen E, Thyme SB, Huang P, Akhmetova L, Pauli A et al (2014) Efficient mutagenesis by Cas9 protein-mediated oligonucleotide insertion and large-scale assessment of single-guide RNAs. PLoS One 9(5):e98186

Gardeitchik T, Mohamed M, Fischer B, Lammens M, Lefeber D, Lace B et al (2014) Clinical and biochemical features guiding the diagnostics in neurometabolic cutis laxa. Eur J Hum Genet 22(7):888–895

Ghali N, Sobey G, Burrows N (2019) Ehlers-Danlos syndromes. BMJ 366:l4966

Gistelinck C, Gioia R, Gagliardi A, Tonelli F, Marchese L, Bianchi L et al (2016a) Zebrafish collagen type I: molecular and biochemical characterization of the major structural protein in bone and skin. Sci Rep 6:21540

Gistelinck C, Witten PE, Huysseune A, Symoens S, Malfait F, Larionova D et al (2016b) Loss of type I collagen Telopeptide Lysyl hydroxylation causes musculoskeletal abnormalities in a zebrafish model of Bruck syndrome. J Bone Mineral Res Off J Am Soc Bone Mineral Res 31(11):1930–1942

Goumans MJ, ten Dijke P (2018) TGF-beta signaling in control of cardiovascular function. Csh Perspect Biol 10(2)

Graul-Neumann LM, Hausser I, Essayie M, Rauch A, Kraus C (2008) Highly variable cutis laxa resulting from a dominant splicing mutation of the elastin gene. Am J Med Genetics Part A 146a(8):977–983

Greally MT, Kalis NN, Agab W, Ardati K, Giurgea S, Kornak U et al (2014) Autosomal recessive cutis laxa type 2A (ARCL2A) mimicking Ehlers-Danlos syndrome by its dermatological manifestations: report of three affected patients. Am J Med Genet A 164A(5):1245–1253

Guernsey DL, Jiang H, Evans SC, Ferguson M, Matsuoka M, Nightingale M et al (2009) Mutation in Pyrroline-5-carboxylate reductase 1 gene in families with cutis Laxa type 2. Am J Hum Genet 85(1):120–129

Hadj-Rabia S, Callewaert BL, Bourrat E, Kempers M, Plomp AS, Layet V et al (2013) Twenty patients including 7 probands with autosomal dominant cutis laxa confirm clinical and molecular homogeneity. Orphanet J Rare Dis 8:36

Hanada K, Vermeij M, Garinis GA, de Waard MC, Kunen MG, Myers L et al (2007) Perturbations of vascular homeostasis and aortic valve abnormalities in fibulin-4 deficient mice. Circ Res 100(5):738–746

Handley MT, Megarbane A, Meynert AM, Brown S, Freyer E, Taylor MS et al (2014) Loss of ALDH18A1 function is associated with a cellular lipid droplet phenotype suggesting a link between autosomal recessive cutis laxa type 3A and Warburg micro syndrome. Molecul Genet Genomic Med 2(4):319–325

Hardin JS, Zarate YA, Callewaert B, Phillips PH, Warner DB (2018) Ophthalmic findings in patients with arterial tortuosity syndrome and carriers: a case series. Ophthalmic Genet 39(1):29–34

Hennekam RC (2006) Hutchinson-Gilford progeria syndrome: review of the phenotype. Am J Med Genet A 140(23):2603–2624

Hennies HC, Kornak U, Zhang H, Egerer J, Zhang X, Seifert W et al (2008) Gerodermia osteodysplastica is caused by mutations in SCYL1BP1, a Rab-6 interacting golgin. Nat Genet 40(12):1410–1412

Hirai M, Ohbayashi T, Horiguchi M, Okawa K, Hagiwara A, Chien KR et al (2007) Fibulin-5/DANCE has an elastogenic organizer activity that is abrogated by proteolytic cleavage in vivo. J Cell Biol 176(7):1061–1071

Horiguchi M, Inoue T, Ohbayashi T, Hirai M, Noda K, Marmorstein LY et al (2009) Fibulin-4 conducts proper elastogenesis via interaction with cross-linking enzyme lysyl oxidase. Proc Natl Acad Sci U S A 106(45):19029–19034

Hosen MJ, Vanakker OM, Willaert A, Huysseune A, Coucke P, De Paepe A (2013) Zebrafish models for ectopic mineralization disorders: practical issues from morpholino design to post-injection observations. Front Genet 4:74

Howe K, Clark MD, Torroja CF, Torrance J, Berthelot C, Muffato M et al (2013) The zebrafish reference genome sequence and its relationship to the human genome. Nature 496(7446):498–503

Hoyer J, Kraus C, Hammersen G, Geppert JP, Rauch A (2009) Lethal cutis laxa with contractural arachnodactyly, overgrowth and soft tissue bleeding due to

a novel homozygous fibulin-4 gene mutation. Clin Genet 76(3):276–281
Hu Q, Shifren A, Sens C, Choi J, Szabo Z, Starcher BC et al (2010) Mechanisms of emphysema in autosomal dominant cutis laxa. Matrix Biol J Int Soc Matrix Biol 29(7):621–628
Huang J, Davis EC, Chapman SL, Budatha M, Marmorstein LY, Word RA et al (2010) Fibulin-4 deficiency results in ascending aortic aneurysms: a potential link between abnormal smooth muscle cell phenotype and aneurysm progression. Circ Res 106(3):583–592
Hucthagowder V, Sausgruber N, Kim KH, Angle B, Marmorstein LY, Urban Z (2006) Fibulin-4: a novel gene for an autosomal recessive cutis laxa syndrome. Am J Hum Genet 78(6):1075–1080
Hucthagowder V, Morava E, Kornak U, Lefeber DJ, Fischer B, Dimopoulou A et al (2009) Loss-of-function mutations in ATP6V0A2 impair vesicular trafficking, tropoelastin secretion and cell survival. Hum Mol Genet 18(12):2149–2165
Hunter AG (1988) Is geroderma osteodysplastica underdiagnosed? J Med Genet 25(12):854–857
Hwang WY, Peterson RT, Yeh JR (2014) Methods for targeted mutagenesis in zebrafish using TALENs. Methods 69(1):76–84
Igoucheva O, Alexeev V, Halabi CM, Adams SM, Stoilov I, Sasaki T et al (2015) Fibulin-4 E57K Knock-in mice recapitulate cutaneous, vascular and skeletal defects of recessive cutis Laxa 1B with both elastic Fiber and collagen fibril abnormalities. J Biol Chem 290(35):21443–21459
Jefferies KC, Cipriano DJ, Forgac M (2008) Function, structure and regulation of the vacuolar (H+)-ATPases. Arch Biochem Biophys 476(1):33–42
Kaler SG (2011) ATP7A-related copper transport diseases-emerging concepts and future trends. Nat Rev Neurol 7(1):15–29
Kaler SG (2013) ATP7A-related copper transport diseases: emerging concepts and future trends. J Trace Elem Med Biol 27:19
Kaler SG, Gallo LK, Proud VK, Percy AK, Mark Y, Segal NA et al (1994) Occipital horn syndrome and a mild Menkes phenotype associated with splice site mutations at the MNK locus. Nat Genet 8(2):195–202
Kantaputra PN, Kaewgahya M, Wiwatwongwana A, Wiwatwongwana D, Sittiwangkul R, Iamaroon A et al (2014) Cutis laxa with pulmonary emphysema, conjunctivochalasis, nasolacrimal duct obstruction, abnormal hair, and a novel FBLN5 mutation. Am J Med Genet Part A 164a(9):2370–2377
Kappanayil M, Nampoothiri S, Kannan R, Renard M, Coucke P, Malfait F et al (2012) Characterization of a distinct lethal arteriopathy syndrome in twenty-two infants associated with an identical, novel mutation in FBLN4 gene, confirms fibulin-4 as a critical determinant of human vascular elastogenesis. Orphanet J Rare Dis 7:61
Kardos GR, Wastyk HC, Robertson GP (2015) Disruption of proline synthesis in melanoma inhibits protein production mediated by the GCN2 pathway. Mol Cancer Res 13(10):1408–1420
Kariminejad A, Afroozan F, Bozorgmehr B, Ghanadan A, Akbaroghli S, Khorshid HRK et al (2017) Discriminative features in three autosomal recessive cutis laxa syndromes: cutis laxa IIA, cutis laxa IIB, and geroderma osteoplastica. Int J Mol Sci 18(3)
Kerr B, Delrue MA, Sigaudy S, Perveen R, Marche M, Burgelin I et al (2006) Genotype-phenotype correlation in Costello syndrome: HRAS mutation analysis in 43 cases. J Med Genet 43(5):401–405
Kimura T, Sakisaka T, Baba T, Yamada T, Takai Y (2006) Involvement of the Ras-Ras-activated Rab5 guanine nucleotide exchange factor RIN2-Rab5 pathway in the hepatocyte growth factor-induced endocytosis of E-cadherin. J Biol Chem 281(15):10598–10609
Kobayashi N, Kostka G, Garbe JH, Keene DR, Bachinger HP, Hanisch FG et al (2007) A comparative analysis of the fibulin protein family. Biochemical characterization, binding interactions, and tissue localization. J Biol Chem 282(16):11805–11816
Kodama H, Okabe I, Yanagisawa M, Kodama Y (1989) Copper deficiency in the mitochondria of cultured skin fibroblasts from patients with Menkes syndrome. J Inherit Metab Dis 12(4):386–389
Kodama H, Sato E, Gu YH, Shiga K, Fujisawa C, Kozuma T (2005) Effect of copper and diethyldithiocarbamate combination therapy on the macular mouse, an animal model of Menkes disease. J Inherit Metab Dis 28(6):971–978
Koppe R, Kaplan P, Hunter A, MacMurray B (1989) Ambiguous genitalia associated with skeletal abnormalities, cutis laxa, craniostenosis, psychomotor retardation, and facial abnormalities (SCARF syndrome). Am J Med Genet 34(3):305–312
Kornak U, Reynders E, Dimopoulou A, van Reeuwijk J, Fischer B, Rajab A et al (2008a) Impaired glycosylation and cutis laxa caused by mutations in the vesicular H+-ATPase subunit ATP6V0A2. Nat Genet 40(1):32–34
Kornak U, Reynders E, Dimopoulou A, Van Reeuwijk J, Fischer B, Rajab A et al (2008b) Impaired glycosylation and cutis laxa caused by mutations in the vesicular H+-ATPase subunit ATP6V0A2. Nat Genet 40(1):32–34
Kretz R, Bozorgmehr B, Kariminejad MH, Rohrbach M, Hausser I, Baumer A et al (2011) Defect in proline synthesis: Pyrroline-5-carboxylate reductase 1 deficiency leads to a complex clinical phenotype with collagen and elastin abnormalities. J Inherit Metab Dis 34(3):731–739
Kurotaki N, Imaizumi K, Harada N, Masuno M, Kondoh T, Nagai T et al (2002) Haploinsufficiency of NSD1 causes Sotos syndrome. Nat Genet 30(4):365–366
Lazoff SG, Rybak JJ, Parker BR, Luzzatti L (1975) Skeletal dysplasia, occipital horns, diarrhea and obstructive uropathy- a new hereditary syndrome. Birth Defects Orig Artic Ser 11(5):71–74
Le Saux O, Beck K, Sachsinger C, Silvestri C, Treiber C, Goring HH et al (2001) A spectrum of ABCC6 muta-

tions is responsible for pseudoxanthoma elasticum. Am J Hum Genet 69(4):749–764

Le VP, Stoka KV, Yanagisawa H, Wagenseil JE (2014) Fibulin-5 null mice with decreased arterial compliance maintain normal systolic left ventricular function, but not diastolic function during maturation. Physiol Rep 2(3):e00257

Lee YC, Huang HY, Chang CJ, Cheng CH, Chen YT (2010) Mitochondrial GLUT10 facilitates dehydroascorbic acid import and protects cells against oxidative stress: mechanistic insight into arterial tortuosity syndrome. Hum Mol Genet 19(19):3721–3733

Lessel D, Ozel AB, Campbell SE, Saadi A, Arlt MF, McSweeney KM et al (2018) Analyses of LMNA-negative juvenile progeroid cases confirms biallelic POLR3A mutations in Wiedemann–Rautenstrauch-like syndrome and expands the phenotypic spectrum of PYCR1 mutations. Hum Genet 137(11–12):921–939

Li Y, Ren P, Dawson A, Vasquez HG, Ageedi W, Zhang C et al (2020) Single-cell transcriptome analysis reveals dynamic cell populations and differential gene expression patterns in control and aneurysmal human aortic tissue. Circulation 142(14):1374–1388

Liang ST, Audira G, Juniardi S, Chen JR, Lai YH, Du ZC et al (2019) Zebrafish carrying pycr1 gene deficiency display aging and multiple behavioral abnormalities. Cell 8(5)

Lin DS, Chang JH, Liu HL, Wei CH, Yeung CY, Ho CS et al (2011a) Compound heterozygous mutations in PYCR1 further expand the phenotypic spectrum of De Barsy syndrome. Am J Med Genet A 155a(12):3095–3099

Lin DS, Yeung CY, Liu HL, Ho CS, Shu CH, Chuang CK et al (2011b) A novel mutation in PYCR1 causes an autosomal recessive cutis laxa with premature aging features in a family. Am J Med Genet A 155a(6):1285–1289

Liu X, Zhao Y, Gao J, Pawlyk B, Starcher B, Spencer JA et al (2004) Elastic fiber homeostasis requires lysyl oxidase-like 1 protein. Nat Genet 36(2):178–182

Loeys B, De Paepe A, Urban Z (1993) EFEMP2-related cutis Laxa. In: Adam MP, Ardinger HH, Pagon RA, Wallace SE, LJH B, Stephens K et al (eds) GeneReviews((R)). Seattle (WA): University of Washington, Seattle University of Washington, Seattle. GeneReviews is a registered trademark of the University of Washington, Seattle. All rights reserved

Loeys B, Van Maldergem L, Mortier G, Coucke P, Gerniers S, Naeyaert JM et al (2002) Homozygosity for a missense mutation in fibulin-5 (FBLN5) results in a severe form of cutis laxa. Hum Mol Genet 11(18):2113–2118

Malakan Rad E, Zeinaloo AA, Kariminejad A, Kornak U, Fischer-Zirnsak B, Mohamadpour M (2016) Novel FBLN5 mutation of congenital autosomal recessive cutis Laxa with isolated right ventricular non-compaction (RVNC): new findings on echocardiographic speckle-tracking strain imaging of RVNC. Iranian J Pediatrics 26(6)

Malan V, Rajan D, Thomas S, Shaw AC, Louis Dit Picard H, Layet V et al (2010) Distinct effects of allelic NFIX mutations on nonsense-mediated mRNA decay engender either a Sotos-like or a Marshall-Smith syndrome. Am J Hum Genet 87(2):189–198

Malfait F, Francomano C, Byers P, Belmont J, Berglund B, Black J et al (2017) The 2017 international classification of the Ehlers-Danlos syndromes. Am J Med Genet C Semin Med Genet 175(1):8–26

Marchegiani S, Davis T, Tessadori F, van Haaften G, Brancati F, Hoischen A et al (2015) Recurrent mutations in the basic domain of TWIST2 cause Ablepharon Macrostomia and barber-say syndromes. Am J Hum Genet 97(1):99–110

Martinelli D, Haeberle J, Colafati S, Giunta C, Hausser I, Goffredo BM et al (2010) Pyrroline-5-carboxylate (P5C) synthase deficiency: novel clinical and biochemical in sights. J Inherit Metab Dis 33:S26

Martinelli D, Häberle J, Rubio V, Giunta C, Hausser I, Carrozzo R et al (2012) Understanding pyrroline-5-carboxylate synthetase deficiency: clinical, molecular, functional, and expression studies, structure-based analysis, and novel therapy with arginine. J Inherit Metab Dis 35(5):761–776

McLaughlin PJ, Chen Q, Horiguchi M, Starcher BC, Stanton JB, Broekelmann TJ et al (2006) Targeted disruption of fibulin-4 abolishes elastogenesis and causes perinatal lethality in mice. Mol Cell Biol 26(5):1700–1709

Mohamed M, Kouwenberg D, Gardeitchik T, Kornak U, Wevers RA, Morava E (2011) Metabolic cutis laxa syndromes. J Inherit Metab Dis 34(4):907–916

Morava E, Lefeber DJ, Urban Z, de Meirleir L, Meinecke P, Gillessen Kaesbach G et al (2008) Defining the phenotype in an autosomal recessive cutis laxa syndrome with a combined congenital defect of glycosylation. Eur J Human Genet: EJHG 16(1):28–35

Morava É, Guillard M, Lefeber DJ, Wevers RA (2009) Autosomal recessive cutis laxa syndrome revisited. Eur J Hum Genet 17(9):1099–1110

Nabatanzi A, Da S, Male M, Chen S, Huang C (2020) Type II acquired cutis laxa associated with recurrent urticarial vasculitis: brief report. Allergy Asthma Clin Immunol 16:1

Nakamura T, Lozano PR, Ikeda Y, Iwanaga Y, Hinek A, Minamisawa S et al (2002) Fibulin-5/DANCE is essential for elastogenesis in vivo. Nature 415(6868):171–175

Nampoothiri S, Paepe AD, Kannan R, Malfait F, Laer LV, Renard M et al (2010) Lethal arteriopathy syndrome of infants associated with a novel mutation of FBLN4 gene - identification of a new syndrome and understanding the critical role of fibulin 4 in human vascular elastogenesis. Ann Pediatr Cardiol 3(2):202

Nasevicius A, Ekker SC (2000) Effective targeted gene 'knockdown' in zebrafish. Nat Genet 26(2):216–220

Niihori T, Aoki Y, Narumi Y, Neri G, Cave H, Verloes A et al (2006) Germline KRAS and BRAF mutations in cardio-facio-cutaneous syndrome. Nat Genet 38(3):294–296

Noda K, Kitagawa K, Miki T, Horiguchi M, TO A, Taniguchi T et al (2020) A matricellular protein fibulin-4 is essential for the activation of lysyl oxidase. Sci Adv 6(48)

Papke CL, Yanagisawa H (2014) Fibulin-4 and fibulin-5 in elastogenesis and beyond: insights from mouse and human studies. Matrix Biol J Int Soc Matrix Biol 37:142–149

Papke CL, Tsunezumi J, Ringuette LJ, Nagaoka H, Terajima M, Yamashiro Y et al (2015) Loss of fibulin-4 disrupts collagen synthesis and maturation: implications for pathology resulting from EFEMP2 mutations. Hum Mol Genet 24(20):5867–5879

Paulsen IF, Bredgaard R, Hesse B, Steiniche T, Henriksen TF (2014) Acquired cutis laxa: diagnostic and therapeutic considerations. J Plast Reconstr Aesthet Surg 67(10):e242–e243

Pepin M, Schwarze U, Superti-Furga A, Byers PH (2000) Clinical and genetic features of Ehlers-Danlos syndrome type IV, the vascular type. N Engl J Med 342(10):673–680

Petris MJ, Camakaris J, Greenough M, LaFontaine S, Mercer JFB (1998) A C-terminal di-leucine is required for localization of the Menkes protein in the trans-Golgi network. Hum Mol Genet 7(13):2063–2071

Petris MJ, Strausak D, Mercer JFB (2000) The Menkes copper transporter is required for the activation of tyrosinase. Hum Mol Genet 9(19):2845–2851

Renard M, Holm T, Veith R, Callewaert BL, Adès LC, Baspinar O et al (2010) Altered TGFB signaling and cardiovascular manifestations in patients with autosomal recessive cutis laxa type i caused by fibulin-4 deficiency. Eur J Hum Genet 18(8):895–901

Reversade B, Escande-Beillard N, Dimopoulou A, Fischer B, Chng SC, Li Y et al (2009) Mutations in PYCR1 cause cutis laxa with progeroid features. Nat Genet 41(9):1016–1021

Robertson SP, Bankier A (1999) Sotos syndrome and cutis laxa. J Med Genet 36(1):51–56

Robertson IB, Horiguchi M, Zilberberg L, Dabovic B, Hadjiolova K, Rifkin DB (2015) Latent TGF-beta-binding proteins. Matrix Biol J Int Soc Matrix Biol 47:44–53

Roche N, Houtmeyers P, Janssens S, Blondeel P (2010) Barber-say syndrome in a father and daughter. Am J Med Genet A 152A(10):2563–2568

Rohner N, Perathoner S, Frohnhofer HG, Harris MP (2011) Enhancing the efficiency of N-ethyl-N-nitrosourea-induced mutagenesis in the zebrafish. Zebrafish 8(3):119–123

Rohrbach M, Vandersteen A, Yis U, Serdaroglu G, Ataman E, Chopra M et al (2011) Phenotypic variability of the kyphoscoliotic type of Ehlers-Danlos syndrome (EDS VIA): clinical, molecular and biochemical delineation. Orphanet J Rare Dis 6:46

Royce PM, Camakaris J, Danks DM (1980) Reduced lysyl oxidase activity in skin fibroblasts from patients with Menkes' syndrome. Biochem J 192(2):579–586

Ryter SW, Rosas IO, Owen CA, Martinez FJ, Choi ME, Lee CG et al (2018) Mitochondrial dysfunction as a pathogenic mediator of chronic obstructive pulmonary disease and idiopathic pulmonary fibrosis. Ann Am Thorac Soc 15(Suppl 4):S266–SS72

Saito A, Horie M, Nagase T (2018) TGF-beta signaling in lung health and disease. Int J Mol Sci 19(8)

Samland AK, Sprenger GA (2009) Transaldolase: from biochemistry to human disease. Int J Biochem Cell Biol 41(7):1482–1494

Sander JD, Cade L, Khayter C, Reyon D, Peterson RT, Joung JK et al (2011) Targeted gene disruption in somatic zebrafish cells using engineered TALENs. Nat Biotechnol 29(8):697–698

Sato F, Wachi H, Starcher BC, Seyama Y (2006) Biochemical analysis of elastic fiber formation with a frameshift-mutated tropoelastin (fmTE) at the C-terminus of tropoelastin. J Health Sci 52(3):259–267

Shafqat S, Velaz-Faircloth M, Henzi VA, Whitney KD, Yang-Feng TL, Seldin MF et al (1995) Human brain-specific L-proline transporter: molecular cloning, functional expression, and chromosomal localization of the gene in human and mouse genomes. Mol Pharmacol 48(2):219–229

Shin SJ, Yanagisawa H (2019) Recent updates on the molecular network of elastic fiber formation. Essays Biochem 63(3):365–376

Sinnige PF, van Ravenswaaij-Arts CMA, Caruso P, Lin AE, Boon M, Rahikkala E et al (2017) Imaging in cutis laxa syndrome caused by a dominant negative ALDH18A1 mutation, with hypotheses for intracranial vascular tortuosity and wide perivascular spaces. Eur J Paediatric Neurol EJPN Off J Eur Paediatric Neurol Soc 21(6):912–920

Skidmore DL, Chitayat D, Morgan T, Hinek A, Fischer B, Dimopoulou A et al (2011) Further expansion of the phenotypic spectrum associated with mutations in ALDH18A1, encoding Delta(1)-pyrroline-5-carboxylate synthase (P5CS). Am J Med Genet A 155A(8):1848–1856

Sousa SB, Jenkins D, Chanudet E, Tasseva G, Ishida M, Anderson G et al (2014) Gain-of-function mutations in the phosphatidylserine synthase 1 (PTDSS1) gene cause Lenz-Majewski syndrome. Nat Genet 46(1):70–76

Sterner-Kock A, Thorey IS, Koli K, Wempe F, Otte J, Bangsow T et al (2002) Disruption of the gene encoding the latent transforming growth factor-beta binding protein 4 (LTBP-4) causes abnormal lung development, cardiomyopathy, and colorectal cancer. Genes Dev 16(17):2264–2273

Stranecky V, Hoischen A, Hartmannova H, Zaki MS, Chaudhary A, Zudaire E et al (2013) Mutations in ANTXR1 cause GAPO syndrome. Am J Hum Genet 92(5):792–799

Su CT, Huang JW, Chiang CK, Lawrence EC, Levine KL, Dabovic B et al (2015) Latent transforming growth factor binding protein 4 regulates transforming growth factor beta receptor stability. Hum Mol Genet 24(14):4024–4036

Syx D, Malfait F, Van Laer L, Hellemans J, Hermanns-Lê T, Willaert A et al (2010) The RIN2 syndrome: a new

autosomal recessive connective tissue disorder caused by deWciency of Ras and Rab interactor 2 (RIN2). Hum Genet 128(1):79–88
Szabo Z, Crepeau MW, Mitchell AL, Stephan MJ, Puntel RA, Yin Loke K et al (2006) Aortic aneurysmal disease and cutis laxa caused by defects in the elastin gene. J Med Genet 43(3):255–258
Tang J, Robertson S, Lem KE, Godwin SC, Kaler SG (2006) Functional copper transport explains neurologic sparing in occipital horn syndrome. Genet Med Off J Am College Med Genet 8(11):711–718
Tartaglia M, Mehler EL, Goldberg R, Zampino G, Brunner HG, Kremer H et al (2001) Mutations in PTPN11, encoding the protein tyrosine phosphatase SHP-2, cause Noonan syndrome. Nat Genet 29(4):465–468
Tartaglia M, Kalidas K, Shaw A, Song X, Musat DL, van der Burgt I et al (2002) PTPN11 mutations in Noonan syndrome: molecular spectrum, genotype-phenotype correlation, and phenotypic heterogeneity. Am J Hum Genet 70(6):1555–1563
Teame T, Zhang Z, Ran C, Zhang H, Yang Y, Ding Q et al (2019) The use of zebrafish (Danio rerio) as biomedical models. Anim Front 9(3):68–77
Tinkle B, Castori M, Berglund B, Cohen H, Grahame R, Kazkaz H et al (2017) Hypermobile Ehlers-Danlos syndrome (a.k.a. Ehlers-Danlos syndrome type III and Ehlers-Danlos syndrome hypermobility type): clinical description and natural history. Am J Med Genet C Semin Med Genet 175(1):48–69
Toriello HV (1990) Wiedemann-Rautenstrauch syndrome. J Med Genet 27(4):256–257
Tsukahara M, Imaizumi K, Kawai S, Kajii T (1994) Occipital horn syndrome: report of a patient and review of the literature. Clin Genet 45(1):32–35
Tümer Z, Møller LB (2010) Menkes disease. Eur J Hum Genet 18(5):511–518
Urban Z (2012) The complexity of elastic Fiber biogenesis: the paradigm of cutis Laxa. J Invest Dermatol 132(Suppl 3):E12–E14
Urban Z, Davis EC (2014) Cutis laxa: intersection of elastic fiber biogenesis, TGFbeta signaling, the secretory pathway and metabolism. Matrix Biol J Int Soc Matrix Biol 33:16–22
Urban Z, Gao J, Pope FM, Davis EC (2005) Autosomal dominant cutis laxa with severe lung disease: synthesis and matrix deposition of mutant tropoelastin. J Invest Dermatol 124(6):1193–1199
Urban Z, Hucthagowder V, Schürmann N, Todorovic V, Zilberberg L, Choi J et al (2009) Mutations in LTBP4 cause a syndrome of impaired pulmonary, gastrointestinal, genitourinary, musculoskeletal, and dermal development. Am J Hum Genet 85(5):593–605
Vallet SD, Ricard-Blum S (2019) Lysyl oxidases: from enzyme activity to extracellular matrix cross-links. Essays Biochem 63(3):349–364
van Bon BW, Gilissen C, Grange DK, Hennekam RC, Kayserili H, Engels H et al (2012) Cantu syndrome is caused by mutations in ABCC9. Am J Hum Genet 90(6):1094–1101
Van Damme T, Gardeitchik T, Mohamed M, Guerrero-Castillo S, Freisinger P, Guillemyn B et al (2017) Mutations in ATP6V1E1 or ATP6V1A cause autosomal-recessive cutis Laxa. Am J Hum Genet 100(2):216–227
van der Pluijm I, Burger J, van Heijningen PM, van Vliet N, Milanese C et al (2018) Decreased mitochondrial respiration in aneurysmal aortas of Fibulin-4 mutant mice is linked to PGC1A regulation. Cardiovasc Res 114(13):1776–1793
Van Maldergem L, Loeys B (1993) FBLN5-Related Cutis Laxa. In: Adam MP, Ardinger HH, Pagon RA, Wallace SE, Bean LJH, Stephens K, et al., editors. GeneReviews((R)). Seattle (WA): University of Washington, Seattle University of Washington, Seattle. GeneReviews is a registered trademark of the University of Washington, Seattle. All rights reserved
Van Maldergem L, Vamos E, Liebaers I, Petit P, Vandevelde G, Simonis-Blumenfrucht A et al (1988) Severe congenital cutis laxa with pulmonary emphysema: a family with three affected sibs. Am J Med Genet 31(2):455–464
Van Maldergem L, Dobyns W, Kornak U (1993) ATP6V0A2-Related Cutis Laxa. In: Adam MP, Ardinger HH, Pagon RA, Wallace SE, Bean LJH, Stephens K, et al., editors. GeneReviews((R)). Seattle (WA)
Van Maldergem L, Yuksel-Apak M, Kayserili H, Seemanova E, Giurgea S, Basel-Vanagaite L et al (2008) Cobblestone-like brain dysgenesis and altered glycosylation in congenital cutis laxa. Debre Type Neurol 71(20):1602–1608
Vanakker OM, Martin L, Gheduzzi D, Leroy BP, Loeys BL, Guerci VI et al (2007) Pseudoxanthoma elasticum-like phenotype with cutis laxa and multiple coagulation factor deficiency represents a separate genetic entity. J Investig Dermatol 127(3):581–587
Vanakker O, Callewaert B, Malfait F, Coucke P (2015) The genetics of soft connective tissue disorders, pp 229–255
Verhoeven NM, Huck JH, Roos B, Struys EA, Salomons GS, Douwes AC et al (2001) Transaldolase deficiency: liver cirrhosis associated with a new inborn error in the pentose phosphate pathway. Am J Hum Genet 68(5):1086–1092
Vodo D, Sarig O, Peled A, Frydman M, Greenberger S, Sprecher E (2015) Autosomal-dominant cutis laxa resulting from an intronic mutation in ELN. Exp Dermatol 24(11):885–887
Wachi H, Nonaka R, Sato F, Shibata-Sato K, Ishida M, Iketani S et al (2008) Characterization of the molecular interaction between tropoelastin and DANCE/fibulin-5. J Biochem 143(5):633–639
Wan W, Gleason RL Jr (2013) Dysfunction in elastic fiber formation in fibulin-5 null mice abrogates the evolution in mechanical response of carotid arteries during maturation. Am J Physiol Heart Circ Physiol 304(5):H674–H686

Wan W, Yanagisawa H, Gleason RL Jr (2010) Biomechanical and microstructural properties of common carotid arteries from fibulin-5 null mice. Ann Biomed Eng 38(12):3605–3617

Wessels MW, Catsman-Berrevoets CE, Mancini GM, Breuning MH, Hoogeboom JJ, Stroink H et al (2004) Three new families with arterial tortuosity syndrome. Am J Med Genet A 131(2):134–143

Willaert A, Khatri S, Callewaert BL, Coucke PJ, Crosby SD, Lee JGH et al (2012) GLUT10 is required for the development of the cardiovascular system and the notochord and connects mitochondrial function to TGFβ signaling. Hum Mol Genet 21(6):1248–1259

Witkos TM, Chan WL, Joensuu M, Rhiel M, Pallister E, Thomas-Oates J et al (2019) GORAB scaffolds COPI at the trans-Golgi for efficient enzyme recycling and correct protein glycosylation. Nat Commun 10(1):127

Wolthuis DFGJ, Van Asbeck E, Mohamed M, Gardeitchik T, Lim-Melia ER, Wevers RA et al (2014) Cutis laxa, fat pads and retinopathy due to ALDH18A1 mutation and review of the literature. Eur J Paediatr Neurol 18(4):511–515

Woods IG, Schier AF (2008) Targeted mutagenesis in zebrafish. Nat Biotechnol 26(6):650–651

Yamauchi Y, Tsuruga E, Nakashima K, Sawa Y, Ishikawa H (2010) Fibulin-4 and -5, but not Fibulin-2, are associated with Tropoelastin deposition in elastin-producing cell culture. Acta Histochem Cytochem 43(6):131–138

Yanagisawa H, Davist EC, Starcher BC, Ouchi T, Yanagisawa M, Richardson JA et al (2002) Fibulin-5 is an elastin-binding protein essential for elastic fibre development in vivo. Nature 415(6868):168–171

Yeo GC, Keeley FW, Weiss AS (2011) Coacervation of tropoelastin. Adv Colloid Interf Sci 167(1–2):94–103

Yildirim Y, Tolun A, Tuysuz B (2011) The phenotype caused by PYCR1 mutations corresponds to geroderma osteodysplasticum rather than autosomal recessive cutis laxa type 2. Am J Med Genet A 155a(1):134–140

Zampatti S, Castori M, Fischer B, Ferrari P, Garavelli L, Dionisi-Vici C et al (2012) De Barsy syndrome: a genetically heterogeneous autosomal recessive cutis laxa syndrome related to P5CS and PYCR1 dysfunction. Am J Med Genet A 158A(4):927–931

Zhang T, Peterson RT (2020) Modeling lysosomal storage diseases in the zebrafish. Front Mol Biosci 7:82

Zhang MC, He L, Giro M, Yong SL, Tiller GE, Davidson JM (1999) Cutis laxa arising from frameshift mutations in exon 30 of the elastin gene (ELN). J Biol Chem 274(2):981–986

Zheng Q, Davis EC, Richardson JA, Starcher BC, Li T, Gerard RD et al (2007) Molecular analysis of fibulin-5 function during de novo synthesis of elastic fibers. Mol Cell Biol 27(3):1083–1095

Zilberberg L, Todorovic V, Dabovic B, Horiguchi M, Courousse T, Sakai LY et al (2012) Specificity of latent TGF-beta binding protein (LTBP) incorporation into matrix: role of fibrillins and fibronectin. J Cell Physiol 227(12):3828–3836

14 Collagen VI Muscle Disorders: Mutation Types, Pathogenic Mechanisms and Approaches to Therapy

Shireen R. Lamandé

Abstract

Mutations in the genes encoding the major collagen VI isoform, *COL6A1*, *COL6A2* and *COL6A3*, are responsible for the muscle disorders Bethlem myopathy and Ullrich congenital muscular dystrophy. These disorders form a disease spectrum from mild to severe. Dominant and recessive mutations are found along the entire spectrum and the clinical phenotype is strongly influenced by the way mutations impede collagen VI protein assembly. Most mutations are in the triple helical domain, towards the N-terminus and they compromise microfibril assembly. Some mutations are found outside the helix in the C- and N-terminal globular domains, but because these regions are highly polymorphic it is difficult to discriminate mutations from rare benign changes without detailed structural and functional studies. Collagen VI deficiency leads to mitochondrial dysfunction, deficient autophagy and increased apoptosis. Therapies that target these consequences have been tested in mouse models and some have shown modest efficacy in small human trials. Antisense therapies for a common mutation that introduces a pseudoexon show promise in cell culture but haven't yet been tested in an animal model. Future therapeutic approaches await new research into how collagen VI deficiency signals downstream consequences.

S. R. Lamandé (✉)
Murdoch Children's Research Institute and Department of Paediatrics, University of Melbourne, Royal Children's Hospital, Parkville, VIC, Australia
e-mail: shireen.lamande@mcri.edu.au

Keywords

Collagen VI · Bethlem myopathy · Ullrich congenital muscular dystrophy · Muscle

Abbreviations

LoF	Loss of function
UCMD	Ullrich congenital muscular dystrophy

14.1 Introduction

The extracellular matrix protein collagen VI forms beaded microfibrils that are found in almost all tissues (Timpl and Chu 1994). Its location, close to basement membranes and in the pericellular matrix, and interactions with many other extracellular matrix proteins, cell surface integrin and anthrax receptors, suggests that collagen VI has tissue anchoring functions and contributes to cell-matrix interactions. In addition to these structural roles, collagen VI can signal to

J. Halper (ed.), *Progress in Heritable Soft Connective Tissue Diseases*, Advances in Experimental Medicine and Biology 1348, https://doi.org/10.1007/978-3-030-80614-9_14

promote cell growth and survival, reduce oxidative damage and alter metabolism.

There are five functional collagen VI genes in humans, *COL6A1*, *COL6A2*, *COL6A3*, *COL6A5*, and *COL6A6* (Fitzgerald et al. 2008; Gara et al. 2008; Cescon et al. 2015). The *COL6A4* gene is present in mice, but in humans, chimpanzees and gorillas it has been interrupted by a pericentric inversion and is non-functional (Fitzgerald et al. 2008; Gara et al. 2008). *COL6A1*, *COL6A2*, and *COL6A3* encode the most abundant collagen VI isoform and mutations in these genes cause muscular dystrophy. A clear role for *COL6A5* and *COL6A6* mutations in inherited disorders is yet to emerge.

This review will briefly summarise the clinical phenotypes, Bethlem myopathy and Ullrich congenital muscular dystrophy (UCMD), and the kinds of mutations in the *COL6A1*, *COL6A2* and *COL6A3* genes that underlie these disorders. The focus will be on updating the information in our 2018 review (Lamandé and Bateman 2018) and highlighting new data about collagen VI assembly and structure, pathogenic mechanisms, and targeted therapies. Our knowledge about endotrophin, a bioactive fragment cleaved from the C-terminus of the α3(VI) chain, and its role in cancer, obesity and diabetes has expanded rapidly in the last few years and will be covered in the final section.

14.2 Collagen VI Structure and Assembly

Understanding collagen VI structure and its complex hierarchical assembly pathway is crucial when interpreting sequence changes in the collagen VI genes, anticipating how they could affect function and interpreting pathogenicity, and so this section will provide a summary of its key features. Further information, about how collagen VI was discovered and the early biochemical and electron microscopy studies, can be found in two detailed reviews (Timpl and Engel 1987; Timpl and Chu 1994).

The major collagen VI isoform contains α1(VI), α2(VI), and α3(VI) protein chains, encoded by *COL6A1*, *COL6A2* and *COL6A3* respectively, in a 1:1:1 ratio. Each chain has a triple helical region – Gly-X-Y amino acid triplet repeats – of 335–336 amino acids. The triple helix is relatively small compared to other collagen types and it has some flexibility because each chain has interruptions to the Gly-X-Y repeats. N- and C-terminal globular regions flank the triple helix and make up the majority of the protein. These globular regions are mostly tandem arrays of ~200 amino acid domains that have homology to the von Willebrand factor A domains (Fig. 14.1a; Whittaker and Hynes 2002). Three different domains are found at the C-terminus of the α3(VI) chain; C3- a unique domain, C4 – a fibronectin type III repeat, and C5 – a Kunitz protease inhibitor motif.

14.2.1 Clinical Features

Collagen VI intracellular assembly begins when the three chains, α1(VI), α2(VI), and α3(VI), associate at the C-terminus (Lamandé et al. 2006) and the central triple helical regions wind together to form the collagen VI monomer. The α4(VI), α5(VI) and α6(VI) chains are similar to the α3(VI) chain and can replace the α3(VI) chain in some low abundance and tissue restricted isoforms (Fitzgerald et al. 2008; Gara et al. 2008). Next, staggered antiparallel dimers with a supercoiled overlap region form and are stabilised by disulphide bonds (Furthmayr et al. 1983) (Fig. 14.1b). This assembly step is initiated by interactions between the α2(VI) C2 domain of one monomer and the triple helix of a second monomer (Ball et al. 2003). Finally, tetramers, the secreted form of collagen VI, are formed when two dimers come together in register, their outer N-terminal regions cross over, and the structure is stabilised by disulphide bonds formed between the α3(VI) cysteine residues (Furthmayr et al. 1983; Timpl and Chu 1994) (Fig. 14.1b).

Following secretion, tetramers associate end-to-end to form the characteristic beaded microfibrils. The N-terminal regions overlap in the microfibrils so that the N-terminal triple helical regions of adjacent tetramers are in

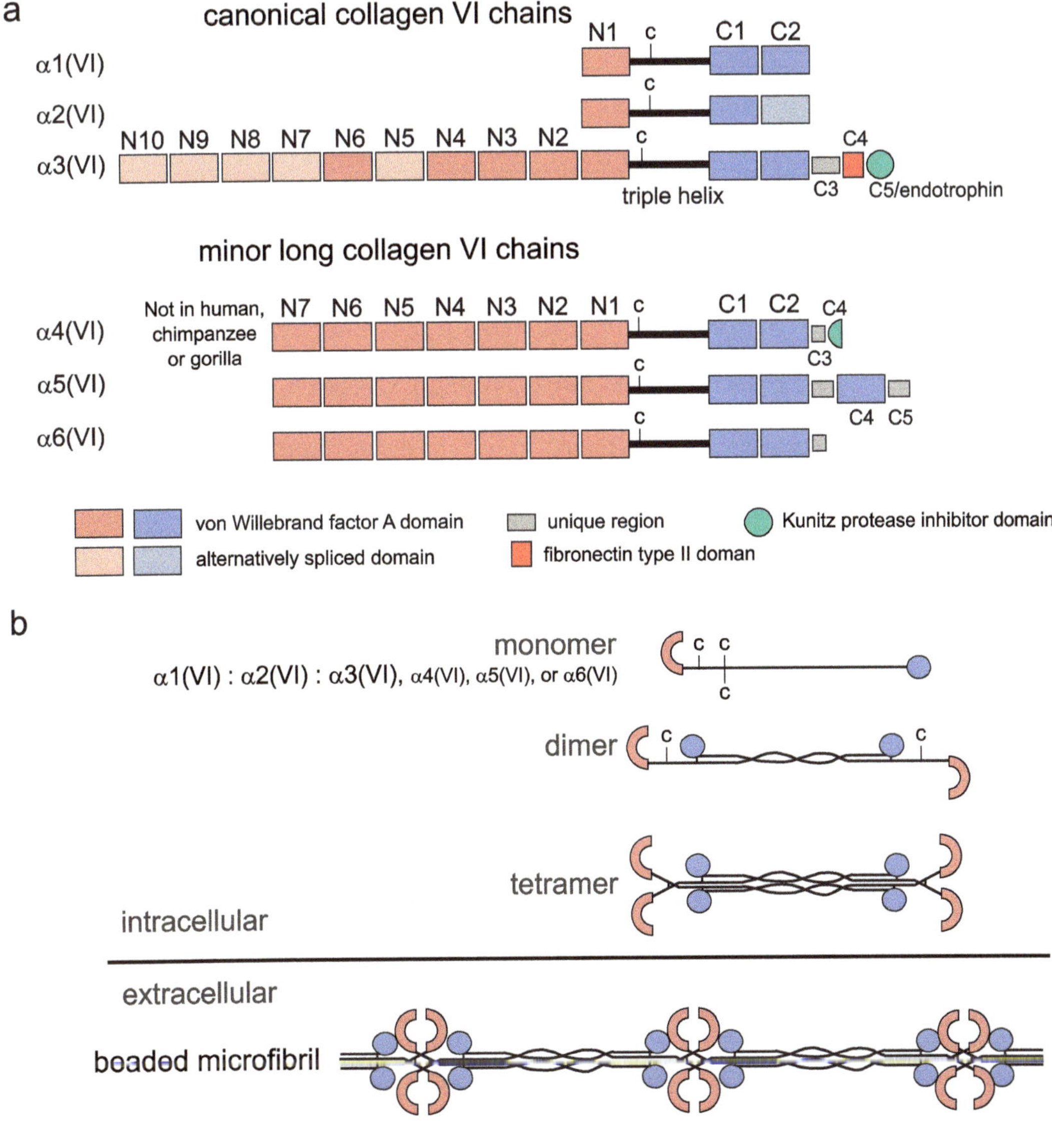

Fig. 14.1 **Collagen VI protein subunits and assembly** (**a**) Schematic showing the three chains that assemble together into the canonical and most abundant collagen VI isoform, α1(VI), α2(VI) and α3(VI), and below, the minor chains, α4(VI), α5(VI) and α6(VI), that can take the place of α3(VI). (**b**) Collagen VI assembles inside the cell into a triple helical monomer, then disulphide bonded dimers, and finally tetramers that are also stabilised by disulphide bonds. Secreted tetramers interact end-to-end and overlap to form beaded microfibrils in the extracellular matrix

close contact with each other and also contact the N- and C-terminal globular regions of adjacent tetramers forming the bead (Chu et al. 1989). This arrangement is beautifully illustrated in 3D reconstructions of the collagen VI microfibril bead region from cryo-TEM images (Godwin et al. 2017). The large N-terminal end of the α3(VI) chain is flexible (Solomon-Degefa et al. 2020) and extends outwards from the beads making it available to interact with other extracellular matrix proteins (Beecher et al. 2011).

14.3 Bethlem Myopathy and Ullrich Congenital Muscular Dystrophy

The collagen VI muscular dystrophies, Bethlem myopathy and Ullrich congenital muscular dystrophy (UCMD), are caused by mutations in *COL6A1*, *COL6A2* and *COL6A3*. These disorders illustrate the important role that collagen VI has in skeletal muscle. In addition to progressive muscle weakness and wasting, patients have skin and tendon abnormalities that reflect a more generalised connective tissue disorder. Recent studies in mouse models document bone, lung and nervous system phenotypes (see later) that suggest changes in these tissues could also contribute to Bethlem myopathy and UCMD. This section will highlight the main features of the collagen VI muscle disorders. Detailed clinical descriptions and patient photographs can be found in (Bönnemann 2011a, b).

14.3.1 Clinical Features

The collagen VI disorders form a continuous disease spectrum; Bethlem myopathy is at the mild end, UCMD is at the severe end and in between are a range of phenotypes that do not fit easily into either Bethlem myopathy or UCMD. UCMD is usually obvious at birth. Patients have muscle weakness, hypotonia and conspicuous joint laxity, particularly involving the distal joints. Joint contractures are sometimes also present at birth. The most severe patients are never able to walk independently, however, most start to walk in early childhood then lose ambulation around their teenage years. The muscle weakness progresses slowly but severe and worsening large joint contractures are a major contributor to losing ambulation. Respiratory insufficiency is common after patients lose the ability to walk and respiratory support, often only at night, is part of standard care allowing patients with UCMD to routinely reach adulthood.

Bethlem myopathy patients have many of the same clinical features as UCMD but the symptoms are significantly milder. Children may have mild muscle weakness and distal joint laxity. Achilles tendon contractures develop around ten years of age and in the teenage years contractures progress to affect fingers, shoulders and sometimes the spine. Progressive muscle weakness and contractures combine to limit mobility and around two thirds of patients over 50 years old need walking aids.

Changes in the skin are common in both Bethlem myopathy and UCMD reflecting collagen VI expression in dermal fibroblasts. Soft velvety skin on the palms, keratosis pilaris – a rough textured skin caused by abnormal keratinisation – on the arms and legs, and abnormal scarring – keloid or atrophic scars – can all be present.

14.3.2 Mouse Models Suggest Broader Tissue Involvement

While muscle weakness and joint contractures have the most profound impact on quality of life for Bethlem myopathy and UCMD patients and have been the most thoroughly studied, collagen VI is found in many other tissues and these are becoming the focus of attention. Detailed studies are difficult in patients because the tissues are not readily accessible and so much of the data comes from studying mouse models.

Collagen VI is closely associated with the basement membrane surrounding peripheral nerves (Marvulli et al. 1996) where it is needed for proper myelination (Chen et al. 2014). The *Col6a1*$^{-/-}$ mouse has hypermyelination, impaired nerve conduction, impaired motor coordination, and delayed response to acute pain (Chen et al. 2014). Collagen VI is found in the neuromuscular junction and the *Col6a1*$^{-/-}$ mouse illustrates that it is needed to stabilise acetylcholine receptor clusters and for efficient neuromuscular synaptic transmission (Cescon et al. 2018). Nerve regeneration following injury is impaired in the *Col6a1*$^{-/-}$ mouse and this appears to be because fewer macrophages are recruited to the injury site than in wild type mice (Chen et al. 2015). In addition, the macrophage population recruited to the injured nerve in the knock out tends to be skewed towards pro-inflammatory (M1) macrophages and contains fewer anti-inflammatory (M2) macrophages than seen in wild type mice with similar injuries (Chen et al. 2015). For more information about collagen VI in the central and peripheral nervous system see (Gregorio et al. 2018).

Collagen VI is found throughout the lung, particularly around basement membranes, and *Col6a1*$^{-/-}$ mice have significant lung abnormalities that include airway space changes – larger and fewer alveoli, increased branching, and increased mucosal thickness – and reduced lung function (Mereness et al. 2020). Collagen VI is also expressed by osteoblasts (Izu et al. 2016). Although trabecular bone formation is normal, *Col6a2*$^{-/-}$ mice have reduced trabecular bone volume and trabecular number (Pham et al. 2020). It turns out that the knockout mice have more osteoclasts than wild type mice suggesting that the bone deficiency is caused by increased breakdown as a consequence of enhanced osteoclastogenesis (Pham et al. 2020). These two studies highlight some of the diverse pathological tissue changes that could also be part of the human collagen VI disease phenotypes and contribute to decreased quality of life.

14.4 Collagen VI Mutations: An Overview and Update

Dominant and recessive mutations in *COL6A1*, *COL6A2* and *COL6A3* are found in patients across the clinical disease spectrum from Bethlem myopathy to UCMD. The mutations have taught us a lot about collagen VI assembly and the protein domains that are key for normal assembly, and conversely, an understanding of collagen VI assembly is helpful when predicting if a sequence change is likely to be pathogenic. Also feeding into the way we interpret sequence changes identified in patients with genetic disorders is the information generated in large scale exome and genome sequencing projects and available in the gnomAD database (gnomad.broadinstitute.org) (Karczewski et al. 2020). Individuals with severe paediatric disorders and their first degree relatives are not included in gnomAD and so disease causing mutations are greatly underrepresented allowing the data to be used to help infer pathogenicity. An earlier review described the mutation types in detail (Lamandé and Bateman 2018) and so the focus here is to provide an overview for the general reader.

14.4.1 Premature Stop Codons and Haploinsufficiency

Mutations that introduce premature stop codons, either point mutations that directly introduce a stop codon or frameshift mutations, are a common cause of genetic disorders (Frischmeyer and Dietz 1999) and in most cases the mutant mRNA is degraded by nonsense-mediated mRNA decay (see reviews (Fang et al. 2013; Nasif et al. 2018)). The end result is protein haploinsufficiency. Homozygous or compound heterozygous premature stop codon mutations, therefore, most commonly result in complete loss of function (LoF). GnomAD consortium researchers have developed a continuous measure of intolerance to LoF variation in which each gene is placed on a spectrum of LoF tolerance (Karczewski et al. 2020). A pLI score reflects the probability of being intolerant to loss of function, the closer to 1, the more intolerant. A pLI >0.9 is considered extremely intolerant and means that loss of a single copy of the gene (haploinsufficiency) is not tolerated. *COL6A2* and *COL6A3* have a pLI score = 0 suggesting that premature stop codon mutations that result in haploinsufficiency are tolerated. In agreement with this, patient and family analyses show that heterozygous carriers of *COL6A2* and *COL6A3* premature stop mutations have no obvious phenotype but patients with homozygous or compound heterozygous premature stop mutations completely lack collagen VI and tend to have severe disease (Camacho Vanegas et al. 2001; Higuchi et al. 2001; Demir et al. 2002). By contrast, *COL6A1* has a pLI score of 1 indicating that loss of a single copy of the gene is not tolerated. This is consistent with several reports showing that *COL6A1* haploinsufficiency can cause pathology, either Bethlem myopathy or mild Bethlem myopathy-like features (Lamandé et al. 1998; Baker et al. 2007; Peat et al. 2007). In the absence of a clear explanation for why *COL6A2* and *COL6A3* haploinsufficiency is tolerated while *COL6A1* haploinsufficiency can cause disease these studies are routinely overlooked but it would be prudent to take these data into account when providing genetic counselling for families with *COL6A1* premature stop codon mutations. Similar to *COL6A2* and *COL6A3*, homozygous and compound heterozygous *COL6A1* loss of function mutations tend to cause severe disease (Giusti et al. 2005; Peat et al. 2007)(Fig. 14.2a).

14.4.2 Triple Helical Glycine Substitutions

The triple helical Gly-X-Y repeat is crucial for collagen folding and structure. Heterozygous glycine substitution mutations that interrupt the triplet repeat are common in collagenopathies, and collagen VI disorders are no exception. The dominant glycine mutations cluster in Gly-X-Y triplets 3–20 at the N-terminal end of the helix (Fig. 14.2b). The mutations can be in any of the three chains and they produce a spectrum of clinical severity (Lampe et al. 2005; Pace et al. 2008; Butterfield et al. 2013; Lamandé and Bateman 2018). Most of the N-terminal glycine mutations that have been studied functionally do not prevent monomer or dimer formation but instead compromise tetramer and microfibril assembly (Pace et al. 2008) confirming that collagen VI assembly relies on the correct triple helical structure in this region. These mutations have a severe dominant negative effect because almost all the secreted collagen VI tetramers contain at least one mutant chain and are structurally abnormal. For a more nuanced evaluation of how N-terminal glycine mutations affect collagen VI assembly see (Pace et al. 2008; Butterfield et al. 2013; Lamandé and Bateman 2018).

Recessive glycine substitution mutations are a relatively rare cause of UCMD but have been reported in *COL6A1*, *COL6A2* and *COL6A3* (Lampe et al. 2005; Briñas et al. 2010; Butterfield et al. 2013)(Fig. 14.2b). The mutations are C-terminal to the dominant glycine mutations and are likely to act by severely impairing triple helix folding and limiting the number of monomers available for dimer formation (Lamandé et al. 2002). Thus patients with homozygous or compound heterozygous recessive glycine substitutions have a severe collagen VI deficiency and heterozygous carriers lack a clinical phenotype.

14.4.3 In-Frame Deletions in the Triple Helix

All the *COL6A1*, *COL6A2* and *COL6A3* exons that code for the triple helix begin and end with complete codons and this means that exon skipping mutations within the triple helix lead to synthesis of chains with in-frame deletions. Exon skipping mutations in the triple helix are common in Bethlem myopathy and UCMD. Most of the reported exon skip mutations are dominant and, like the dominant glycine substitutions, they are found towards the N-terminal end of the helix. They generally compromise tetramer and microfibril formation and cause severe disease (Baker et al. 2005). There are two exceptions to this. Mutations that cause *COL6A1* exon 14 or *COL6A2* exon 11 (equivalent exons) skipping cause Bethlem myopathy. The milder phenotype is because these mutations delete the cysteine residues that stabilise collagen VI dimers. The mutant monomers are almost completely excluded from further assembly into dimers, and so about half the normal amount of collagen VI is secreted (Lamandé et al. 1999; Baker et al. 2007). In-frame deletions in the C-terminal two thirds of the triple helix are recessive mutations that prevent the mutant chains being incorporated into triple helical monomers (Fig. 14.2b) (Camacho Vanegas et al. 2001).

14.4.4 Mutations in the N- and C-terminal Globular Regions

Proving that a sequence variation outside of the triple helix is pathogenic or otherwise remains an enormous diagnostic challenge. The N- and

a **loss of function mutations**

loss of both alleles of either of the three genes not tolerated

α1(VI) - loss of one allele not tolerated, pLI =1

α2(VI) - loss of one allele tolerated, pLI = 0

α3(VI) - loss of one allele tolerated, pLI = 0

b **glycine substitution mutations**

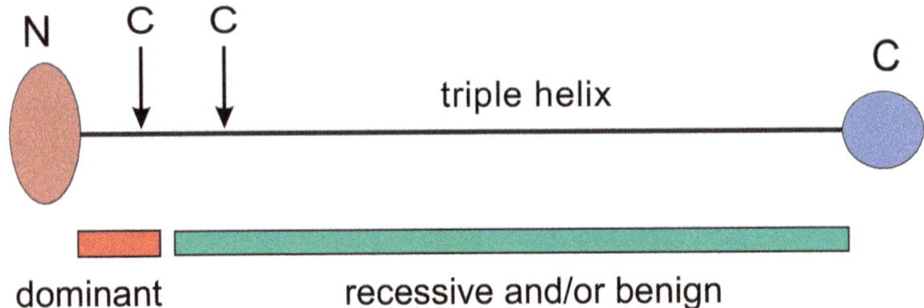

c **exon skipping mutations**

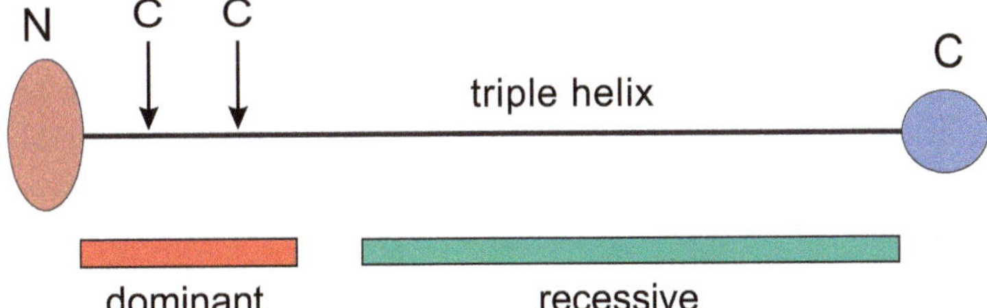

Fig. 14.2 Overview of the main collagen VI mutations types (**a**) loss of function mutations, most commonly mutations that introduce a premature stop codon and lead to nonsense mediated mRNA decay cause disease in patients with homozygous or compound heterozygous mutations. The gnomAD algorithm, pLI, predicts the probability of being loss of function intolerant with a score > 0.9 indicating that loss of a single copy of the gene is not tolerated. The algorithm predicts loss of one α1(VI) allele is not tolerated, but loss of one α2(VI) or α3(VI) is tolerated. This fits with patient data showing that α1(VI) haploinsufficiency causes Bethlem myopathy (see text). (**b**) Glycine substitutions towards the N-terminal end of the triple helix are dominant mutations while more C-terminal glycine substitutions are recessive. Some C-terminal glycine substitutions could be benign. (**c**) Exon skipping mutations in the N-terminal third of the triple helix cause dominant disease while exon skipping mutations in the C-terminal half of the helix are recessive

C-terminal globular regions are highly polymorphic – the gnomAD data base contains hundreds of sequence variants that cause amino acid substitutions in these non-collagenous regions. Some of these are likely to be recessive pathogenic variants but without detailed functional studies it is not possible to determine which ones are benign and which are deleterious. Adding to the complexity, some patients have multiple rare sequence variations in one or more of the collagen VI genes (Lampe and Bushby 2005).

There is strong data supporting the pathogenicity of some recessive mutations in the α2(VI) C1 and C2 domains (reviewed in (Lamandé and Bateman 2018)). There is also strong evidence that supports the pathogenicity of two dominant mutations in the α3(VI) N2 domain reported to cause Bethlem myopathy (Pan et al. 1998; Baker et al. 2007). When α3(VI) N2 domains containing p.G1679E or p.L1726R mutations are expressed recombinantly they are retained entirely within the cell in contrast to the wild type N2 domain which is efficiently secreted (Sasaki et al. 2000; Solomon-Degefa et al. 2020). The structural model of the N2 domain shows that both mutations are located in the core of the domain and so they are likely to have a detrimental effect on domain folding (Solomon-Degefa et al. 2020).

A number of amino acid changes in the collagen VI globular regions that were reported to cause Bethlem myopathy and UCMD do not meet the more stringent requirements for determining pathogenicity that currently apply (for examples see (Lamandé and Bateman 2018)). In the absence of functional data, a cautious approach should be taken when reporting and assessing amino acid changes, particularly if possible recessive mutations are described that expand the phenotypes associated with collagen VI mutations.

14.4.5 A Common Deep Intronic Mutation Causes UCMD

Using muscle RNAseq, an in-frame non-collagenous insertion towards the N-terminal end of the α1(VI) triple helix was discovered in four unrelated UCMD patients who were previously negative for directed collagen VI mutation screening (Cummings et al. 2017). Parallel whole genome sequencing identified the same *de novo* heterozygous deep intronic variant in all four patients (*COL6A1* c.930 + 189C > T). The mutation creates a donor splice site that leads to the 72 nucleotide pseudoexon insertion. This mutation has since been found in 31 additional patients and was *de novo* in all the cases where parents were available (Cummings et al. 2017; Bolduc et al. 2019). Two factors contributed to these mutations being missed previously. Firstly, whole exome sequencing does not usually capture fragments that are located far from exon boundaries. Secondly, many directed collagen VI mutation screens PCR-amplify and sequence patient fibroblast mRNA and the pseudoexon is only rarely incorporated into mRNA in fibroblasts (~10–18% of total α1(VI) mRNA) meaning that it could easily be overlooked in Sanger sequencing profiles (Bolduc et al. 2019). By contrast, the pseudoexon was present in ~40–48% of total α1(VI) mRNA in muscle (Bolduc et al. 2019). It is likely that the mutant chains are incorporated into monomers, dimers and tetramers, but that the tetramers are abnormal, because end-to-end association into microfibrils is reduced. All patients with this mutation had mild disease in their early years but rapidly progressed in severity and lost ambulation on average by 9 years of age. This mutation appears to be the most common UCMD mutation and should be included in all patient mutation detection and screening protocols.

14.5 Pathogenic Mechanisms and Potential Targeted Therapies

14.5.1 Genetic Approaches

In theory, gene replacement strategies could be therapeutically useful in patients with homozygous or compound heterozygous premature stop codon mutations, potentially turning severe UCMD into the milder Bethlem myopathy. Dominant negative mutations, however, are much more common and pose a greater challenge because the mutant mRNA would need to be very efficiently targeted for breakdown to have therapeutic benefit. In addition, single base changes are very hard to target with current technologies. Antisense oligonucleotide splice modulation is the leading therapeutic strategy for Duchenne muscular dystrophy, and three oligonucleotides have been approved by the FDA for testing in Duchenne patients (Hanson et al. 2021). They are designed to induce skipping of one or more exons to restore the dystrophin reading frame and produce internally deleted but partially functional dystrophin. The *COL6A1* deep intronic mutation described above provides an exciting opportunity to develop similar exon skipping approaches using antisense oligonucleotides to prevent incorporation of the pseudoexon into the α1(VI) mRNA. The mutation is common, and if the pseudoexon is efficiently excluded, collagen VI production could be restored to almost normal levels. The utility of this approach was recently demonstrated using human fibroblast cultures and two antisense oligonucleotides were identified that reduced the mutant mRNA by more than 90% (Aguti et al. 2020). A second proof-of-principle antisense oligonucleotide therapy selectively targeted an 18 nucleotide heterozygous genomic deletion in *COL6A3* (Marrosu et al. 2017). The mutant mRNA was selectively degraded and more collagen VI was deposited into the extracellular matrix of patient fibroblasts. Unfortunately, mouse models with the *COL6A1* c.930 + 189C > T or its equivalent mouse version, or the *COL6A3* genomic deletion are not available for testing *in vivo* efficacy and safety, so translating this approach to clinical trials is some way off.

14.5.2 Targeting Mitochondrial Abnormalities, Apoptosis and Autophagy

Abnormal mitochondria – swollen, disorganised cristae and paracrystalline inclusions – are a feature of Bethlem myopathy and UCMD (Tagliavini et al. 2013; Zamurs et al. 2015). The *Col6a1*$^{-/-}$ mouse has similar mitochondrial abnormalities as well as dilated sarcoplasmic reticulum and an increase in apoptotic nuclei (Irwin et al. 2003). The discovery that dysregulation of the mitochondrial permeability transition pore, a mitochondrial inner membrane high-conductance channel, leads to mitochondrial dysfunction, and that the defective organelles are not adequately removed by autophagy has been the subject of a detailed review (Bernardi and Bonaldo 2013). The mitochondrial abnormality in the *Col6a1*$^{-/-}$ mouse can be rescued with cyclosporin A (Irwin et al. 2003), and two similar drugs that, unlike cyclosporin A, do not induce immunosuppression – Debio 025 (Tiepolo et al. 2009) and NIM811 (Zulian et al. 2014). A small trial treating UCMD patients with cyclosporine A for 1–3 years showed significant improvement in limb muscle strength in 5 out of 6 patients but respiratory function declined (Merlini et al. 2011). Cyclosporin A treatment corrected the mitochondrial abnormalities, increased muscle regeneration and decreased apoptosis in the trial patients (Merlini et al. 2011). These results raise hope that cyclosporin A or similar drugs could be useful but the improvements were relatively small and unlikely to significantly change the course of the disease.

Autophagy is an essential mechanism that degrades unneeded and damaged organelles and proteins and recycles essential nutrients (Mizushima and Komatsu 2011; Ohsumi 2014). Autophagy is defective in the *Col6a1*$^{-/-}$ mouse, and Bethlem and UCMD patients, and is unable to efficiently remove the damaged mitochondria (Grumati et al. 2010)(for a detailed review of autophagy and collagen VI disorders see (Castagnaro et al. 2020). Possible therapeutic approaches emerged with the finding that reactivating autophagy in the *Col6a1*$^{-/-}$ mouse with fasting, a low protein diet, rapamycin or cyclosporin A, was able to eliminate the abnormal mitochondria, and rescue the structural and functional muscle abnormalities (Grumati et al. 2010). Seven adults with collagen VI disorders were treated for one year with a low protein, normal calorie diet to see if autophagy could be reactivated (Castagnaro et al. 2016). Encouragingly, autophagy was reactivated, apoptosis reduced, the expected worsening of muscle function was counteracted, and some motor and respiratory parameters improved (Castagnaro et al. 2016). This clinical study supports the hypothesis that reactivating autophagy could be an effective therapeutic approach although a long term dietary approach is challenging and patient compliance difficult to maintain. Rapamycin also stimulates autophagy, and is an effective treatment for the

Col6a1$^{-/-}$ mouse (Grumati et al. 2010) but unfortunately it is not suitable for long term treatment in patients because it has potent immunosuppressive side effects (Madeo et al. 2014). Other potential therapies that stimulate autophagy in the *Col6a1*$^{-/-}$ mouse and are safe for human use include spermidine (Chrisam et al. 2015) and pterostilbene, a chemical similar to resveratrol but with higher bioavailability and stability (Metti et al. 2020).

14.6 Endotrophin: A Bioactive Collagen VI Peptide

Matrikines, protein fragments cleaved from extracellular matrix proteins and with biological activities unrelated to their parental protein, are being increasingly recognised as key regulators of important biological functions (Gaggar and Weathington 2016; Sivaraman and Shanthi 2018; Papadas et al. 2020). One such matrikine is endotrophin, a fragment cleaved from the C-terminus of the α3(VI) chain. While not a direct cause of the inherited collagen VI disorders, endotrophin has a profound influence on obesity, metabolic disease and cancer and patients with severely reduced or absent collagen VI will also have reduced endotrophin and the biological activities endotrophin possesses.

The α3(VI) C5 domain is required for collagen VI microfibril assembly (Lamandé et al. 2006) but is later cleaved off (Aigner et al. 2002; Heumüller et al. 2019). Until recently, the cleavage sites and the responsible proteases had not been identified. Fragments from 8–100 kDa that contain the C5 domain are present in human serum and tissue, and we now know that BMP-1 can cleave off just the C5 endotrophin domain, and a furin-like proprotein convertase releases a large C2-C5 fragment (Heumüller et al. 2019). This larger fragment is significantly more abundant than free endotrophin. Other enzymes, for example MMPs, could also be responsible for releasing fragments containing endotrophin (Li et al. 2020).

Collagen VI is highly expressed in a range of cancers and the role of endotrophin in promoting cancer has been recently reviewed (Lamandé and Bateman 2018; Wang and Pan 2020). Endotrophin promotes tumour growth, inflammation, angiogenesis and resistance to the chemotherapy drug cisplatin (Iyengar et al. 2005; Park et al. 2013). *Col6a1*$^{-/-}$ mice show dramatically reduced primary mammary tumour growth (Iyengar et al. 2005), and significantly, when wild type mice are treated with an endotrophin neutralising antibody as well as cisplatin-thiazolidinedione therapy mammary tumour growth is completely inhibited (Park et al. 2013). This suggests that targeting endotrophin could improve chemotherapy outcomes in patients (Bu et al. 2019). It also suggests that UCMD patients with homozygous or compound heterozygous premature stop codon mutations, effectively collagen VI knockouts, could be less susceptible to malignant tumour growth.

There is strong evidence from mouse studies that endotrophin also influences insulin sensitivity and diabetes. Transgenic mice that express endotrophin just in adipose tissue have increased inflammation and fibrosis and become insulin resistant (Sun et al. 2014). Treating wild type mice with an endotrophin neutralising antibody reverses the metabolic consequences of a high fat diet and improves insulin sensitivity (Sun et al. 2014).

14.7 Conclusions

Dominant negative mutations in the collagen VI triple helical regions, glycine substitution and exon skip mutations, are the most common cause of Bethlem myopathy and UCMD. These mutations have highlighted the protein features that are important for collagen VI assembly, and understanding the way mutations affect assembly has provided some insights into genotype phenotype relationships and the likely disease course for patients. A major deficiency in our understanding is how to judge whether sequence changes outside of the triple helix are deleterious or benign. Detailed structural and biochemical studies of these domains are shedding light but there is still a long way to go.

Dominant negative mutations are challenging to target therapeutically with the genetic strategies that are being developed for other muscular dystrophies; however, one common mutation that introduces a novel pseudoexon is an exception and an exon skipping oligonucleotide is able to dramatically reduce use of the pseudoexon *in vitro*. How effective this strategy will turn out to be in patients is still unclear because patient muscle will contain a background of mutant collagen. Therapeutically targeting the cellular consequences of collagen VI mutations is a major focus. Dysfunctional mitochondria and ineffective removal of the damaged organelles by autophagy can be targeted with cyclosporin A and autophagy stimulating drugs, respectively, and are promising approaches. There are still major gaps in our understanding about how mutant collagen VI signals to produce the downstream cellular consequences and this reinforces the importance of research that focuses on the basic biochemistry and cell biology of collagen VI.

References

Aguti S et al (2020) Exon-skipping oligonucleotides restore functional collagen VI by correcting a common COL6A1 mutation in Ullrich CMD. Mol Ther Nucleic Acids 21:205–216. https://doi.org/10.1016/j.omtn.2020.05.029

Aigner T et al (2002) The C5 domain of Col6A3 is cleaved off from the Col6 fibrils immediately after secretion. Biochem Biophys Res Commun 290(2):743–748. https://doi.org/10.1006/bbrc.2001.6227

Baker NL et al (2005) Dominant collagen VI mutations are a common cause of Ullrich congenital muscular dystrophy. Hum Mol Genet 14(2):279–293. https://doi.org/10.1093/hmg/ddi025

Baker NL et al (2007) Molecular consequences of dominant Bethlem myopathy collagen VI mutations. Ann Neurol 62(4):390–405. https://doi.org/10.1002/ana.21213

Ball S et al (2003) Structural basis of type VI collagen dimer formation. J Biol Chem 278(17):15326–15332. https://doi.org/10.1074/jbc.M209977200

Beecher N et al (2011) Collagen VI, conformation of A-domain arrays and microfibril architecture. J Biol Chem 286(46):40266–40275. https://doi.org/10.1074/jbc.M111.265595

Bernardi P, Bonaldo P (2013) Mitochondrial dysfunction and defective autophagy in the pathogenesis of collagen VI muscular dystrophies. Cold Spring Harb Perspect Biol 5(5):a011387. https://doi.org/10.1101/cshperspect.a011387

Bolduc V et al (2019) A recurrent COL6A1 pseudoexon insertion causes muscular dystrophy and is effectively targeted by splice-correction therapies. JCI Insight 4(6). https://doi.org/10.1172/jci.insight.124403

Bönnemann CG (2011a) The collagen VI-related myopathies: muscle meets its matrix. Nat Rev Neurol 7(7):379–390. https://doi.org/10.1038/nrneurol.2011.81

Bönnemann CG (2011b) The collagen VI-related myopathies Ullrich congenital muscular dystrophy and Bethlem myopathy. Handb Clin Neurol 101:81–96. https://doi.org/10.1016/B978-0-08-045031-5.00005-0

Briñas L et al (2010) Early onset collagen VI myopathies: genetic and clinical correlations. Ann Neurol 68(4):511–520. https://doi.org/10.1002/ana.22087

Bu D et al (2019) Human endotrophin as a driver of malignant tumor growth. JCI Insight 5. https://doi.org/10.1172/jci.insight.125094

Butterfield RJ et al (2013) Position of glycine substitutions in the triple helix of COL6A1, COL6A2, and COL6A3 is correlated with severity and mode of inheritance in collagen VI myopathies. Hum Mutat 34(11):1558–1567. https://doi.org/10.1002/humu.22429

Camacho Vanegas O et al (2001) Ullrich scleroatonic muscular dystrophy is caused by recessive mutations in collagen type VI. Proc Natl Acad Sci U S A 98(13):7516–7521. https://doi.org/10.1073/pnas.121027598

Castagnaro S et al (2016) Autophagy activation in COL6 myopathic patients by a low-protein-diet pilot trial. Autophagy 12(12):2484–2495. https://doi.org/10.1080/15548627.2016.1231279

Castagnaro S et al (2020) Autophagy in the mesh of collagen VI. Matrix Biol. https://doi.org/10.1016/j.matbio.2020.12.004

Cescon M et al (2015) Collagen VI at a glance. J Cell Sci 128(19):3525–3531. https://doi.org/10.1242/jcs.169748

Cescon M et al (2018) Collagen VI is required for the structural and functional integrity of the neuromuscular junction. Acta Neuropathol 136(3):483–499. https://doi.org/10.1007/s00401-018-1860-9

Chen P et al (2014) Collagen VI regulates peripheral nerve myelination and function. FASEB J 28(3):1145–1156. https://doi.org/10.1096/fj.13-239533

Chen P et al (2015) Collagen VI regulates peripheral nerve regeneration by modulating macrophage recruitment and polarization. Acta Neuropathol 129(1):97–113. https://doi.org/10.1007/s00401-014-1369-9

Chrisam M et al (2015) Reactivation of autophagy by spermidine ameliorates the myopathic defects of collagen VI-null mice. Autophagy 11(12):2142–2152. https://doi.org/10.1080/15548627.2015.1108508

Chu ML et al (1989) Sequence analysis of alpha 1(VI) and alpha 2(VI) chains of human type VI collagen reveals internal triplication of globular domains similar to the A domains of von Willebrand factor and two alpha

2(VI) chain variants that differ in the carboxy terminus. EMBO J 8(7):1939–1946
Cummings BB et al (2017) Improving genetic diagnosis in Mendelian disease with transcriptome sequencing. Sci Transl Med 9(386). https://doi.org/10.1126/scitranslmed.aal5209
Demir E et al (2002) Mutations in COL6A3 cause severe and mild phenotypes of Ullrich congenital muscular dystrophy. Am J Hum Genet 70(6):1446–1458. https://doi.org/10.1086/340608
Fang Y et al (2013) Nonsense-mediated mRNA decay of collagen -emerging complexity in RNA surveillance mechanisms. J Cell Sci 126(Pt 12):2551–2560. https://doi.org/10.1242/jcs.120220
Fitzgerald J et al (2008) Three novel collagen VI chains, alpha4(VI), alpha5(VI), and alpha6(VI). J Biol Chem 283(29):20170–20180. https://doi.org/10.1074/jbc.M710139200
Frischmeyer PA, Dietz HC (1999) Nonsense-mediated mRNA decay in health and disease. Hum Mol Genet 8(10):1893–1900. https://doi.org/10.1093/hmg/8.10.1893
Furthmayr H et al (1983) Electron-microscopical approach to a structural model of intima collagen. Biochem J 211(2):303–311. https://doi.org/10.1042/bj2110303
Gaggar A, Weathington N (2016) Bioactive extracellular matrix fragments in lung health and disease. J Clin Invest 126(9):3176–3184. https://doi.org/10.1172/JCI83147
Gara SK et al (2008) Three novel collagen VI chains with high homology to the alpha3 chain. J Biol Chem 283(16):10658–10670. https://doi.org/10.1074/jbc.M709540200
Giusti B et al (2005) Dominant and recessive COL6A1 mutations in Ullrich scleroatonic muscular dystrophy. Ann Neurol 58(3):400–410. https://doi.org/10.1002/ana.20586
Godwin ARF et al (2017) Defining the hierarchical organisation of collagen VI microfibrils at nanometre to micrometre length scales. Acta Biomater 52:21–32. https://doi.org/10.1016/j.actbio.2016.12.023
Gregorio I et al (2018) Collagen VI in healthy and diseased nervous system. Dis Model Mech 11(6). https://doi.org/10.1242/dmm.032946
Grumati P et al (2010) Autophagy is defective in collagen VI muscular dystrophies, and its reactivation rescues myofiber degeneration. Nat Med 16(11):1313–1320. https://doi.org/10.1038/nm.2247
Hanson B, Wood MJA, Roberts TC (2021) Molecular correction of Duchenne muscular dystrophy by splice modulation and gene editing. RNA Biol:1–15. https://doi.org/10.1080/15476286.2021.1874161
Heumüller SE et al (2019) C-terminal proteolysis of the collagen VI α3 chain by BMP-1 and proprotein convertase(s) releases endotrophin in fragments of different sizes. J Biol Chem 294(37):13769–13780. https://doi.org/10.1074/jbc.RA119.008641
Higuchi I et al (2001) Frameshift mutation in the collagen VI gene causes Ullrich's disease. Ann Neurol 50(2):261–265. https://doi.org/10.1002/ana.1120
Irwin WA et al (2003) Mitochondrial dysfunction and apoptosis in myopathic mice with collagen VI deficiency. Nat Genet 35(4):367–371. https://doi.org/10.1038/ng1270
Iyengar P et al (2005) Adipocyte-derived collagen VI affects early mammary tumor progression in vivo, demonstrating a critical interaction in the tumor/stroma microenvironment. J Clin Invest 115(5):1163–1176. https://doi.org/10.1172/JCI23424
Izu Y et al (2016) Collagens VI and XII form complexes mediating osteoblast interactions during osteogenesis. Cell Tissue Res 364:623–635. https://doi.org/10.1007/s00441-015-2345-y
Karczewski KJ et al (2020) The mutational constraint spectrum quantified from variation in 141,456 humans. Nature 581(7809):434–443. https://doi.org/10.1038/s41586-020-2308-7
Lamandé SR et al (1998) Reduced collagen VI causes Bethlem myopathy: a heterozygous COL6A1 nonsense mutation results in mRNA decay and functional haploinsufficiency. Hum Mol Genet 7(6):981–989. https://doi.org/10.1093/hmg/7.6.981
Lamandé SR et al (1999) Bethlem myopathy and engineered collagen VI triple helical deletions prevent intracellular multimer assembly and protein secretion. J Biol Chem 274(31):21817–21822. https://doi.org/10.1074/jbc.274.31.21817
Lamandé SR et al (2002) Kinked collagen VI tetramers and reduced microfibril formation as a result of Bethlem myopathy and introduced triple helical glycine mutations. J Biol Chem 277(3):1949–1956. https://doi.org/10.1074/jbc.M109932200
Lamandé SR et al (2006) The C5 domain of the collagen VI alpha3(VI) chain is critical for extracellular microfibril formation and is present in the extracellular matrix of cultured cells. J Biol Chem 281(24):16607–16614. https://doi.org/10.1074/jbc.M510192200
Lamandé SR, Bateman JF (2018) Collagen VI disorders: insights on form and function in the extracellular matrix and beyond. Matrix Biol 71–72:348–367. https://doi.org/10.1016/j.matbio.2017.12.008
Lampe AK et al (2005) Automated genomic sequence analysis of the three collagen VI genes: applications to Ullrich congenital muscular dystrophy and Bethlem myopathy. J Med Genet 42(2):108–120. https://doi.org/10.1136/jmg.2004.023754
Lampe AK, Bushby KMD (2005) Collagen VI related muscle disorders. J Med Genet 42(9):673–685. https://doi.org/10.1136/jmg.2002.002311
Li X et al (2020) Critical role of matrix metalloproteinase 14 in adipose tissue remodeling during obesity. Mol Cell Biol 40(8). https://doi.org/10.1128/MCB.00564-19
Madeo F et al (2014) Caloric restriction mimetics: towards a molecular definition. Nat Rev Drug Discov 13(10):727–740. https://doi.org/10.1038/nrd4391
Marrosu E et al (2017) Gapmer antisense oligonucleotides suppress the mutant allele of COL6A3 and restore functional protein in Ullrich muscular dystrophy. Mol Ther Nucleic Acids 8:416–427. https://doi.org/10.1016/j.omtn.2017.07.006

Marvulli D, Volpin D, Bressan GM (1996) Spatial and temporal changes of type VI collagen expression during mouse development. Dev Dynamics 206(4):447–454. https://doi.org/10.1002/(SICI)1097-0177(199608)206:4<447::AID-AJA10>3.0.CO;2-U

Mereness JA et al (2020) Collagen VI deficiency results in structural abnormalities in the mouse lung. Am J Pathol 190(2):426–441. https://doi.org/10.1016/j.ajpath.2019.10.014

Merlini L et al (2011) Cyclosporine a in Ullrich congenital muscular dystrophy: long-term results. Oxidative Med Cell Longev 2011:139194. https://doi.org/10.1155/2011/139194

Metti S et al (2020) The polyphenol Pterostilbene ameliorates the Myopathic phenotype of collagen VI deficient mice via autophagy induction. Frontiers in Cell and Developmental Biology 8:580933. https://doi.org/10.3389/fcell.2020.580933

Mizushima N, Komatsu M (2011) Autophagy: renovation of cells and tissues. Cell 147(4):728–741. https://doi.org/10.1016/j.cell.2011.10.026

Nasif S, Contu L, Mühlemann O (2018) Beyond quality control: the role of nonsense-mediated mRNA decay (NMD) in regulating gene expression. Semin Cell Dev Biol 75:78–87. https://doi.org/10.1016/j.semcdb.2017.08.053

Ohsumi Y (2014) Historical landmarks of autophagy research. Cell Res 24(1):9–23. https://doi.org/10.1038/cr.2013.169

Pace RA et al (2008) Collagen VI glycine mutations: perturbed assembly and a spectrum of clinical severity. Ann Neurol 64(3):294–303. https://doi.org/10.1002/ana.21439

Pan TC et al (1998) Missense mutation in a von Willebrand factor type a domain of the alpha 3(VI) collagen gene (COL6A3) in a family with Bethlem myopathy. Hum Mol Genet 7(5):807–812. https://doi.org/10.1093/hmg/7.5.807

Papadas A et al (2020) Versican and Versican-matrikines in Cancer progression, inflammation, and immunity. The Journal of Histochemistry and Cytochemistry: Official Journal of the Histochemistry Society 68(12):871–885. https://doi.org/10.1369/0022155420937098

Park J, Morley TS, Scherer PE (2013) Inhibition of endotrophin, a cleavage product of collagen VI, confers cisplatin sensitivity to tumours. EMBO Mol Med 5(6):935–948. https://doi.org/10.1002/emmm.201202006

Peat RA et al (2007) Variable penetrance of COL6A1 null mutations: implications for prenatal diagnosis and genetic counselling in Ullrich congenital muscular dystrophy families. Neuromuscular disorders: NMD 17(7):547–557. https://doi.org/10.1016/j.nmd.2007.03.017

Pham HT et al (2020) Collagen VIα2 chain deficiency causes trabecular bone loss by potentially promoting osteoclast differentiation through enhanced TNFα signaling. Sci Rep 10(1):13749. https://doi.org/10.1038/s41598-020-70730-7

Sasaki T et al (2000) A Bethlem myopathy Gly to Glu mutation in the von Willebrand factor a domain N2 of the collagen alpha3(VI) chain interferes with protein folding. FASEB journal: official publication of the Federation of American Societies for Experimental Biology 14(5):761–768. https://doi.org/10.1096/fasebj.14.5.761

Sivaraman K, Shanthi C (2018) Matrikines for therapeutic and biomedical applications. Life Sci 214:22–33. https://doi.org/10.1016/j.lfs.2018.10.056

Solomon-Degefa H et al (2020) Structure of a collagen VI α3 chain VWA domain array: adaptability and functional implications of myopathy causing mutations. J Biol Chem 295(36):12755–12771. https://doi.org/10.1074/jbc.RA120.014865

Sun K et al (2014) Endotrophin triggers adipose tissue fibrosis and metabolic dysfunction. Nat Commun 5:3485. https://doi.org/10.1038/ncomms4485

Tagliavini F et al (2013) 'Ultrastructural changes in muscle cells of patients with collagen VI-related myopathies', *muscles*. Ligaments and Tendons Journal 3(4):281–286

Tiepolo T et al (2009) The cyclophilin inhibitor Debio 025 normalizes mitochondrial function, muscle apoptosis and ultrastructural defects in Col6a1−/− myopathic mice. Br J Pharmacol 157(6):1045–1052. https://doi.org/10.1111/j.1476-5381.2009.00316.x

Timpl R, Chu M-L (1994) Microfibrillar collagen type VI. In: Yurchenco PD, Birk DE, Mecham RP (eds) Extracellular matrix assembly and structure. Academic Press (Biology of Extracellular Matrix), San Diego, pp 207–242. https://doi.org/10.1016/B978-0-12-775170-2.50012-3

Timpl R, Engel J (1987) Type VI Collagen. In: Mayne R, Burgeson RE (eds) Structure and function of collagen types. Academic, Orlando, pp 105–143. https://doi.org/10.1016/B978-0-12-481280-2.50008-8

Wang J, Pan W (2020) The biological role of the collagen Alpha-3 (VI) chain and its cleaved C5 domain fragment Endotrophin in Cancer. Onco Targets Ther 13:5779–5793. https://doi.org/10.2147/OTT.S256654

Whittaker CA, Hynes RO (2002) Distribution and evolution of von Willebrand/integrin a domains: widely dispersed domains with roles in cell adhesion and elsewhere. Mol Biol Cell 13(10):3369–3387. https://doi.org/10.1091/mbc.e02-05-0259

Zamurs LK et al (2015) Aberrant mitochondria in a Bethlem myopathy patient with a homozygous amino acid substitution that destabilizes the collagen VI α2(VI) chain. J Biol Chem 290(7):4272–4281. https://doi.org/10.1074/jbc.M114.632208

Zulian A et al (2014) NIM811, a cyclophilin inhibitor without immunosuppressive activity, is beneficial in collagen VI congenital muscular dystrophy models. Hum Mol Genet 23(20):5353–5363. https://doi.org/10.1093/hmg/ddu254

Connective Tissue Disorders in Domestic Animals

15

Jennifer Hope Roberts and Jaroslava Halper

Abstract

Though soft tissue disorders have been recognized and described to some detail in several types of domestic animals and small mammals for some years, they remain uncommon. Because of their low prevalence, not much progress has been made not only in improved diagnosis but also in our understanding of the biochemical basis and pathogenesis of these diseases in animals. Ehlers-Danlos syndrome (EDS) described in dogs already in 1943 and later in cats has only minor impact on the well-being of the dog as its effects on skin of these animals are rather limited. The involved skin is thin and hyperextensible with easily inflicted injuries resulting in hemorrhagic wounds and atrophic scars. Joint laxity and dislocation common in people are less frequently found in dogs. No systemic complications, such as organ rupture or cardiovascular problems which have devastating consequences in people have been described in cats and dogs. The diagnosis is based on clinical presentation and on light or electron microscopic features of disorganized and fragmented collagen fibrils. Several case of bovine and ovine dermatosparaxis analogous to human Ehlers-Danlos syndrome type VIIC were found to be caused by mutations in the procollagen I N-proteinase (pnPI) or *ADAMTS2* gene, though mutations in other sites are likely responsible for other types of dermatosparaxis. Cattle suffering from a form of Marfan syndrome (MFS) were described to have aortic dilatation and aneurysm together with ocular abnormalities and skeletal involvement. As in people, mutations at different sites of bovine *FBN1* may be responsible for Marfan phenotype. Hereditary equine regional dermal asthenia (HERDA), or hyperelastosis cutis, has been recognized in several horse breeds as affecting primarily skin, and, occasionally, tendons. A mutation in cyclophilin B, a chaperon involved in proper folding of collagens, has been identified in some cases. Warmblood fragile foal syndrome (WFFS) is another Ehlers-Danlos-like disorder in horses, affecting primarily Warmbloods who present with skin fragility and joint hyperextensibility. Degenerative suspensory ligament desmitis (DSLD) affects primarily tendons and ligaments of certain horse breeds. Data from our laboratory showed excessive accumulation of

J. H. Roberts
Department of Pathology, College of Veterinary Medicine, The University of Georgia, Athens, GA, USA
e-mail: Jennifer.Roberts@uga.edu

J. Halper (✉)
Department of Pathology, College of Veterinary Medicine, and Department of Basic Sciences, AU/UGA Medical Partnership, The University of Georgia, Athens, GA, USA
e-mail: jhalper@uga.edu

J. Halper (ed.), *Progress in Heritable Soft Connective Tissue Diseases*, Advances in Experimental Medicine and Biology 1348, https://doi.org/10.1007/978-3-030-80614-9_15

proteoglycans in organs with high content of connective tissues. We have identified increased presence of bone morphogenetic protein 2 (BMP2) in active foci of DSLD and an abnormal form of decorin in proteoglycan deposits. Our most recent data obtained from next generation sequencing showed disturbances in expression of genes for numerous proteoglycans and collagens.

Keywords

Ehlers-Danlos syndrome · Dogs · Cats · Dermatosparaxis · Cattle · Sheep · HERDA · Hyperelastosis cutis · WFFS · *PLOD1* · DSLD · Horses

Abbreviations

ADAMTS	a-Disintegrin-and-metallo-proteinase-with-thrombospondin-like-motifs
BMP	Bone morphogenetic protein
cbEGF-like	Calcium-binding epidermal growth factor-like
CypB	Cyclophilin B
DSLD	Degenerative suspensory ligament desmitis
EDS	Ehlers-Danlos syndrome
ESPA	Equine systemic proteoglycan accumulation
HERDA	Hereditary equine regional dermal asthenia
MFS	Marfan syndrome
PLOD1	Procollagen-lysine-2-oxoglutarate-5-dioxygenase 1
TGFβ	Transforming growth factor β
WFFS	Warmblood fragile foal syndrome

Soft tissue disorders have been recognized and described to some extent in several types of domestic animals and small mammals such as rabbit and cat (Minor 1980; Sinke et al. 1997; Weingart et al. 2014). Whether the relative paucity of cases in the literature is due to naturally low incidence of this group of diseases in domestic animals, or whether there is little incentive for studies in these animals is not clear. Though Ehlers-Danlos syndrome (EDS) was first described in dogs as cutaneous asthenia already in 1947 (Arlein 1947), and though it is recognized that dogs are very suitable to serve as models for numerous human hereditary diseases and cancer (Tsai et al. 2007; Switonski 2014) not many comprehensive studies have been done on connective tissue diseases in this species. Ehlers-Danlos syndrome affects preferentially certain breeds such as dachshunds, boxers, German shepherds and St. Bernards (Bellini et al. 2009; Scott et al. 2001) with a dominant pattern of inheritance (Hegreberg et al. 1969). More recently it has been reported in Doberman pinschers as well (Jaffey et al. 2019). Unlike in people, skin carries the brunt of the disease in dogs. It is thin and hyperextensible with easily inflicted injuries resulting in hemorrhagic wounds and atrophic scars (Bellini et al. 2009; Paciello et al. 2003; Barrera et al. 2004). Skin folds in the extremities, downy "chicken-like" skin, and "cigarette paper" scars are features found in dogs as well as in people suffering from Ehlers-Danlos syndrome, whereas joint laxity and dislocation common in people are less frequently found in dogs (Bellini et al. 2009; Paciello et al. 2003; Malfait et al. 2010). Organ rupture and/or dissecting aneurysm, two dramatic and catastrophic events in people, have not been identified in dogs and other species (Paciello et al. 2003; Malfait et al. 2010). However, a 2015 case study found spontaneous vascular rupture in a dog diagnosed with Ehlers-Danlos syndrome (Uri et al. 2015). Microscopic examination of the affected canine dermis usually shows fragmented and disorganized collagen fibers and fibrils and low number of poorly organized elastic fibers (Bellini et al. 2009; Paciello et al. 2003).

Cats are occasionally affected by cutaneous asthenia or Ehlers-Danlos as well (Patterson and Minor 1977; Sequeira et al. 1999). As in dogs their skin is hyperextensible and fragile because of decreased tensile strength. Dominant pattern of inheritance has been documented in some, but not all afflicted cats (Patterson and Minor 1977; Fitchie 1972; Butler 1975; Scott 1974). Electron microscopic examination reveals disorganized packing of collagen fibrils into fibers in the dermis, and the presence of fibrils with abnormally large or small diameters (Patterson and Minor 1977). The diagnosis of cutaneous asthenia or Ehlers-Danlos in dogs and cats is based on clinical presentation and electron microscopic features of collagen fibrils rather than on the detection of a specific, as of yet to be identified in most cases, underlying biochemical or genetic defect. It is likely that the diagnosis and prevalence of this disorder is underreported as EM examination would not be done on a routine basis.

The limited extent of the disease in cats makes it more similar to so called dermatosparaxis described in cattle (Hanset and Ansay 1967) and sheep though the mode of inheritance of the latter has been determined to be autosomal recessive (Fjølstad and Helle 1974; Van Halderen and Green 1988). Dermatosparaxis has been characterized to more detail in White Dorper and Dorper sheep where it has been recognized to be analogous to type VIIC Ehlers-Danlos syndrome in people, including defect in *ADAMTS2* (Vaatstra et al. 2011; Zhou et al. 2012; Malfait et al. 2017). The disorder is limited to skin which is quite fragile due to subcutaneous accumulation of gelatinous fluid (Vaatstra et al. 2011). Most recent data identified a premature stop codon in the *ADAMTS2* gene in these sheep (Zhou et al. 2012). The a-Disintegrin-and-metalloproteinase-with-thrombospondin-like-motifs 2 (ADAMTS2) is an extracellular metalloproteinase involved in removal of propeptides of type I-III procollagens which is necessary for proper assembly of collagen fibrils in the extracellular matrix (Wang et al. 2006; Tang 2001).

A case of bovine dermatosparaxis analogous to human EDS type VIIC and ovine dermatosparaxis was found to be caused also by mutations in the procollagen I N-proteinase (pnPI) or *ADAMTS2* gene (Colige et al. 1999; Holm et al. 2008). As a consequence, severe skin fragility is a prominent feature of both EDS type VIIC and bovine dermatosparaxis (Tang 2001; Colige et al. 1999). Dermatosparaxis affecting a herd of Drakensberger cattle in South Africa was characterized by only mild skin pathology suggesting that this cattle carried a mutation in a site different from the one involved described by Colige et al. (1999; Holm et al. 2008). A recent case study involving two siblings with dermatosparaxis had determined that there was a 17 bp deletion in the coding region of the *ADAMTS2* gene in two sibling Limousine calves. The clinical presentation of both calves was severe and resembled similar cases in other cattle species (Carty et al. 2016).

A form of Marfan syndrome was also recognized in cattle (Singleton et al. 2005; Hirano et al. 2012; Besser et al. 1990). The first report of a known case described aortic dilatation and aneurysm together with ocular abnormalities (microspherophakia, ectopia lentis, and lens opacities) and skeletal involvement characterized by long thin limbs, joint and tendon laxity, and postural kyphosis in several related calves (Besser et al. 1990). Disruption of aortic elastic fibers and decreased incorporation of fibrillin in the extracellular matrix were identified by light microscopy (Potter et al. 1993; Potter and Besser 1994). Similarly, Gigante et al. have found decrease in elastic fibers in all of the examined tissues (ligaments, periosteum, joint capsule and peripheral arteries) (Gigante et al. 1999). As in humans, mutations at different sites of bovine *FBN1* may be responsible for Marfan phenotype. For example, a mutation in one of the calcium-binding epidermal growth factor-like (cbEGF-like) domains of fibrillin-1 was responsible for the Marfan phenotype in cattle described by Singleton et al. (2005). Calcium binding by cbEGF-like domains contributes to proper conformation and stability of fibrillin-1 (Gigante et al. 1999; Downing et al. 1996; Reinhardt et al. 1997). The mutations in *FBN1* lead to decrease in levels of functional fibrillin-1, and thus to increased levels of active TGFβ (Gigante et al. 1999). Hirano et al. detected a mutation on the post-furin cleavage site sequence of the C-terminal domain of fibrillin-1 (Hirano et al. 2012).

Though standing somewhat apart from Ehlers-Danlos and Marfan syndromes, two other hereditary canine disorders are worth mentioning because of known underlying biochemical defects. Cutaneous mucinosis was described only recently in shar-pei dogs (Lopez et al. 1999) and is characterized by hyaluronic acid accumulation limited to the dermis (Zanna et al. 2008). This gives shar-pei dogs their characteristic creased appearance consisting of thick skin folds on their heads. Increased hyaluronan synthase-2 expression is considered the culprit of this disorder (Zanna et al. 2009; Docampo et al. 2011). The second disorder is primary lens luxation or isolated ectopia lentis afflicting mainly terrier breeds (Farias et al. 2010). A mutation in a splice donor recognition site in regional candidate gene *ADAMTS17* has been recognized as the cause by Farias et al. (2010) and Gould et al. (2011). This has been confirmed in later papers which observed this phenomenon in several different canine breeds, the majority of which were terriers (Gharahkhani et al. 2015). Ectopia lentis is a recognized component of Marfan and Weill-Marchesani syndromes. Interestingly, Morales et al. have described truncating mutations in human *ADAMTS17* gene in a form of Weill-Marchesani syndrome in people with ocular abnormalities and short stature (Morales et al. 2009) which appears to be very similar to association of glaucoma and short stature in several canine breeds carrying mutation in *ADAMTS17* gene (Jeanes et al. 2019). Mutations in one or more other *ADAMTS* genes causes Ehlers-Danlos syndrome VIIC in people and dermatosparaxis in cattle and sheep (see above) (Zhou et al. 2012; Colige et al. 1999), and possibly a form of Ehlers-Danlos in horses as well (see below) (Plaas et al. 2011).

Hereditary equine regional dermal asthenia (HERDA), also known under the name of hyperelastosis cutis, was first described in Quarter Horses (and later in other breeds) as a disease not dissimilar to cutaneous asthenia of dogs and Ehlers-Danlos in people: it affects primarily skin which is hyperelastic, fragile and thin with atrophic scars (Hardy et al. 1988; Gunson et al. 1984; White et al. 2004; Borges et al. 2005; Brounts et al. 2001). Though the lesions occur mostly on the dorsum, legs are often involved as well (White et al. 2004). Manifestations of the disease usually appear around 2 years of age when horses begin training, and saddling leads to skin trauma (Grady et al. 2009). Gunson et al. noted fragmentation and disorganization of collagen fibers both by light and electron microscopy, and suggested that a non-inflammatory degradation and phagocytosis of collagen are relevant in the pathogenesis of HERDA (Gunson et al. 1984). Subsequent investigators have noted changes in the skin of horses with HERDA which confirmed the findings of Gunson: thin dermis with sparse, fragmented collagen in affected horses by light microscopy and variation in diameter of collagen fibrils by electron microscopy, fragile skin associated with poor wound healing (Hardy et al. 1988; White et al. 2004; Borges et al. 2005). Moreover, biomechanical analysis revealed lower tensile strength and modulus of elasticity of affected skin (Grady et al. 2009). Ishikawa et al. have described the presence of varying diameters of collagen fibrils in HERDA tendons with high proportions of very small fibrils (Ishikawa et al. 2012). Similarly, faulty arrangement of collagen fibers leads to thinning and ulceration of the corneas of affected horses (Mochal et al. 2010). It has also been proposed that UV exposure can increase collagenase degradation of faulty collagen in affected horses. One study found that areas of skin exposed to sunlight were more susceptible to enzymatic degradation (Rashmir-Raven et al. 2015). Whether HERDA is an autosomal dominant (Hardy et al. 1988) or, more likely, an autosomal recessive disorder (White et al. 2004; Borges et al. 2005; Grady et al. 2009) is not quite clear. A homozygous missense mutation in the sixth residue of mature cyclophilin B protein was found in HERDA horses (Ishikawa et al. 2012; Tryon et al. 2007). Cyclophilin B (CypB or PPIB), a member of the peptidyl-propyl isomerase, acts as a chaperon involved in proper folding of collagens and in several other functions directed at modulating extracellular matrix.

The CypB mutation leads to substitution of glycine to arginine. The consequence is delayed folding of type I procollagen in the endoplasmic reticulum of HERDA horses, and a decrease in hydroxylysine and glucosyl-galactosyl hydroxylysine content of nascent collagen (Ishikawa et al. 2012). The ratio of pyridinoline to deoxypyridinoline is increased in HERDA horses and may serve as a marker for HERDA (Swiderski et al. 2006). However, it is clear from the literature that not all HERDA horses have a mutation in the cyclophilin B gene (Rüfenacht et al. 2010).

Warmblood fragile foal syndrome (WFFS) is another equine disorder bearing similarity to Ehlers-Danlos syndrome. It is an autosomal recessive disorder caused usually by a nucleotide variant in position 678 in the procollagen-lysine-2-oxoglutarate-5-dioxygenase 1 (*PLOD1*) gene where Gly is replaced by Arg (Dias et al. 2019). So far only Warmblood horses were found to be homozygous, though the WFFS allele has been identified in Thoroughbreds as well (Reiter et al. 2020). Several Warmblood breeds have fairly high carrier frequency of 17%. Vast majority of homozygous equines die *in utero*, the few documented foals born alive died not long after birth due to severe skin abnormalities, such as fragility and hyperextensibility leading to large open wounds, and hyperextensibility of digital joints (Reiter et al. 2020).

Degenerative suspensory ligament desmitis (DSLD) is a chronic, debilitating disease affecting primarily Peruvian Pasos and Peruvian Paso crosses (Mero and Pool 2002a; Mero and Scarlett 2005a). However, many other breeds, including Quarter horses are afflicted as well (Halper et al. 2006, 2011). It tends to run in families though no clear pattern of heritability has been established. Likewise, the pathogenesis and biochemical defects are not well understood. DSLD is characterized by an insidious onset of bilateral or quadrilateral lameness without a history of trauma or performance related injury. Typically, horses present with one or more dropped fetlocks (fetlock is a metacarpophalangeal joint between the cannon bone and the pastern (Fig. 15.1b). Fetlock effusion, static and dynamic hyperextension and degenerative joint disease are hallmarks on physical examination. Ultrasonography of affected suspensory ligaments reveals diffuse loss of echogenicity, and an irregular fiber pattern (Mero and Scarlett 2005b; Dyson et al. 1995; Dyson 2007; Gibson and Steel 2002). Treatment for DSLD-affected horses is empirical and directed at minimizing musculoskeletal pain and providing support exercise for the suspensory apparatus (Xie et al. 2011). Though originally DSLD was considered a collagen disorder strictly limited to suspensory ligaments (Mero and Pool 2002b), our data show that it is a systemic disease involving tissues and organs with high content of connective tissue components. Abnormal accumulations of proteoglycans were identified not only in the suspensory ligaments, but also in the superficial and deep digital flexor tendons, patellar and nuchal ligaments, aorta, coronary arteries and sclerae of DSLD-affected horses, and to minor extent in the skin (Halper et al. 2006; Haythorn et al. 2020). In mild or incipient cases proteoglycans are present only intracellularly, surrounding nuclei of tenocytes (Fig. 15.2b). However, with the progression of the disease this material is deposited extracellularly (Fig. 15.2c), and, with time, it displaces collagen fibers (in tendons) and elastic fibers (ligaments), and in many cases it is interspaced with newly formed cartilage foci (Fig. 15.2d). In light of these observations, a more appropriate term for this disease process may be equine systemic proteoglycan accumulation (ESPA) (Halper et al. 2006).

The occasional presence of whirls of active exuberant fibroblasts which contain very little collagen in tendons from DSLD-affected horses represents most likely an early stage of the disease. With time these foci eventually progress to a less cellular (and finally acellular) phase characterized by increasing proteoglycan content. Typically, no inflammatory or fibrotic changes accompany deposits of proteoglycans or proliferative lesions at any stage, early or late (Halper et al. 2006). We hypothesize that the proliferating fibroblasts (tenoblasts) secrete proteoglycans which, as the disease progresses, then accumulate in tissues. The stimulus for the proliferation of fibroblasts and the subsequent production of proteoglycans is unknown. Miller and Juzwiak

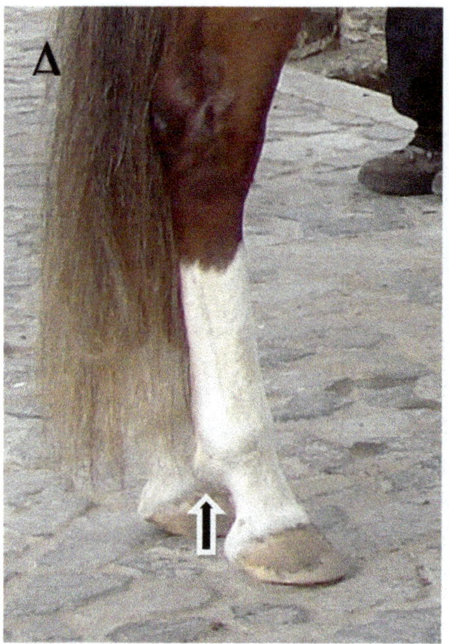

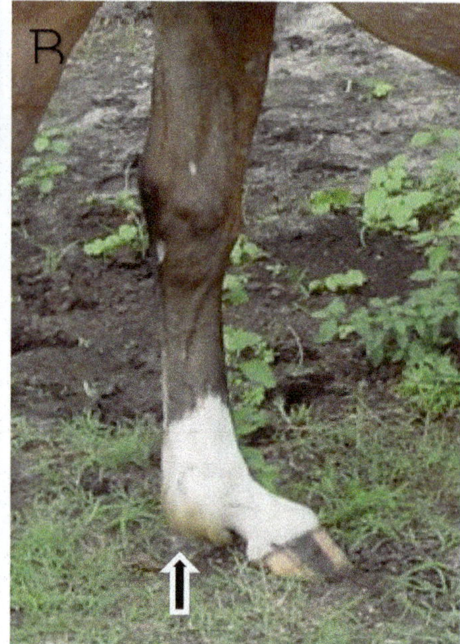

Fig. 15.1 Fetlock of healthy and DSLD-affected horses. (**a**) The photo shows a fetlock (metacarpophalangeal joint, ↑) of a back leg of a healthy Peruvian Paso horse. (**b**) This is in sharp contrast with a fetlock of a back leg of a DSLD-affected horse, note that the joint has dropped almost to the ground (↑). This is due to the weakness of the suspensory ligaments

have described a case of a 3-month-old foal suffering from acute rupture of suspensory ligaments in the hindlimbs. On necropsy they found multifocal fibrovascular proliferation with only minimal inflammation and proteoglycan accumulation (Miller and Juzwiak 2010). This would be similar to the proliferative lesions described above. More recently we have identified the presence of bone morphogenetic protein 2 (BMP2) in the cytoplasm of these proliferating fibroblasts, but not in normal tissue or in areas with accumulating proteoglycans or cartilage metaplasia (Fig. 15.3) (Young et al. 2018).

Other investigators have also observed an increased accumulation of proteoglycans in suspensory ligaments and superficial and deep digital flexor tendons in DSLD, but not in other organs (Plaas et al. 2011; Schenkman et al. 2009), or they found lesions similar to those seen in DSLD but present only in few tissues or organs. For example, arterial medial calcification with elastin fiber disorganization in multiple arteries was described in a horse as an isolated phenomenon limited to the blood vessels (Fales-Williams et al. 2008). However, it is not clear whether any tendons or ligaments were examined as well. In our experience calcifications within the blood vessel walls were rare in DSLD, though whether affected horses have propensity for sudden death due to coronary artery involvement or rupture of the aorta is a matter of some controversy (Halper et al. 2006). Whether there is a relationship between DSLD and isolated aortic rupture syndrome occurring with higher frequency in Friesian horses remains to be seen (Holmes et al. 1973; van der Linde-Sipman et al. 1985; Ploeg et al. 2013), as these horses do not exhibit other problems. The presence of periaortic hemorrhages, media necrosis and fibrosis of the media and fibrosis of the adventitia of the aorta and pulmonary trunk at least in some horses suggests the possibility of a soft tissue disorder akin to Marfan or Ehlers-Danlos (van der Linde-Sipman et al. 1985).

Hui et al. have described a horse with calcifying ligament desmopathy resembling DSLD (Hui et al. 2018), including the presence of cartilaginous metaplasia. However, extensive calcifications described in their cases are considered to be

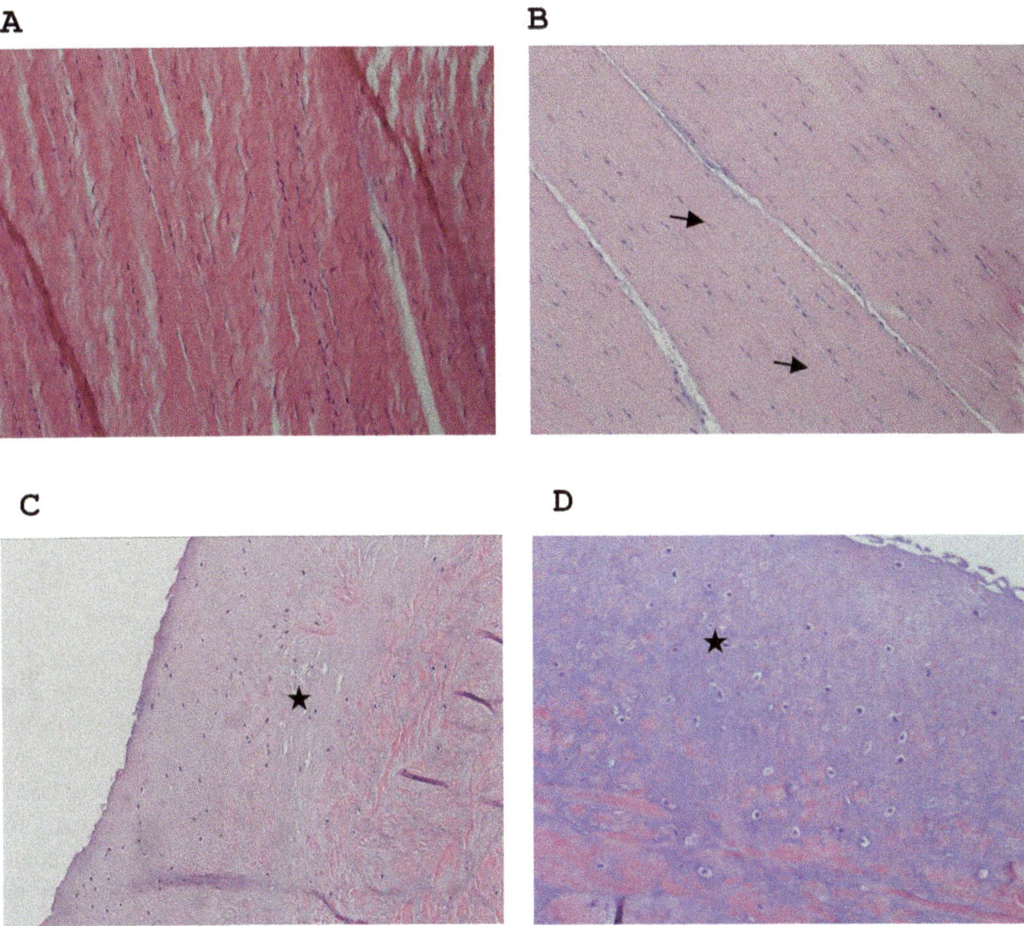

Fig. 15.2 Proteoglycan presence in DSLD tendons. (**a**) Control, healthy tendon stained with hematoxylin and eosin, original magnification × 200. (**b**) DSLD tissue, lightly infiltrated with proteoglycans (→), stained with hematoxylin and eosin, original magnification × 200. (**c**) DSLD tissue with heavy accumulation of proteoglycans (*) stained with hematoxylin and eosin. (**d**) Incipient cartilage formation in tendon heavily infiltrated with proteoglycans (*). Original magnification × 100. (Portions of this figure originally appeared in Kim et al. 2010)

an unusual, if any, feature of DSLD (Halper and Mueller 2018).

Several aspects make the presence of cartilaginous metaplasia in the sclera of sheep as described by Smith et al. different from scleral involvement in horses with DSLD: no involvement of other tissues and organs, and high incidence of scrapies in these sheep (Smith et al. 2011). In addition, we have not seen real cartilaginous metaplasia in any of the examined sclera in our cases so far.

The pathogenesis and underlying biochemical/genetic defect of DSLD are under investigation. Our data point to the presence of an abnormal form of decorin with altered biological activity in these proteoglycan deposits. The abnormal form of decorin is the result of increased sulfation of at 6-position of N-acetylgalactosamine, and an increased ratio of glucuronic to iduronic acid in DSLD-affected tissues. As a consequence, chondroitin sulfate replaces dermatan sulfate at least partially during synthesis of the modified decorin (Kim et al. 2010).

Plaas et al. found abundant aggrecan content in proteoglycan accumulations of suspensory ligaments of DSLD-affected horses. This aggrecan appears to be of two types: one is a high molecular weight aggrecan identical to the form

A

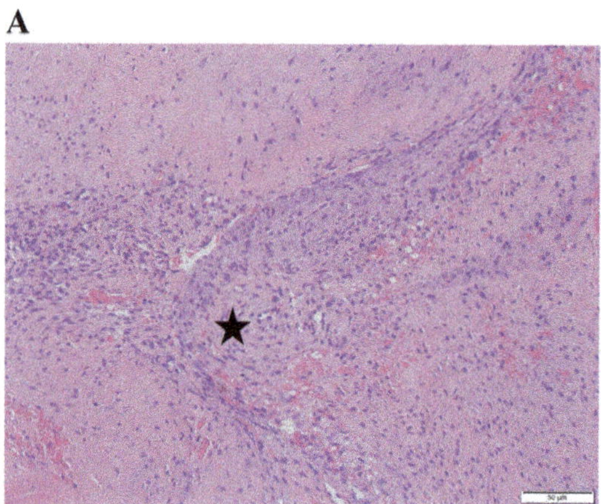

B

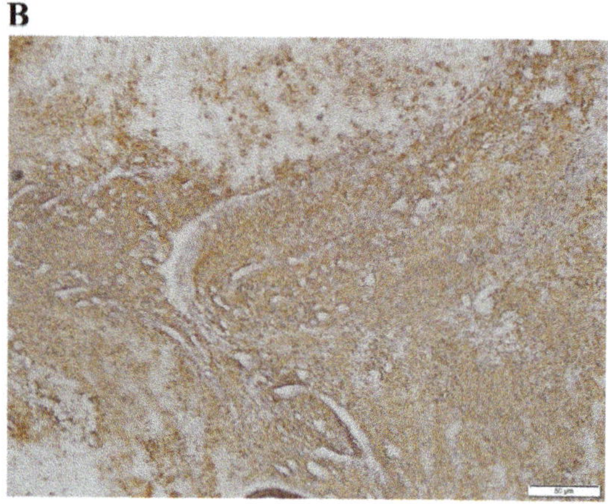

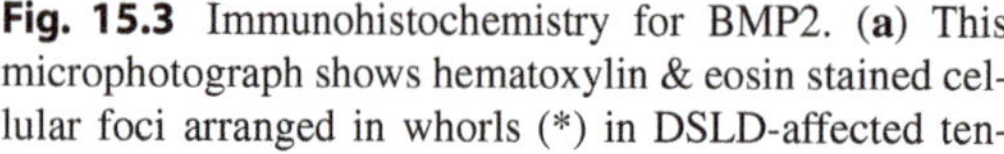

Fig. 15.3 Immunohistochemistry for BMP2. (**a**) This microphotograph shows hematoxylin & eosin stained cellular foci arranged in whorls (*) in DSLD-affected tendons. (**b**) The cells in these whorls immunostained strongly for BMP2. (This figure originally appeared in Young et al. 2018)

produced by equine mesenchymal cells and similar to the form synthesized by articular cartilage. This high molecular species of aggrecan might be a substrate for ADAMTS5. ADAMTS5 would then generate aggrecan fragments constituting the second type of aggrecan found in DSLD. The authors found elevated levels of ADAMTS5 and ADAMTS4 in these ligaments. Their findings bear similarity to findings in ADAMTS5 knockout $TS5^{-/-}$ mice (Plaas et al. 2011). $TS5^{-/-}$ mice have problems in skin wound healing (Velasco et al. 2011) and develop biomechanically compromised tendons due to aggrecan accumulation in the pericellular matrix of fibroblasts. In turn, as Wang et al. point out, the aggrecan accumulation may lead to tendon dysfunction through impairment of TGFβ signaling ALK5/SMAD3 pathway (Wang et al. 2012). ADAMTS5 cleavage of aggrecan occurs as one of numerous responses to TGFβ1stimulation of formation of connective tissue and fibrosis during physiological and pathological reparative processes, including perhaps even DSLD (Plaas et al. 2011; Velasco et al. 2011). The absence of ADAMTS5 (for example in $TS5^{-/-}$ mice) leads to assembly of pericellular CD-44 bound hyaluronan-aggrecan complexes into larger cell aggregates which disrupt repair of healing tissues such as dermis and tendons (Velasco et al. 2011; Wang et al. 2012). This assembly is mediated by increased SMAD1/5/8 signaling as the absence of ADAMTS5 also disrupts TGFβ1 signaling pathways (Velasco et al. 2011). Our most recent data suggest a profound disturbance of proteoglycan and collagen metabolism (Haythorn et al. 2020). Unexpectedly, next generation sequencing of skin RNA from DSLD (and control) horses revealed decreased gene expression for protein cores of many proteoglycans (including aggrecan), most collagens and some growth factors. Whether the increased expression genes for hyaluronan synthase 2 and cell-migration inducing hyaluronidase indicate that it is hyaluronan constituting the bulk of the proteoglycan or rather glycosaminoglycan accumulation in DSLD tissues (Haythorn et al. 2020). The increased expression of *BMP2* gene confirms our previous finding of elevated BMP2 presence in active DSLD foci (Young et al. 2018). The overexpression of keratin genes was somewhat surprising but it would explain the skin changes reported by owners of affected horses (Haythorn et al. 2020).

Naturally occurring diseases in domestic animals have a potential to serve as good models for analogous human diseases, and at times they might be better suited for this task than the commonly used rodent models. However, unless the pathogenesis and biochemistry of the syndromes and disorders are better defined and described for each species the usefulness of serving of these naturally occurring disorders as models for human disease will be limited. This is the case particularly relevant to dogs as a half out of the 450 naturally occurring hereditary canine diseases has the potential to serve as a model for human disease (Gunson et al. 1984).

References

Arlein MS (1947) Generalized acute cutanous asthenia in a dog. J Am Vet Med Assoc 111:57

Barrera R, Mañe C, Duran E, Vives MA, Zaragoza C (2004) Ehlers-Danlos in a dog. Can Vet J 45:355–356

Bellini MH, Caldini ETEG, Scapinelli MP, Simões MJ, Machado DB, Nürmberg R (2009) Increased elastic microfibrils and thickening of fibroblastic nuclear lamina in canine cutaneous asthenia. Vet Dermatol 20:139–143

Besser TE, Potter KA, Bryan GM, Knowlen GG (1990) An animal model of Marfan syndrome. Am J Med Genet 29:581–594

Borges AS, Conceição LG, Alves ALG, Fabris VE, Pessoa MA (2005) Hereditary equine regional dermal asthenia in three related quarter horses in Brazil. Vet Dermatol 16:125–130

Brounts SH, Rashmir-Raven AM, Black SS (2001) Zonal dermal separation: a distinctive histopathological lesion associated with hyperelastosis cutis in a quarter horse. Vet Dermatol 12:219–224

Butler WF (1975) Fragility of the skin in a cat. Res Vet Sci 19:213

Carty CI, Lee AM, Wienandt NA, Stevens EL, Alves DA, Browne JA et al (2016) Dermatosparaxis in two Limousin calves. Ir Vet J 69:15

Colige A, Sieron AL, Li SW, Schwarze U, Petty E, Wertelecki W, Wilcox W, Krakow D, Cohn DH, Reardon W, Byers PH, Lapière CM, Prockop DJ, Nusgens BV (1999) Human Ehlers-Danlos syndrome type VII C and bovine dermatosparaxis are caused by mutations in the procollagen I N-proteinase gene. Am J Hum Genet 65:308–317

Dias NM, de Andrade DGA, Teixeira-Neto AR, Trinque CM, Oliveira-Filho JP, Winand NJ et al (2019) Warmblood fragile foal syndrome causative single nucleotide polymorphism frequency in warmblood horses in Brazil. Vet J 248:101–102

Docampo MJ, Zanna G, Fondevila D, Cabrera J, López-Iglesias C, Carvalho A et al (2011) Increased HAS2-driven hyaluronic acid synthesis in Shar-Pei dogs with hereditary cutaneous hyaluronosis (mucinosis). Vet Dermatol 22:535–545

Downing AK, Knott V, Werner JM, Cardy CM, Campbell ID, Handford PA (1996) Solution structure of a pair of calcium-binding epidermal growth factor-like domains: implications for the Marfan syndrome and other genetic disorders. Cell 85:597–605

Dyson S (2007) Diagnosis and management of common suspensory lesions in the forelimbs and hindlimbs of sport horses. Clin Tech Equine Pract 6:179–188

Dyson SJ, Arthur RM, Palmer SE, Richardson D (1995) Suspensory ligament desmitis. Vet Clin North Am 11:172–215

Fales-Williams A, Sponseller B, Flaherty H (2008) Idiopathic arterial medial calcification of the thoracic arteries in an adult horse. J Vet Diagn Investig 20:692–697

Farias GHG, Johnson GS, Taylor JF, Giuliano E, Katz ML, Sanders DN, Schnabel RD, McKay SD, Khan S, Gharahkani P, O'Leary CA, Pettitt L, Forman OP, Boursnell M, McLaughlin B, Ahonen S, Lohi H, Hernandez-Merino E, Gould DJ, Sargan DR, Mellersh C (2010) An ADAMTS17 splice donor siten mutation in dogs with primary lens luxation. Investig Ophthalmol 51:4716–4721

Fitchie P (1972) The Ehlers-Danlos syndrome (in a cat). Vet Res 90:165

Fjølstad M, Helle O (1974) A hereditary dysplasia of collagen tissues in sheep. J Pathol 112:183–188

Gharahkhani P, O'Leary CA, Duffy DL, Kyaw-Tanner M (2015) Potential modifying loci associated with primary Lens luxation, pedal hyperkeratosis, and ocular phenotypes in miniature bull terriers. Invest Ophthalmol Vis Sci 56(13):8288–8296

Gibson KT, Steel CM (2002) Conditions of the suspensory ligament causing lameness in horses. Equine Vet Educ 4:50–64

Gigante A, Chillemi C, Potter KA, Bertoni-Freddari C, Greco F (1999) Elastic fibers of musculoskeletal tissues in bovine Marfan syndrome: a morphometric study. J Orthop Res 17:624–628

Gould D, Pettitt L, McLaughlin B, Holmes N, Forman O, Thomas A et al (2011) ADAMTS17 mutation associated with primary lens luxation is widespread among breeds. Vet Ophthalmol 14:378–384

Grady JJ, Elder SH, Ryan PL, Swiderski CE, Rashmir-Raven AM (2009) Biomechanical and molecular characteristics of hereditary equine regional dermal asthenia in quarter horses. Vet Dermatol 20:591–599

Gunson DE, Halliwell RE, Minor RR (1984) Dermal collagen degradation and phagocytosis. Occurrence in a horse with hyperextensible fragile skin. Arch Dermatol 120:599–604

Halper J, Mueller POE (2018) Dystrophic mineralization in degenerative suspensory ligament desmitis. Equine Vet Educ 30:424–426

Halper J, Kim B, Khan A, Yoon JH, Mueller POE (2006) Degenerative suspensory ligament desmitis as a systemic disorder characterized by proteoglycan accumulation. BMC Vet Res 2:12

Halper J, Khan A, Mueller POE (2011) Degenerative suspensory ligament desmitis – a new reality. Pak Vet J 31:1–8

Hanset R, Ansay M (1967) Dermatosparaxie (peau dechiree) ceha la veau un degaut general de tissu conjonctef, de nature hereditaire. Ann Med Vet 11:451

Hardy MH, Fisher KR, Vrablic OE, Yager JA, Nimmo-Wilkie JS, Parker W, Keeley FW (1988) An inherited connective tissue disease in the horse. Lab Investig 59:253–262
Haythorn A, Young M, Stanton J, Zhang J, Mueller POE, Halper J (2020) Differential gene expression in skin RNA of horses affected with degenerative suspensory ligament desmitis. J Orthop Surg Res 15(1):460
Hegreberg GA, Padgett GA, Gorham JA, Henson JB (1969) A connective tissue disease of dogs and mink resembling the Ehlers Danlos syndrome of man II. Mode of inheritance. J Hered 60:249–254
Hirano T, Matsuhashi T, Kobayashi N, Watanabe T, Sugimoto Y (2012) Identification of an FBN1 mutation in bovine Marfan syndrome-like disease. Anim Genet 43:11–17
Holm DE, van Wilpe E, Harper CK, Duncan NM (2008) The occurrence of dermatosparaxis in a commercial Drakensberger cattle herd in South Africa. J S Afr Vet Assoc 79:19–24
Holmes JR, Rezakhani A, Else RW (1973) Rupture of a dissecting aortic aneurysm into the left pulmonary artery in a horse. Equine Vet J 5:65–70
Hui SKY, Turner SJ, Leaman TR, de Brot S, Barakzai SZ (2018) Quadrilateral suspensory and straight sesamoidean ligament calcifying desmopathy in an Arabian mare. Equine Vet Educ 30:419–423
Ishikawa Y, Vranka JA, Boudko SP, Pokidysheva E, Mizuno K, Zientek K, Keene DR, Rashmir-Raven AM, Nagata K, Winand NJ, Bächinger HP (2012) The mutation in cyclophilin B that causes hyperelastosis cutis in the American quarter horse does not affect peptidyl-prolyl cis-trans isomerase activity, but shows altered cyclophilin B-protein interactions and affects collagen folding. J Biol Chem 287:22253–22265
Jaffey JA, Bullock G, Teplin E, Guo J, Villani NA, Mhlanga-Mutangadura T et al (2019) A homozygous ADAMTS2 nonsense mutation in a Doberman Pinscher dog with Ehlers Danlos syndrome and extreme skin fragility. Anim Genet 50:543–545
Jeanes EC, Oliver JAC, Ricketts SL, Gould DJ, Mellersh CS (2019) Glaucoma-causing ADAMTS17 mutations are also reproducibly associated with height in two domestic dog breeds: selection for short stature may have contributed to increased prevalence of glaucoma. Canine Genet Epidemiol 6:5
Kim B, Yoon JH, Zhang J, Eric Mueller PO, Halper J (2010) Glycan profiling of a defect in decorin glycosylation in equine systemic proteoglycan accumulation, a potential model of progeroid form of Ehlers-Danlos syndrome. Arch Biochem Biophys 501(2):221–231
Lopez A, Spracklin D, McConkey S, Hanna P (1999) Cutaneous mucinosis and mastocytosis in a shar-pei. Can Vet J 40:881–883
Malfait F, Wenstrup RJ, De Paepe A (2010) Clinical and genetic aspects of Ehlers-Danlos syndrome, classic type. Genet Med 12:597–605
Malfait F, Francomano C, Byers P, Belmont J, Berglund B, Black J et al (2017) The 2017 international classification of the Ehlers Danlos syndromes. Am J Med Genet 175:8–26
Mero JL, Pool, R (ed) (2002a) Twenty cases of degenerative suspensory ligament desmitis in Peruvian Paso horses. In: Proceedings, Annual Convention of AAEP ; Orlando, FL
Mero JL, Pool R (2002b) Twenty cases of degenerative suspensory ligament desmitis in Peruvian Paso horses. Abstract for AAEP, Orlando 48:329–334
Mero J, Scarlett J (2005a) Diagnostic criteria for degenerative suspensory ligament desmitis in Peruvian Paso horses. J Equine Vet Sci 25:224–228
Mero J, Scarlett J (2005b) Diagnostic criteria for degenerative suspensory ligament desmitis in Peruvian Paso horses. J Equine Vet Sci 25:224–228
Miller KD, Juzwiak JS (2010) Bilateral degenerative suspensory desmitis with acute rupture in a Standardbred colt. Equine Vet Educ 22:267–270
Minor RR (1980) Collagen metabolism. Am J Pathol 98:226–280
Mochal CA, Miller WW, Cooley AJ, Linford RL, Ryan PL, Rashmir-Raven AM (2010) Ocular findings in quarter horses with hereditary equine regional dermal asthenia. J Am Vet Med Assoc 237(3):304–310
Morales J, Al-Sharif L, Khalil DS, Shinwari JM, Bavi P, Al-Mahrouqi RA, Al-Rahji A, Alkuraya FS, Meyer BF, Al Tassan N (2009) Homozygous mutations in ADAMTS10 and ADAMTS17 cause lenticular myopia, ectopia lentis, glaucoma, spherophakia, and short stature. Am J Hum Genet 85:558–568
Paciello O, Lamagna F, Lamagna B, Papparella S (2003) Ehlers-Danlos-like syndrome in 2 dogs: clinical, histologic, and ultrastructural findings. Vet Clin Pathol 32:13–18
Patterson DF, Minor RR (1977) Hereditary fragility and hyperextensibility of the skin of cats. Lab Investig 37:170–179
Plaas A, Sandy JD, Liu H, Diaz MA, Schenkman D, Magnus RP, Bolam-Bretl C, Kopesky PW, Wang VM, Galante JO (2011) Biochemical identification and immunolocalization of aggrecan, ADAMTS5 and inter-alpha-trypsin-inhibitor in equine degenerative suspensory ligament desmitis. J Orthop Res 29:900–906
Ploeg M, Saey V, de Bruijn CM, Gröne A, Chiers K, van Loon G, Ducatelle R, van Weeren PR, Back W, Delesalle C (2013) Aortic rupture and aorto-pulmonary fistulation in the Friesian horse: characterisation of the clinical and gross post mortem findings in 24 cases. Equine Vet J 45:101–106
Potter KA, Besser TE (1994) Cardiovascular lesions in bovine Marfan syndrome. Vet Pathol 31:501–509
Potter KA, Hoffman Y, Sakai LY, Byers PH, Besser TE, Milewicz DM (1993) Abnormal fibrillin metabolism in bovine Marfan syndrome. Am J Pathol 142:803–810
Rashmir-Raven A, Lavagnino M, Sedlak A, Gardner K, Arnoczky S (2015) Increased susceptibility of skin from HERDA (hereditary equine regional dermal asthenia)-affected horses to bacterial collagenase degradation: a potential contributing factor to the clinical signs of HERDA. Vet Dermatol 26:476–e111

Reinhardt DP, Ono RN, Sakai LY (1997) Calcium stabilizes fibrillin-1 against proteolytic degradation. J Biol Chem 272:1231–1236
Reiter S, Wallner B, Brem G, Haring E, Hoelzle L, Stefaniuk-Szmukier M et al (2020) Distribution of the warmblood fragile foal syndrome type 1 mutation (PLOD1 c.2032G>a) in different horse breeds from Europe and the United States. Genes (Basel) 11(12)
Rüfenacht S, Straub R, Steinmann B, Winand N, Bidaut A, Stoffel MH, Gerber V, Wyder M, Müller E, Roosje PJ (2010) Swiss warmblood horse with symptoms of hereditary equine regiomal asthena without mutation in the cyclophylin B gene (PPIB). Schweiz Arch Tierheilkd 152:188–192
Schenkman D, Armien A, Pool R, Williams JM, Schultz RD, Galante JO (2009) Systemic proteoglycan deposition is not a characteristic of equine degenerative suspensory ligament desmitis (DSLD). J Equine Vet Sci 29:748–752
Scott DV (1974) Cutaneous asthenia in a cat resembling Ehler-Danlos syndrome in man. Vet Med Small Anim Clin 69:1256
Scott DW, Miller VH, Griffin CE (2001) Congenital and hereditary diseases. Muller and Kirk's small animal dermatology. WB Saunders, Philadelphia, pp 913–1003
Sequeira JL, Rocha NS, Bandarra EP, Figueiredo LM, Eugenio FR (1999) Collagen dysplasia (cutaneous asthenia) in a cat. Vet Pathol 36:603–606
Singleton AC, Mitchell AL, Byers PH, Kathleen A, Potter KA, Pace JM (2005) Bovine model of Marfan syndrome results from an amino acid change (c.3598G4A, p.E1200K) in a calcium-binding epidermal growth factor-like domain of fibrillin-1. Hum Mutat 25:348–352
Sinke JD, van Dijk JE, Willemse T (1997) A case of Ehlers-Danlos-like syndrome in a rabbit with a reviwe of the disease in other species. Vet Q 19:182–186
Smith JD, Hamir AN, Greenlee JJ (2011) Cartilaginous metaplasia in the sclera of Suffolk sheep. Vet Pathol 48:827–829
Swiderski C, Pasquali M, Schwarz L, Boyle C, Read R, Hopper R, Ryan P, Rashmir-Raven A (2006) The ratio of urine deoxypyridinoline to pyridinoline identifies horses with hyperelastosis cutis (a.K.a. hereditary equine regional dermal asthenia or HERDA). J Vet Intern Med 20:756
Switonski M (2014) Dog as a model in studies on human hereditary diseases and their gene therapy. Reprod Biol 14:44–50
Tang BL (2001) ADAMTS: a novel family of extracellular matrix proteases. Int J Biochem Cell Biol 33:33–44
Tryon RC, White SD, Bannasch DL (2007) Homozygosity mapping approach identifies a missense mutation in equine cycophilin B (PPIB) associated with HERDA in the American quarter horse. Genomics 90:93–102
Tsai KL, Clark LA, Murphy KE (2007) Understanding hereditary diseases using the dog and human as companion model systems. Mamm Genome 18:444–451
Uri M, Verin R, Ressel L, Buckley L, McEwan N (2015) Ehlers-Danlos syndrome associated with fatal spontaneous vascular rupture in a dog. J Comp Pathol 152:211–216
Vaatstra BL, Halliday WD, Waropastrakul S (2011) Dermatosparaxis in two white Dorper lambs. N Z Vet J 59:258–260
van der Linde-Sipman JS, Kroneman J, Meulenaar H, Vos JH (1985) Necrosis and rupture of the aorta and pulmonary trunk in four horses. Vet Pathol 22:51–53
Van Halderen A, Green JR (1988) Dermatosparaxis in White Dorper sheep. J South Afr Vet Assoc 59:45–46
Velasco J, Li J, DiPietro L, Stepp MA, Sandy JD, Plaas A (2011) Adamts5 deletion blocks murine dermal repair through CD44-mediated aggrecan accumulation and modulation of transforming growth factor β1 (TGFβ1) signaling. J Biol Chem 286:26016–26027
Wang WM, Ge G, Nagase H, Greenspan DS (2006) TIMP-3 inhibits the procollagen N-proteinase ADAMTS-2. Biochem J 398:515–519
Wang VM, Bell RM, Thakore R, Eyre DR, Galante JO, Li J, Sandy JD, Plass A (2012) Murine tendon function is adversely affected by aggrecan accumulation due to the knockout of ADAMTS5. J Orthop Res 30:620–626
Weingart C, Haußer I, Kershaw O, Kohn B (2014) Ehlers-Danlos-like-Syndrom bei einer Katze [Ehlers-Danlos like syndrome in a cat]. Schweiz Arch Tierheilkd 156:543–548
White SD, Affolter VK, Bannasch DL, Schultheiss PC, Hamar DW, Chapman PL, Naydan D, Speir SJ, Rosychuk RAW, Rees C, Veneklasen GO, Martin A, Revier D, Jackson HA, Bettenay S, Matousek J, Campbell KL, Ihrke PJ (2004) Hereditary equine regional dermal asthenia ('hyperelastosis cutis') in 50 horses: clinical, histological, immunohistological and ultrastructural findings. Vet Dermatol 15:207–217
Xie L, Spencer ND, Beadle RE, Gaschen L, Buchert MR, Lopez MJ (2011) Effects of athletic conditioning on horses with degenerative suspensory ligament desmitis: a preliminary report. Vet J 189:49–57
Young M, Moshood O, Zhang J, Sarbacher CA, Mueller POE, Halper J (2018) Does BMP2 play a role in the pathogenesis of equine degenerative suspensory ligament desmitis? BMC Res Notes 11:672
Zanna G, Fondevila D, Bardagi M, Docampo MJ, Bassols A, Ferrer L (2008) Cutaneous mucinosis in shar-pei dogs is due to hyaluronic acid deposition and is associated with high levels of hyaluronic acid in serum. Vet Dermatol 18:314–318
Zanna G, Docampo, M.J., Fondevila, D., Bardagi, M., , Bassols, A., Ferrer, L. Hereditary cutaneous mucinosis in shar pei dogs is associated with increased hyaluronan synthase-2 mRNA transcription by cultured dermal fibroblasts. Vet Dermatol 2009;20:377–382
Zhou H, Hickford JG, Fang Q (2012) A premature stop codon in the ADAMTS2 gene is likely to be responsible for dermatosparaxis in Dorper sheep. Anim Genet 43:471–473

Correction to: Clinical and Molecular Delineation of Cutis Laxa Syndromes: Paradigms for Elastic Fiber Homeostasis

Aude Beyens, Lore Pottie, Patrick Sips, and Bert Callewaert

Correction to:
Chapter 13 in: J. Halper (ed.), *Progress in Heritable Soft Connective Tissue Diseases*, Advances in Experimental Medicine and Biology 1348, https://doi.org/10.1007/978-3-030-80614-9_13

The book was inadvertently published with incorrect Chapter title. The title has been corrected in Chapter 13 now.

The updated version of the chapter can be found at https://doi.org/10.1007/978-3-030-80614-9_13

J. Halper (ed.), *Progress in Heritable Soft Connective Tissue Diseases*, Advances in Experimental Medicine and Biology 1348, https://doi.org/10.1007/978-3-030-80614-9_16

Index

J. Halper (ed.), *Progress in Heritable Soft Connective Tissue Diseases*, Advances in Experimental Medicine and Biology 1348, https://doi.org/10.1007/978-3-030-80614-9

GPSR Compliance
The European Union's (EU) General Product Safety Regulation (GPSR) is a set of rules that requires consumer products to be safe and our obligations to ensure this.

If you have any concerns about our products, you can contact us on

ProductSafety@springernature.com

In case Publisher is established outside the EU, the EU authorized representative is:

Springer Nature Customer Service Center GmbH
Europaplatz 3
69115 Heidelberg, Germany

www.ingramcontent.com/pod-product-compliance
Ingram Content Group UK Ltd.
Pitfield, Milton Keynes, MK11 3LW, UK
UKHW050139280726
14058UKWH00006B/736

* 9 7 8 3 0 3 0 8 0 6 1 6 3 *